RENEWALS 458-4574

Individual Susceptibility to Genotoxic Agents in the Human Population

ENVIRONMENTAL SCIENCE RESEARCH

Editorial Board

Alexander Hollaender
Associated Universities, Inc.
Washington, D.C.

Ronald F. Probstein
Massachusetts Institute of Technology
Cambridge, Massachusetts

Bruce L. Welch
Environmental Biomedicine Research, Inc.
and
The Johns Hopkins University School of Medicine
Baltimore, Maryland

Recent Volumes in this Series

Volume 21 MEASUREMENT OF RISKS
Edited by George G. Berg and H. David Maillie

Volume 22 SHORT-TERM BIOASSAYS IN THE ANALYSIS OF COMPLEX ENVIRONMENTAL MIXTURES II
Edited by Michael D. Waters, Shahbeg S. Sandhu, Joellen Lewtas Huisingh, Larry Claxton, and Stephen Nesnow

Volume 23 BIOSALINE RESEARCH: A Look to the Future
Edited by Anthony San Pietro

Volume 24 COMPARATIVE CHEMICAL MUTAGENESIS
Edited by Frederick J. de Serres and Michael D. Shelby

Volume 25 GENOTOXIC EFFECTS OF AIRBORNE AGENTS
Edited by Raymond R. Tice, Daniel L. Costa, and Karen Schaich

Volume 26 HUMAN AND ENVIRONMENTAL RISKS TO CHLORINATED DIOXINS AND RELATED COMPOUNDS
Edited by Richard E. Tucker, Alvin L. Young, and Allan P. Gray

Volume 27 SHORT-TERM BIOASSAYS IN THE ANALYSIS OF COMPLEX ENVIRONMENTAL MIXTURES III
Edited by Michael D. Waters, Shahbeg S. Sandhu, Joellen Lewtas, Larry Claxton, Neil Chernoff, and Stephen Nesnow

Volume 28 UTILIZATION OF MAMMALIAN SPECIFIC LOCUS STUDIES IN HAZARD EVALUATION AND ESTIMATION OF GENETIC RISK
Edited by Frederick J. de Serres and William Sheridan

Volume 29 APPLICATION OF BIOLOGICAL MARKERS TO CARCINOGEN TESTING
Edited by Harry A. Milman and Stewart Sell

Volume 30 INDIVIDUAL SUSCEPTIBILITY TO GENOTOXIC AGENTS IN THE HUMAN POPULATION
Edited by Frederick J. de Serres and Ronald W. Pero

A Continuation Order Plan is available for this series. A continuation order will bring delivery of each new volume immediately upon publication. Volumes are billed only upon actual shipment. For further information please contact the publisher.

Individual Susceptibility to Genotoxic Agents in the Human Population

Edited by

Frederick J. de Serres

National Institute of Environmental Health Sciences
Research Triangle Park, North Carolina

and

Ronald W. Pero

University of Lund
Lund, Sweden

Technical Editor

William Sheridan

National Institute of Environmental Health Sciences
Research Triangle Park, North Carolina

PLENUM PRESS • NEW YORK AND LONDON

Library of Congress Cataloging in Publication Data

Main entry under title:

Individual susceptibility to genotoxic agents in the human population.

(Environmental science research; v. 30)
"Proceedings of a workshop on individual susceptibility to genotoxic agents in the human population, held May 10–12, 1982, at the National Institute of Environmental Health Sciences, in Research Triangle Park, North Carolina."
Includes bibliographical references and index.
1. Genetic toxicology – Congresses. 2. Toxicity testing – Congresses. 3. Mutagenicity testing – Congresses. I. de Serres, Frederick J. II. Pero, Ronald W. III. National Institute of Environmental Health Sciences. IV. Series.
RA1224.3.I53 1984 616′.042 84-11467
ISBN 0-306-41679-4

Proceedings of a workshop on Individual Susceptibility to Genotoxic Agents in the Human Population, held May 10–12, 1982,
at the National Institute of Environmental Health Sciences,
in Research Triangle Park, North Carolina

© 1984 Plenum Press, New York
A Division of Plenum Publishing Corporation
233 Spring Street, New York, N.Y. 10013

All rights reserved

No part of this book may be reproduced, stored in a retrieval system, or transmitted in any form or by any means, electronic, mechanical, photocopying, microfilming, recording, or otherwise, without written permission from the Publisher

Printed in the United States of America

PREFACE

As a result of the industrial revolution, man's technological achievements have been truly great, increasing the quality of life to almost unimagined proportions; but all this progress has not been accomplished without equally unimagined health risks. Sufficiently diagnostic short-term assay procedures have been developed in recent years for us to determine that there are mutagenic agents among thousands of chemicals to which the human population is exposed today. These chemicals were not significantly present prior to the industrial revolution. As of today, there are no procedures available which have been adequately demonstrated to assess individual susceptibility to genotoxic exposures, and as a result we have had to rely on extrapolating toxicological data from animal model systems. The question is can we afford to allow such an increased environmental selection pressure via mutagenic exposures to occur without expecting adverse long-term effects on our health. It is apparent from this line of reasoning that what is lacking and immediately needed are test procedures that can be applied to humans to assess genotoxic exposure as well as individual susceptibility to it.

There have already been two conferences which have focused attention on this research area. "Guidelines for studies of human populations exposed to mutagenic and reproductive hazards" (A. D. Bloom, ed., March of Dimes Birth Defects Foundation, White Plains, New York, 1981) and "Indicators of genotoxic exposure in humans" (Banbury Report 13, B. A. Bridges, B. E. Butterworth, and I. B. Weinstein, eds., Cold Spring Harbor Laboratory, 1982) were important beginnings in approaching the need for examining this field. During this same period, The Swedish Council for Planning and Coordination of Research established a program entitled "Chemical Health Risks in our Environment" to stimulate a better organization of research in this area for Sweden.

In May 1982 we organized the first American-Swedish Workshop dealing with "Individual Susceptibility to Genotoxic Agents in the Human Population" at the National Institute of Environmental Health Sciences in the Research Triangle Park, North Carolina. Our intentions were to pick up the momentum produced by the two earlier conferences, by focusing more directly on the "State-of-the-Art" assay procedures already being developed or planned in the United States and Sweden for use on humans. It is hoped that publication of these proceedings will emphasize the importance of international collaboration and place into a better perspective the biochemical and genetic methods that can be developed and assessed in the near future as epidemiological tools for the determination of mutagenic risk to man.

Ronald W. Pero
Lund, Sweden

CONTENTS

INVESTIGATION OF GENETIC HAZARDS:

GUIDANCE FROM OCCUPATIONAL AND ENVIRONMENTAL STUDIES

Irving J. Selikoff

Environmental Sciences Laboratory
Mount Sinai School of Medicine of
The City University of New York
New York, New York 10029

There is a paradox in discussing the use of occupational experience in relation to individual susceptibility to genotoxic agents. First, disease that is seen occurs in individuals, and reflects their individual susceptibility. The research that is done, however, is focused on populations. Therefore, in this presentation I will move from one foot to the other. Secondly, there's long been a curious dichotomy with regard to reproductive experience and occupational and environmental health matters. Curious, because there's been a concordance in many ways. In each, both somatic and genetic material are included, many species can be affected, many different organs, different tissues, and both are very selective within populations. A large variety of agents can affect individuals in each, both males and females can be involved. Yet this dichotomy has continued.

Several recent examples may be given. For instance, when it became clear a decade ago that bis-chloromethyl ether was capable of producing lung cancer and nasal cancer in humans, there were no studies started among the populations involved to see whether there might also be adverse reproductive experience. Our own laboratory was very much involved in 1974 in vinyl chloride research [1] and it was clear that this chemical could produce a good deal of cancer. There were some 6000 to 10,000 heavily exposed polymerization workers and probably more than a million people involved in chronic exposure at much lower levels. There have been few studies with regard to reproductive experience, even though the occupational experience was very clear and even though chromosome aberrations had been demonstrated [2]. I am not aware of concurrent studies with regard to aflotoxins in East Africa and there isn't a single nickel smelter to my knowledge in which the workers are followed for lung cancer and the offspring simultaneously considered in relation to adverse reproductive experience. This discordance has lasted for many decades. In part, perhaps it's been structural. There's been compartmentalization of research. Geneticists, pediatricians, and obstetricians have been very much involved in one area, and environmental scientists, pathologists and biostatisticians in another. This is beginning to change. Not only is this meeting an example, but also the perspectives of NIEHS, where such interdigitation is very much encouraged.

It may be useful to review how we came upon many of our current concepts in environmental and occupational disease, to see whether some guidance might be provided for an increase in such interdigitation and joint studies.

We both began with description in our earliest decades, and also enumeration. Perhaps immediately postwar, we began to somewhat diverge in our approaches. In occupational and environmental health, we began to have a much greater focus with regard to etiology.

How did this come about? [3]. In environmental and occupational disease (and cancer is a good example) data collection was primarily in terms of registries, very much as were reproductive experience studies. This was just before, during and after World War II. Data began to come in from cancer registries; data of different quality from different parts of the world. Variations in incidence were found. Stomach cancer was high in Chile, Iceland, and Japan, low in the United States, Roumania, etc. For cancer of the esophagus, very much the opposite; low in the United States and high in France and China. Colon-rectum cancer was high in Scotland and the United States, and low in Roumania. Explanation was needed for these differences. One that was quite obvious was that people were genetically different, that people in France were different from those in Egypt, etc. That was a possibility.

However, when geographical pathologists examined what happened when people moved from one country to another, they found that when people in Japan, who had high stomach cancer risk, moved to the United States, their children here had the low rates of other Americans. In contrast, the low colon cancer rates in their home country changed to the high colon cancer rates of all other Americans when they came to this country. It was surely unlikely that there was enough genetic drift in one generation to account for this.

Secondly, even within one country, it was noticed that rates were changing. For example, there's been a sharp decrease in stomach cancer in the United States. We don't know why, although we are very thankful. But it hasn't been equally true for all people in the

United States. It's true for whites, but not so much for black people. With regard to colon cancer, the rates have stayed the same from 1930 to 1980 among whites. However, they have been increasing for our black population. Cancer of the esophagus has very much remained the same for whites, although it has been sharply rising for blacks. Genetic differences seemed an incomplete explanation.

Explanations were sought. Hints were available. One came from radiation studies. Beginning in the Schneeburg area a century ago, it was observed that what was then thought to be sarcoma of the lung (now known to be lung cancer) was very common among the miners [4]. Fifty years later the same was found in Joachimstal, on the Czech side of the mountains, with autopsies of the miners showing 50% with lung cancer. The explanation that the cancers might be exogenous in origin was suggested by a young woman journalist, who opined that it could be due to radiation since these were the mines from which the Curies obtained pitchblende for their radium studies.

The second thread that has been woven into our current understanding came from our greatest public health error, our failure in the 1920s, 1930s, and 1940s to predict what was going to happen in the 1960s, 1970s, and 1980s, as a result of the extraordinary increase in cigarette smoking that began at that time. Fortunately, this met with new advances that were being made in chronic disease epidemiology and population studies. The American Cancer Society, for example, registered one million people in one-third of the counties of the United States 1959-1960. A vast amount of data were collected. People were asked, "Where were you born?", "Where was your father born?", "How much fried food do you eat?", "How many hours do you sleep at night?", "How much schooling have you had?", "How far down do you smoke your cigarettes?", and a wide variety of other questions. All were then followed prospectively. Within five years, by the mid-60s, the results were clear. In two groups of more than 37,000 men each, carefully matched for many variables except for cigarette smoking, it was found that people who smoked cigarettes had much more lung cancer than those who did not. This was true for a number of neoplasms and, of course, for a number of other diseases, as well [5].

The use of large scale population studies was obviously very valuable. This was true in identifying not only a direct effect on the lung in cigarette smoking, but for such diseases as cancer of the bladder and cancer of the pancreas, which were not the result of a direct effect but a diffuse chemical influence. Individuals who smoked cigarettes had much greater risk of developing cancer of the bladder, compared to those who did not.

The third thread that allowed us to reach our present understanding came from a series of unnatural, unwanted, unplanned circumstances in which human beings were exposed to a variety of toxic agents. This began about 200 years ago, with Percivall Pott's original description of cancer of the scrotum in chimney sweeps, in 1775 [6]. One hundred or so years later, Dr. Rehn reported cancer of the bladder in aniline dye workers [7]. Yamagiwa (1915) followed and produced experimental cancer of skin in rabbits with coal tar eventually leading to the chemical studies of Ernest Kennaway and the isolation of a variety of carcinogenic chemicals. Rehn's work was germinally important. He described three cases of bladder cancer in 1895 and suggested that the three workers he studied, since they worked in the same plant making fuchsin, might have gotten their bladder cancer because of the chemical in their environment. This was lésè majesté since Connheim had noted decades before that it was due to epithelial cell rests in the bladder mucosa. Yet Rehn proposed exogenous etiology. Scientists soon published additional case reports. Once again population studies were utilized. In Basel, for example, of the twelve deaths of cancer of the bladder in the first decade of this century, six were in the dye workers of the city, although only 2% of the males worked in the dye industry. The same was true in Frankfurt, where approximately 25% of all bladder cancers recorded at the turn of the century were in dye workers. Thus, the use of population studies, albeit primitive, allowed us to define an association between a specific exposure and disease.

Our failure in the United States to appreciate this led to a tragedy. When in the first World War we could no longer import aniline dyes from Switzerland and Germany, the duPont Company set up its own works in Salem County, New Jersey, the great Chambers Works. By 1931, the first cases of bladder cancer were seen. Dr. Gehrman, the Medical Director of duPont, did a magnificent study, cystoscoping 532 men. Even at that early time, 4-1/2% had bladder tumors, and he was able to define that some had been exposed to β-naphthylamine, to benzidine or to both chemicals [8].

Another agent that has been very well studied - asbestos - has again given evidence in this regard. In 1898, Dr. Montague Murray, a physician at the Charing Cross Hospital in London, saw a man who was very short of breath. He was told that the man worked in a local asbestos factory that had opened up only 10 years or so before. When the man died the following year, an autopsy was done. Diffuse interstitial fibrosis was found with what we now know as asbestos bodies.

Since then, we have learned a good deal [9]. First, during the 30s, 40s, and 50s, we learned that, unlike other dusts (coal, silica, diatomaceous earth, aluminum), with asbestos the pleura was often involved, although the mechanism is still not clear. This might be simply discrete plaques, with no particular signs or symptoms, or more diffuse, sometimes causing pulmonary insufficiency.

In the mid-30s, Dr. Kenneth Lynch, at that time Professor of Pathology at the Medical University of South Carolina, described a man who had both interstitial fibrosis (asbestosis) and cancer of the lung. He proposed there might be an association [10]. Ten years after that, Dr. Weiss in Germany reported another unusual event. A man who had worked with asbestos was found to have a malignancy of the mesothelial lining of the chest (mesothelium). Since a primary tumor of the mesothelium is termed a mesothelioma, this was the first case of mesothelioma related to asbestos work. The description then was very much like what we have been seeing since - diffuse, malignant, often fatal within six months or a year. The following decade saw other people who have been exposed to asbestos suffer malignancy in the lining of the abdomen. Again the mesothelial lining, therefore, mesothelioma. And again diffuse and invariably fatal. Such cases began to increase so much that scientists began to urgently study the problem. This included studies of insulation workers. Population studies were once more used.

In the New York-New Jersey metropolitan area, where there were a total of 1249 insulation workers on January 1, 1963, 1117 were X-rayed. It was found that about half had abnormal X-rays. But it was also found that of the 725 with less than 20 years from onset of exposure, most had normal X-rays. It was only after that point that the X-rays tended to become abnormal (Table 1). This demonstrated that we were dealing with an important time factor [11].

In the same area, in 1943, there had been 632 men on the rolls of this union. Nine died before 20 years and 623 were still alive, 20 years or more after the onset of their work. It was found, examining their mortality experience as a population from January 1, 1943 to January 1, 1963, that there was an extraordinary increase in the number of deaths observed compared to the number expected. Twelve were due to asbestosis. That was no surprise. What was unusual was that instead of 32 deaths of cancer, 95 had occurred (Table 2) [12]. Instead of 6 deaths of cancer of the lung, there

TABLE 1. X-Ray Changes in Asbestos Insulation Workers [11]

Onset of exposure (yrs.)	No.	% Normal	% Abnormal	Asbestosis (grade) 1	2	3
40+	121	5.8	94.2	35	51	28
30-39	194	12.9	87.1	102	49	18
20-29	77	27.2	72.8	35	17	4
10-19	379	55.9	44.1	158	9	0
0-9	346	89.6	10.4	36	0	0
	1,117	51.5	48.5	366	126	50

TABLE 2. Observed and Expected Number of Deaths Among 632 Asbestos Workers Exposed to Asbestos Dust 20 Years or Longer [12]

	Years				Total,
Cause of death	1943-1947	1948-1952	1953-1957	1958-1962	1943-1962
Total, all causes..............	28	54	85	88	255
Observed (asbestos workers)					
Expected (US white males).....	39.7	50.8	56.6	54.4	203.5
Total cancer, all sites.........	13	17	26	39	95
Observed (asbestos workers)					
Expected (US white males	5.7	8.1	13.0	9.7	36.5
Cancer of lung and pleura.......	6	8	13	18	45
Observed (asbestos workers)					
Expected (US white males).....	0.8	1.4	2.0	2.4	6.6
Cancer of stomach, colon, and rectum......................	4	4	7	14	29
Observed (asbestos workers)					
Expected (US white males)....	2.0	2.5	2.6	2.3	9.4
Cancer of all other sites combined..................	3	5	6	7	21
Observed (asbestos workers)					
Expected (US white males)....	2.9	4.2	8.4	5.0	20.5
Asbestosis.....................	0	1	4	7	12
Observed (asbestos workers)					

were 42. Dr. Lynch at the postmorten table in 1935 had been correct.

Mesothelioma in the past had been found, in general, in approximately 1 out of 10,000 deaths, in autopsy series. Here it was almost 1 of 10. There was also increased cancer at a number of other sites.

This cohort has now almost reached extinction, giving virtually their total experience. By 1977, instead of 329 expected deaths, there were 478. Once again, some were due to asbestosis. However, the majority of excess deaths were due to cancer where instead of 57, there were 210. Instead of 13 deaths of lung cancer, there were 93. One out of very 5 of these people died of cancer of the lung. Instead of no deaths of mesothelioma, there were 38. And increases were seen in cancer of the gastrointestinal tract and at a number of other sites.

Therefore, with radiation studies, chemical studies, studies of particles, with smoking studies, a very important advance was made. It was demonstrated that we can identify things that are carcinogenic to humans. This has been coupled with the understanding that "every cancer has a cause," and the causes are generally exogeneous - itself a remarkable step forward. If you hear of a person with cancer of

TABLE 3. Membership of Asbestos Insulation Workers' Union*, January 1, 1967, Classified by Age and by Years from First Exposure to Asbestos Dust [15]

Age, years	Total number of members	Number of years since first exposure to asbestos								
		0-9	10-14	15-19	20-24	25-29	30-34	35-39	40-49	50+
15-19	244	244								
20-24	1,695	1,695								
25-29	2,412	2,066	345	1						
30-34	2,762	1,065	1,356	341						
35-39	2,988	313	1,141	1,342	192					
40-44	2,260	79	424	1,026	591	139	1			
45-49	1,589	49	131	433	442	487	47			
50-54	1,297	27	88	214	332	377	182	77		
55-59	984	13	49	129	206	176	146	193	72	
60-64	703	1	21	59	131	126	87	99	179	
65-69	419	---	6	18	41	58	45	29	201	21
70-74	255	---	---	6	14	22	21	16	105	71
75-79	111	---	1	---	4	8	4	7	37	50
80-84	52	---	---	---	---	2	1	2	16	31
85+	29	---	---	---	---	---	---	2	7	20
Total	17,800	5,552	3,562	3,569	1,953	1,395	534	425	617	193

* Membership in the United States and Canada of the International Association of Heat and Frost Insulators and Asbestos Workers, AFL-CIO, CLC.

TABLE 4. Deaths Among 17,800 Asbestos Insulation Workers in the United States and Canada January 1, 1967-December 31, 1976 (Number of Mean 17,800, Man-Years of Observation 166,853)

Underlying cause of death	Expected*	Observed		Ratio o/e	
		(BE)	(DC)	(BE)	(DC)
Total deaths, all causes	1,658.9	2,271	2,271	1.37	1.37
Total cancer, all sites	319.7	995	922	3.11	2.88
Cancer of lung	105.6	486	429	4.60	4.06
Pleural mesothelioma	†	63	25	---	---
Peritoneal mesothelioma	†	112	24	---	---
Mesothelioma, n.o.s.	†	0	55	---	---
Cancer of esophagus	7.1	18	18	2.53	2.53
Cancer of stomach	14.2	22	18	1.54	1.26
Cancer of colon-rectum	38.1	59	58	1.55	1.52
Cancer of larynx	4.7	11	9	2.34	1.91
Cancer of pharynx, buccal	10.1	21	16	2.08	1.59
Cancer of kidney	8.1	19	18	2.36	2.23
All other cancer	131.8	184	252	1.40	1.91
Noninfectious pulmonary diseases, total	59.0	212	188	3.59	3.19
Asbestosis	†	168	78	---	---
All other causes	1,280.2	1,064	1,161	0.83	0.91

* Expected deaths are based upon white male age-specific U.S. death rates of the U.S. National Center for Health Statistics, 1967-1976.

† Rates are not available, but these have been rare causes of death in the general population.

(BE): Best evidence. Number of deaths categorized after review of best available information (autopsy, surgical, clinical).

(DC): Number of deaths as recorded from death certificate information only

the brain, or cancer of the colon, or kidney, one of the thoughts that crosses your mind is "I wonder what caused it." That would not have been the case before World War II.

Since these early studies, a variety of agents have been identified as increasing cancer risk to humans, ranging from bis-chloromethyl ether and chromium, to auramine and diethylstilbesterol, β-naphthylamine and nickel smelting.

When did this understanding take hold? It is difficult to date intellectual advances of this type [13]. Dr. Doll last year made an estimate [14]. I think he is right. He said that in 1951 most people believed that cancer was a "degenerative process"; inextricably confounded with age. As we grew older, we had increased risk of dying of cancer. Sixteen years later, in the mid-60s, it was generally accepted that most cancers were attributable to exogeneous causes and that therefore 4 out of 5 were preventable.

TABLE 5. Deaths Among 17,800 Asbestos Insulation Workers in the United States and Canada January 1, 1967-December 31, 1976 [15] (Number of Men 17,800, Man-Years of Observation 166,853)

		Observed		Ratio o/e	
Underlying cause of death	Expected*	(BE)	(DC)	(BE)	(DC)
Total deaths, all causes	1,658.9	2,271	2,271	1.37	1.37
Cancer, all sites	319.7	995	922	3.11	2.88
Deaths of less common malignant neoplasms					
Pancreas	17.5	23	49	1.32	2.81
Liver, biliary passages	7.2	5	19	0.70	2.65
Bladder	9.1	9	7	0.99	0.77
Testes	1.9	2	1	--	--
Prostate	20.4	30	28	1.47	1.37
Leukemia	13.1	15	15	1.15	1.15
Lymphoma	20.1	19	16	0.95	0.80
Skin	6.6	12	8	1.82	1.22
Brain	10.4	14	17	1.35	1.63

* Expected deaths are based upon white male age-specific U.S. death rates of the U.S. National Center for Health Statistics, 1967-1976.

(BE): Best evidence. Number of deaths categorized after review of best available information (autopsy, surgical, clinical).

(DC): Number of deaths as recorded from death certificate information only.

This intellectual leap is still very much being developed. Thus, in these studies - the smoking studies, the occupational studies - we have been able to give not merely "yes" or "no" answers, but also to tease out some of the subtleties and specific attributes of these experiences. For example, with the insulation workers, we enrolled all 17,800 members of this Union in the United States and Canada on January 1, 1967 (Table 3). We have been observing them since. In the first ten years, by 1977, instead of 1659 deaths, there were 2271 [15]. Again, some deaths were due to asbestosis, but the great excess were due to cancer. Instead of 320 expected, 995 occurred (Table 4). Once more, cancer of the lung, mesothelioma, cancer of the gastrointestinal tract, the oropharynx, the larynx and cancer of the kidney. We did not find an excess of cancer of the liver, cancer of the bladder, of leukemia, or lymphoma, and a number of other sites (Table 5). Each tissue, we found, had its own predilection for response, or lack of response, to this agent.

We knew when each of these men began work. Figures 1-3 are graphic descriptions of when these people died, compared to when they began work. For asbestosis, very few deaths were seen in less than 15 years from onset of work exposure. Most occur at 35, 40, 45 years from onset. This is also true for cancer of the lung, where the dif-

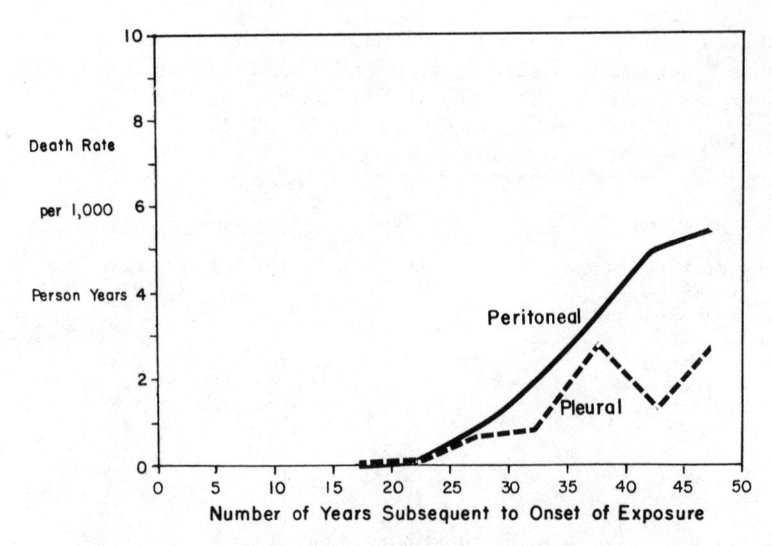

Fig. 1. Deaths per thousand person years of experience of pleural and peritoneal mesothelioma among 17,800 asbestos insulation workers from 1967-1976, analyzed by duration from onset of employment in five-year periods. Ratios between observed and expected deaths cannot be computed, since expected rates are not available for the general population. At least as of 45 years from onset, a decline in rates was not noted. This is consistent with the finding that cigarette smoking did not play a role in the risk of developing mesothelioma in contrast to lung cancer [17].

ference between expected and observed deaths does not really show much until 15 years from onset and does not reach a peak until approximately 30-35 years from onset. For mesothelioma, exactly the same. Little is seen in less than 15 or 20 years; then it begins to take off. So far it has not plateaued. These people begin exposure when they are 17, 18, 19 as apprentices and do not die until they are 55, 60, or more. Two-thirds are dead before the age of 65. This emphasizes the importance of a long period of observation. It should be a minimum of 15 to 40 years for cancer. Shorter observations can be useful, but they are likely to be incomplete [16].

We have not yet been able to take the next step of determining the cellular and molecular basis for latency, what happens over years. There is progression on X-ray of asbestosis. We do not know the cellular and molecular basis for this long-term event. Latency gives us pause when we consider other things. For example, the observations in 1974 concerning vinyl chloride indicated that the first angiosarcomas of the liver occurred, on the average, 18-1/2 years

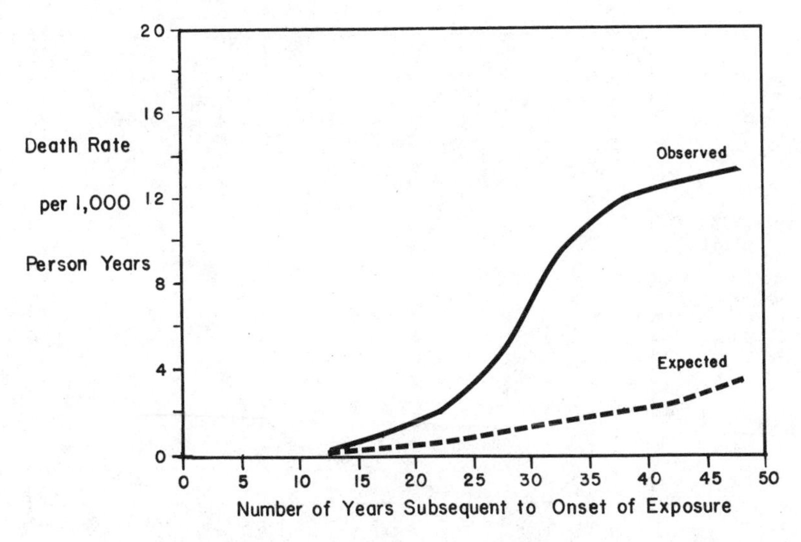

Fig. 2. To facilitate comparison with death rates for mesothelioma and asbestosis, death rates per thousand person years of lung cancer are shown, in five-year periods of duration from onset.

following onset of exposure. For us, as public health scientists, the import of latency had to be considered with the knowledge that in the United States in 1950 in the early years of the vinyl industry, we only produced 250 million pounds of vinyl chloride whereas in 1977, production reached 6 billion pounds. This was true also for synthetic organic chemicals in general. In 1940, we made 1 billion pounds; in 1976, 300 billion pounds. With long latency, our experience is still ahead of us.

Our next lesson came from an old factory in Paterson, New Jersey. In 1940, the U.S. Navy helped set up a plant here (the Union Asbestos and Rubber Company) to manufacture asbestos insulation for our new ships. To rebuild our Navy, they set up factories for many things, electronic gear, motors, steel plates, etc. The plant produced the Navy's insulation. From 1941 to 1945, it employed 933 men. We have been following these 933 men since. They are dying of the usual things that people who were exposed to asbestos die of; i.e., lung cancer, mesothelioma, gastrointestinal cancer, etc. (Table 6) [17]. Here we had an unusual opportunity of studying dose-response. Because of wartime conditions, some of the men worked for a week, two months, often while waiting to go into service. Others worked for the full thirteen years, until the plant closed in 1954. These men have been traced and are still being followed. Men who had a large dose, i.e., more than two years, have

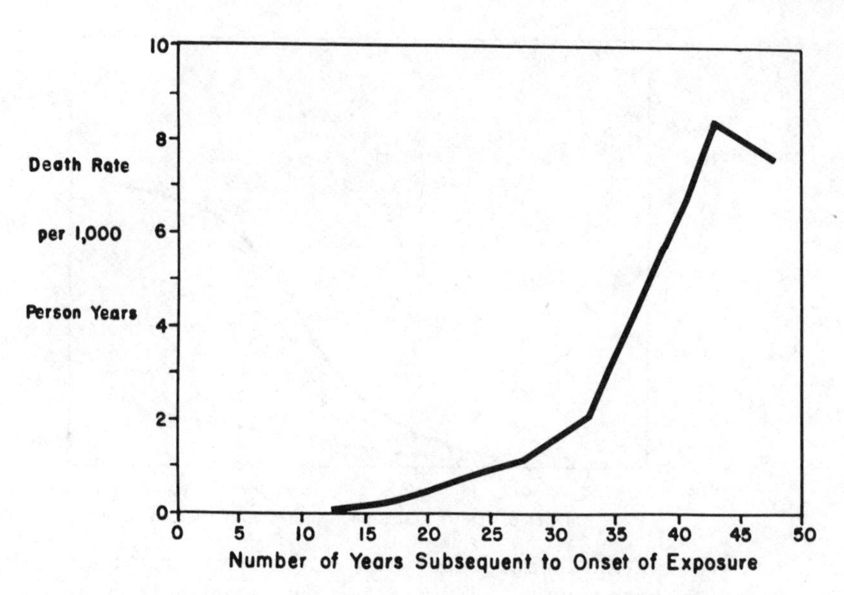

Fig. 3. Death rates of asbestosis among 17,800 asbestos insulation workers 1967-1976, analyzed in five-year periods of duration from onset of employment. Some decline is seen after 45 years from onset, possibly related to the added burden of smoking-induced lung disease superimposed upon asbestosis (at least in some cases) with selective survival of nonsmokers [17].

been compared to men who had a small dose, a month or less [18]. They had much more cancer (Fig. 4). The greater exposure, the greater the risk. What was particularly interesting, however, was that even one month or less of exposure was sufficient, if people were followed for 30 to 35 years, to more than double the cancer risk. These men had the same exposure, to the same material, in the same city, in the same years, making the same products with the same machinery, but differed in dose. It was clear that in human populations, as in animal studies, there is an important dose-response relationship. The same is true with smoking. Too, we learned that brief exposure, if excessive, can be hazardous. We are not sure what this means in molecular terms.

Our next question, which is certainly the subject of this meeting, concerns individual susceptibility. I have told you that among some groups exposed to asbestos, 1 out of 15 or so dies of mesothelioma. This is another way of saying that 14 out of 15 do not. If you are a heavy smoker and smoke 2 or 3 packs a day you have a 1 in 10 chance of dying of lung cancer; another way of saying that 9 out of 10 do not die of cancer of the lung. Why? There is this vast

TABLE 6. Categorization of 304 Deaths, 1961-77, 20 or More Years from Onset of Employment, 1941-45, for 582 Amosite Asbestos Factory Workers [17]

Cause of death	Based on DC only	Based on BE	Difference between BE and DC
All causes	304	304	0
Cancer, all sites	103	116	+13
Lung	52	60	+8
Pleural mesothelioma	1	7	+6
Peritoneal mesothelioma	0	7	+7
Mesothelioma not specified above	3	0	-3
Larynx, oral cavity, and pharynx	3	5	+2
Esophagus	1	1	0
Kidney	2	2	0
Colon-rectum	11	11	0
Stomach	4	4	0
Prostate gland	4	4	0
Bladder	2	2	0
Pancreas	5	4	-1
Other and unspecified	15	9	-6
Noninfectious pulmonary diseases, total	29	24	-5
Asbestosis	8	18	+10
Cardiovascular diseases	130	122	-8
Other and unspecified causes	42	42	0

individual susceptibility which we so far cannot explain. But we are beginning to nibble away at the edges of the problem. For example, with our knowledge concerning cigarette smoking we have been studying the interaction between two carcinogens, i.e., asbestos and cigarette smoking. In our first group of men, there were 370 alive on January 1, 1963 who had been in the union in 1963. They were at least 20 years from onset of their work; many, much longer. These were people, therefore, who were very much at risk. Among these 370 men, there were 87 who never smoked cigarettes and 283 who had. We did not expect much lung cancer in the nonsmokers. If you do not smoke you tend not to die of lung cancer. Among the 283 smokers, we anticipated three deaths of cancer of the lung, given their smoking habits. There was not one lung cancer by 1967 among the nonsmokers [19]; when they died of other things, their lungs contained much asbestos. But they did not die of cancer of the lung. On the other hand, among the 283 smokers, where there should have been three deaths of cancer of the lung, there were 24. With this suggestion of interaction of asbestos and cigarette smoking, we very much enlarged our study with the 17,800 men (Table 3). We used as controls our American Cancer Society study of people regis-

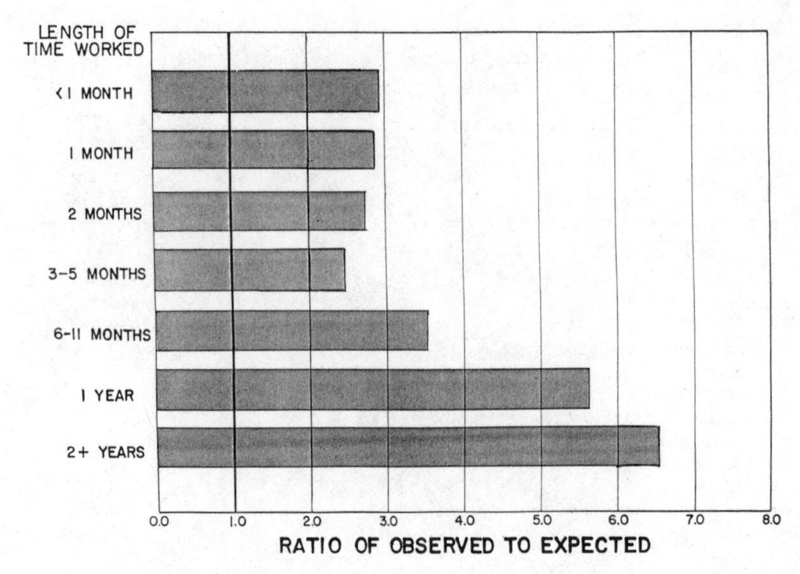

Fig. 4. Ratio of cumulative observed to expected probabilities of dying from lung cancer from 5 through 35 elapsed years since onset of work in an amosite asbestos factory, 1941-45, according to length of time worked. Observed probabilities shown are those for lung cancer classified according to best evidence available [18].

tered in 1959. We had among them more than 73,000 men who were very much like the insulation workers - they were white, they were not farmers, they had stated on registration that they were exposed at their work to dust, fumes, vapors, chemicals, radiation, etc., and they had no more than a high school education. They were alive on January 1, 1967 when we began our other study, and we have traced them thereafter. We knew their smoking habits. We watched both of these groups for the next 10 years. This is what we found: for the American Cancer Society men who neither worked with asbestos nor smoked cigarettes, their death rates for lung cancer per 100,000 per year was 11.3; for the asbestos workers who did not smoke it was five times as much, 58. Of course, five times a very small number is still not very much. On the other hand, for the American Cancer Society men who smoked but did not work with asbestos the rate was 122. For those who had both asbestos work and cigarette smoking the rate was an extraordinary 601 per 100,000 per year (Table 7) [20]. This demonstrates the importance of multiple factor interaction. However, this was not true for all tissues. We found it so for lung cancer, for the esophagus, the larynx and the pharynx but not for mesothelioma. That occurs equally among smokers and nonsmokers. It was not true for cancer of the stomach, the colon, the rectum or the

TABLE 7. Age-Standardized Lung Cancer Death Rates* for Cigarette Smoking and/or Occupational Exposure to Asbestos Dust Compared with No Smoking and No Occupational Exposure to Asbestos Dust [20]

Group	Exposure to asbestos	History cigarette smoking?	Death rate	Mortality difference	Mortality ratio
Control	No	No	11.3	0.0	1.00
Asbestos workers	Yes	No	58.4	+47.1	5.17
Control	No	Yes	122.6	+111.3	10.85
Asbestos workers	Yes	Yes	601.6	+590.3	53.24

* Rate per 100,000 man-years standardized for age on the distribution of the man-years of all the asbestos workers. Number of lung cancer deaths based on death certificate information.

kidneys. Again, each tissue responds differently to multiple factor interaction. We do not know why.

There have been other observations, of equal interest. We examined the experience of men who stopped smoking. The risk of those who stopped smoking, after 5 to 10 years, went to one-half or one-third compared to their colleagues who continued to smoke. This observation demonstrates that while the effect (cancer) is not reversible, the risk of the disease is.

While we are beginning to understand something about selection within high risk groups, we again do not know its cellular basis, as smoking-particle interaction. It is there, obviously of considerable importance, and we do not understand it.

Could immunomodification have an important influence of individual susceptibility? We are very interested in this in our Laboratory [21]. While it is likely to be essential in explaining much with regard to cancer, I do not know whether this is going to be true for reproductive effects.

Mesothelioma As a "Signal" Tumor

We have been very interested in mesothelioma at Mount Sinai since the first modern paper on the subject was written at our hospital in 1931 [22]. Dr. Klemperer and Dr. Rabin reported two cases that they had seen, collected three from the German literature, published the five. It is still, I think, one of the best pathology papers on the subject. We had a surgical pathologist at the time, Dr. Otani, a very brilliant man who did not believe this. He said that with two cases, it was probably a variant of something. So we had a debate. When I first came to Mount Sinai in the early

1940s I was wondering what these people were arguing about. Here was a disease few ever saw, yet they would spend hours discussing the histopathology. With the debate, we kept looking for it. Yet our records show that from 1931 to 1960 we saw only three more cases. Thus, it was a very rare disease in the past. You can image our consternation then when, in 1960, Dr. J. C. Wagner in Johannesburg reported 47 cases of mesothelioma that he had seen in the previous five years [23]. They came from a part of South Africa, the Northwestern portion of the Cape Province, with many small crocidolite asbestos mines. He had visited the relatives of these patients. He found that there had been opportunity in 45 of the 47, 30-35 or so years before, for some asbestos contact. Often, they had as children played on the tailings heaps outside of the mines. This demonstrated for the first time that one need not have much asbestos exposure to suffer neoplastic disease. Asbestosis is different. A fair amount of asbestos is needed to cause significant asbestosis. For cancer, one needs just enough to start neoplastic change at one site, with independent growth thereafter. There will be other things found that cause mesothelioma [24, 25]. It is inconceivable that a tissue would respond to only one agent. But at least in studies thus far this seems to be largely so. Cochrane and Webster reported last year that in a series of 107 patients with mesothelioma, there was history of asbestos exposure in 106 [26]. When a positive biopsy was seen in the Pathology Department, they spoke to the patients and thus were able to get accurate histories. (In mortality series, one may get a history concerning possible exposure, 30 years before from the second wife or third cousins.) Mesothelioma is a very good marker, a very good signal for exposure to a toxic agent. A number of such signals have been identified, as with angiosarcoma of the liver, scrotal cancer, nasal cancer in woodworkers and nickel smelting, etc. These signals are not population-based but are of things so rare that when they occur they are striking. If one encounters an angiosarcoma of the liver, a history of whether there has been vinyl chloride or arsenic exposure would certainly be taken. Such signals are not, however, always heeded. I was reminded of this because registries failed to demonstrate the growing problem of mesothelioma. During the 40s, 50s, and 60s, mesotheliomas were occurring with increasing frequency, and they were collected in cancer registries. Biological bookkeepers were ticking them off. For example, Dr. Clemmensen published in 1981 the data on mesothelioma in the Danish Cancer Registry which was initiated in 1943. Incidence steadily rose both in males and females. Yet we were not alerted to what was happening in the Registry. Rather, we were alerted by an alert physician.

With Wagner's observation, Molly Newhouse, a very capable epidemiologist at the London School of Hygiene, then studied the 76 cases of mesothelioma in the files of the London Hospital [27]. She also visited the relatives and found that 31 had worked with asbestos. That came as no surprise. But of the 45 who had not worked with the mineral, 9 had simply lived in the household of an asbestos worker. These were women who had washed their husbands' clothes when they came home from work (household contact mesothelioma). Of the 36 who had neither worked with asbestos nor lived with someone who worked with asbestos, 11 had lived within a half mile of one of the asbestos plants in London, at Barking (neighborhood mesothelioma). This again demonstrated that it did not take much of the agent to produce the disease.

We have been studying this, in quantitative terms. In the factory in Paterson, we have been X-raying the wives and children (household contacts) who lived with the employees from 1941 to 1945. In the first 679 that we examined one-third had abnormal X-rays. They felt well but their X-rays were abnormal, albeit insufficient to result in shortness of breath. In an ongoing mortality study of these family contacts, it appears that lung cancer will be about doubled in incidence (observed vs. expected rates) and that approximately 1% of deaths, more than 20 years from onset of residential exposure, will be due to mesothelioma. What Dr. Wagner told us, what Dr. Newhouse has been telling us, has been the importance of low-level exposure. This can be translated into the vast questions of consumer exposures, trace contamination of the environment with toxic chemical wastes, of indirect (bystander) occupational exposure. These can be immensely complex, with permutations that include multiple simultaneous or sequential agents at varying levels, intermingling occupational circumstances, environmental contamination, life-style, food, genetic, and other influences. Each can be further modified by persistence both in the environment and, sometimes, in our tissues. To the ecologists' environmental burdens, we juxtapose human tissue burdens (PCBs, PBBs, inorganic particles, other).

Our age of innocence is over. We now know that there are many exogenous agents that can cause disease, as a result of our environmental and occupational studies. A major problem is how to use this knowledge. There is both opportunity and responsibility [28]. What to do with tissue burdens? Should there be research concentration on removal of chemicals from body tissues? In our studies in Michigan, we found that in 1978 approximately 97% of that State's general population had residual PBB in serum/depot fat - as a result of contamination of Michigan's dairy food supply in 1973-1974 [29]. What are the consequences likely to be? Should prophylactic measures be advocated?

Should we use animal study results in selecting our response to these questions [30]. There is at the moment reluctance to use the results of laboratory studies for public health and regulatory purposes. This, despite knowledge of at least seven chemicals which were shown to be carcinogenic in animals before man. These include such important agents as diethylstilbestrol, vinyl chloride, bis-chloromethyl ether, estrogens, aflotoxins, and 4-aminobiphenyl. The reason is a structural one. In the United States, and in Sweden as well, we do not ban agents that are shown to be toxic. We have laws which say they are to be controlled. Control means regulation, dose-response, debate, setting of levels, concern with large populations, rather than yes or no scientific answers [31].

Control also converts a scientific problem into one with ethical terms [32]. If, with a straight-line dose relationship, we will allow some exposure, we will also allow some disease. How much disease will be a public decision, with interdigitation of science and public policy. We know now something of how to define the toxicity of agents, in humans and in animals. How and whether we use these methods, is a matter of public policy. Once the information is available, it is a social decision whether, with what degree of rigor, it is to be applied. And many of the decisions that have to be made will reflect the fact that our scientific generation is the first to therefore have the opportunity - and responsibility - of preventing preventable disease.

REFERENCES

1. I. J. Selikoff and E. C. Hammond, eds., Toxicity of Vinyl Chloride and Polyvinyl Chloride, Annals of The New York Academy of Sciences, New York, Vol. 246 (1975).
2. A. Ducatman, K. Hirschhorn, and I. J. Selikoff, Vinyl chloride exposure and human chromosome aberrations, Mutat. Res., 31: 163-168 (1975).
3. I. J. Selikoff, Scientific basis for control of environmental health hazards, in: Public Health and Preventive Medicine, 11th edition (J. M. Last, ed.), Appleton-Century-Crofts, New York (1980), pp. 529-542.
4. M. Hartung, H.-H. Schaller, and E. Brand, On the question of the pathogenetic importance of cobalt for hard metal fibrosis of the lung, Intl. Arch. Occup. Environ. Health, 50:53-57 (1982).
5. E. C. Hammond, Smoking in relation to the death rates of one million men and women, in: Epidemiological study of cancer and other chronic diseases, Monograph 19, Bethesda, Maryland, National Cancer Institute (1966).
6. P. Pott, Cancer of the scrotum. Chirurgical Observations, Hawes, Clark, and Collings, London, 1775, pp. 63-68.
7. L. Rehn, Blasengeschwulste bei Fuchsin in Arbritern, Arch. Klin. Chirugie, 50:588-600, 1895.
8. G. H. Gehrmann, The carcinogenetic agent - chemistry and industrial aspects, J. Urology, 31(2):126-137 (1934).
9. I. J. Selikoff, Asbestos-associated disease, in: Maxcy-Rosenau Public Health and Preventive Medicine, 11th ed. (J. Last, ed.), Appleton-Century-Crofts (1980), pp. 568-644.

10. K. M. Lynch and W. A. Smith, Pulmonary asbestosis. III. Carcinoma of lung in asbestos-silicosis, Am. J. Cancer, 24:56 (1935).
11. I. J. Selikoff, J. Churg, and E. C. Hammond, The occurrence of asbestosis among insulation workers in the United States, Ann. N.Y. Acad. Sci., 132:139-155 (1965).
12. I. J. Selikoff, J. Churg, and E. C. Hammond, Asbestos exposure and neoplasia, JAMA, 188:22-26 (1964).
13. E. Mayr, The Growth of Biological Thought, Harvard University Press, Cambridge (1982), pp. 974.
14. R. Doll, Avoidable cancer: attribution of risk, So. Afr. Ca. Bull., 25(3): 125-126 (1981).
15. I. J. Selikoff, E. C. Hammond, and H. Seidman, Mortality experience of insulation workers in the United States and Canada, Ann. N.Y. Acad. Sci., 330:91-116 (1979).
16. I. J. Selikoff, E. C. Hammond, and H. Seidman, Latency of asbestos disease among insulation workers in the United States and Canada, Cancer 46:2736-2740 (1980).
17. I. J. Selikoff, H. Seidman, and E. C. Hammond, Mortality effects of cigarette smoking among amosite asbestos factory workers, J. Natl. Cancer Inst., 65(3):507-513 (1980).
18. H. Seidman, I. J. Selikoff, and E. C. Hammond, Short-term asbestos work exposure and long-term observations, Ann. N.Y. Acad. Sci., 330:61-89 (1979).
19. I. J. Selikoff, E. C. Hammond, and J. Churg, Asbestos exposure, smoking and neoplasia, JAMA, 204(2):106-112 (1968).
20. E. C. Hammond, I. J. Selikoff, and H. Seidman, Asbestos exposure, cigarette smoking and death rates, Ann. N.Y. Acad. Sci., 330:473-490 (1979).
21. J. G. Bekesi, J. P. Roboz, A. Fischbein, and I. J. Selikoff, Immune dysfunction due to xenobiotics, Presented at the Second Intl. Conf. on Immunopharmacology, Washington, D.C. (July 5-10, 1982).
22. P. Klemperer and C. B. Rabin, Primary neoplasms of the pleura, A report of five cases, Arch. Path., 11:385-412 (1931).
23. J. C. Wagner, C. A. Sleggs, and P. Marchand, Diffuse pleural mesothelioma and asbestos exposure in the North Western Cape Province, Brit. J. Indust. Med., 17:260 (1960).
24. Y. I. Baris, L. Simonato, R. Saracci, et al., Malignant mesothelioma and radiological chest abnormalities in two villages in central Turkey, Lancet, 2:984-997 (1981).
25. A. N. Rohl, A. M. Langer, G. Moncure, I. J. Selikoff, and A. Fischbein, Endemic pleural disease associated with exposure to mixed fibrous dust in Turkey, Science, 216:518-520 (1982).
26. J. C. Cochrane and I. Webster, Mesothelioma in relation to asbestos fibre exposure, So. Afr. Med. J., 59(24):848 (1981).
27. M. L. Newhouse and H. Thompson, Mesothelioma of pleura and peritoneum following exposure to asbestos in the London area. Brit. J. Indus. Med., 22:261 (1965).

28. I. J. Selikoff, Practical questions for practical politicians, Ann. N.Y. Acad. Sci., 387:1-9 (1982).
29. M. S. Wolff, H. A. Anderson, and I. J. Selikoff, Human tissue burdens of halogenated aromatic chemicals in Michigan, JAMA, 247:2112-2116 (1982).
30. D. Rall, Relevance of results from laboratory animal toxicology studies, in: Maxcy-Rosenau, Public Health and Preventive Medicine, 11th Edition, Appleton-Century-Crofts (1980), pp. 543-549.
31. I. J. Selikoff, The regulations debate: deciding whether people live or die, CUNY Graduate School Magazine, 1(2):12-19 (1982).
32. I. J. Selikoff, Multiple factor interaction in environmental disease: potential for risk modification and risk reversal. Presented at the Symposium on Assessment and Perception of Health Risk, Royal Society of Canada and the Science Council of Canada, Toronto, Canada, October 18, 1982 (in press).

EPIDEMIOLOGY AND POPULATION MONITORING

IN GENETIC RISK ASSESSMENT

Ernest B. Hook

Bureau of Maternal and Child Health
New York State Department of Health
Albany, New York 12237

and

Department of Pediatrics
Albany Medical College
Albany, New York 12208

INTRODUCTION

There are at the outset two quite separate approaches to genetic risk assessment and population monitoring that while often confused, must be carefully distinguished [1, 2, 3]. The first may be termed "surveillance": the review and analysis of fluctuations in background rates of mutation in various populations to determine if there has been a mutagen introduced into the population or an increase in one already present. The second, to which the term "monitoring" may be applied and which is often of greater social concern is the evaluation over time of a population which has been exposed to a suspected or known mutagen. The process of monitoring thus addresses a specific problem, although the precise population at risk and the substances involved may or may not be known exactly. Because of the generality of the goals of surveillance - the review of background rates - it is possible to consider surveillance in a rather broad way and present an overview (see Ref. 2). With regard to monitoring however, the topic Dr. DeSerres has asked me to address, so much depends upon specific aspects of the population and the substances involved that it is practically impossible to deal with the issue usefully in the general way. In this discussion I will focus on a few examples that serve a heuristic purpose, address some methodological and policy issues, and call attention to a new approach that looks promising.

Rationale

If a population has been exposed to a putative toxin, usually the last consequence of social concern is germinal mutagenic effects. This may appear almost blasphemous to a group of mutation scientists, but on reflection I think it is clear that it is the immediate and short term effects upon mortality and morbidity of the exposed population (including those in utero) that is the primary concern of society. There are already enough reasons to remove toxins with such acute effects from the population without seeking a rationale from study of mutation. (For long term effects the main social concern of course is the development of degenerative disease, malignancy or other chronic illness.)

It is however, in populations exposed to dosages of toxins so low that they don't have any obvious morbidity or mortality that the concern with germinal mutation usually arises. The substances which are exemplars in lower organisms are the so called "super mutagens"— compounds which are highly mutagenic at doses which otherwise result in little or no toxicity to cells or organs, at least in some experimental systems [4].

Consistency in Human Studies

A major problem in studies of putative environmental hazards upon human populations is consistency of evidence. It is often assumed by many in the regulatory area that the results of a single investigation in humans are sufficient to warrant inferences about putative hazards and can justify immediate decisive social action. Certainly in some circumstances all the data that may be available are the results of a single accidental population exposure and one would not urge the repetition of any catastrophe in order to repeat the observations. But what is impressive in reviewing investigations of environmental hazards is how often reliance upon a single study or even a few apparently convincing ones could lead to false inferences. A notable example - albeit not from the field of human mutation - was the simultaneous appearance of three reports from different centers that reserpine use predisposes to breast cancer [5, 6, 7]. It was realized this was incorrect only after massive efforts to confirm the association were unsuccessful. It was eventually concluded even by the original workers themselves that the first observations were probably attributable to an unlikely concatenation of chance events. (For references and further discussion see Refs. 8 and 9.) Of course this can occur in studies of experimental organisms as well. But here at least one has the opportunity to control precisely the associated variables. For studies of human populations, there is the additional complexity of numerous variables which may affect the outcome, of which the observer often has little knowledge and control. It is disturbing how often those with adminstrative regulatory responsibility appear unaware of this, and assume that single unconfirmed observations in human populations provide grounds for rash public action or statements.

Chromosome Breakage Episodes

One such example is the action of the Consumer Product Safety Commission some years ago which banned abruptly certain spray adhesives because of a single poorly controlled study claiming that exposure to these increased human chromosome breakage [10]. Subsequent studies [11, 12, 13] with one belated and overlooked exception [14], could not confirm this. (The exception appears of some interest in that it was written when the episode was over and authors interpreted their result as negative. Yet as I analyze their data, the proportion of cells with abnormalities tender to be higher in cases than controls. I suspect this was a chance result and that the authors interpreted their data as negative, perhaps subconsciously, in the light of their knowledge that the ban had been lifted and others had concluded the substance did not damage chromosomes.) The best evidence for the absence of effect came from the blind reexamination of the original slides by other investigators with conclusion that they had been misinterpreted [11, 15]. The ban and resultant publicity resulted in massive public anxiety and expense, wasted time of medical geneticists, and most tragically led to elective abortions by unnecessarily frightened pregnant women [16]. This was in 1973. One would have hoped that because of this episode and the involvement of individuals from six U.S. agencies that were immune for such further action.

The subsequent E.P.A. study of chromosome breakage in Love Canal residents and the lack of appropriate controls in their study was an astonishing and disturbing "deja-vu" experience for workers in human chromosome breakage [17, 18]. (Actually, the data in the original report, even using historical not simultaneous controls found a higher overall rate of breakage in controls than in cases. It was however, the claim that some type of unusual events - supernumerary acentrics - were over represented in the exposed group that led to the public concern.) (See also Addendum.)

But these episodes are atypical in that the original poorly controlled studies which precipitated the problems were widely, indeed almost unanimously condemned by all scientists knowledgeable in the area of human chromosome breakage [18]. It was only the willingness of those outside this scientific field, but within parts of the regulatory establishment, to accept the protocols and self-appointed credentials of the original investigators that provoked the problem. But usually deficiencies in human studies are not so clear cut, nor is there agreement among experts as in these episodes. Often what occurs at least in human mutageneis is that similar and apparently acceptable approaches to investigation of a human hazard come up with opposite findings, providing enough room for adherents on both sides of an issue to argue different viewpoints.

Radiation and Down Syndrome

In the field of human mutation, this is well illustrated by the diverse data on irradiation and germinal cell chromosome mutations resulting in such defects as trisomy 21 (Down syndrome). One would imagine such a question could be answered relatively straightforwardly by studies in human populations. But even in experimental animals, there is disagreement.

Positive studies in humans including both case-control studies, and retrospective cohort studies, are impressive [19, 20, 21]. (One positive study which almost certainly is biased by poor ascertainment in the control group is unfortunately, widely cited but should not be considered pertinent [22]. See also discussion in Ref. 23.) Yet data from other centers have not been confirmatory. For example, studies at Hiroshima and Nagasaki, a retrospective cohort study, revealed if anything a negative association of Down's syndrome and exposure to ionizing irradiation there [24]. Stevenson's et al., retrospective cohort study also found insignificant results [25]. (Uchida's cohort study had reported 8 observed compared to 1.95 expected, Stevenson's 3 observed vs. 2.1. expected. But if expected values are calculated from more recently available data [26] which are pertinent to populations studied by Uchida and Stevenson then the number expected are increased by about 33%.) Negative results by others have also been reported [27, 28].

There are several different ways of trying to reconcile differences among these studies. One obvious possibility is that there is a modest effect which by chance is found in some but not in other studies. And differences in the application of radiation, the dose, the interval between irradiation and pregnancy under study, etc., conceivably could account for the observed variation. (A study by Alberman et al. [21] has suggested the increased risk of a Down's syndrome is only manifest 10 years after exposure to radiation, although their data have yet to be confirmed by other studies.) If such factors do account for the difference then at least under some circumstances maternal preconceptual ionizing irradiation is associated with a relative risk significantly greater than one of a subsequent Down's syndrome livebirth. The available data however, still don't allow firm inferences as to just what those special circumstances are. Moreover, a convinced skeptic might find strong grounds to challenge some of the positive studies. For example, some years ago when contemplating replicating studies I asked one prominent worker in the field for a copy of the interview form used in eliciting the radiation history of cases and controls. This person indicated that no standard fixed form had been used but all individuals had been interviewed in meticulous detail about radiation exposure, without reference to a standard written guide. This makes it difficult for others to replicate exactly this particular case-control study. Moreover, many students of population surveys would question if bias had arisen because it is conceivable that, without the conscious awareness of the investigator, the controls were queried less intensively than cases. (There is in addition a separate type of memory bias that can occur as mothers of cases may be more likely to recall putative causal factors during or prior to gestation than parents of control children.) I should point out however, that while this objection could be raised regarding case-control studies, it is not applicable to a cohort design study since such a study starts with a group with defined exposure and notes only which pregnancy outcomes end in a chromosome abnormality, an objective endpoint. Nevertheless, at least some defined written protocol for eliciting information should be used.

Studies searching for radiation induced non-disjunction in mammals show about as much consistency as those in humans. Of interest the only positive study that has not been challenged on statistical grounds is from Uchida's group, the same group from which the most consistent positive studies in humans have come. (See references and discussion in Ref. 29.)

There is no question that ionizing irradiation in insects increases non-disjunction, but the doses required to produce any significant effect are several hundred R, much higher than those humans are likely to be exposed to under most circumstances [30].

Thus despite much work, we still cannot make strong statements about the putative effect of preconceptual ionizing irradiation in moderate doses upon Down's syndrome, a well defined endpoint. Perhaps the safest statement to make is that if such radiation does have an effect it is unlikely to be a strong one.

One might infer from the discussion to this point that I am pessimistic about the value of monitoring human populations for effects of exposure to putative mutagens. This is not entirely the case. Such studies, are worthwhile if "positive" results in observations in humans are interpreted cautiously and confirmation, if possible, is sought in other sources before taking some precipitate public action. And such studies, if necessary, can at least provide reassurance about the apparent absence of effect of a putative mutagen or at least put a likely confidence interval about its magnitude of effect if any.

Given a population exposed to a potentially mutagenic substance, what types of investigations may be taken from the perspective of monitoring for alteration of human mutation rates? Obviously investigation of chromosome breakage and rearrangement is one approach to this problem in somatic tissues, but there is already extensive commentary upon this approach in the literature [31, 32, 33] and this provides no direct data on germinal mutations.

Mutation in Sperm

For germinal mutation rates, I believe the most exciting possibility now on the horizon is the assay of chromosome abnormalies in human sperm, originally reported by Rudak et al. [34], and further developed to the point of achieving consistent success, albeit with difficulty, by Martin et al. [35, 36].

Martin (personal communication) had studied by May 5, 1982 686 haploid figures from sperm of over 18 men and found 7.9% abnormal cytogenetically. Of interest, none of the sperm are YY, i.e., have 2 Y chromosomes. (This suggests incidentally that the double sparkling body, the so called YFF body present in about 2% to 3% of sperm and thought by some to represent Y chromosomal non-disjunction during spermatogeneis, is probably an artifact of some sort. At least it does not appear likely to indicate the presence of 2 Y chromosome in sperm and probably cannot be regarded as a mutational event. The observation of no YY sperm in 686 studied excludes a rate of 0.5% or greater with 95% confidence.)

Martin's work [35, 36] suggests that if all sperm have equal chance of fertilization - which admittedly is not known to be the case - then around 8 to 10% of all zygotes would be cytogenetically abnormal because of a paternal error. If abnormalies attributable to maternal error are also included then the proportion of cytogenetically abnormal zygotes probably is at least 16 to 20% and perhaps even larger. Yet 3 weeks later (at 5 weeks after the LMP) the proportion of cytogenetically abnormal recognized conceptuses has been estimated at 5% [37]. Either there is a great loss of cytogenetically abnormal zygotes between conception and the 3 week embryo, or else something is wrong with the data or assumptions in reaching either of the estimates compared above! (See also Addendum.)

In any event it is obvious that study of the haploid complement of sperm of those exposed to toxic substances may provide the most direct evidence of human mutagenic effect of exposure.

Other Approaches

For investigation where only a few males are exposed to substances of concern this approach may be the optimal one for evaluating mutagenic effect (barring more invasive procedures such as testicular biopsies). But for investigations of any larger scale on women, this is not feasible and one is left with the use of other markers. Unfortunately the rate of mutagenic outcomes in the human livebirths resulting from germinal mutations is so small relatively speaking that one must study a relatively large population to detect an association with environmental hazard, particularly if doing a cohort study, that is starting with a group of exposed individuals and investigating outcomes.

The proportion of all livebirths with a cytogenetics abnormality is about 6 per 1000 (about half of which are clinically significant) [38]. About 80% of clinically cytogenetic abnormalities are attributable it is likely to a germinal mutation. The proportion of livebirths with a clinically significant single gene disorder may be estimated at about 11.5 to 12.5 per 1000 [39]. (Somewhere between 0.2 to 2 per 1000 of these have resulted from mutation in the most recent generation.) This of course does not consider specific locus mutations not associated with clinically significant lesions. (See derivations of rates and references in Ref. 39.)

Because of the relative scarcity of these mutations and the difficulty and expense of detecting them on any scale, cohort studies of populations for germinal mutation in livebirths are likely to be difficult and expensive. (See also reference 44 in Addendum.)

Cytogenetic Mutations Detected Prenatally

For mutations detectable at prenatal cytogenetic diagnosis this may not present such a problem because here at least the study is conducted upon the fetus for clinical reasons. There is no cost to the investigator for detection of the mutation, only for collection of the data. A certain fraction of women undergo prenatal cytogenetic testing because of anxiety due to parental exposure to putative mutagens. Sufficient data from this group may eventually accumulate to provide direct data about cytogenetic consequences of exposure to some of these suspected hazardous substances. As one example, in a recent analysis of data reported to two chromosome registries, there was a small group of women - 136 - studied because of exposure to a putative mutagen in whom the observed rate of mutant (or potentially mutant) structural abnormalities was 1.5%, significantly greater than in the remainder (31/27.089 = 0.1%), a different that remained after adjustment for parental age [40]. Clearly these are miniscule numbers, but we are just at the beginning of harvesting the extensive data that are accumulating.

Of course we could also use cases reported to this source for a case-control investigation if one had reason to suspect an environmental substance unrelated to the parental reason for undergoing amniocentesis.

Lastly, the rate of mutant cytogenetic abnormality is very high, about 30%, in spontaneous embryonic and fetal death [37]. However, collecting and evaluating such mutations systematically is extraordinarily difficult and costly. Indeed to my knowledge there are only two such ongoing studies in the world [41, 42]. Some use of data from these sources for investigation of putative environmental effects upon chromosome abnormalities is discussed elsewhere in this volume by Stein.

REFERENCES

1. J. W. Flynt, Jr., in: Congenital Defects - New Directions in Research (D. T. Janerich, R. G. Skalko, and I. H. Porter, eds.), pp. 119-127, Academic Press, New York (1974).
2. E. B. Hook, in: Progress in mutation research (K. C. Bora, G. R. Douglas, E. R. Nestmann, eds.), Vol. 3, pp. 9-38, Elsevier North Holland, Amsterdam (1982). (See also Addendum, Ref. 43.)
3. E. B. Hook and P. K. Cross, in: Environmental Mutagens and Carcinogens (T. Sugimura, S. Kondo, H. Takebe, eds.), pp. 613-621, University of Tokyo and Alan R. Liss, Tokyo and New York (1982).
4. F. J. DeSerres, Prospects for a revolution in the methods of toxicilogical evaluation, Mutation Research, 38:165-172 (1976).
5. Boston Collaborative Drug Surveillance Program, Reserpine and breast cancer, Lancet, 2:669-671 (1974).
6. B. Armstrong, N. Stevens, and R. Doll, Retrospective study of the association between use of ravwolfia derivatives and breast cancer in English women, Lancet, 2:672-675 (1974).
7. O. P. Heinonen, S. Shapiro, L. Tvominen, and M. I. Turunen, Reserpine use in relation to breast cancer, Lancet, 2:675-677 (1974).
8. S. Shapiro and D. Slone, Comment, J. Chr. Dis., 32:105-107 (1979).
9. S. Shapiro, Discussion, J. Chr. Dis., 32:108-113 (1979).
10. Anonymous, Was this scare necessary? Med. World News, 14, No. 39, 15-17 (1973).
11. H. A. Lubs, R. S. Verma, R. L. Summitt, and F. Hecht, Reevaluation of the effect of spray adhesives in human chromosomes, Clin. Genet., 9:302-306 (1976).
12. E. B. Hook, N. H. Harcher, P. A. Brinson, O. S. Stanecky, L. Fisher, G. Feck, and P. Greenwald, Negative outcome of a blind assessment of the association between spray adhesives exposure and human chromosome breakage, Nature, 249:165-166 (1974).
13. B. T. Gong and P. W. Smith, Normal chromosome findings in spray-adhesive users (Letters), JAMA, 227:1259-1260 (March 18) (1974).
14. D. W. Bianchi, C. H. Donaldson, and P. S. Moorhead, Absence of damage to human chromosomes by spray adhesives, Mutation Research, 26:545-551 (1974).
15. Anonymous, Spray glue ban has come unstuck, Med. World News, 15, (7):17.
16. Anonymous, Sticking to the spray glue band, Med. World News, 15(4), 17, 18 (1974).
17. Biogenics Corporation, Pilot cytogenetic study of the residents of Love Canal, New York (unpublished report to U.S. Environmental Protection Agency, May 14, 1980), pp. 1-20 (1980).
18. G. B. Kolata, Love Canal: False alarm caused by botched study, Science, 208:1239-1242 (1980).
19. I. A. Uchida and E. J. Curtis, A possible association between maternal radiation and mongolism, Lancet, 2:848-850 (1960).

20. I. A. Uchida, R. Holunga, and C. Lawler, Maternal radiation and chromosomal aberrations, Lancet, 2:1045-1049 (1968).
21. E. Alberman, P. E. Polani, J. A. Fraser Roberts, C. C. Spicer, M. Elliot, and E. Armstrong, Parental exposure to x-irradiation and Down's syndrome, Ann. Hum. Genet., 36:195-208 (1972).
22. N. Kuchupillai, C. Verma, M. S. Grewal, and V. Ramalingas, Down's syndrome and related abnormalities in an area of high background radiation in coastal Kenala, Nature, 262:60 (1976).
23. E. B. Hook and I. H. Porter, in: Population cytogenetics: Studies in Humans (E. B. Hook and I. H. Porter, eds.), pp. 353-365, Academic Press, New York (1977).
24. W. J. Schull and J. V. Neel, Maternal radiation and mongolism, Lancet, 1:537-538 (1962).
25. A. C. Stevenson, R. Mason, and K. D. Edwards, Maternal diagnostic x-irradiation before conception and the frequency of mongolism in children subsequently born, Lancet, 2:1335-1337 (1970).
26. E. B. Hook, in: Proceedings of the National Foundation-March of Dimes Symposium, Memphis, 1977 Birth Defects Original Article Series (D. Bergsma and R. L. Summitt, eds.), 14 (6C), pp. 249-267, Alan R. Liss, New York (1978).
27. B. H. Cohen, A. M. Lilienfield, S. Kramer, and L. C. Hyman, in: Population Cytogenetics - Studies in Humans (E. B. Hook and I. H. Porter, eds.), pp. 301-352, Academic Press, New York (1977).
28. C. O. Carter, K. A. Evans, and A. M. Stewart, Maternal radiation and Down's syndrome (mongolism), Lancet, 2:1042 (1961).
29. Committee on the Biological effects of ionizing radiation, The effects on populations of exposure to low levels of ionizing radiation 1980, pp. 1-524, National Academy Press, Washington (1980).
30. J. W. Mavor, The production of non-disjunction by x-rays, J. Exp. Zoology, 39:381-432 (1924).
31. H. J. Evans and D. C. Lloyd, eds., Mutagen induced chromosome damage in man, pp. 1-354, Yale University Press, New Haven (1978).
32. P. G. Archer, M. Bender, A. D. Bloom, J. G. Brewer, A. V. Carrano, and R. J. Preston, in: Guideline for studies of human populations exposed to mutagenic and reproductive hazards (A. D. Bloom, ed.), pp. 1-35, March of Dimes Birth Defects Foundation, White Plains (1981).
33. E. B. Hook, Epidemiologic aspects of studies of somatic chromosome breakage and sister chromatid exchange, Mutation Research, 99:372-382 (1982).
34. E. Rudak, P. A. Jacobs, and R. Yanagimachi, Direct analysis of the chromosome constitution of human spermatozoa, Nature, 274: 911-913 (1978).
35. R. H. Martin, W. Balkan, K. Burns, and C. C. Lin, Direct chromosomal analysis of human spermatozoa: Results from 15 normal men. Sixth International Congress of Human Genetics, p. 46, Jerusalem (1981) abstracts (see also Addendum, Ref. 46).

36. R. H. Martin, C. H. Lin, W. Balkan, and K. Burns, Direct chromosomal analysis of human spermatozoa: preliminary results from 18 normal men, American Journal Human Genetics, 34:459-468 (1982).
37. E. B. Hook, Prevalence rate of chromosome abnormalities during human gestation and implications for studies of environmental mutagens, Lancet, 2:169-172 (1981).
38. E. B. Hook and J. L. Hamerton, in: Population Cytogenetics: Studies in Humans (E. B. Hook and I. H. Porter, eds.), pp. 63-79, Academic Press (1977).
39. E. B. Hook, in: Banbury Report 13: Indicators of genotoxic exposure in man and animals (I. B. Weinstein, B. E. Butterworth, B. A. Bridges, eds.), Banbury Conference Series, Cold Spring Harbor Press, Cold Spring Harbor, New York, 19-30 (1982).
40. E. B. Hook, D. M. Schreinemachers, A. M. Willey, and P. K. Cross, Rates of mutant structural chromosomal rearrangements in human fetuses: Data from prenatal cytogenetic studies and associations with maternal age and parental mutagen exposure, American Journal Human Genetics, 35:96-109 (1983).
41. D. Warburton, Z. Stein, J. Kline, and M. Susser, in: Human Embryonic and Fetal Death (I. H. Porter and E. B. Hook, eds.), pp. 261-287, Academic Press, New York (1980).
42. T. Hassold, T. Chen. J. Funkhouser, T. Jouss, B. Manuel, J. Matsura, A. Matsuyama, C. Wilson, J. A. Yamane, and P. A. Jacobs, A cytogenetic study of 1000 spontaneous abortions, Ann. Hum. Genet., 44:151-178 (1980).

ADDENDUM (March 2, 1984)

Since this manuscript was submitted some additional work has been published pertinent to the themes of its topic. A more up to date review of the contribution of chromosome abnormalities to human morbidity and mortality has appeared [43]. A general review of the methods, achievements and limitations in mutation epidemiology has also been published [44]. A "blind" restudy of the population at Love Canl has been carried out, using simultaneous controls, by the Center for Disease Control, which has indicated no evidence for increased chromosome breakage in this population [45]. Further studies of human sperm have appeared both by Martin and by others at the Lawrence Laboratory in Livermore [47]. Martin finds a rate of 8.5% abnormal (2.7% hypohaploid, 2.4% hyperhaploid and 3.3% structurally abnormal in 1000 sperm from 33 men). The Livermore group also found 8.5% abnormal but different proportions in each category: 1.0% hypohaploid, 1.0% hyperhaploid, and 6.5% structurally abnormal in 1133 sperm from five men. They also have suggestive evidence for interindividual variation in structural abnormalities observing a range in these from 1.3% to 10.4%. The precise reasons for the differences between laboratories are uncertain but may be related to interindividual or geographic (i.e., environmental) variation, or perhaps some as yet unappreciated technical differences.

REFERENCES FOR ADDENDUM

43. E. B. Hook, Perspectives in mutation epidemiology: Contribution of chromosome abnornalities to human morbidity and mortality and some comments upon surveillance of chromosome mutation rates, Mutation Research, 114:389-423 (1983).
44. J. R. Miller and nine coauthors, Mutation epidemiology: review and recommendations, Mutation Research, 123:1-11 (1983).
45. C. E. Heath and 17 coauthors, A study of cytogenetic patterns in persons living near the Love Canal, Center for Disease Control (CDC) report issued May 1983, Atlanta, Georgia.
46. R. H. Martin, W. Balkan, K. Burns, A. W. Rodemaker, C. C. Lin, and N. L. Rudd, The chromosome constitution of 1000 human spermatozoa, Human Genetics, 63:305-309 (1983).
47. B. Brandriff, L. Gordon, L. Ashworth, G. Watchmaker, A. Wyrobek, and A. Carrano, Background frequencies of chromosome aberrations in human sperm, Program and abstracts, 15th annual meeting, Environmental Mutagen Society, February 19-23, p. 164 (1984).

AN OVERVIEW OF APPROACHES FOR GENETIC MONITORING OF HUMANS

H. Eldon Sutton

The Genetics Institute and Department of Zoology
The University of Texas at Austin
Austin, Texas 78712

The subject of monitoring human populations for mutation has been much discussed in recent years and, I am pleased to say, some progress has been made in developing techniques and in implementing programs. Both the techniques and the programs are still very limited, and this review must still give more attention to possibilities than to existing programs.

I shall divide the monitoring systems into two major groups (Table 1). The first group includes those useful for monitoring the general population, a population in which exposure to mutagens is presumed to be low. Furthermore, the purpose of such monitoring is primarily to detect an unexpected increase in mutation rates due to agents not known to be mutagenic or not recognized to be in the environment. The population base would be large.

The second major group consists of specific populations or cohorts known to be exposed to a real or potential mutagen. The purpose in this instance is to measure the risk from that exposure, which may be high or low. Examples of such groups include occupational exposures to specific agents and exposures of nonworker residential or other geographic cohorts to chemical agents from industrial plants or to residues from chemical dumps. In the latter case, the nature of the agents may be well defined, as in the accident at Seveso, or they may be poorly defined, as at Love Canal. The size of the populations would be rather small typically.

A serious problem in any epidemiological study is the selection of proper controls or, as must often be the case, comparison groups, since true control groups may not be available. For occupational or similar exposure, it may be possible to set baselines on individuals

TABLE 1. Human Populations for Which the Question of Monitoring for Environmental Mutagenesis Arises

I. The general population:
Low exposure
Agents unknown
Very large numbers of persons
II. Specific exposed groups
Low to high exposure
Agents known or unknown
Small numbers of persons

prior to the exposure. For many other situations, especially accidental exposures to the public, extreme care must be taken to select valid comparison groups. It should not be necessary to point out the importance of contemporaneous controls, but recent history indicates otherwise. For large studies of the general population, the only controls possible may be the same population at different times. This places enormous importance on the use of reliable test systems, systems that have been thoroughly studied and whose results are not subject to temporal variations in laboratory procedure. Many of these points are discussed in greater detail in Bloom [1].

1. MONITORING THE GENERAL POPULATION

Let us consider first the prospects and problems of monitoring the general population for increased mutation. We would like to be able to measure germinal effects directly, since this would give the best estimate of risk to future generations. But somatic systems will also be reviewed, since these may be more feasible, and they are of intrinsic interest as well because of the probable relationship between somatic mutation and cancer risk. A summary of monitoring systems is given in Table 2.

1.1. Mutations in Offspring of Exposed Persons

The most direct demonstration of a genetic effect would be among the offspring of exposed persons. Estimation of risk in these cir-

TABLE 2. Tests Currently Available or under Development for Mutational Risk in Human Populations

Test	General population (G) or exposed (E)	Germinal (G) or somatic (S)	Measure of genetic effect (G) or exposure (E)	Time for positive effect[a]	Duration of positive effect[a]
Observations on offspring of exposed					
Sentinel phenotypes	G	G	G	long	long
Mutant proteins	G	G	G	long	long
Chromosomes	G	G	G	long	long
Observations on sperm of exposed					
YFF sperm	G,E	G	G	short	short/medium?
Mutant proteins	G,E	G	G	short?	?
Morphology	E	G	G	short	?
Observations on (potentially) exposed persons					
Lymphocyte chromosomes	E	S	G	short	long
Sister chromatid exchange	E	S	G	short	medium
Micronuclei	E	S	G	short	medium?
Body fluid mutagenicity: amniotic fluid	G	S	E	short	short
urine/blood	E	S	E	short	short
Fertility	E	G	G?	short	medium?
DNA repair	E	S	E	short	short
DNA alkylation	E	G,S	E	short	short
Mutant proteins in somatic cells	G,E	S	G	short	long
Protein alkylation	E	S	E	short	short
Cancer epidemiology	G,E	S	G	long	long

[a] Short = days or weeks; medium = months; long = years.

cumstances is direct, with many fewer assumptions than for other measures.

1.1.1. Mendelian Phenotypes

Sometimes called "sentinal phenotypes," these are simply inherited traits whose transmission to offspring is well understood. In practice, only dominant traits are considered because the great majority of recessive mutations will be expressed in later generations. Any dominant trait not present in one of the parents would be considered to result from mutation, provided there has been adequate verification of biological relationships. To be useful, a trait should be rare, associated with substantial defect, readily diagnosed in the neonatal period, and identifiable in a screening program. Few such traits are available, and the effort and population size required for an effective program would both be enormous [2, 3]. For that reason, there has been no serious effort to implement such a surveillance program.

A second approach to monitoring dominant traits is simply to record the frequency of all dominant traits without regard to which individuals may be new mutations. The arguments on which such an approach is based are summarized in the BEIR reports [4, 5]. Briefly, the frequencies of detrimental dominant traits are presumed to be maintained by mutation, and a certain proportion, perhaps 20%, are replaced each year by new mutations. Thus the overall frequency is a simple function of the mutation rate. There is disagreement on the current frequency of dominant traits [6, 7], a disagreement that reflects the inadequacy of the present systems of data collection. There are indeed enormous problems in setting up a system that would be effective for the very large populations required. At present, the Centers for Disease Control collects information on a limited basis on congenital defects in newborns. This is obviously a place to conduct pilot studies to improve the recognition of Mendelian dominant traits. I believe, however, that an effective program would take years of experience before it could be perfected.

1.1.2. Mutant Proteins in Offspring

Some of the cleanest examples of Mendelian dominant traits are proteins that vary in their amino acid sequences. Since the first discoveries of such variants of human hemoglobin, many hundreds have been found in human proteins, mostly electromorphs. For dozens of human loci, variants can be reliably and efficiently detected by gel electrophoresis. Therefore, one should be able to detect mutant offspring by their possession of a protein variant that is not inherited from either parent. Neel [8, 9] has discussed the various aspects of this system in detail. Some five hundred thousand loci have been tested using this approach in Japan, with one probable mutant detected [10].

The principal shortcoming of the mutant protein system is the very large population required and the extensive laboratory effort. Although the number of persons decreases as more loci become testable, it is still difficult to foresee this as an effective monitoring system in the near future. The use of two-dimensional gels makes the problem of numbers less severe, provided the gels can be automated [8]. Still, one is talking only about detecting mutagen exposures to millions of persons, and the effect might not be detected for years after the exposure. On the positive side is the fact that this is a germinal mutation test. While electromorphs would only result from nucleotide substitutions, the extensions of the test to activity variants would also pick up deletions and frame shift mutations.

1.1.3. Chromosome Studies in Offspring

A substantial health burden is presented by abnormalities in chromosome structure and number. Structural changes in chromosomes due to breakage and abnormal rejoining are caused by the same agents that may cause small deletions and point mutations. We may view them as part of the general spectrum of mutational response. Changes in chromosome number - genomic mutations - occur by different mechanisms, most likely by interference with spindle function, and may be caused by quite different agents.

A number of centers have reported observations on unselected newborn karyotypes, with a frequency of abnormal karyotypes occurring in approximately one in 200 births. Roughly half of these are structural and half genomic abnormalities, with the great majority of genomic changes being new mutations rather than having been transmitted. The frequency of chromosomal changes is thus sufficiently large so that a doubling could be detected in some thousands of newborns, a not unreasonable number to monitor. A monitoring program could be put in place now by organizing the centers that already have the capability to do karyotyping. (For example, see Hecht et al. [11].) This would involve doing karyotypes on unselected newborns, the great majority of whom have normal phenotypes. Considerable care would have to be exercised to avoid biases in ascertainment.

Another possible system would be to monitor through a centralized agency the approximately 40,000 prenatal karyotypes now done annually in the United States, as well as the large numbers done in other countries. These are done primarily on high risk pregnancies, especially if the mother is in an older age group. Results would have to be corrected for age, and care would be necessary to avoid other biases of ascertainment. But the prenatal karyotypes are being done without regard to the fetal phenotype. Most are being done in medical centers by qualified cytogeneticists. Implementation of a screening program would therefore require no additional laboratory studies. The cost would be only that associated with collecting the information from the laboratories and assuring that epidemiological and laboratory standards are maintained.

It has also been suggested that spontaneous abortions could be used to monitor chromosome abnormalities [12]. Certainly the frequency of chromosome abnormalities is much higher in spontaneous abortions. However, it would seem that the problems of obtaining satisfactory samples of spontaneous abortions jeopardize the validity of the results. For example, it would be difficult to assure that only spontaneous abortuses are examined.

In considering the karyotypes of offspring of exposed persons, one must note some drawbacks. The time between exposure and effect may be many years. It would likely be difficult to identify an agent responsible for increasing the rate of abnormalities, although that should not prevent the program being carried out. On the positive side is the fact that the effects are clearly germinal and are related to health. No substantial extrapolations are required to estimate health risks. Many of the chromosomal aberrations persist in the gonial cells and would thus give a measure of the cumulative exposure of the father. The effect in mothers is complicated by the entrance of all oogonia into meiotic prophase during the fetal period, with arrest in the dictyotene stage for many years. These are factors that are both an advantage and a disadvantage.

1.2. Genotypes of Sperm

1.2.1. Y Bodies in Sperm

The heterochromatic long arm of the human Y chromosome fluoresces brightly either in interphase somatic cells or in Y-bearing sperm. A promising test for genomic mutations is the presence of two Y bodies in sperm (YFF sperm), presumed to represent nondisjunction of the Y chromosome during spermatogenesis, although other mechanisms have not been ruled out. In the sperm of normal males, less than 2% of sperm have two Y bodies. Kapp [13] has shown that patients under treatment for cancer with mutagenic agents have a marked increase in sperm with two Y bodies. The frequency eventually returned to normal after cessation of treatment.

The Y-body test needs much more study before it can be accepted as a monitor of human mutagenesis. The test is sufficiently promising to justify development of pilot programs. Among the advantages is the fact that it is based on germ cells. A positive response to a specific exposure means at the very least that the agent reached the germ cells. An increase can be detected in just one person. Although changes in postgonial cells would be short-lived, gonial changes would be long lasting and cumulative. As a monitor for the general population, a sample of relatively few persons would be required. Furthermore, if an increase were suspected, the number of human subjects could be increased virtually without limit.

Present drawbacks are based primarily on our ignorance about the factors that influence the test. Mice do not have a heterochromatic Y chromosome and hence cannot be used experimentally to answer many of the questions that exist. Much additional experience is necessary before the presence of two Y bodies can be equated with nondisjunction and before the results can be used to estimate health risk.

I will note here some of the excellent work on sperm karyotypes that has been done in several laboratories using the hamster egg fusion technique [14, 15]. These methods are not yet ready for use in screening, but the applications are obvious. At present the effort required for assessment of just one person preclude use in monitoring.

1.2.2. Proteins in Sperm

An interesting possible test has been developed for mice and should, with development of new antibodies, be extended to humans. In brief, Ansari et al. [16] were able to use antibodies to rat lactate dehydrogenase-X (LDH-X), found only in sperm, to detect variants of mouse sperm LDH-X. Normal mouse LDH-X does not react strongly with antibodies to the rat form, but some mutant forms apparently can react.

As yet no enzymes are known in human sperm for which antibodies exist to mutant forms. The advantages of such a system have already been enumerated for the Y-body test.

1.3. Measures of Exposure in Germinal Systems

The previous test systems were all based on a genetic change. The agent responsible not only got to the germ cells but also caused a genetic effect that was measured. We may also measure exposure rather than effect, on the assumption that where there is exposure there is effect. Translation of exposure into effect would require careful studies to bridge the two types of events.

In the case of screening of the general population for germinal exposures, the only testing that comes to mind would be to test semen with a microbial test such as the Ames test. Because the components of semen are mixed only at the moment of ejaculation, a positive result would not mean that the sperm had been exposed more than transitorially to the mutagen. If one were conducting tests on sperm as part of a general screening program, testing the semen in this way would seem an inexpensive and worthwhile additional system. I am unaware that anyone has done studies of this possibility.

1.4. Somatic Mutations in the General Population

Although of less direct interest, increased mutations in somatic cells are presumed to reflect increased risk of mutation in germ cells. Not all chemicals are able to cross the blood-testis barrier, but many do. If it were shown that the population is experiencing an increase in somatic mutations, the assumption must be that germinal mutations also are increased until proven otherwise [17, 18].

Somatic mutations also have the advantage, shared only by examination of individual sperm in germinal systems, that each cell is a unit of observation. Thus the individual is equivalent to an entire population. One can assess individual risks. This is especially important if the population is a small exposed group, but it is also important in identifying persons in the general population who may be very sensitive to mutagens. For example, do most mutations occur in persons with repair defects? Or are there other genetic bases for mutational sensitivity?

Several systems are available or are under active development that detect genetic changes in somatic cells. For the most part, the sensitivity of these tests, or lack of sensitivity, makes them useful only on populations with high exposures. Furthermore, most are not yet sufficiently reliable so that results at one time can be compared with results at another. This makes it difficult to look for small changes over time in a population - the most likely situation to arise. Except for cancer incidence in the general population, most tests will likely be applied first to cohorts of potentially high risk persons. The tests are therefore discussed individually under that heading. They can of course be applied to samples of the general population as well.

1.4.1. Cancer Epidemiology

The most effective system now available for monitoring the general population is the occurrence of cancer. Although we cannot equate cancer with somatic mutation in a simplistic way, there are convincing arguments and evidence that mutation is usually an important part of carcinogenesis [19-21]. Cancer would seem to have many handicaps as a monitor. It has a long latent period, often many years. A large population base is required. And because of the much higher frequency in older persons, it is one of many competing causes of death. Nevertheless, because it is such an important health problem, and because it can be reliably identified as a cause of death, the incidence of cancer should be a very useful monitor of mutational risk in the general population. Cancer epidemiologists are experienced in separating out the many variables that confound the reported frequencies of cancer. They can identify geographic areas of high and low risk for various types of cancer. A change in cancer incidence over time is likely to be valid and significant for mutational risk. At present, cancer epidemiology is our most effective system for monitoring exposure to mutagens. It is the only system that has led to identification of unsuspected exposure to mutagens [22].

1.5. Measures of Exposure in Somatic Systems

Only one type of test is available at present that measures exposure to unidentified mutagens and that can be applied on a large scale - a microbial test such as the Ames test. It would be possible to use the Ames test to monitor the general population, using blood and urine from a limited sample of persons selected to be representative. These could be the same persons used for other tests, such as Y bodies in sperm. In any case, careful consideration must be given to smoking habits and other featues of life style that might be relevant.

One special application of the Ames test has been suggested by Dr. Margery Shaw (personal communication). The 40,000 amniocenteses done each year in the United States provide a rich source of material for population monitoring. Specifically, the Ames test could be applied both to the amniotic fluid and to urine collected at the same time from the mother. This should give an excellent picture of the exposure of fetuses to mutagens at a time when they may be extremely sensitive. Such a test program could be instituted now and could be very informative.

2. MONITORING SPECIAL EXPOSURE GROUPS

Let us now consider tests that may be useful for detecting genetic changes or risk of genetic changes in small populations with known or suspected exposure to mutagens. Examples of such populations are workers in a particular industry or those living in the vicinity of a particular industrial plant. Or the population might be all persons who have used a particular medication. In any case, only those tests are useful that can be applied to a small population. Furthermore, I shall assume that the exposure is substantially larger, as is the anticipated effect, than would be experienced by the general population. A corollary is that more intensive laboratory testing and epidemiological investigations can be applied to each individual in the cohort.

2.1. Germinal Effects in Special Exposure Groups

Some of the tests already listed for the general population also can be applied to special populations. Tests on offspring would not be useful. All require very large numbers. The absence of a detectable effect in the children of Japanese atomic bomb survivors [23] and of other heavily irradiated parents underscores the need for enormous populations if such effects are to be detected directly. At this point, we can only attempt to measure effects and exposures in the parental generation and extrapolate to later generations.

2.1.1. YFF Sperm

One of the tests that can be applied to small groups is the presence of two Y bodies in sperm (YFF sperm), described above for the general population. Although much more experience is required with this test, it is potentially an important part of our test battery. The interpretation remains to be established, but the limited evidence suggests that mutagens cause an increase in YFF bodies.

2.1.2. Sperm Morphology

One test that may be especially useful on occupationally exposed males is sperm morphology [24, 25]. (See also p.427.) For reasons that are not entirely clear, mutagens generally cause aberrant morphology in sperm in experimental animals [26]. But whatever the reason, the relationship seems sound experimentally. More needs to be known about dose-response and the spectrum of mutagens able to induce the effect. This could be a fairly simple and useful test in some situations.

2.1.3. Fertility

Interference in fertility is an effect that is often, though not necessarily, a result of treatment with mutagens. Monitoring the general population for fertility would seem a pointless exercise. Any minor changes in fertility due to environmental agents would be swamped by changing social customs and economic influences. Only careful personal interviews can distinguish these from biological effects. Such interviews are possible on smaller populations, however. The effects sought would necessarily be large, but such effects have already been reported in employees of several manufacturing plants [27]. For the most part, loss of fertility seems to be a function of recent exposure and could well presage other delayed biological effects.

2.1.4. DNA Alkylation

In experimental organisms, mutagens such as alkylating agents that form stable DNA adducts can be measured in sperm DNA to demonstrate not only that the agent reached the relevant constituents of the relevant cells but also the extent to which the reaction occurred. Ordinarily this is done with radioactively labeled mutagens [28]. However, the test can be based on chemical analysis of DNA, although this is likely to be less sensitive and more time consuming. The success of the test depends on knowledge of the agent to which the individuals are exposed, with substantial prior studies with experimental animals characterizing the actions of the agent. Also, the alkylation of a nucleotide does not necessarily lead to mutation. Thus the test is more a measure of exposure than of genetic effect. Given sufficient experimental studies, there should be fewer weak links in extrapolating to health risks than in the case of many other measures.

2.2. Somatic Effects in Special Exposure Groups

Several tests are available or are substantially developed that are especially useful for small populations. This is due in part to the possibility for contemporaneous controls, an impossibility if the entire population is the experimental group.

2.2.1. Chromosome Aberrations in Cultured Lymphocytes

At the present time, this is probably the test that has been most often used to detect effects in populations exposed to mutagens. Structural changes in cultured lymphocyte chromosomes have been shown to increase in several groups exposed to ionizing radiation [29-35]. The structural alterations persist for many years, and small numbers of persons may be sufficient to show a positive effect.

There is much less experience with induction of chromosome aberrations by chemicals in mammals, especially humans, although some chemicals, such as epichlorhydrin, have been shown to increase such aberrations [36]. Many, perhaps most, chemical lesions can be repaired. The extent to which such repair is accompanied by a reduction in risk is unknown. The chromosome aberrations themselves are of unknown health significance. Possibly they are more a measure of exposure than of genetic effect. These are questions that must be answered before we know how to interpret chromosome aberrations.

2.2.2. Sister Chromatid Exchange

Currently the most sensitive test for chemical effects on chromosomes is sister chromatid exchange (SCE) [37-42]. Many chemicals cause an increase in SCE in test animals, often at levels that are too low to give responses in other test systems. This sensitivity is illustrated in x ray exposures as well. Evans and Vijayalaxmi [43] reported the frequency of SCE's to be 30 times greater than chromosome aberrations.

The mechanism for SCE production is not understood. At this time, one cannot equate SCE's with mutation. Some agents that increase SCE's in cultured lymphocytes appear to cause long lasting changes in the DNA. And many mutagens are extremely effective in

SCE induction. It seems safe, and certainly wise, to assume that an increase in SCE's signifies an increased risk of mutation.

There has been some use of SCE in testing special exposure groups [44-49]. These have shown the feasibility of this method as a sensitive indicator. The chromosome damage, as reflected in increased SCE, may last for months. There are still many variables to be investigated before SCE results can be extrapolated to human risk. Smoking is known to increase SCE's. Is this a measure of the mutagenic effects of smoking? In spite of these questions, measures of SCE appear ready to take their place on the short list of tests avail able to monitor the mutational risk to small numbers of persons. Care must be taken with interpretation, but only by attempting to use the test can we learn its strengths and weaknesses.

2.2.3. Micronucleus Test

Acentric chromosome fragments do not segregate properly at mitosis but often appear as cytoplasmic inclusions - so-called micronuclei. Their number reflects the extent to which chromosomes have been broken and not repaired. Schmid [50, 51] has used micronuclei as a test of chromosome breakage in experimental animals, using dividing bone marrow cells. Heddle [52, 53] adapted the test to circulating human lymphocytes, raising the possibility of its use in screening humans for mutational exposure. Positive responses were obtained both with X-rays and mitomycin C.

Much study remains to be done before the utility of the micronucleus test can be evaluated. Although it should be more efficient to score than karyotypes in detecting chromosome breakage, more extensive comparative studies are required. In the meantime, it is a promising lead to follow.

2.2.4. DNA Alkylation

Measure of DNA adducts has already been mentioned as a test of germ cell exposure. It can also be done on other tissues, such as white cells, and of course can be carried out on both females and males. Again, it is likely to be useful only with exposure to specific known agents.

2.2.5. DNA Repair

Many, perhaps most, mutagenic alterations in DNA are repaired. The extent of repair can be measured by the incorporation of radioactive nucleotides during "unscheduled DNA synthesis" [54, 55]. As yet, no adequate test system has been developed to measure DNA repair *in vivo* in man, although there are very promising starts in that direction [56]. (See pages 333 and 349). Our general knowledge of repair systems in mammals is still very limited. It is thus premature to start extensive pilot testing of repair systems to measure mutagenic exposure. But this is a promising direction for research.

2.2.6. Mutant Proteins in Somatic Cells

It has been assumed that the same amino acid substitutions that occur as a result of mutation in germ cells can also arise in somatic cells. In this case, rare cells would have the mutant genotype and would not likely influence the overall phenotype. But it should be possible to detect such variant cells if efficient screening methods can be devised [17, 57]. The validity of this approach has been demonstrated by Stamatoyannopoulos et al. [58], who found rare hemoglobin S- and C-positive red cells in normal persons. There remain major technical problems in detecting such cells by direct staining, and application to population monitoring is a prospect that must be delayed until the technical problems are solved [59]. Such variant cells should be directly equatable with mutation, although somatic, and should therefore be of direct relevance to mutagenic risk.

Mutant enzymes offer the possibility of selection in cultured cells. X-linked loci are especially useful, since all cells are effectively hemizygous for such loci. The locus that has been most studied is the *hprt* (hypoxanthine phosphoribosyltransferase) locus. Mutations that involve loss of activity are resistant to certain pure analogs, such as 8-azaguanine and 6-thioguanine. Such resistance was used to detect variant cells in freshly cultured fibroblasts and lymphocytes from humans [60, 61].

Albertini [62, 63] has been especially concerned with adapting the *hprt*$^-$ mutants for monitoring. The frequency of 6-thioguanine-resistant cells rises from a normal value of about 10^{-4} among lymphocytes to values some 20-fold higher in cancer patients under treatment. Evans and Vijayalaxmi [43] found an increase in azaguanine-resistant cells in lymphocytes treated *in vitro* with mitomycin C or X-rays. The increase paralleled that in SCE's, although there were 7000 SCE's for each *Ag*^{r} cell recovered. Such a ratio need not characterize all mutagens, however, since the mutational mechanisms may be different for different agents. Loss of activity of *hprt* could result either from point mutations or deletions [64]. A related locus under investigation is that for adenine phosphoribosyltransferase [65]. Although many questions remain about the translation of these results into genetic risk, the system is very promising and deserves careful study.

2.2.7. Body Fluid Analysis

The previous systems have measured genetic effects of one type or another in somatic cells. As we discussed earlier in the case of germ cells, risk estimates may also be based, although less securely, on measures of exposure. Demonstrated lack of exposure is at least a reassuring situation. The most extensively used test for exposure is analysis of body fluids for mutagenicity using microbial test organisms, most often the Ames Salmonella test. The tests and their variations are too well known to extend the discussion here. Blood and urine ordinarily are not mutagenic in these tests. Positive results have been demonstrated with a number of agents, such as medication [66], epichlorhydrin [67], and industrial agents in the rubber industry [68].

One characteristic of such tests is their dependence on recent exposure. As mutagenic metabolities are cleared from the body, the tests become negative. Therefore, analysis of body fluids complements those tests whose effects are lasting but where the agent may be difficult to pinpoint.

2.2.8. Protein Alkylation

Another test of exposure is the formation of protein derivatives by alkylating agents that also attack DNA. This method obviously has severe limitations, but it should be useful in appropriate circumstances. Hemoglobin is the human protein most easily isolated, and ethylated hemoglobin has been used to measure exposure to ethylene [69-71]. (see also page 315.) An advantage of this approach is its sensitivity to recent exposures and the ability to discriminate a specific industrial or therapeutic exposure from the potential mutagens.

3. CURRENT STATUS AND RECOMMENDATIONS

The many tests enumerated above and summarized in Table 2 have too often been characterized as requiring more research. Is nothing available that can be recommended right now? I think the answer is yes. Some, of course, are already being used to monitor workers in industry. Others could be used, both in special populations and in the general population.

As noted, cancer epidemiology is currently the best established test of mutagenic risk for the general population. Others could be used, however, and we should certainly seek tests that are more responsive to recent exposures. Central surveillance of cytogenic analyses of amniocenteses would provide valuable information without any additional laboratory effort. The biases of such a population can be evaluated and corrected. Any general increase in structural or genomic mutations should be recognizable. It might be very difficult to identify the agent responsible for such an increase, but that is no excuse for not knowing whether an increase has occurred. A beginning has been made by the Centers for Disease Control in implementing this system, and we should support it strongly.

Certain other tests could more appropriately be viewed as pilot projects at this time. These could be applied to a representative sample of the general population to gain experience with the tests and determine their utility for monitoring purposes. Such tests include SCE, chromosomes in cultured lymphocytes, micronuclei in lymphocytes, and YFF bodies in sperm. Tests at a much more experimental stage include $\underline{hprt}^-$ lymphocytes and DNA repair.

Monitoring of special exposure groups, especially those with chronic exposure to known mutagens, should be promptly instituted where this is not already the case. All the tests listed above with the exception of those based on amniocentesis are appropriate. Additional tests, such as analysis of body fluids, should also be added, and specific exposures may suggest others of the tests.

When this review was prepared, it was interesting to consider where we are today as compared to ten years ago [72-79]. We have made progress. We still emphasize the need for more research, but it is possible to start monitoring now. Indeed, modest beginnings already exist. There are test systems not yet thought of in 1972. I am therefore optimistic that the next decade will solve many of the problems we face today, and of course it will present us with some new problems.

REFERENCES

1. A. D. Bloom (ed.), Guidelines for Studies of Human Populations Exposed to Mutagenic and Reproductive Hazards, March of Dimes Birth Defects Foundation, White Plains, New York (1981), 163 pp.
2. H. E. Sutton, Workshop on monitoring human mutagenesis, Teratology 4:103-108 (1971).
3. H. B. Newcombe, Techniques for monitoring and assessing the significance of mutagenesis in human populations, Chemical Mutagens, Vol. 3 (A. Hollaender, ed.), pp. 57-77 (1973).
4. BEIR (Advisory Committee on the Biological Effects of Ionizing Radiation), The Effects on Populations of Exposure to Low Levels of Ionizing Radiation, National Academy of Sciences - National Research Council, Washington, D.C. (1972), 217 pp.
5. BEIR (Advisory Committee on the Biological Effects of Ionizing Radiation), The Effects on Populations of Exposure to Low Levels of Ionizing Radiation, National Academy of Sciences - National Research Council, Washington, D.C. (1981).
6. A. C. Stevenson, The load of hereditary defects in human populations, Radiat. Res., Suppl. 1:306-325 (1959).
7. B. K. Trimble and M. E. Smith, The incidence of genetic disease and the impact on man of an altered mutation rate, Canad. J. Genet. Cytol., 19:375-385 (1977).

8. J. V. Neel, T. O. Tiffany, and N. G. Anderson, Approaches to monitoring human populations for mutation rates and genetic disease, Chemical Mutagens, Vol. 3 (A. Hollaender, ed.), pp. 105-150 (1973).
9. J. V. Neel, Monitoring for genetic effects in man, Environmental Monitoring, National Academy of Sciences, Washington, D.C. (1977), pp. 113-129.
10. J. V. Neel, C. Satoh, H. B. Hamilton, M. Otake, K. Goriki, T. Kageoka, M. Fujita, S. Neriishi, and J. Asakawa, Search for mutations affecting protein structure in children of atomic bomb survivors: Preliminary report, Proc. Nat. Acad. Sci. (U.S.), 77:4221-4225 (1980).
11. F. Hecht, B. K. McCaw, D. Peakman, and A. Robinson, New translocations in human lymphocytes: A mutagen monitoring system, Environ. Health Persp., 31:19-22 (1979).
12. P. A. Jacobs, Population surveillance: A cytogenetic approach, Genetic Epidemiology (N. E. Morton and C. S. Chung, eds.), Academic Press, New York (1978), pp. 463-481.
13. R. W. Kapp, Jr., Detection of aneuploidy in human sperm, Environ. Health Persp., 31:27-31 (1979).
14. E. Rudak, P. A. Jacobs, and R. Yanagimachi, Direct analysis of the chromosome constitution of human spermatozoa, Nature, 274: 911-913 (1978).
15. R.H. Martin, C. C. Lin, W. Balkan, and K. Burns, Direct chromosomal analysis of human spermatozoa: Preliminary results from 18 normal men, Amer. J. Human Genet., 34:459-468 (1982).
16. A. A. Ansari, M. A. Baig, and H. V. Malling, In vivo germinal mutation detection with "monospecific" antibody against lactate dehydrogenase-X, Proc. Nat. Acad. Sci. (U.S.), 77:7352-7356 (1980).
17. H. E. Sutton, Prospects of monitoring environmental mutagenesis through somatic mutations, Monitoring, Birth Defects and Environment (E. B. Hook, D. T. Janerich, and I. H. Porter, eds.), pp. 237-248, Academic Press, New York (1971).
18. H. E. Sutton, Somatic mutation in human populations, Human Genetics, Part A: The Unfolding Genome, Alan R. Liss, New York, pp. 289-298 (1982).
19. B. N. Ames, W. E. Durston, E. Yamasaki, and F. D. Lee, Carcinogens are mutagens: A simple test system combining liver homogenates for activation and bacteria for detection, Proc. Nat. Acad. Sci. (U.S.), 70:2281-2285 (1973).
20. A. G. Knudson, Jr., Genetics and etiology of human cancer, Adv. Human Genetics, 8:1-66 (1977).
21. D. S. Straus, Somatic mutation, cellular differentiation, and cancer causation, J. Nat. Cancer Inst., 67:233-241 (1981).
22. R. W. Miller, The discovery of human teratogens, carcinogens, and mutagens: Lessons for the future, Chemical Mutagens, Vol. 5 (A. Hollaender and F. J. de Serres, eds.), pp. 101-126 (1978).
23. W. J. Schull, M. Otake, and J. V. Neel, Genetic effects of the atomic bombs: A reappraisal, Science, 213, 1220-1227 (1981).

24. A. J. Wyrobek and W. R. Bruce, The induction of sperm-shape abnormalities in mice and humans, Chemical Mutagens, Vol. 5 (A. Hollaender and F. J. de Serres, eds.), pp. 257-285 (1978).
25. A. J. Wyrobek, G. Watchmaker, L. Gordon, K. Wong, D. Moore II, and D. Whorton, Sperm shape abnormalities in carbarbyl-exposed employees, Environ. Health Persp., 40:255-265 (1981).
26. J. C. Topham, The detection of carcinogen-induced sperm head abnormalities in mice, Mutat. Res., 69:149-155 (1980).
27. R. W. Miller, Pollutants and children: Lessons from case histories, in (1), pp. 155-163.
28. W. R. Lee, Dosimetry of chemical mutagens in eukaryote germ cells, Chemical Mutagens, Vol. 5 (A. Hollaender and F. J. de Serres, eds.), pp. 177-202 (1978).
29. A. A. Awa, S. Neriishi, T. Honda, M. C. Yoshida, T. Sofuni, and T. Matsui, Chromosome-aberration frequency in cultured blood-cells in relation to radiation dose of A-bomb survivors, Lancet, 2:903-905 (1971).
30. T. Sofuni, H. Shimba, K. Ohtaki, and A. A. Awa, G-banding analysis of chromosome aberrations in Hiroshima atomic bomb survivors, Radiation Effects Research Foundation Technical Report 13-77 (1978).
31. D. C. Lloyd, The problems of interpreting aberration yields induced by in vivo irradiation of lymphocytes, Mutagen-Induced Chromosome Damage in Man (H. J. Evans and D. C. Lloyd, eds.), pp. 77-88, Yale Univ. Press, New Haven (1978).
32. W. F. Brandom, G. Saccomanno, V. E. Archer, P. G. Archer, and A. D. Bloom, Chromosome aberrations as a biological dose-response indicator of radiation exposure in uranium miners, Radiat. Res., 76:159-171 (1978).
33. W. F. Brandom, P. G. Archer, A. D. Bloom, V. E. Archer, R. W. Bistline, and G. Saccomanno, Chromosome changes in somatic cells of workers with internal depositions of plutonium Biological Implications of Radionuclides Released from Nuclear Industries, Vol. 2, pp. 195-210, International Atomic Energy Agency, Vienna (1979).
34. K. E. Buckton, G. E. Hamilton, L. Paton, and A. O. Langlands, Chromosome aberrations in irradiated ankylosing spondylitis patients, Mutagen-Induced Chromosome Damage in Man (H. J. Evans and D. C. Lloyd, eds.), pp. 142-150, Edinburgh Univ. Press, Edinburgh (1978).
35. H. J. Evans, K. E. Buckton, G. E. Hamilton, and A. Carothers, Radiation-induced chromosome aberrations in nuclear-dockyard workers, Nature, 277:531-534 (1979).
36. R. J. Srám, Z. Zudová, and N. P. Kuleshov, Cytogenetic analysis of peripheral lymphocytes in workers occupationally exposed to epichlorohydrin, Mutat. Res., 70:115-120 (1980).
37. S. A. Latt, Sister chromatid exchanges, indices of human chromosome damage and repair: Detection by fluorescence and induction by mitomycin C, Proc. Nat. Acad. Sci. (U.S.), 71:3162-3166 (1974).

38. S. A. Latt, Sister chromatid exchange formation, Ann. Rev. Genet., 15:11-55 (1981).
39. S. Wolff, Sister chromatid exchange, Ann. Rev. Genet., 11:183-201 (1977).
40. S. Wolff, Sister chromatid exchange: The most sensitive mammalian system for determining the effects of mutagenic carcinogens, Genetic Damage in Man Caused by Environmental Agents (K. Berg, ed.), pp. 229-246, Academic Press, New York (1979).
41. A. V. Carrano, L. H. Thompson, P. A. Lindl, and J. L. Minkler, Sister chromatid exchange as an indicator of mutagenesis, Nature, 271:551-553 (1978).
42. P. E. Perry, Chemical mutagens and sister-chromatid exchange, Chemical Mutagens, Vol. 6 (F. J. de Serres and A. Hollaender, eds.), pp. 1-39, Plenum Press, New York (1980).
43. H. J. Evans and Vijayalaxmi, Induction of 8-azaguanine resistance and sister chromatid exchange in human lymphocytes exposed to mitomycin C and X-rays _in vitro_, Nature, 292:601-605 (1981).
44. F. Funes-Cravioto, C. Zapata-Gayon, B. Kolmodin-Hedman, B. Lambert, J. Lindsten, E. Norberg, M. Nordenskjöld, R. Olin, and Å. Swensson, Chromosome aberrations and sister chromatid exchange in workers in chemical laboratories and a rotoprinting factory and in children of women laboratory workers, Lancet, 2: 322-325 (1977).
45. D. H. Hollander, M. S. Tockman, Y. W. Liang, D. S. Borgaonkar, and J. K. Frost, Sister chromatid exchanges in the peripheral blood of cigarette smokers and in lung cancer patients; and the effect of chemotherapy, Hum. Genet., 44:165-171 (1978).
46. T. Raposa, Sister chromatid exchange studies for monitoring DNA damage and repair capacity after cytostatics _in vitro_ and in lymphocytes of leukaemic patients under cytostatic therapy, Mutat. Res., 57:241-251 (1978).
47. V. F. Garry, J. Hozier, D. Jacobs, R. L. Wade, and D. G. Gray, Ethylene oxide: Evidence of human chromosomal effects, Environ. Mutagen., 1:375-382 (1979).
48. B. Lambert, U. Ringborg, and A. Lindblad, Prolonged increase of sister-chromatid exchanges in lymphocytes of melanoma patients after CCNU treatment, Mutat. Res., 59:295-300 (1979).
49. L. G. Littlefield, S. P. Colyer, and R. J. DuFrain, Comparison of sister-chromatid exchanges in human lymphocytes after G_0 exposure to mitomycin _in vivo_ vs. _in vitro_, Mutat. Res., 69: 191-197 (1980).
50. W. Schmid, The micronucleus test, Mutat. Res., 31:9-15 (1975).
51. W. Schmid, The micronucleus test for cytogenetic analysis, Chemical Mutagens, Vol. 4 (A. Hollaender, ed.), pp. 31-53 (1976).
52. P. I. Countryman and J. A. Heddle, The production of micronuclei from chromosome aberrations in irradiated cultures of human lymphocytes, Mutat. Res., 41:321-332 (1976).

53. J. A. Heddle, R. D. Benz, and P. I. Countryman, Measurement of chromosomal breakage in cultured cells by the micronucleus technique, Mutagen-Induced Chromosome Damage in Man (H. J. Evans and D. C. Lloyd, eds.), pp. 191-200, Yale Univ. Press, New Haven (1978).
54. J. D. Regan and R. B. Setlow, Repair of chemical damage to human DNA, Chemical Mutagens, Vol. 3 (A. Hollaender, ed.), pp. 151-170 (1973).
55. J. E. Cleaver, Methods for studying excision repair of DNA damaged by physical and chemical mutagens, Handbook of Mutagenicity Test Procedures (B. J. Kilbey, M. Legator, W. Nichols, and C. Ramel, eds.), pp. 19-48, Elsevier, Amsterdam (1977).
56. R. W. Pero and F. Mitelman, Another approach to in vivo estimation of genetic damage in humans, Proc. Nat. Acad. Sci. (U.S.), 76:462-463 (1979).
57. H. E. Sutton, Somatic cell mutations, Birth Defects (A. G. Motulsky and W. Lenz, eds.), pp. 212-214, Excerpta Medica, Amsterdam (1974).
58. G. Stamatoyannopoulos, P. E. Nute, T. Papayannopoulou, T. McGuire, G. Lim, H. F. Bunn, and D. Rucknagel, Development of a somatic mutation screening system using Hb mutants. IV. Successful detection of red cells containing the human frameshift mutants Hb Wayne and Hb Cranston using monospecific fluorescent antibodies, Am. J. Hum. Genet., 32:484-496 (1980).
59. W. L. Bigbee, E. W. Branscomb, H. B. Weintraub, T. Papayannopoulou, and G. Stamatoyannopoulos, Cell sorter immunofluorescence detection of human erythrocytes labelled in suspension with antibodies specific for hemoglobin S and C, J. Immunol. Methods, 45:117-127 (1981).
60. R. De Mars and K. R. Held, The spontaneous azaguanine-resistant mutants of diploid human fibroblasts, Humangenetik, 16:87-110 (1972).
61. K. Sato, R. S. Slesinski, and J. W. Littlefield, Chemical mutagenesis at the phosphoribosyltransferase locus in cultured human lymphoblasts, Proc. Nat. Acad. Sci. (U.S.), 69:1244-1248 (1972).
62. G. H. Strauss and R. J. Albertini, Enumeration of 6-thioguanine-resistant peripheral blood lymphocytes in man as a potential test for somatic cell mutations arising in vivo, Mutat. Res., 61:353-379 (1979).
63. R. J. Albertini, Drug-resistant lymphocytes in man as indicators of somatic mutation, Teratogenesis, Carcinogenesis, and Mutagenesis, 1:25-48 (1980).
64. R. Cox and W. K. Masson, Do radiation-induced thioguanine-resistant mutants of cultured mammalian cells arise by HGPRT gene mutation or X-chromosome rearrangement?, Nature, 276:629-630 (1978).
65. R. De Mars, Suggestions for increasing the scope of direct testing for mutagens and carcinogens in intact humans and animals, Mammalian Cell Mutagenesis: The Maturation of Test Systems, Banbury Report No. 2, pp. 329-340, Cold Spring Harbor Laboratory (1979).

66. V. Minnich, M. E. Smith, D. Thompson, and S. Kornfeld, Detection of mutagenic activity in human urine using mutant strains of Salmonella typhimurium, Cancer, 38:1253-1258 (1976).
67. M. S. Legator, L. Truong, and T. H. Connor, Analysis of body fluids including alkylation of macromolecules for detection of mutagenic agents, Chemical Mutagens, Vol. 5 (A. Hollaender and F. J. de Serres, eds.), pp. 1-23 (1978).
68. K. Falck, M. Sorsa, and H. Vainio, Mutagenicity in urine of workers in rubber industry, Mutat. Res., 79:45-52 (1980).
69. S. Osterman-Golkar, L. Ehrenberg, D. Segerbäck, and I. Hällström, Evaluation of genetic risks of alkylating agents. II. Haemoglobin as a dose monitor, Mutat. Res., 34:1-10 (1976).
70. L. Ehrenberg, S. Osterman-Golkar, D. Segerbäck, K. Svensson, and C. J. Calleman, Evaluation of genetic risks of alkylating agents. III. Alkylation of haemoglobin after metabolic conversion of ethene to ethene oxide in vivo, Mutat. Res., 45: 175-184 (1977).
71. C. J. Calleman, L. Ehrenberg, B. Jansson, S. Osterman-Golkar, D. Segerbäck, K. Svensson, and C. A. Wachmeister, Monitoring and risk assessment by means of alkyl groups in hemoglobin in persons occupationally exposed to ethylene oxide, J. Environ. Path. Toxicol., 2:427-442 (1978).
72. J. F. Crow, Chemical risk to future generations, Scientist and Citizen, 10:113-117 (1968).
73. J. V. Neel and A. D. Bloom, The detection of environmental mutgens, Med. Clin. N. Amer., 53, 1243-1256 (1969).
74. J. F. Crow, Human population monitoring, Chemical Mutagens, Vol. 2 (A. Hollaender, ed.), pp. 591-605 (1970).
75. J. V. Neel, Evaluation of the effects of chemical mutagens on man: The long road ahead, Proc. Nat. Acad. Sci. (U.S.), 67: 908-915 (1970).
76. F. Vogel and G. Röhrborn (eds.), Chemical Mutageneis in Mammals and Man, 519 pp., Springer-Verlag, New York-Berlin-Heidelberg (1970).
77. E. B. Hook, D. T. Janerich, and I. H. Porter (eds.), Monitoring, Birth Defects and Environment: The Problem of Surveillance, 308 pp., Academic Press, New York (1971).
78. H. E. Sutton and M. I. Harris (eds.), Mutagenic Effects of Environmental Contaminants, 195 pp., Academic Press, New York (1972).
79. C. Ramel (ed.), Evaluation of Genetic Risks of Environmental Chemicals, Ambio Special Report No. 3, Royal Swedish Academy of Sciences/Universitetsforlaget, 27 pp. (1973).

EVALUATION OF GENOTOXIC EFFECTS IN HUMAN POPULATIONS*

M. D. Waters[1], J. W. Allen[1], L. D. Claxton[1],
N. E. Garrett[2], S. L. Huang[2], M. M. Moore[1],
Y. Sharief[2], and G. H. Strauss[1]

[1]Genetic Toxicology Division
Health Effects Research Laboratory
U.S. Environmental Protection Agency
Research Triangle Park, North Carolina 27711

[2]Northrop Services, Inc.
Environmental Sciences
Research Triangle Park, North Carolina 27709

1. INTRODUCTION

1.1. Need to Detect Chemically Induced Genetic Damage in Humans

There are demonstrable associations in experimental animals between DNA damage in somatic cells and the development of cancer, and between DNA damage in germ cells and the incidence of genetic disease in offspring. Thus, there is substantial evidence of the need to detect and to quantitate chemically induced genetic damage in humans in order to assess the potential for cancer and genetic disease. Current methods for estimating human risk of cancer have been based on knowledge of human exposure and epidemiological data. An alternative approach would base such estimates of risk on knowledge of exposure and of damage to the DNA of human cells and tissues. The

---

*This manuscript has been reviewed by the Health Effects Research Laboratory, U.S. Environmental Protection Agency, and approved for publication. Approval does not signify that the contents necessarily reflect the views and policies of the Agency, nor does mention of trade names or commercial products constitute endorsement or recommendation for use.

principal difficulty in performing such assessments lies in the uncertainty of translating information from tests for genetic damage in human cells and tissues into reliable estimates of risk for cancer or genetic disease.

This paper concerns: (1) the development of mammalian cell methods that may be used ultimately to evaluate genotoxic effects in humans and (2) our initial attempts to implement an approach whereby such methods can be used, collectively, to relate quantitative information on genetic damage to quantitative estimates of risk for cancer or genetic disease. Such an approach is not new, having been advocated previously by Sobels [1, 2], Malling [3], and Lyon [4].

The "parallelogram method" is an approach involving somatic- and germ-cell monitoring techniques whereby quantitative dose-response relationships are developed among component test systems. With the parallelogram method, illustrated in Fig. 1, estimates of chemical (external) exposure and cellular and molecular (internal) dose provide the quantitative basis for extrapolating data from whole animal systems to human populations. The use of animal and human cells *in vitro* enables direct comparisons of induced effects under rather carefully controlled dosages. Such comparisons have conventionally been based on measurements of total dose to the DNA or common molecular products in the DNA determined after exposure of the organism to the agent in question.

The relative mutagenic effects of individual agents may be subject to further quantitation if the kinds of DNA adducts formed and their distribution within the molecule after exposure can be characterized in the various test systems and tissues. A vast literature already exists concerning the products of chemical interactions with DNA and the initiation of carcinogenesis. Singer [5], in classifying the reaction products formed by the alkylation of nucleic acids and polynucleotides, found the correlation with carcinogenicity to be good for certain alkylation products. Singer also reported that the relative amount of specific alkylated derivatives of DNA is similar, even though total alkylation may vary a thousandfold for DNA. This observation holds for analyses made in the test tube, in intact cells, or in the whole animal [5]. In the final analysis, the reaction products formed by the interaction of chemicals with DNA are of consequence for assessment purposes only if they relate directly to genetically significant targets.

1.2. Biological and Dosimetric Methods Available

Under certain conditions, genotoxic damage can be measured directly in acutely exposed humans. However, if the genetic damage is repaired, little or no evidence may exist that can be related to heritable mutations or tumors in the exposed population. Studies

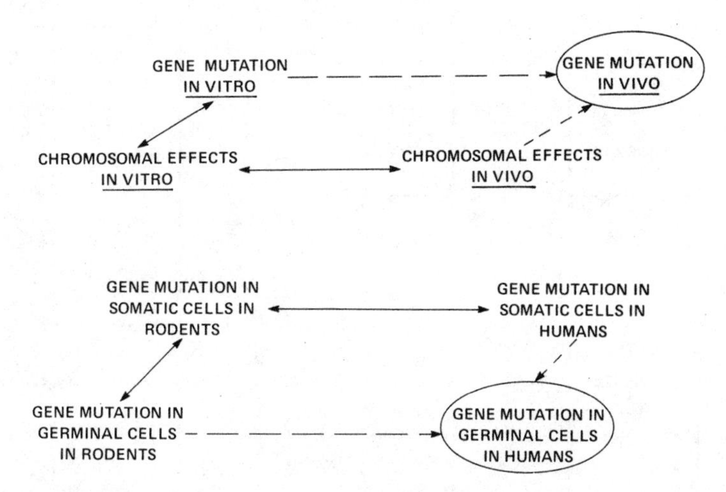

Fig. 1. a) Parallelogram for extrapolation to gene mutation *in vivo*. b) Parallelogram for extrapolation to gene mutation in germinal cells in humans.

using low chronic exposures are also limited, because they may not produce a measurable genotoxic endpoint or evidence of an increased disease frequency. Thus, within the limits of current methodology, some combination of animal and short-term genetic tests must be used to supplement any available human exposure data.

The relationships presented in Fig. 1 and expanded in Fig. 2 illustrate the parallelogram method of predicting genotoxic effects in humans. As illustrated, external exposure (e.g., ambient air) or internal exposure (e.g., intraperitoneal injection) can be functionally related to mutation or tumor incidence in whole animal or short-term test systems. Data from these test systems can, in turn, be used to estimate effects in humans under similar exposure conditions. Estimates are made in a stepwise fashion and are corrected by factors that account for differences in test system sensitivity, size differences between species, etc. If uncertainties in this approach can be held to a minimum, one or more test systems then can be used to assess possible carcinogenicity or heritable mutation in humans.

As indicated by the arrows in Fig. 2, one species system (e.g., mouse) can be used to relate external exposure to internal exposure (blood level) and internal exposure to the level of DNA alkylation in germ tissue. However, this species may not be the best one for linking the level of DNA alkylation to mutation in germ tissue. This may be effectively achieved with a different species (e.g., Drosophila), which allows DNA alkylation to be functionally related to mutation in germ tissue. The surrogate species is useful, pro-

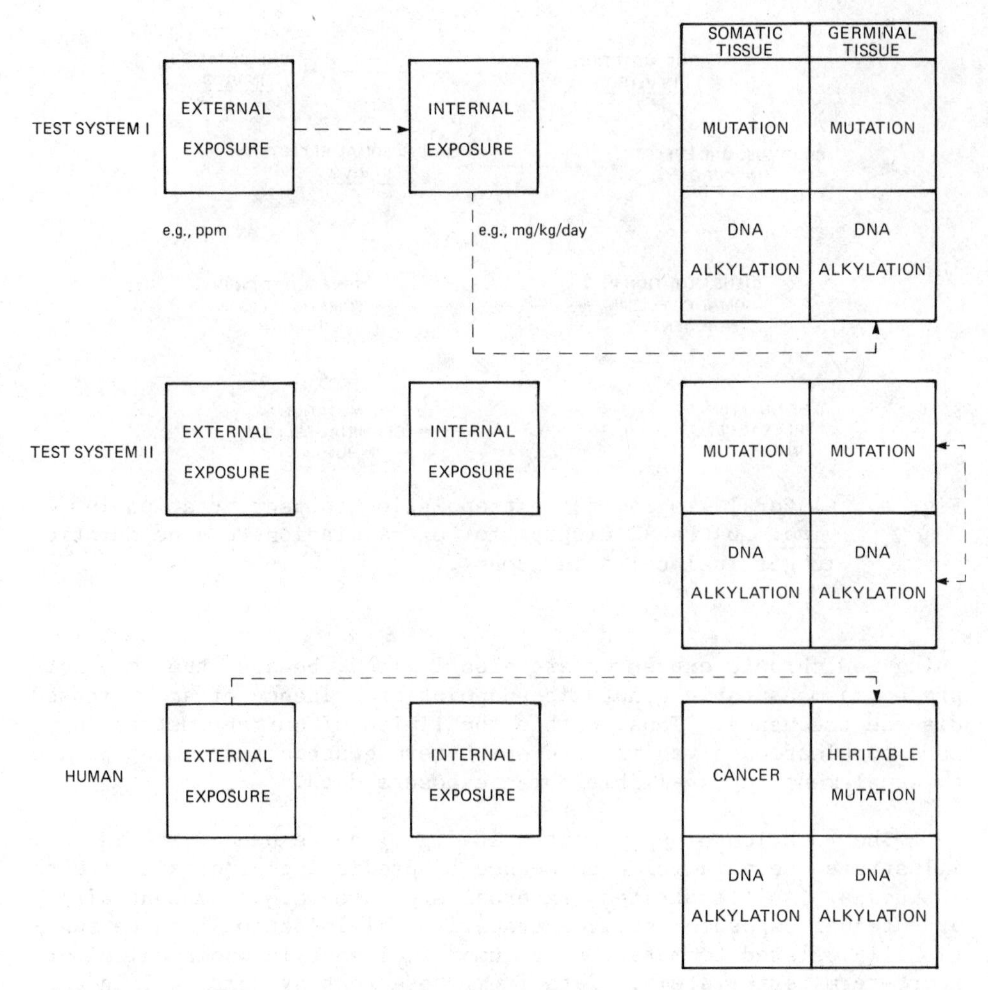

Fig. 2. Expanded concept of parallelogram method considering external and internal exposure.

viding that DNA alkylation and gene mutation are parallel in the two species over the dose range considered. Such comparative dosimetry forms the quantitative basis of the assessment process referred to as the parallelogram method [6-10].

1.3. Chemicals under Investigation

Several chemical agents are under investigation in this laboratory for which parallel studies in animal and human assay systems *in vitro* and *in vivo* are feasible. These include the cancer chemotherapeutic agent, cyclophosphamide, and the industrial intermediate, ethyl carbamate (urethane). Each of these compounds presents a useful model for the purposes of our investigations. The former agent is of substantial clinical relevance and may be useful in further development of the parallelogram approach. The latter agent is of perhaps greater theoretical significance in terms of our lack of understanding of its metabolism, especially in short-term tests, and with regard to the difficulties in extrapolating data from such tests to experimental animals and humans.

Cyclophosphamide is an alkylating nitrogen mustard; its reactive moiety is the bis(2-chloroethyl) group. The compound is widely used clinically for treatment of Hodgkin's Disease and in the treatment of lymphoma and leukemia, neuroblastoma, carcinoma of the breast, and a variety of other malignant diseases. The agent is also used as an immunosuppressant in several types of nonmalignant diseases, including rheumatoid arthritis, nephrotic syndrome in children, and chronic hepatitis. Hence, there are substantial opportunities to obtain human exposure and effects data with this compound.

Ethyl carbamate (EC) is considered a "problem carcinogen" because of erratic or generally negative genetic test results [12]. EC was one of the earliest known chemical carcinogens [13], and it has continued to be a public health concern because of its former use in medicine and current widespread use in industry [14]. Of several "model rodent carcinogens" recently studied, EC alone also proved to be carcinogenic for Old and New World species of monkey [15]. Despite 40 yr of various toxicological studies with EC, it still is not understood in terms of mechanistic pathways leading to pathology. Covalent binding to DNA is known to occur [16]; however, EC has proven relatively inactive in most short-term tests for carcinogens [12, 17]. Therefore, EC provides a useful, albeit difficult, "model carcinogen" for comparative studies in genetic test systems.

2. METHODOLOGY

2.1. Introduction to Component Tests

The following section describes the component test methods being employed in this laboratory to measure chromosomal effects and gene mutation, and to provide biological evidence of exposure (urine screening). Results obtained from such biological measurements will be integrated using appropriate dosimetric techniques. The specific test methods to be described are:

- Methods for measuring chromosomal effects
  - Sister chromatid exchange, *in vitro* and *in vivo*
  - Aberrations, *in vitro* and *in vivo*

- Methods for measuring (putative) gene mutation
  - HGPRT, fibroblasts, in vitro
  - HGPRT, lymphocytes, in vitro and in vivo
  - TK, lymphoma cells, in vitro
- Methods for measuring exposure
  - Urine screening with *Salmonella typhimurium*
- Dosimetry methods
  - Total DNA adducts

2.2. Description of Test Methods

2.2.1. Methods for Measuring Chromosomal Damage

Major areas of cytogenetic investigation include human monitoring, and in vitro and in vivo mammalian cell studies. These areas are discussed below.

2.2.1a. Human Monitoring. Two major cytogenetic endpoints are evaluated in human monitoring studies - chromosome aberrations and sister chromatid exchange (SCE) induction. Peripheral lymphocytes are obtained from individuals exposed to a chemical agent and cultured in an appropriate growth medium containing phytohemagglutinin (PHA) to stimulate cell cycling. Standard procedures [11, 18, 19] for cell harvest, slide preparation, and aberration analysis are followed.

In the SCE studies, 5×10^6 lymphocytes are cultured in 10 ml of growth medium, to which 25 μM 5-bromodeoxyuridine (BrdU) is added after 24 hr [22] (Fig. 3). Cells are harvested after 72 hr in culture (2 hr after a colchicine treatment to collect metaphases), by which time a high proportion of cells have cycled twice and become unequally substituted with BrdU in the sister chromatids. Cells are then stained with a BrdU-sensitive dye and examined for differentiated sister chromatids in which SCEs manifest themselves as reciprocal exchanges. First metaphase division cells can also be harvested at earlier times (e.g., 48-54 hr) and analyzed for chromosome aberrations. The alternative methods for aberration analysis (cells with or without BrdU substitution), the appropriate conditions under which aberration and SCE analyses are applied, and the individual strengths and weaknesses of the two endpoints have been reviewed [11, 18-21]. For further details concerning the procedure described here, see Allen et al. [22].

2.2.1b. In vitro Analyses. The in vitro studies routinely conducted involve the use of lymphocytes (human, rodent) or Chinese hamster V79 cells. Details concerning these analyses can be found in [22, 23]. As shown in Fig. 4, alternative activation sources and multiple genotoxic endpoints can be accommodated by the V79 system;

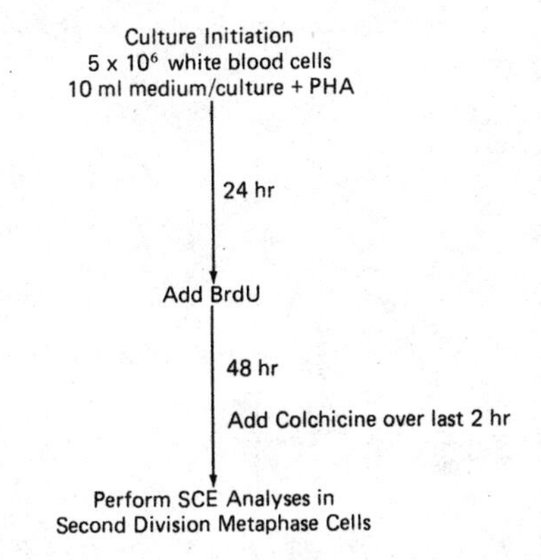

Fig. 3. Protocol for culturing human lymphocytes for SCE analysis.

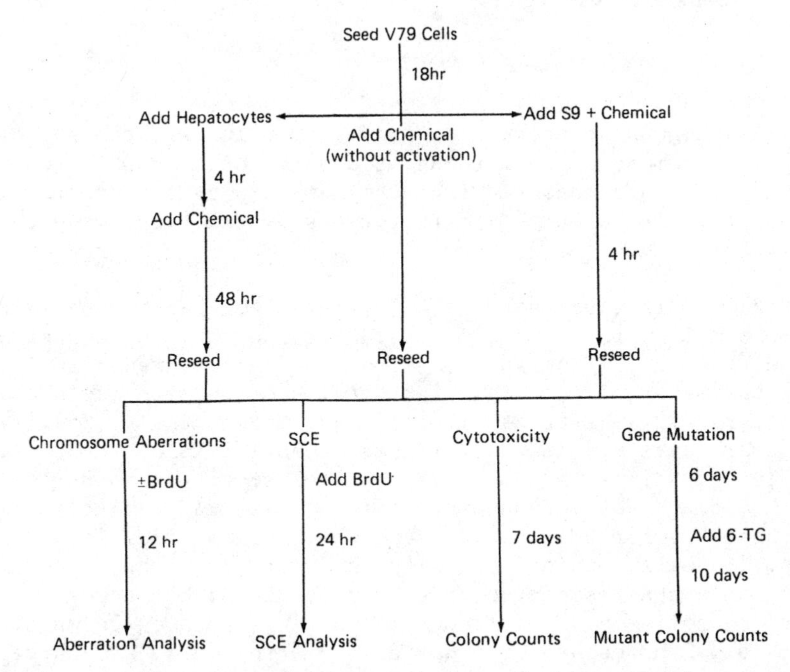

Fig. 4. Protocol for in vitro analysis of chromosomal, gene-mutagenic, and cytotoxic effects from mutagens/carcinogens.

the number of cells seeded varies with the activation system employed and the particular endpoint(s) evaluated. For activation, liver S9 enzymes or intact cells (e.g., hepatocytes) may be used. After test chemical treatment, cells are reseeded under separate growth conditions for each endpoint being assayed, i.e., gene muta-

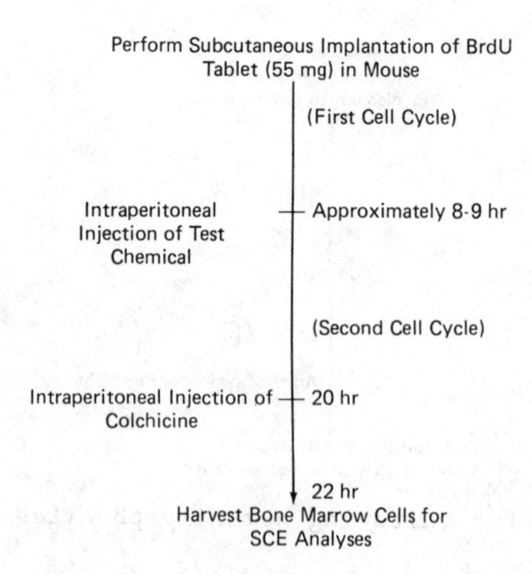

Fig. 5. Protocol for in vivo analysis of SCE induction in rodent bone marrow cells.

tions (6-thioguanine resistance and/or ouabain resistance), cytotoxicity, chromosome aberrations, and SCE. With this approach, several endpoints are measured from the same exposure population, and the sensitivities of the different types of genetic damage can be readily compared.

2.2.1c. In vivo Analyses. In vivo cytogenetic studies involve the use of mice, rats, or hamsters. SCE and chromosome aberration analyses are usually performed using bone marrow, regenerating liver, and/or spermatogonial cells. For details of these procedures, see Allen et al. [24], Latt et al. [11], and McFee et al. [25]. Aberrations are analyzed at the first metaphase division (approximately 12 hr in somatic cells, 26 hr in germ cells) after exposure to the test chemical. For SCE analyses (Fig. 5), a BrdU tablet is implanted subcutaneously for DNA labeling and the test chemical is administered intraperitoneally, coincident with the start of either the first or second DNA synthesis period or just prior to BrdU treatment. Metaphases are collected by intraperitoneal injection of colchicine (approximately 2 hr before cell harvest); marrow cells are harvested and analyzed for SCE approximately 22-24 hr after the start of labeling. BrdU tablet methodology can also be used for studies of transplacental SCE induction in specific fetal rodent tissues [26].

2.2.2. Methods for Measuring Gene Mutation

Presumptive gene mutation is being measured in a variety of mammalian and human lymphocyte and fibroblast culture systems. Exposures are in vitro as well as in vivo.

2.2.2a. HGPRT, Fibroblasts - in vitro. The method employed to detect gene mutation at the HGPRT locus in normal human fibroblasts was developed by Albertini and DeMars [27] using foreskin fibroblasts. In this method, the induction of 8-azaguanine (AG^r) or 6-thioguanine (TG^r) resistant cells is studied in the presence or absence of S9 activation.

Guanine, hypoxanthine, and 6-mercaptopurine are good substrates for HGPRTase, an enzyme that generally plays a salvage role in mammalian purine metabolism. HGPRTase plays a key role in the anabolism of 6-oxypurines and their analogs, converting 8-azaguanine (AG) or 6-thioguanine (TG) to nucleotides that inhibit the growth of normal fibroblasts. Because of their reduced ability to produce these nucleotides, HGPRT-deficient cells proliferate in the presence of AG or TG. It thus becomes possible to select these cells from a mutagenized cell population; the mutants resemble the naturally occurring Lesch-Nyhan (LN) cells.

The HGPRT locus is known to be on the X chromosome. Since males have a single X chromosome and females only one active X chromosome, mutations at the HGPRT locus should be readily recoverable as AG or TG resistance in cells of male or female origin.

Previous studies indicated that only S-phase cells were sensitive to mutagens, thereby implying that DNA replication was required for the induction of mutation [28-34]. Experiments have been conducted using synchronized human cells to detect mutagenic activity of several chemical agents, including cyclophosphamide. Unfortunately, experiments using synchronized cells receiving only a single treatment of cyclophosphamide have failed to produce mutations. Studies using a different protocol, i.e., nonsynchronized cells receiving multiple treatments, have indicated that cyclophosphamide is mutagenic to human cells. The protocols used in these studies [single treatment in synchronized cells and multiple treatments in nonsynchronized cells] are discussed below.

As illustrated in Fig. 6, in tests involving a single treatment, human fibroblasts are synchronized according to the method of Milo and DiPaolo [35]. With this method, 1.5×10^6 S-phase cells per 100 mm dish are exposed to the chemical test agent for 4 hr. The treated cells are then grown in FCS-F10 medium (10% fetal calf serum in Ham's medium) for a 7 day expression period with two passages. After 7 days, 10^4 cells are plated per 60-mm dish, and each dish receives TG at a final concentration of 7.5×10^{-5} moles. The selective medium is renewed once every week. In order to determine cloning efficiency, cells are plated (100 cells per 60-mm dish) and grown in FCS-F10 medium. Dishes used to estimate the cell survival are stained at the end of the second week; dishes used to determine mutagenesis are stained at the end of the third week. Mutant colonies are then counted, and the mutation frequencies are determined using a Poisson distribu-

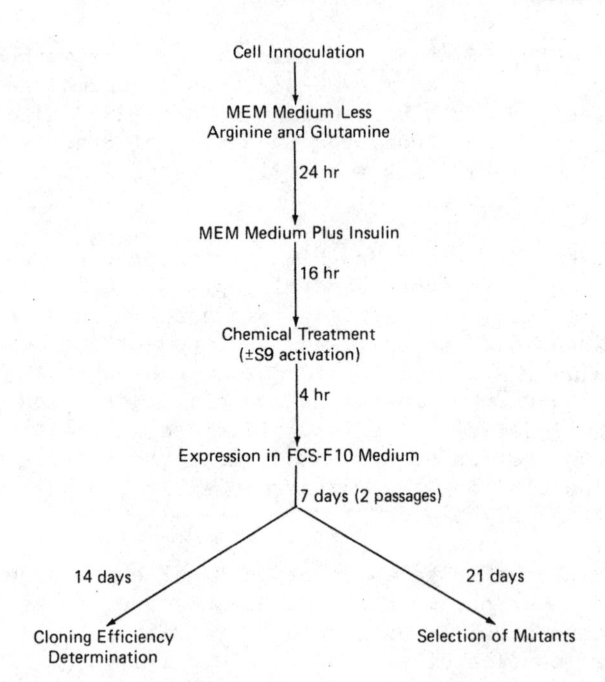

Fig. 6. Induction of 6-thioguanine resistance studied in synchronized human fibroblasts in the presence or absence of S9 activation.

tion [27, 36-39]. In order to test for promutagens, a reaction mixture consisting of a balanced salt solution and cofactors, 1 ml S9 fraction, mutagen dissolved in 0.1 ml dimethyl sulfoxide, and 8.5 ml F10 medium is prepared. S-phase cells are exposed to the mixture and are incubated at 37°C for 4 hr. After a 7-day expression period, cells are plated and mutation frequencies are determined.

For tests involving multiple treatments (Fig. 7), 2.5×10^5 nonsynchronized cells per 60 mm dish are exposed to the test agent for 4 hr daily for 3 to 4 days. Five ore more dishes are used for each concentration. The treated cells are then expressed in FCS-F10 medium for 7 days, with two passages. Because of the toxic effect of S9, cell exposures to the reaction mixture and promutagen are limited to 4 hr each day with a reduced amount of S9, for 4 consecutive days. After each treatment, the S9 and mutagens are removed and fresh medium is introduced to the cultures. After a 7-day expression period, cells are plated and mutation frequencies are determined.

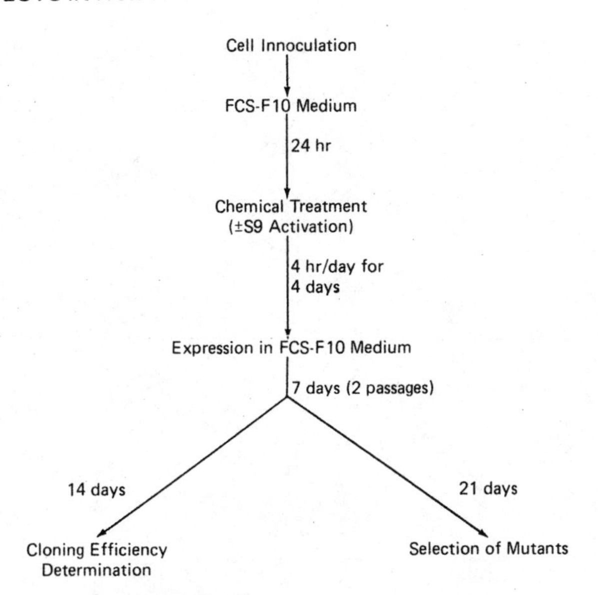

Fig. 7. Induction of 6-thioguanine resistance studied in nonsynchronized human fibroblasts in the presence or absence of S9 activation.

2.2.2b. HGPRT, Lymphocytes - *in vivo*. The Strauss-Albertini test was originally developed as a direct technique for assessing the *in vivo* incidence of mutant cells arising in circulating lymphocytes in animals and man [40, 41]. In the test, rare TG' peripheral blood lymphocyte (PBL) variants that incorporate ^{3}H-TdR in response to PHA, despite the presence of 6-thioguanine, can be enumerated by light microscopy after autoradiography (Fig. 8). The 42-hr culture period is of insufficient length to allow all responding lymphocytes to divide such that *in vitro* TG^r PBL frequencies correspond to *in vivo* frequencies. Conditions for testing humans are those shown; however, rat lymphocytes are cultured 60 hr prior to adding ^{3}H-TdR. The test includes a cryopreservation step to eliminate "phenocopies," PBLs that are artifically TG' owing to the resistance capacity of cells already cycling at the moment of challenge.

Variants are assumed to be $HGPRT^-$ mutants by virtue of various characterizations and analogy with the naturally occurring LN mutation. The test has been suggested as a means to diagnose LN heterozygosity [42]. Thioguanine-resistant PBLs have been found at frequencies ranging from 4 to 6×10^{-6} in normal individuals and at significantly higher frequencies in humans and animals receiving iatrogenic and experimental mutagen exposures, respectively [40-41]. The Strauss-Albertini test is well suited to long-term repeated monitoring due to the brevity of both latency and patency periods in-

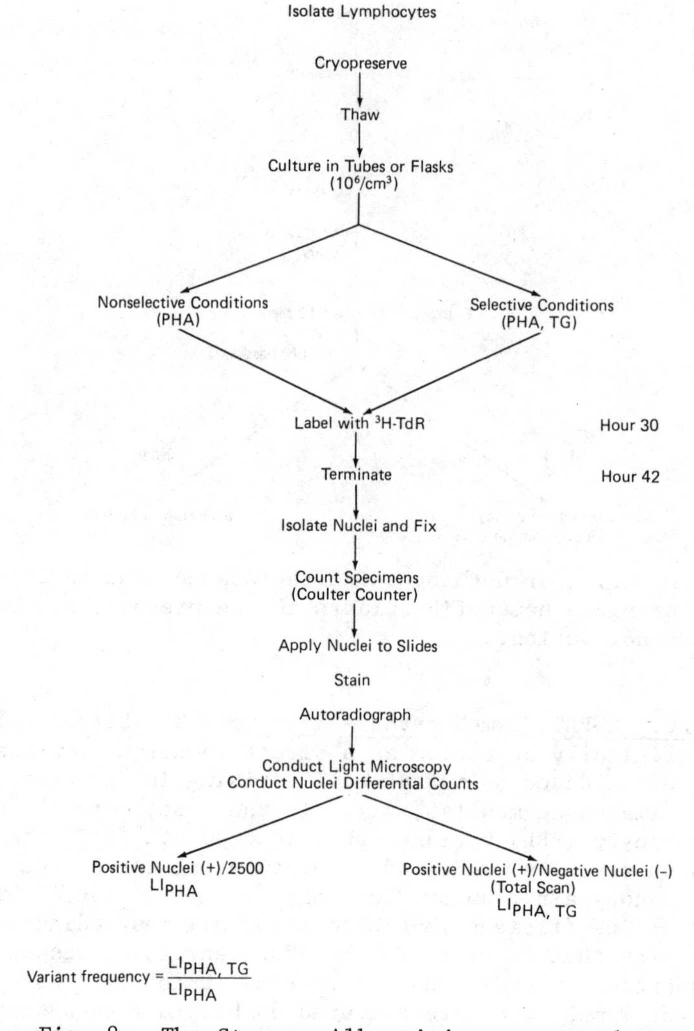

Fig. 8. The Strauss-Albertini test procedure.

fluencing variant frequencies after exposures to mutagens. Confirmation of the mutant nature of variants is now possible using the new lymphocyte clonal assay as follows:

The lymphocyte clonal assay is an adaptation of an immunological technique for cloning nontransformed T-lymphocytes from blood and other tissues [43]. By using T-cell growth factor, continuously growing T-lymphocytes (CTL) can be grown for months with maintenance of normal functions. Rare drug-resistant precursors cells can be isolated and expanded to form visible colonies in appropriate selective media containing T-cell growth factor. As depicted in

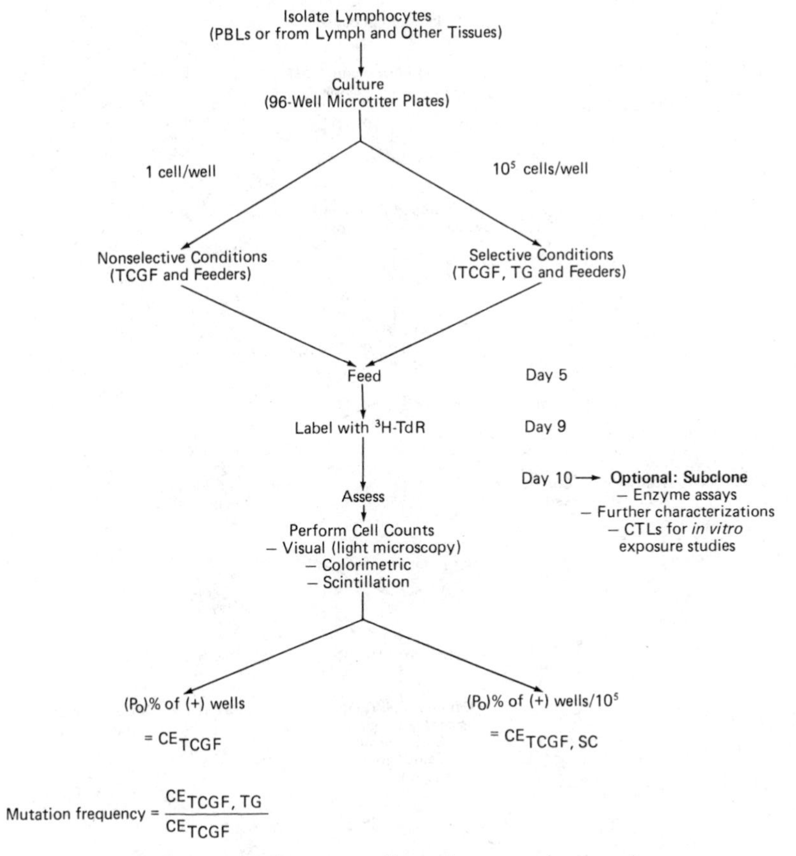

Fig. 9. Lymphocyte clonal assay procedure.

Fig. 9, lymphocytes are diluted in growth factor medium containing 0.05% X-irradiated sheep red blood cells, which serve as feeders; cells are then plated in paired 96-well microtiter plates at 10^5 cells per well in the presence of TG (final concentration = 2×10^{-5} moles) and at 1 cell/well in the absence of TG. Spent media are replaced with fresh media (at 5 and 9 days with labeling ^{3}H-TdR occurs on day 9). On day 10, activity per well is assessed by colorimetry, light microscopy, and scintillation spectrophotometry. Cloning efficiencies (CE) are calculated for each plate, assuming a Poisson distribution, such that the ratio $CE_{TCGF,TG}/CE_{TCGF}$ approximates in vivo or in vitro (for CTLs) mutation frequencies. The high sensitivity of this method can be confirmed by reconstruction experiments using mixtures of minority LN PBLs (and in further experiments, TG^r CTLs cloned from normal individuals) in a majority of normal cells. The observed recoveries are satisfactory for detecting mutant cells in normal persons, i.e., persons not at risk from mutagen

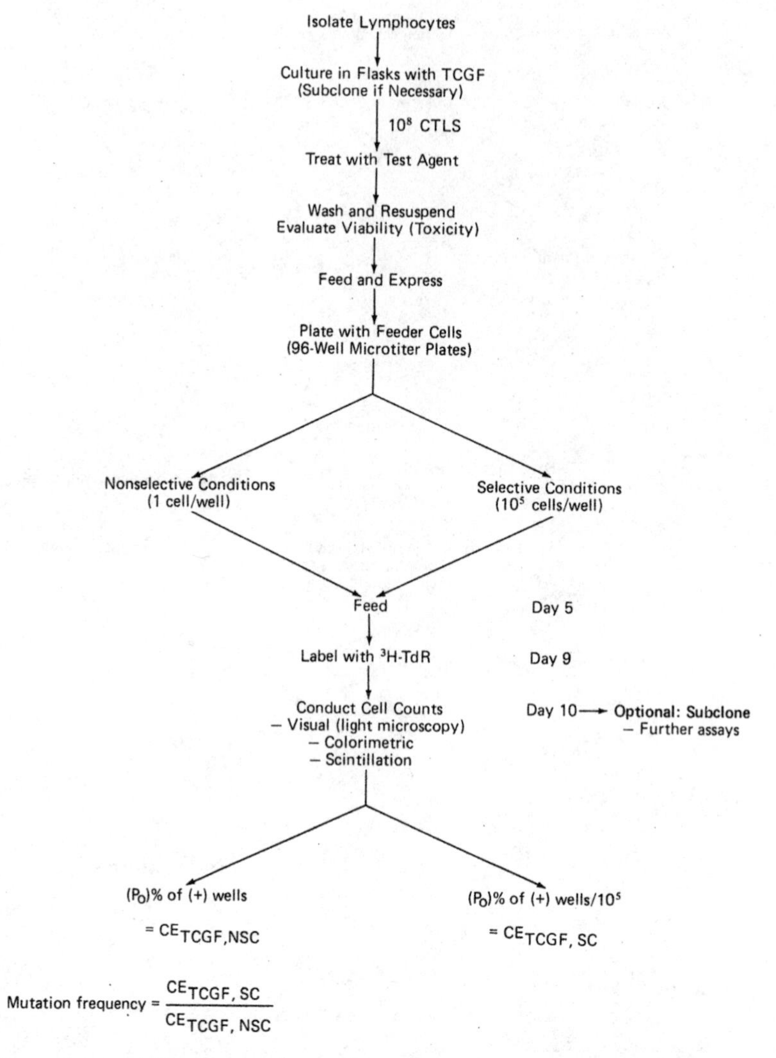

Fig. 10. A possible procedure for the adaptation of the clonal assay to the evaluation of genotoxic damage after in vitro exposures to an agent.

exposures (mutagen frequency is approximately 4 to 6 × 10^{-6}). Strauss-Albertini and lymphocyte clonal assays conducted in parallel yielded comparable results for frozen/thawed PBLs and CTLs [43].

2.2.2c. HGPRT, Lymphocytes - in vitro. The lymphocyte clonal assay may be adapted to the evaluation of genotoxic damage after in vitro exposures to various agents (Fig. 10). Individual susceptibility to mutagens and the relevancy of lymphocyte specific locus

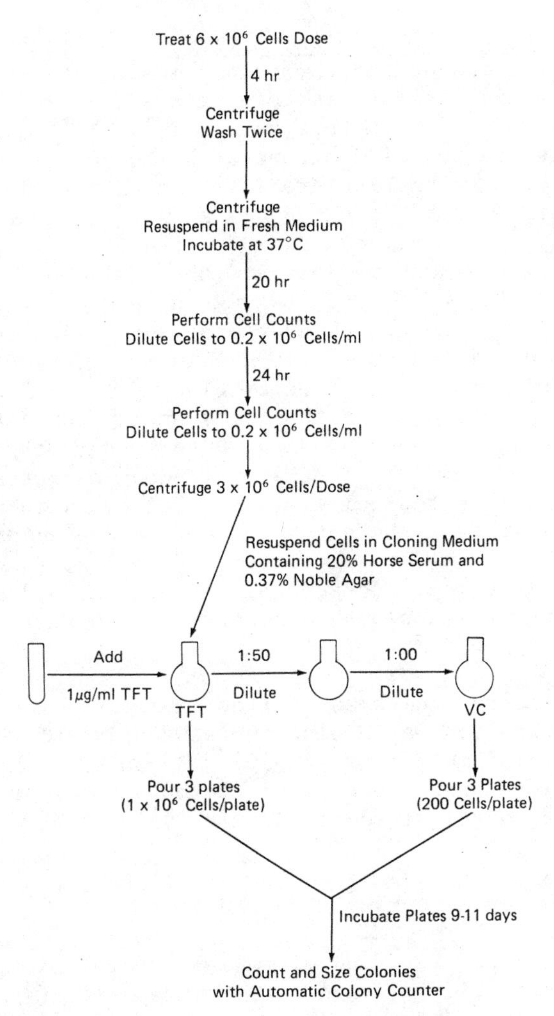

Fig. 11. Protocol for the analysis of mutation at the thymidine kinase locus of L5178Y mouse lymphoma cells.

assays for mutation may be questioned in terms of detectable short-term effects. In addition, the assay may be used for mutagenicity testing in mice, because mice were used in the first immunological cloning studies. The response of target lymphocytes from diverse tissues to a variety of chemical agents may be assayed in mice (and in humans, to a lesser extent). Such studies would help to develop the means to extrapolate the results from experiments with rodents to humans.

2.2.2d. TK, Lymphoma Cells in vitro. The method for determining gene mutations at the thymidine kinase (TK) locus involves the use of a special isolated heterozygous cell line of L5178Y mouse lymphoma cells [44-46]. In this assay, TK^{+-}-3.7.2C cells (6×10^6 cells/dose) are treated with either a pure chemical test agent or a complex mixture for 4 hr. An Aroclor-induced rat liver S9 system can also be incorporated if exogenous activation is required. Following the 4-hr treatment, cells are washed free of test agent, resuspended in fresh growth medium, and the newly induced mutant cells allowed to express for 2 days. During expression, the cells are maintained in log-phase growth by performing cell counts and daily adjusting cell density to 0.2×10^6 cells/ml. Thymidine kinase-deficient mutants are quantitated by cloning 3×10^6 cells/dose in 1 μg/ml trifluorothymidine, a thymidine analog toxic to TK-competent cells. Cells are incubated for 9 to 11 days to allow the formation of colonies that can be quantitated on an automatic colony counter. Because treated cells are also cloned without selective agent (600 cells/dose), the mutant number can be adjusted for plating efficiency and the data expressed as a mutant frequency (Fig. 11). Survival is calculated according to the method of Clive and Spector [47] and includes both a measure of growth in suspension and growth in cloning medium.

Recent research indicates that the detectable mutants include chromosomal aberrations affecting the expression of the TK locus as well as single gene mutations [48-50]. Because of the type of mutant may correlate with the size of the mutant colony (Fig. 11), it may be possible to quickly determined relative clastogenic as well as mutagenic potential [50].

2.2.3. Methods for Measuring Exposure

Body fluid analysis using microbial mutagenesis tests provides a biological indication of exposure to mutagenic or promutagenic substances.

2.2.3a. Urine Screening with Salmonella typhimurium. A major difficulty in environmental epidemiology is the measurement or estimation of exposure of individuals. Under certain conditions, urine screening allows investigators to determine whether individuals have been exposed to mutagens or promutagens and, in some circumstances, to make a semiquantitative estimate of individual exposure. The methodology used to screen urine for mutagens with bacterial systems was published first by Yamasaki and Ames [51]. Legator et al. [52] recently reviewed this method and other methods for screening various tissues and fluids for the presence of mutagens. The reader should refer to this review for a basic understanding of the techniques commonly used.

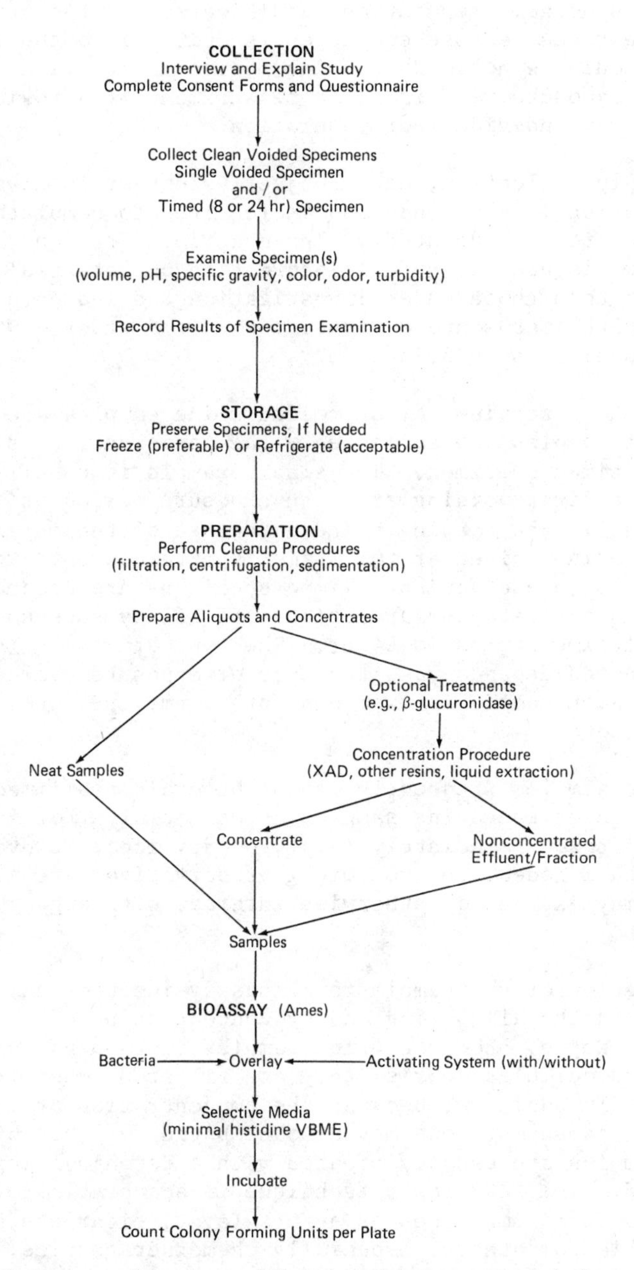

Fig. 12. Urine monitoring with microbial mutagenicity screening.

Urine monitoring (Fig. 12) involves four major procedures: collection, storage, preparation, and bioassay of the urine sample. Care must be taken because each step is critical to the final result. It should be noted that urine monitoring provides only an estimate of exposure, and is not a measurement of a toxicological effect upon any individual or population.

In sample collection, each subject volunteer is given a thorough explanation of the study and is required to complete a questionnaire related to the study. The questionnaire, completed during an interview, is used to obtain information on potential exposure to agents other than those under investigation and indicates the volunteer's health status and attitude toward the study. When possible, the questionnaire is quantitative.

After the interview, clean voided urine samples are collected. Two types of samples are collected - the occasional or single specimen and the timed specimen. The single sample is usually collected either at the first voiding after an exposure period or immediately after arising in the morning. The advantage of the morning specimen is that urine voided at this time is more concentrated and in a better state of preservation. Timed specimens are needed when attempting to quantitate the excretion rate of any substance. The more typical time period is 24 hr. The type of sample (single or timed) collected depends upon the type of exposure evaluated. After collection, each sample is examined for volume, pH, specific gravity, odor, color, etc.

Because samples frequently cannot be evaluated immediately, proper storage of the urine samples is essential. For most studies, samples are frozen immediately (e.g., on dry ice); however, simple refrigeration is adequate. Chemical preservatives are avoided because they may extract or otherwise interact with organic mutagens present.

Preparation of the sample for bioassay includes any "clean up" procedures and the aliquoting and/or concentrating of organics within each sample. Because urine samples contain bacteria, fibrous material, and cellular debris, they are filtered, centrifuged, or sedimented. In addition, because the concentration of mutagens may be low and because mutagens may also be bound to other biochemical species, samples are usually treated with a deconjugating enzyme and concentrated. The XAD2 resin technique is acceptable for concentrating mutagens from some types of agents (e.g., cigarette smoke) but less desirable for others, especially chemotherapeutics. The original (neat) sample is evaluated as a check on mutagen concentration and the concentration method.

The bioassay method presently employed is the Ames *Salmonella typhimurium* plate incorporation assay. The urine sample is plated in the same manner as any liquid test sample. It should be recognized that some urine samples can contain enough histidine to give increased numbers of spontaneous revertants per plate and that false positive results are possible. Histidine may be removed from neat or concentrated samples by treating the samples with histidase. Data are analyzed and interpreted as described by Stead et al. [53].

2.2.4. Dosimetry Methods

The following methods are applicable to animals dosed with radiolabeled chemical and to other cell types after the initial cell separation step.

Lymphocytes are separated from blood serum and lysed in a buffer containing Tris Cl, pH 8, EDTA, and 1% SDS. The concentration of radiolabeled chemical is determined in whole blood, lymphocyte fraction, and lysate. The lysate is incubated 18 to 24 hr with 1 mg/ml pronase to digest protein. The lysate is then treated with an equal volume of phenol saturated with 50 mM Tris Cl and 1 mM EDTA. The aqueous phase containing the nucleic acid is extracted two additional times with phenol; RNA and DNA are then precipitated with alcohol. The nucleic acid is resuspended and treated with ribonuclease to obtain DNA. The purity of the fraction can be improved and ribonuclease removed by a subsequent treatment with pronase, phenol, and alcohol.

Purified DNA is degraded by acid hydrolysis [54, 55] or by enzymatic degradation [56]. The DNA constituents - bases, nucleosides, and nucleotides - may be analyzed for total and specific DNA adducts.

3. RESULTS AND DISCUSSION

3.1. Comparative Evaluation of a Chemotherapeutic Agent - Cyclophosphamide

The data in Table 1 summarize preliminary evaluations of cyclophosphamide performed in this laboratory. The results are positive in the Salmonella TA100 strain in the strain in the presence of an S9 metabolic activation system. Positive results in this strain usually are indicative of base-pair substitution mutations. The compound is positive following multiple treatments in human fibroblasts at the HGPRT locus. Similarly, the agent is positive in *in vivo* rat lymphocytes at the HGPRT locus [41] and for SCE induction in rodent cells *in vivo* [26] and *in vitro* (with activation).

As described in Table 1, positive effects were generated at a dose equivalent to 4 to 5 logarithmic dose units (LDUs), the most sensitive of these tests being gene mutation in the human fibroblasts.

TABLE 1. Effects of Cyclophosphamide in Short-Term Tests

	Ames/Salmonella TA98	Ames/Salmonella TA100	Human BFU[a] (*in vivo*) TA100	Human Fibroblasts HGPRT (*in vitro*)	Rat Lymphocytes HGPRT
Point/Gene Mutation					
Qualitative results	-/-[b]	-/+	+/+	-/+	+/0
Quantitative results[c]				1 μg/ml	3 mg/kg
-log (10^{-5} x dose)[d]	---	4.1[e]	---	5	4.5
SCE Induction			Maternal Bone Marrow (*in vivo*)	Embryo (*in vivo*)	V79 Chinese Hamster Cell (*in vitro*)
Qualitative results			+	+	-/+
Quantitative results[c]			5 mg/kg	5 mg/kg	5 μg/ml
-log (10^{-5} x dose)[d]			4.3	4.3	4.3

[a]BFU = body fluid (urine test).
[b]Test results are shown without/with S9 activation.
[c]Approximate minimum dose required.
[d]Measure of sensitivity, 100,000 to 1 ppm equals 0 to +5 log dose units.
[e]Calculations based on a total plate volume of 32 ml.

TABLE 2. Comparison of Results from Direct Testing of Cyclophosphamide with Those Obtained Using Urine from Treated Patients

Sample	Activation[a]	Strain	Result
Urine: Cyclophosphamide Treatment	+	TA100	+
	-	TA100	+
	+	TA98	Neg
	-	TA98	Neg
Direct Testing: Cyclophosphamide	+	TA100	+
	-	TA100	Neg
	+	TA98	Neg
	-	TA98	Neg

[a]S9 activation used (+); S9 activation not used (-).

TABLE 3. Human Data for Cyclophosphamide

Chromosomal effects	Total Dose (g)	Logarithmic Dose Units -Log (10^{-5} mg/kg)[a]
SCE, In Vivo		
100 mg/day E = 22 days[b] Raposa[61]	2.2	3.5
Chromosome Aberration, In Vivo		
E = variable Schmid and Bauchinger[62]	1.5 to 1.7	3.7
250 mg/day E = 20 days Arrighi et al.[63]	5	3.1
E = 6 to 8 weeks Schmid and Bauchinger[64]	1.2 to 18.2	3.8
E = variable Bauchinger and Schmid[65]	1.5 to 18.8	3.7
3 to 5 mg/day E = 6 to 8 months Dobos et al.[66]	0.6	4.1

[a]Assuming a 70-kg human and 100,000 to 1 ppm equals 0 to +5 logarithmic dose units; the calculations are based on the lowest cumulative dose.
[b]E = duration of exposure.

The results of urine screening and direct testing of cyclophosphamide are shown in Table 2. When in vitro data from direct compound testing using the plate incorporation assay are compared with data obtained in screening the urine of a patient treated with cyclophosphamide, it may be seen that the exposure is detected and that metabolism of the compound has occurred to yield a direct-acting metabolite. It is thought that phosphoramide mustard represents one of the more potent cytotoxic metabolites of this compound [57].

The data for gene mutation and SCE can be compared with experimental data on chromosomal effects observed in humans. Adequate dosimetry information on cyclophosphamide exposure to patients is available, as shown in Table 3. Since the LDU values in the table were calculated assuming the cumulative dose necessary for producing chromosomal damage, and since the activated metabolites of cyclophosphamide are excreted [58], the lower LDU values better reflect the lowest effective drug concentration. On a logarithmic scale, the dose required for effects in humans is similar to that used in animal and in vitro bioassays (Table 1).

Although the data summarized in Table 3 were obtained from patients undergoing chemotherapy with cyclophosphamide, information was not available on subsequent tumor induction. The International Agency for Research on Cancer (IARC) has reported that more than 100 cases of cancer exist for patients receiving cyclophosphamide, although many patients are concomitantly given other cytostatic drugs and/or radiation [59]. In 1975, IARC reported that at least 10 cases of malignant tumors developed in patients treated with cyclophosphamide where there was no previous clinical evidence of cancer [60]. In view of the large number of people treated with cyclophosphamide, the data base for risk estimation is very small. Although the agent is considered a human carcinogen [59], its effectiveness in prolonging life in a variety of disorders is well established.

The significance of the bioassay data with respect to possible cancer induction in humans can be investigated by examining the data of Wall and Clausen [67] on patients receiving very large doses of cyclophosphamide. Fatal urinary bladder carcinomas developed in four of the five patients treated with high doses. The patients received from 114 to 295 g of cyclophosphamide over a period of several years. These doses are generally 100-fold higher than those reported in Table 3 for chromosomal effects.

Thus, for cyclophosphamide, the bioassays detect effects occurring at similar doses (mg/kg) as do tests of chromosomal damage occurring in humans. However, these doses may be 100-fold less than the cumulative doses required for an 80% tumor incidence in humans. Whether the frequency of cancer induction is significant at these lower doses is unknown.

3.2. A Problem Carcinogen - Ethyl Carbamate (EC)

EC has been consistently negative for *in vitro* SCE induction and/or gene mutation in various mammalian cell systems [23, 68]. However, it has tested clearly positive for *in vivo* SCE induction in several rodent species [69]. These discrepant findings illustrate the difficulties that may be encountered in testing common environmental chemicals under the parallelogram approach. Exogenous activation systems currently employed cannot always simulate *in vivo* metabolic processing. Thus, testing with *in vitro* methods exclusively can lead to misleading results.

Two related carcinogens, ethyl N-hydroxycarbamate (ENHC) and vinyl carbamate (VC), are believed to be metabolically derived from EC and, either directly or indirectly, to account for the carcinogenic activity associated with the parent compound. Early interest centered on ENHC, and potent teratogenic and genotoxic activities were revealed for this substance [70]. However, ENHC has been shown to possess considerably less carcinogenic potential than EC [70].

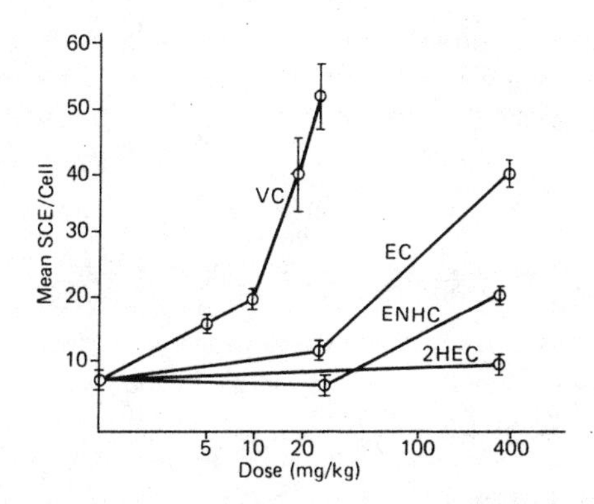

Fig. 13. SCE induction in mouse bone marrow cells after in vivo exposure to vinyl carbamate, ethyl carbamate, ethyl N-hydroxycarbamate, and 2-hydroxyethyl carbamate. Each point represents the mean SCE frequency (± the standard error) usually calculated from 20 to 40 cells/animal, 3 or more animals/treatment. From Allen et al. [23].

On the other hand, VC is a chemical of much current interest, and the few studies to date support its theoretical implication in EC carcinogenesis. Dahl et al. [71, 72] have shown that VC is considerably more potent than EC for the induction of EC-type tumors in rats and mice and for gene mutations in Salmonella. They suggested that VC may be a metabolic intermediate of EC, with VC epoxide possibly representing the ultimate carcinogen. Specific adducts expected from this line of reasoning have been found [73].

Studies have shown that VC is far more reactive than EC or ENHC for inducing SCEs and gene mutations in a variety of in vivo and/or in vitro mammalian cell systems [74]. For example, Fig. 13 illustrates SCE induction levels in mouse bone marrow cells after in vivo exposures to these carbamate chemicals. In this particular study, VC caused effects that were several times greater than EC or ENHC (up to 8 times the baseline SCE level) at less than one-tenth the dose of EC or ENHC. Negative results with 2-hydroxyethyl carbamate (2-HEC), a noncarcinogen, were also seen. Tables 4 and 5 reveal the relative in vitro activities of these carbamate chemicals in human lymphocytes and in V79 cells. Again, VC is clearly the most reactive; EC is essentially negative with or without S9 activation. In the V79 cell system, VC was direct-acting for the production of SCEs, although gene mutations were not induced. Increases up to 8 times the baseline level of SCEs resulted. Yet, with the addition of Aroclor-induced hamster liver S9 enzymes, the doses required to cause this effect greatly decreased, and extensive gene mutagenesis for 6-TG resistance resulted as well. Perhaps the coincident SCE and gene mutation effects stem from VC epoxide lesions, which are

TABLE 4. Toxicity, Gene Mutation, and SCE Induced by Ethyl Carbamate, Vinyl Carbamate, Ethyl N-Hydroxycarbamate, and 2-Hydroxyethyl Carbamate in Chinese Hamster V79 Cells*

Chemical	Dose (mg/ml)	Toxicity (%) +S9[b]	Toxicity (%) -S9	Mutagenesis (6-TG) Mutants/10^6 cells +S9	Mutagenesis (6-TG) Mutants/10^6 cells -S9	Mean SCE/cell[c] +S9	Mean SCE/cell[c] -S9
Ethyl carbamate	20.0	81	64	<10	<10	13 ± 4.4	9 ± 2.5
	10.0	81	77	<10	<10	12 ± 3.5	9 ± 2.7
	5.0	100	84	<10	<10	11 ± 4.0	9 ± 2.6
Vinyl carbamate	1.0	0.4	54	Toxic	<10	Toxic	82 ± 12.4
	0.2	6	90	591	<10	Toxic	30 ± 8.4
	0.01	52	53	236	<10	67 ± 8.5	15 ± 5.0
	0.001	48	58	69	<10	25 ± 5.9	14 ± 4.6
Ethyl N-hydroxy-carbamate	7.5	29	11	<10	62	22 ± 9.0	Toxic
	5.0	45	22	<10	55	18 ± 7.5	39 ± 14.2
	1.0	56	48	<10	51	15 ± 3.6	15 ± 3.6
	0.2	52	50	<10	72	20 ± 8.4	18 ± 9.1
	0.012	69	66	<10	20	14 ± 4.6	14 ± 5.0
2-Hydroxyethyl carbamate	10	86	77	29	<10	13 ± 4.4	13 ± 3.8
	5	68	87	<10	<10	16 ± 5.0	11 ± 4.0
	1	59	61	31	<10	15 ± 4.7	11 ± 3.0
	0.1	123	105	16	<10	16 ± 4.9	12 ± 3.7
Control	0	100	97	<10	<10	13 ± 3.5	11 ± 2.9
DMBA	0.015	65	-	222	-	46 ± 12.2	-
MNNG	0.0001	-	48	-	82	-	52 ± 14.3

[a]From Allen et al.[23].
[b]Aroclor-induced Syrian hamster S9.
[c]Mean ± standard deviation of 30 to 40 cells/treatment.

different from those caused by VC itself. ENHC was less reactive, and 2-HEC was essentially negative.

It has also been observed that Aroclor-induced S9 enzymes obtained from a mouse strain (A) with a high susceptibility to EC carcinogenesis can metabolically convert VC to more reactive genotoxic products than Aroclor-induced S9 enzymes from a mouse strain (C57BL/6) characterized by a low susceptibility to EC carcinogenesis (Fig. 14) [69]. Significantly higher levels of SCE induction and gene mutagenesis result in V79 cells when the former mouse strain S9 preparation is used. That strain-specific differences in S9 metabolic conversions of VC parallel strain-specific differences in carcinogenesis susceptibilities to EC lends some support to the proposition [71, 72] that VC may be a critical intermediate in EC

TABLE 5. *In vitro* SCE Frequencies in Human Lymphocytes Treated with Ethyl Carbamate or Vinyl Carbamate*

Test Chemical	Dose (mg/ml)	Number of Cells	SCE/Cell Ranges	SCE/cell[b]
Control	–[c]	54	3-14	7.4 % 2.7
Ethyl Carbamate	1.0	65	2-15	8.3 % 3.0
	5.0	54	4-14	7.9 % 2.7
	10.0	34	3-20	8.2 % 4.2
Vinyl Carbamate	0.01	31	3-17	9.1 % 3.8
	0.10	27	17-67	41.4 % 12.4
	0.20	28	17-66	39.3 % 12.5

[a]From Allen et al.[22].
[b]Mean value % standard deviation/sample of cells analyzed.
[c]All control and treatment group cells were grown with 25 mmoles BrdU.

pathways to tumor formation. Additional experimentation is aimed at confirming the metabolic relationship between EC and VC and at further clarifying the role and fate of the latter substance.

VC has also been tested in the L5178Y/TK$^{+/-}$ mouse lymphoma assay. In this system, VC was found to give positive mutagenicity at the TK locus, both with and without an Aroclor-induced rat liver S9 (Fig. 15). As with the V79 assay, the dose required to give a genotoxic effect was lower with S9 activation. An analysis of the size of the mutant colonies induced (Fig. 16) indicates that VC does produce a significant proportion of small colony mutants, which have, as a class, been shown to represent chromosome aberrations affecting chromosome 11 [48-50], which carries the TK gene [75]. VC has also been shown to be an efficient inducer of gross aberrations in these L5178Y/TK$^{+/-}$ mouse lymphoma cells [76].

3.3. Dosimetry

A common denominator in our methodology that has not yet been considered is the cellular or molecular (DNA) dose determined for the component test systems. Dose-response data are used to estimate the effects that might occur under similar exposure conditions in humans. These extrapolations are facilitated by using peripheral blood lymphocytes in a number of the component procedures. This

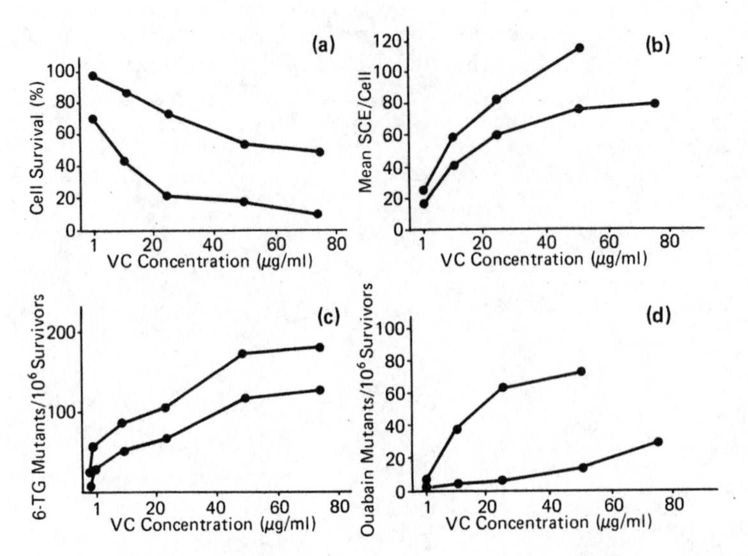

Fig. 14. Comparison of capabilities of A vs. C57BL/6 mouse strain liver enzyme preparations (Aroclor-induced S9 fractions) to activate vinyl carbamate to induce various genotoxic effects in V79 cells. Cytotoxicity (a) is shown as percent cell survival of the control value (100%), SCE frequencies as mean per cell (b), gene mutation frequency for 6-thioguanine resistance (c), and gene mutation frequency for ouabain resistance (d) as number of mutants/10^6 survivors minus control value. These values represent an average of 3 experiments. From Sharief et al. [69].

cell type can be studied directly in _in vitro_ cultures, and in _in vivo_ systems where the cells are obtained from exposed animals or exposed humans. However, a number of complicating factors exist in basing extrapolation procedures on cellular or molecular dose.

After exposure in a human or an experimental animal, the concentration of an agent may change rapidly under physiological conditions. Many chemicals are rapidly metabolized and are removed from the blood plasma within minutes to hours; the concentration in other tissues and the concentration at the molecular targets is similarly time dependent. Perhaps more important with regard to the present investigation is the fact that various repair mechanisms are operative to remove the agent or its metabolites, once their targets have been reached. In addition, it is possible that the cellular or molecular damage sustained will not be expressed as an altered phenotype. Thus, careful experiments must be planned to allow deductions to be made concerning the effects of cellular metabolism and repair

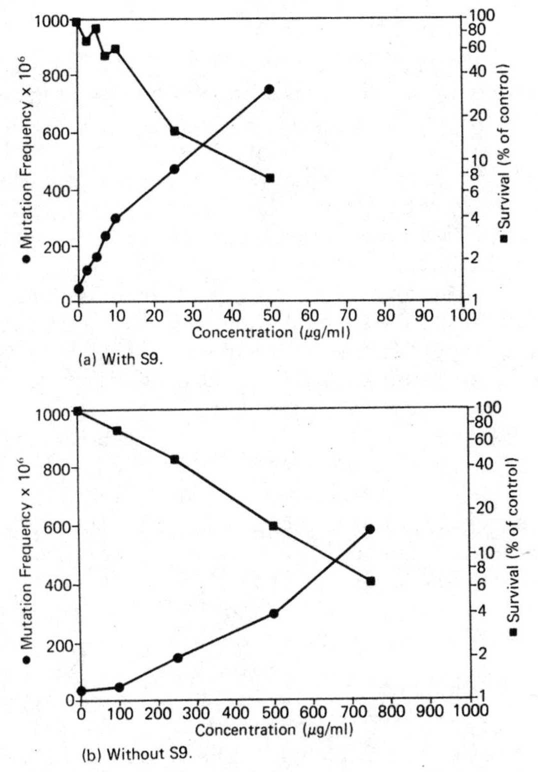

Fig. 15. Mutagenicity of vinyl carbamate at the TK locus of L5178Y mouse lymphoma cells with S9 (a) and without S9 (b) activation.

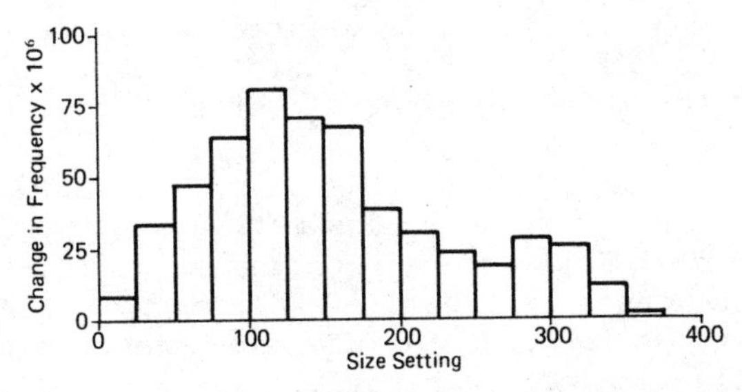

Fig. 16. Colony size of trifluorothymidine-resistant mutants following treatment with 750 μg/ml vinyl carbamate (without S9 activation). Small colony peak is to the left; large colony peaks is to the right.

processes on the relationship between dose and the phenotypic expression of genetic damage in various tissues and species.

More specifically, the feasibility of basing dosimetry efforts on DNA adducts formed from labeled material may be questionable since the modified target may be unstable and the label may be liberated to the intracellular environment, even though a stable base change in DNA or other modification in the molecular target may have been produced. These considerations may be important in understanding recent experiments [77] where comparisons of SCE induction and adduct formation revealed that no demonstrable DNA product was formed at the same ratio as SCE, and, in fact, differences were 7- to 25-fold. Other experiments have shown that substantial alkylation of messenger RNA does not alter specific gene products [78]. Covalent binding data are not readily translated into information regarding the amount of target modification that occurs or may be transcribed after replication of DNA. Finally, isotopically labeled material cannot be used to monitor human exposure.

A recent report by Topal and Baker [79] indicates that methylation of the DNA nucleotide pool may be an important consideration in assessing molecular dose. Methylation of the DNA nucleotide precursor pool was shown to be 190 to 13,000 times that of the equivalent residue in the DNA helix itself.

Nevertheless, determining cellular or total dose and adducts in cell cultures and in experimental animals is relatively straightforward, with several limitations. The agent must be readily available and appropriately labeled isotopically. The specific radioactive activity of the compound and its active metabolite must be sufficient to permit DNA or adducts to be isolated and counted. Ideally, quantitation would be based on relatively stable adducts. For example, alkyl derivatives of oxygen atoms are stable for more than 10 days for DNA treated _in vitro_ [5]. Alkyl phosphotriesters are very stable and the half-life of other alkyl derivatives may be several days. Quantitation of such relatively stable derivatives of DNA is essential in relating genotoxic effects to exposure doses in human populations.

In summary, the question of dosimetry is very complex, and further experimental efforts are required before conclusions can be reached about the appropriateness of various possible measurements or about the most useful ways to integrate these measurements into the overall evaluation of the parallelogram approach.

4. SUMMARY AND CONCLUSIONS

Current methods of risk assessment for chemicals found in the environment are based on estimates of human exposure and associated effects. Agents are defined as cancer-causing only if exposure and effects information exist from case reports or clinical findings. Otherwise, such agent definition depends upon descriptive epidemiological studies where cancer incidence has varied with exposure to the agent or where there are similar findings in case-control and cohort epidemiological studies [59]. Unfortunately, such data can be assembled only after inadvertent exposure of individuals or certain populations. Clearly, other prospective procedures are necessary for dealing with potentially hazardous agents found in the environment.

Data from animal experiments or short-term tests do not have the limitations of case reports in epidemiology, although problems are encountered in extrapolating data from in vitro and in vivo test systems to the human situation. It should be possible, however, to construct a laboratory model that would closely approximate the case for human exposure and effects.

The purpose of this paper has been to describe preliminary experiments and methodology directed toward the development of such an experimental model. This model should be useful in integrating somatic- and germ-cell testing or monitoring data with information on exposure. The rationale for such a model is based upon the parallelogram method described previously. The component parts of the model that we have assembled consist of in vitro and in vivo methods for measuring point/gene mutation and chromosomal effects. Obviously, there are many other genetic endpoints and related effects that could contribute to the detection and quantitation of chemically induced genetic damage in humans. Our research program is at its inception and lacks a number of components before a complete integration of our efforts will be feasible. Proper dosimetry is needed, and these measurements must be standardized (e.g., against direct radiation exposures). Genetic test systems must be made as uniform as possible and internal controls must be added. Despite these difficulties, our preliminary results may be subjected to an integrated assessment that is interpretable within the overall parallelogram concept.

REFERENCES

1. F. H. Sobels, in: Comparative Chemical Mutagenesis (F. J. de Serres and M. D. Shelby, eds.), Plenum Press, New York (1981).
2. F. H. Sobels, in: Progress in Mutation Research, Elsevier/North Holland, Amsterdam (1981).
3. H. V. Malling, Perspectives in mutagenesis, Environ. Mutag., 3:103-108 (1981).
4. M. E. Lyon, in: Utilization of Mammalian Specific Locus Studies in Hazard Evaluation and Estimation of Genetic Risk (F. J. de Serres and W. Sheridan, eds.), in press.
5. B. Singer, N-nitroso alkylating agents: Formation and persistence of alkyl derivarives in mammalian nucleic acids as contributing factors in carcinogenesis, J. Natl. Cancer Inst., 62: 1329-1339 (1979).

6. G. A. Sega, R. B. Cumming, and M. F. Walton, Dosimetry studies on the ethylation of mouse sperm DNA after in vivo exposure to [^{3}H]ethyl methanesulfonate, Mutat. Res., 24:317-333 (1974).
7. W. R. Lee, Molecular dosimetry of chemical mutagens: Determination of molecular dose to the germ line, Mutat. Res., 38: 311-316 (1976).
8. C. S. Aaron, Molecular dosimetry of chemical mutagens: Selection of appropriate target molecules for determining molecular dose to the germ line, Mutat. Res., 38:303-310 (1976).
9. W. R. Lee, Dosimetry of chemical mutagens in eukaryotic germ cells, in: Chemical Mutagens (A. Hollaender and F. J. de Serres, eds.), Vol. 5, pp. 117-202, Plenum Press, New York (1978).
10. C. S. Aaron, A. A. von Zeeland, G. R. Mohn, A. T. Natarajan, A. G. A. C. Knapp, A. D. Tates, and B. W. Glickman, Molecular dosimetry of the chemical mutagen ethyl methanesulfonate: Quantitative comparison of mutation induction in Escherichia coli, V79 Chinese hamster cells and L5178Y mouse lymphoma cells, and some cytological results in vitro and in vivo, Mutat. Res., 69:201-216 (1980).
11. S. A. Latt, J. Allen, S. E. Bloom, A. Carrano, E. Falke, D. Kram, E. Schneider, R. Schreck, R. Tice, B. Whitfield, and S. Wolff, Sister-chromatid exchanges: A report of the Gene-Tox Program, Mutat. Res., 87:17-62 (1981).
12. F. D. de Serres and J. Ashby, eds., Short-Term Tests for Carcinogens: Report of the International Collaboration Program, Elsevier/North Holland, Amsterdam (1981).
13. A. Nettleship, P. S. Henshaw, and H. L. Meyer, Induction of pulmonary tumors in mice with ethyl carbamate (urethane), J. Natl. Cancer Inst., 4:309-319 (1943).
14. International Agency for Research on Cancer, Some antithyroid and related substances, nitrofurans, and industrial chemicals, in: IARC Monograph on the Evaluation of the Carcinogenic Risks of Chemicals to Man, Urethane, 7:111-140, Lyon, France (1974).
15. R. H. Adamson and S. M. Sieber, in: Organ and Species Specificity in Chemical Carcinogenesis (R. Langenbach, S. Nesnow, and J. Rice, eds.), Vol. 24, pp. 129-156, Plenum Press, New York (1983).
16. A. W. Pound, F. Franke, and T. A. Lawson, The binding of ethyl carbamate to DNA of mouse liver in vivo: The nature of the bound molecule and the site of binding, Chem. Biol. Interact., 14:149-163 (1976).
17. J. W. Allen, Y. Sharief, and R. J. Langenbach, in: Genotoxic Effects of Airborne Agents (R. Tice, D. Costa, and K. Schaich, eds.), pp. 443-460, Plenum Press, New York (1982).
18. R. J. Preston, W. Au, M. A. Bender, J. G. Brewen, A. V. Carrano, J. A. Heddle, A. F. McFee, S. Wolff, and J. S. Wassom, Mammalian in vivo and in vitro cytogenetic assays: A report of the U.S. EPA's Gene-Tox Program, Mutat. Res., 87:143-188 (1981).

19. A. D. Bloom, ed., Guidelines for Studies of Human Populations Exposed to Mutagenic and Reproductive Hazards, March of Dimes Birth Defects Foundation, White Plains, New York (1981).
20. A. A. Sandberg, ed., Sister Chromatid Exchange, Alan R. Liss, Inc., New York (1982).
21. S. Wolff, ed., Sister Chromatid Exchange, John Wiley and Sons, New York (1982).
22. J. W. Allen, K. Brock, J. Campbell, and Y. Sharief, in: Single-Cell Mutation Monitoring Systems: Methodologies and Applications (A. A. Ansari and F. J. de Serres, eds.), Plenum Press, New York (in press).
23. J. W. Allen, R. Langenbach, S. Nesnow, K. Sasseville, S. Leavitt, J. Campbell, K. Brock, and Y. Sharief, Comparative genotoxicity studies of ethyl carbamate and related chemicals: Further support for vinyl carbamate as a proximate carcinogenic metabolite, Carcinogenesis, 3:1437-1441 (1982).
24. J. W. Allen, C. F. Shuler, and S. A. Latt, Bromodeoxyuridine tablet methodology for in vivo studies of DNA synthesis, Somatic Cell Genet., 4:393-405 (1978).
25. A. F. McFee, K. W. Lowe, and J. R. San Sebastian, Improved sister-chromatid differentiation using paraffin-coated bromodeoxyuridine tablets in mice, Mutat. Res., 119:83-88 (1983).
26. J. W. Allen, E. El-Nahass, M. K. Sanyal, R. L. Dunn, B. Gladden, and R. L. Dixon, Sister-chromatid exchange analysis in rodent maternal, embryonic, and extra-embryonic tissues: Transplacental and direct mutagen exposures, Mutat. Res., 80:297-311 (1981).
27. R. J. Albertini and R. DeMars, Detection and quantification of X-ray induced mutation in cultured diploid human fibroblasts, Mutat. Res., 18:199-224 (1973).
28. J. H. Taylor, Asynchronous duplication of chromosomes in cultured cells of Chinese hamster, J. Biophys. Biochem. Cytol., 7:455-467 (1960).
29. S. Bader, O. T. Miller, and B. B. Mukherjee, Observations on chromosome duplication in cultured human leucocytes, Exp. Cell Res., 31:100-112 (1963).
30. Y. Kikuchi and A. A. Sanberg, Chronology and pattern of human chromosome replication, 1., Blood leucocytes of normal subjects, J. Nat'l. Cancer Inst., 32:1109-1143 (1964).
31. G. C. Mueller and K. Kajiwara, Early-and late- replicating deoxyribonucleic acid complexes in HeLa nuclei, Biochim. Biophys. Acta, 114:108-119 (1966).
32. B. B. Mukherjee, W. C. Wright, S. K. Ghosal, G. D. Burkhalder, and K. E. Mann, Further evidence for the simultaneous initiation of DNA repliation in both X chromosomes of bovine female, Nature (London), 220:714-716 (1968).
33. R. Braun and H. Willi, Time sequence of DNA replication in phyarium, Biochim. Biophys. Acta, 174:246-252 (1969).
34. S. L. Huang, S. M. S. Huang, C. Casperson, and M. D. Waters, Induction of 6-thioguanine resistance in synchronized human fibro-

blast cells treated with methyl methanesulfonate, N-acetoxy-2-acetylaminofluorene and N-methyl-N'-nitro-N-nitrosoguanidine, Mutat. Res., 83:251-260 (1981).

35. G. E. Milo and J. A. DiPaolo, Neoplastic transformation of human diploid cells in vitro after chemical carcinogen treatment, Nature (London), 275:130-132 (1978).

36. V. M. Maher and J. E. Wessel, Mutation to azaguanine resistance induced in cultured diploid human fibroblasts by the carcinogen, N-acetoxy-2-acetylaminofluorene, Mutat. Res., 28:277-284 (1975).

37. L. Jacobs and R. DeMars, Quantification of chemical mutagenesis in diploid human fibroblasts: Induction of azaguanine-resistant mutants by N-methyl-N'-nitro-N-nitrosoguanidine, Mutat. Res., 53:29-53 (1978).

38. L. Jacobs and R. DeMars, in: Handbook of Mutagenicity Test Procedures (B. J. Kilbey, ed.), Elsevier, Amsterdam (1977).

39. S. L. Huang and M. W. Lieberman, Induction of 6-thioguanine resistance in human cells treated with N-acetoxy-2-acetylaminofluorene, Mutat. Res., 57:349-358 (1978).

40. G. H. Strauss and R. J. Albertini, Enumeration of 6-thioguanine-peripheral blood lymphocytes in man as a potential test for somatic cell mutations arising in vivo, Mutat. Res., 61:353-379 (1979).

41. G. H. S. Strauss, in: The Use of Human Cells for the Assessment of Risk from Physical and Chemical Agents (A. Castellani, ed.), Plenum Press, New York (1982).

42. G. H. Strauss, R. J. Albertini, and B. J. Allen, An enumerative assay of purine analog resistant lymphocytes in women heterozygous for the Lesch-Nyhan mutation, Biochem. Genet., 18:529-547 (1980).

43. G. H. S. Strauss, in: Indicators of Genotoxic Exposures in Man and Animals (B. A. Bridges, B. E. Butterworth, and I. B. Weinstein, eds.), Banbury Report 13, Cold Spring Harbor Laboratory (1982).

44. D. Clive, W. G. Flamm, M. R. Machesko, and N. J. Bernheim, A mutational assay system using the thymidine kinase locus in mouse lymphoma cells, Mutat. Res., 16:77-87 (1972).

45. D. Clive and P. Voytek, Evidence for chemically-induced structural gene mutations at the thymidine kinase locus in cultured L5178Y mouse lymphoma cells, Mutat. Res., 44:69-278 (1977).

46. N. T. Turner, A. G. Batson, and D. Clive (in preparation).

47. D. Clive and J. F. S. Spector, Laboratory procedure for assessing specific locus mutations at the TK locus in cultured L5178Y mouse lymphoma cells, Mutat. Res., 31:17-29 (1975).

48. J. Hozier, J. Sawyer, M. Moore, B. Howard, and D. Clive, Cytogenetic analysis of the L5178Y/TK$^{+/-}$ → TK$^{-/-}$ mouse lumphoma mutagenesis assay system, Mutat. Res., 84:169-181 (1981).

49. J. Hozier, J. Sawyer, D. Clive, and M. Moore, Cytogenetic distinction between the TK^{+} and TK^{-} chromosome in the L5178Y TK$^{+/-}$ mouse lymphoma mutagenesis assay system, Mutat. Res., 105:451-456 (1982).

50. M. M. Moore, D. Clive, J. Hozier, B. E. Howard, A. G. Batson, N. T. Turner, and J. Sawyer, Analysis of trifluorothymidine-resistant (TFT') mutants of L5178Y/TK$^{+/-}$ mouse lymphoma cells, Mutat. Res. (in press).
51. E. Yamasaki and B. N. Ames, Concentration of mutagens from urine by adsorption with the non-polar resin XAD-2: Cigarette smokers have mutagenic urine, Proc. Natl. Acad. Sci. (U.S.A.), 74:3555-3559 (1977).
52. M. S. Legator, E. Bueding, R. Butzinger, T. H. Conner, E. Eisenstadt, M. G. Farrow, G. Ficsor, A. Hsie, J. Seed, and R. S. Stafford, An evaluation of the host-mediated assay and body fluid analysis: A report of the U.S. Environment Protection Agency Gene-Tox Program, Mutat. Res., 98:319-374 (1982).
53. A. G. Stead, V. Hasselblad, J. P. Creason, and L. Claxton, Modeling the Ames test, Mutat. Res., 85:12-27 (1981).
54. D. E. Jenson, Reaction of DNA with alkylating agents, Differential alkylation of poly DA-DT by methylnitrosourea and ethylnitrosourea, Biochemistry, 17:5108-5113 (1978).
55. P. Brooks and P. D. Lawley, in: Chemical Mutagens, Principles, and Methods for Their Detection (A. Hollaender, ed.), Vol. 1, pp. 121-144, Plenum Press, New York.
56. B. Singer, W. J. Bodell, J. E. Cleaver, G. H. Thomas, M. F. Rajewsky, and W. Thon, Oxygens in DNA are main targets for ethylnitrosourea in normal and xeroderma pigmentosum fibroblasts and fetal rat brain cells, Nature, 276:85-88 (1978).
57. A. V. Connors, P. J. Cox, P. B. Farmer, A. B. Foster, and M. Jarman, Some studies of the active intermediates formed in the microsomal metabolism of cyclophosphamide and isophosphamide, Biochem. Pharmacol., 23:115 (1974).
58. L. S. Goodman and A. Gilman, The Pharmacological Basis of Therapeutics, Macmillan, New York (1975).
59. International Agency for Research on Cancer, Chemicals and Industrial Processes Associated with Cancer in Humans, IARC Monographs, Supplement 4 to Vols. 1-29, Lyon, France (1982).
60. International Agency for Research on Cancer, IARC Monographs, Vol. 9, Lyon, France, 135-156 (1975).
61. T. Raposa, Sister chromatid exchange studies for monitoring DNA damage and repair capacity after cytostatics _in vitro_ and in lymphocytes of leukaemic patients under cytostatic therapy, Mutat. Res., 57:241 (1978).
62. E. Schmid and M. Bauchinger, Chromosomenaberrationen in menschlichen peripheren lymphozyten nach endoxanstosstherapie gynakologischer tumoren, Deutsche Medizinische Wochenschrift, 93: 1149 (1968).
63. F. E. Arrighi, T. C. Hsu, and D. E. Bergsagel, Chromosome damage in murine and human cells following cytoxan therapy, Tex. Rep. Biol. Med., 20:545 (1962).
64. E. Schmid and M. Bauchinger, Comparison of the chromosome damage induced by radiation and cytoxan therapy in lymphocytes of patients with gynaecological tumors, Mutat. Res., 21:271 (1973).

65. M. Bauchinger and E. Schmid, Cytogenetische veranderungen in weissen blutzellen nach cyclophosphamidtherapie, Z. Krebsforsch., 72:77 (1969).
66. M. Dobos, D. Schuler, and G. Fekete, Cyclophosphamide-induced chromosomal aberrations in nontumorous patients, Humangenetik, 22:221 (1974).
67. R. L. Wall and K. P. Clausen, Carcinoma of the urinary bladder in patients receiving cyclophosphamide, New England J. Med., 293:271-273 (1975).
68. G. T. Roberts and J. W. Allen, Tissue-specific induction of sister chromatid exchanges by ethylcarbamate in mice, Environ. Mutag., 2:17-26 (1980).
69. Y. Sharief, J. Campbell, S. Leavitt, R. Langenbach, and J. W. Allen, Rodent species and strain specificities for sister-chromatid exchange induction and gene mutagenesis effects from ethyl carbamate, ethyl-N-hydroxycarbamate and vinyl carbamate, Mutat. Res. (in press).
70. S. S. Mirvish, The carcinogenic action and metabolism of urethane and N-hydroxyurethane, Adv. Cancer Res., 11:1-42 (1968).
71. G. A. Dahl, J. A. Miller, and E. C. Miller, Vinyl carbamate as a promutagen and a more carcinogenic analysis of ethyl carbamate, Cancer Res., 38:3793-3804 (1978).
72. G. A. Dahl, E. C. Miller, and J. A. Miller, Comparative carcinogenicities and mutagenicities of vinyl carbamate, ethyl carbamate, and ethyl N-hydroxycarbamate, Cancer Res., 40:1194-1203 (1980).
73. M. L. Ribovich, J. A. Miller, E. C. Miller, and L. G. Timmins, Labeled 1, N^6-ethenoadenosine and 3,N^4-ethenocytidine in hepatic RNA of mice given [ethyl-1,2-^{3}H or ethyl-1-^{14}C] ethyl carbamate (urethan), Carcinogenesis, 3:539-546 (1982).
74. J. W. Allen, R. Langenbach, S. Leavitt, Y. Sharief, J. Campbell, and K. Brock, in: Indicators of Genotoxic Exposure in Man and Animals (B. A. Bridges, B. E. Butterworth, and I. B. Weinstein, eds.), Banbury Report 13, Cold Spring Harbor Laboratory (1982).
75. C. A. Kozak and F. H. Ruddle, Assignment of the genes for thymidine kinase and galactokinase to Mus musculus chromosome 11 and the preferential segregation of this chromosome in Chinese hamster/mouse lymphoma somatic cell hybrids, Somat. Cell Genet., 3:121-133 (1977).
76. C. Doerr and M. Moore, unpublished.
77. S. M. Morris, R. H. Heflich, R. L. Kodell, and D. T. Beranek, Induction of sister-chromatid exchanges in Chinese hamster ovary cells by simple methylating agents: Relationship to specific DNA adducts, Abstracts of the Fourteenth Annual Meeting of the Environmental Mutagenic Society, p. 98, San Antonio, Texas, March 3-6, 1983.
78. H. Fraenkel-Conrat and B. Singer, Effect of introduction of small alkyl groups on mRNA function, Proc. Natl. Acad. Sci. (U.S.A.), 77:1983-1985 (1980).

79. M. D. Topal and M. S. Baker, DNA precursor pool: A significant target for N-methyl-N-nitrosourea in C3H/10T1/2 clone 8 cells, Proc. Natl. Acad. Sci. (U.S.A.), 79:2211-2215 (1982).

ON THE POSSIBLE SIGNIFICANCE OF TCDD RECEPTOR BASED ASSAYS IN ATTEMPTS TO ESTIMATE ENVIRONMENTAL HEALTH HAZARDS

Jan-Åke Gustafsson, Jan Carlstedt-Duke,
Mikael Gillner, Lars-Arne Hansson, Bertil Högberg,
Johan Lund, †Göran Löfroth, Lorenz Poellinger
and Rune Toftgård

Department of Medical Nutrition
Karolinska Institute
Huddinge University Hospital F69
S-141 86 Huddinge

and

†Department of Radiobiology
Wallenberg Laboratory
University of Stockholm
S-106 91 Stockholm, Sweden

INTRODUCTION

Several attempts have been made and are being made to correlate individual responsiveness to certain health hazards with individual differences in inducibility of aryl hydrocarbon hydroxylase (AHH) activity [1-5]. Our approach to this problem is to attempt to measure the concentration of the cytosolic receptor (the 2,3,7,8-tetrachlorodibenzo-p-dioxin or TCDD receptor) believed to be involved in induction of AHH [6], rather than to study the enzyme induction [7-13]. From the field of steroid hormonal receptors we know that we can predict the responsiveness of individual cancer patients to hormonal therapy by monitoring their tumor content of steroid receptors [14-18]. It does not seem unreasonable to suggest that, by analogy, we might be able to predict an individual's responsiveness to certain toxic agents by assaying that individual's levels of receptor for the particular toxic chemical under study.

In the following, a summary of some studies we have performed on the TCDD receptor will be presented. First, as a background, a brief account will be given of our current knowledge in the field of glucocorticoid receptors. Also, a few words will be mentioned about use of steroid receptor assays in prediction of responsiveness of certain malignancies to hormonal therapy.

The Glucocorticoid Receptor

The glucocorticoid receptor is a soluble protein with a molecular weight of about 90,000 and a Stokes radius of 6.1 nm [19]. In its free, unliganded form it is mainly present in the cytoplasm of glucocorticoid target cells but following complex formation with a glucocorticoid hormone arriving from the blood over the plasma membrane (probably via passive diffusion) it is "activated" and the steroid-receptor complex translocates to the cell nucleus where it somehow stimulates transcription of specific mRNA's [20].

The glucocorticoid receptor may be proteolytically digested to smaller fragments. One such fragment has a molecular weight of about 45,000 (Stokes radius 3.6 nm); the molecular weight of the smallest fragment, the so called "mero-receptor", is about 20,000 (Stokes radius 1.9 nm) [21-23]. Experiments performed with these proteolytically induced receptor forms have been helpful in defining different domains of the receptor molecule. The steroid-binding domain resides within the mero-receptor whereas the DNA-binding domain (responsible for the non-specific interaction between the steroid-glucocorticoid receptor complex and DNA) is present on both the 3.6 and native 6.1 nm forms but not on the 1.9 nm form (which, consequently, is not taken up by cell nuclei).

Experiments with rabbit antisera against the glucocorticoid receptor have helped to elucidate the receptor structure further. Several antisera, raised against highly purified native glucocorticoid receptor from rat liver cytosol [24], were found to react with the 6.1 nm receptor but not with the 3.6 or 1.9 nm forms. In summary, therefore, the glucocorticoid receptor appears to consist of three separate parts: a steroid-binding domain, a DNA-binding domain and an immunologically active domain. The biological function of the latter part of the receptor is not known but it is interesting to note that it appears to be lacking in glucocorticoid receptors from certain glucocorticoid insensitive tumor cells [25, 26].

Very recently highly exciting results have been obtained in collaboration between our own laboratory and that of Dr. Keith Yamamoto at the Department of Biochemistry, University of California, San Francisco. Cloned DNA restriction fragments from the genome of murine mammary tumor virus have been shown to bind purified glucocorticoid receptor specifically [27]. Glucocorticoid hormones are known to stimulate transcription of the genome of this tumor virus [20] and it is particularly interesting that not only the promotor region but also a few other regions of the genome appear to bind the purified receptor specifically. These findings may help to deepen our understanding of receptor control of gene expression and may lead to a somewhat modified concept of hormone mechanism of action at the gene level.

Steroid Receptor Assays in Prediction of Responsiveness to Hormonal Therapy

Determination of estrogen and progestin receptors in breast cancer tissue has since long been used both to predict hormonal responsiveness and to evaluate the prognosis of the disease [14-18, 28-34]. Several receptor assays have been used in this work including dextran-coated charcoal assay, sucrose density gradient centrifugation and isoelectric focusing. In the years to come it is possible that immunoassays based on monoclonal receptor antibodies will replace the conventional tracer-based methods [35, 36].

Besides breast cancer also other malignancies have been considered with regard to the clinical usefulness of steroid receptor assays, e.g., prostatic carcinoma (androgen, progestin, and estrogen receptors) [37-40], endometrial carcinoma (estrogen and progestin receptors) [41, 42], leukemia (glucocorticoid receptors) [43-45] and hypernephroma (progestin receptors) [46, 47]. It is possible that the practical clinical importance of receptor assays will increase significantly when simpler immunoassays become available.

The TCDD Receptor

TCDD is the most potent inducer of various enzyme systems including AHH, a cytochrome P-450-dependent microsomal monooxygenase system [48]. Other enzymes induced by TCDD are δ-aminolevulinic acid synthetase [49, 50], UDP-glucuronyl transferase [51] and glutathione-S-transferase B [52]. The induction of AHH by TCDD and other polycyclic aromatic hydrocarbons has been studied in detail.

Nebert and Gelboin [53] first described the induction of AHH in fetal hamster cells. This induction could be blocked by the simultaneous addition of actinomycin D or cycloheximide with the inducing polycyclic aromatic hydrocarbon [54]. This has also been shown to be the case with the specific induction of AHH by TCDD [55]. Haugen et al. [56] showed that the induction of cytochrome P_1-450 in mouse liver involves *de novo* synthesis of the cytochrome. Thus, the induction of AHH activity by polycyclic aromatic hydrocarbons involves both transcription and translation following the addition of the inducer.

Studies on the AHH inducibility in different strains of mice further characterized the genetic regulation of the induction. Certain strains of mice, such as C57BL, could respond to polycyclic aromatic hydrocarbons with regard to AHH induction whereas other strains, such as DBA, could not. By crossing the strains and studying the back-crosses, it was suggested that the induction was regulated genetically by a single locus [57-59]. This genetic regulatory locus for AHH induction has been designated the "*Ah* locus" [60].

Poland et al. [6] were the first to report high-affinity, stereospecific binding of [^{3}H]TCDD by mouse and rat liver cytosol. The binding was stereospecific for those polycyclic aromatic hydrocarbons capable of inducing AHH. The binding could only be found in the cytosol from strains of mice that are responsive [61]. This binding was postulated to be binding to a receptor for TCDD, an important product of the *Ah* locus.

Detection of the TCDD Receptor by Isoelectric Focusing

Using isoelectric focusing in polyacrylamide gel, a single binding species for [^{3}H]TCDD with the same affinity and stereospecificity as the binding described by Poland et al. [6] could be found [7]. The binding of TCDD to this binder was competed for by 2,3,7,8-tetrachlorodibenzofuran (TCDBF), 3-methylcholanthrene (3-Mc), benzo(a)-pyrene and β-naphthoflavone, four inducers of AHH, but not by phenobarbital or 16α-cyanopregnenolone, inducers of other forms of cytochrome P-450 [7]. Neither was the binding competed for by dexamethasone, progesterone, estradiol, testosterone, 2-hydroxyestradiol, retinol, retinoc acid, α-tocopherol, menadione or vitamin K_1. Poland et al. [6] have also shown that thyroxine does not compete for TCDD binding. Thus, the TCDD-binding protein is not similar to any other soluble receptor previously described.

Interaction of the TCDD Receptor with DNA

Chromatography of liver cytosol labelled with [^{3}H]TCDD on DNA-cellulose demonstrated that the TCDD receptor was capable of interacting with DNA [11]. In contrast to the steroid receptors, no activation step was required *in vitro* before the TCDD-receptor complex could interact with DNA. The TCDD-receptor complex bound to DNA immediately following the binding of TCDD to the receptor. However, the free receptor itself could not interact with DNA. The DNA-binding domain of the TCDD receptor is only available for binding when the ligand-binding site is occupied. Partial proteolysis of the TCDD-receptor complex with the trypsin concentration used to obtain the form with pI 5.2 resulted in a complex of apparently unchanged size but lacking the DNA-binding domain [11]. This is analogous to the conversion of the 3.6 nm glucocorticoid receptor complex into the 1.9 nm complex [21, 22]. However, it would seem that the DNA-binding domain of the TCDD receptor is much more accessible to proteolysis with trypsin.

A hypothesis for the mechanism of action of TCDD is presented in Fig. 1.

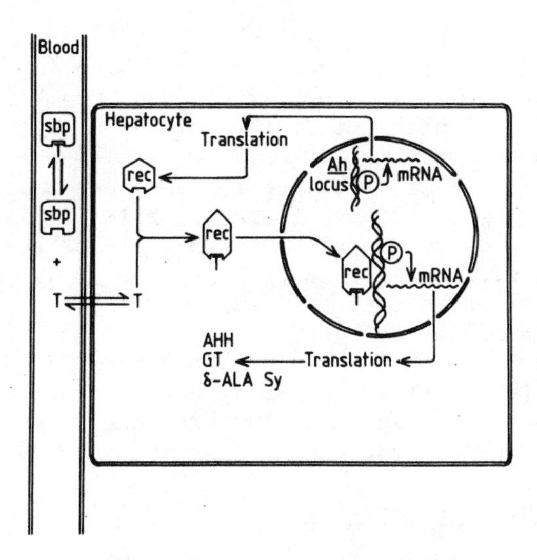

Fig. 1. Current concept of mechanism of action of TCDD. Abbreviations: T, TCDD; sbp, serum TCDD-binding protein; rec, TCDD receptor protein; P, DNA-dependent RNA polymerase; AHH, aryl hydrocarbon hydroxylase; GT, UDP-glucuronyl transferase; δ-ALA Sy, δ-aminolevulinic acid synthetase.

Tissue Distribution of the TCDD Receptor

Using isoelectric focusing in polyacrylamide gel, the tissue distribution of the TCDD receptor was investigated [9]. High levels of the receptor were found in the lung, liver, and kidneys. Low levels were found in brain, muscle, and testis [9]. Inducible AHH activity has been described in all these tissues. Nebert and Gelboin [53] reported inducible AHH activity in fetal hamster cells derived from liver, lung, limb, or brain. They also found that AHH could be induced *in vivo* in the liver, lung, and kidney in rat, monkey, or hamster [62]. In addition to the induction of AHH and glucuronyl transferase, ultrastructural changes were seen in the kidney following the administration of TCDD [63].

Interestingly, the organ containing the highest concentration of TCDD receptor, the thymus, has not been reported to contain any inducible AHH activity. However, TCDD suppresses cell-mediated immune functions in rats [64].

Lee and Dixon [65] found that the induction of AHH in the rat ventral prostate was much greater and longer lasting than in the liver. No TCDD receptor could be found in the ventral prostate but another TCDD-binding species was seen. This was not affected by trypsin and could not be saturated by a 100-fold excess concentration of unlabelled TCDBF. The prostatic TCDD-binder did not interact with DNA. Thus, it does not demonstrate any of the characteristics of an intracellular receptor protein. We have previously shown that the rat ventral prostate contains a major secretory protein which makes up about 20% of the total protein content in the prostate [66, 67]. We have called this protein prostatic secretion protein (PSP) or estramustine-binding protein (EMBP) since it binds estramustine, the dephosphorylated metabolite of the drug Estracyt^R, with a high affinity and a high capacity [66, 67]. PSP is responsible for the selective uptake of estramustine in the ventral prostate of the rat and may possibly also play a similar role in the human [68]. We have now identified the prostatic TCDD-binding protein as PSP [69]. It is possible that this binder masks the occurrence of the TCDD receptor in the prostate as it focuses very closely to the receptor and occurs in such large amounts.

Regulation of the TCDD Receptor

The induction of AHH activity varies greatly with age. In rats, the nuclear AHH activity is better characterized than the microsomal form with regard to ontogeny [70]. In that study, two peaks in the change of AHH activity following induction were found. The first occurs at birth and the second in 3-4-week-old rats. Similar findings have been found in responsive strains of mice [71].

The TCDD receptor concentration in the rat liver was found to follow the inducibility of AHH activity. Peaks of receptor concentraction were seen in neonatal rats and 21-day-old rats. No sexual differences were seen in the receptor concentrations at any of the ages studied [8].

Removal of the pituitary, testes, ovaries or adrenals had no significant effect on the concentration of the hepatic TCDD receptor in adult rats [8]. This means that the TCDD receptor is not under direct endocrine control by any of these organs in the adult rat. In addition, if any of these glands secreted an endogenous ligand for the receptor, there should have been an increase in the amount of free receptor after removal of that organ. No such increase was seen and therefore it is concluded that the endogenous ligand, if there is one, is not secreted by any of these organs.

Uptake of TCDD into Hepatic Cell Nuclei

The intranuclear binding of [^{3}H]TCDD in rat liver has been studied both *in vivo* and *in vitro*. Following the intravenous administration of [^{3}H]TCDD, a maximum uptake by cell nuclei could be observed at 2 h after injection with a concurrent decrease in the cytosolic uptake [12]. Using linear sucrose density gradient centrifugation, dextran-coated charcoal adsorption assay, DEAE-Sepharose ion-exchange chromatography, competition, enzymatic and saturation studies, a high-affinity binding protein for TCDD in liver cell nuclei could be demonstrated both *in vivo* and after an exchange *in vitro* of intravenously administered unlabelled 2,3,7,8-tetrachlorodibenzofuran (TCDBF) for [^{3}H]TCDD. Sucrose density gradient analysis showed a size of 4-5 S for both the cytosolic and nuclear TCDD binding entity. The specific binding of [^{3}H]TCDD to nuclear components was heat labile and saturable and had an equilibrium dissociation constant of 1.05 nM. Based on a differential susceptibility to specific hydrolases, i.e., DNAase, RNAase, trypsin and pronase, the binding entity appears to be a 4-5 S salt-extractable protein [12].

Cytosolic and Nuclear Binding Proteins for TCDD in the Thymus

Since the thymus was found to contain relatively large amounts of the TCDD receptor (cf. above) cytosolic and nuclear binding of TCDD in the thymus was studied in some detail [13]. The high-affinity, low-capacity TCDD-binding species was sensitive to heat and to pronase, trypsin or chymotrypsin but not to DNAase or RNAase, indicating that it was a protein. An excess of unlabelled TCDBF or β-naphthoflavone displaced [^{3}H]TCDD from the binder whereas phenobarbital, prenenolone-16α-carbonitrile or dexamethasone did not compete. Using a dextran-coated charcoal assay, the apparent dissociation constant (K_d) of the [^{3}H]TCDD-binder complex was determined to 0.36 nM and the apparent maximum amount of binding sites (B_{max}) to 68 fmol/mg of cytosolic protein. When analyzed by sucrose density-gradient centrifugation at high ionic strength, the [^{3}H]TCDD-binder complex sedimented at 4-5 S; at low ionic strength the complex sedimented more rapidly, probably due to aggregation. All these data support the interpretation that the demonstrated cytosolic TCDD-binder represents the true receptor protein for TCDD. Following intravenous administration of [^{3}H]TCDD, a low-capacity [^{3}H]TCDD-macromolecule complex was extractable from thymic cell nuclei; this complex behaved identically to the cytosolic [^{3}H]TCDD-receptor complex when exposed to heat or to hydrolytic enzymes and was therefore also identified as a protein. The nuclear [^{3}H]TCDD-protein complex sedimented at 4-5 S at high ionic strength. Furthermore, a maximum uptake of [^{3}H]TCDD in thymic nuclei was observed simultaneously with a decline in cytosolic radioactivity (at 3 h post-injection). These findings suggest that the nuclear [^{3}H]TCDD-protein complex represented [^{3}H]TCDD-receptor complex translocated from the cytoplasm [13]. In conclusion, the rat thymus contains a cytosolic TCDD receptor at a concentration similar to that of the rat hepatic receptor [7]. However, *in vivo* experiments showed that the nuclear uptake of [^{3}H]TCDD (expressed as dpm/mg DNA) in the thymus was only about 6% of that in liver. Further studies are needed for an understanding of the mechanism behind this discrepancy.

The TCDD Receptor in Rat Intestinal Mucosa and Its Possible Dietary Ligands

As mentioned above, upon removal of the pituitary, adrenals, or gonads from the rat, no increase in the number of unoccupied specific TCDD-binding sites could be observed in the liver cytosol [8]. This lack of increase indicates that the pituitary, adrenals, and gonads do not secrete any factor serving as a ligand for the receptor. The existence of (an) exogenous, i.e., dietary, receptor ligand(s) must therefore be considered.

In early experiments, pellet diets were shown to induce xenobiotic metabolism in laboratory animals [72], the responsible factors being mainly plant indoles and flavonoids [73]. Since these two classes of compounds induce AHH, a 3-MC- and TCDD-inducible enzyme system, they could have some affinity for the TCDD receptor. If so, they might represent the physiological ligands of the receptor. To investigate this possibility, we measured the affinities of some AHH-inducers of dietary origin for the TCDD receptor.

In previous studies on the induction of AHH in rats by dietary constituents, the greatest induction was noted in the lung and intestinal mucosa [2]. If the TCDD receptor is the mediator of this induction, it should be present in the intestinal mucosa as well. Using isoelectric focusing in polyacrylamide gel it was possible to detect and to study this receptor [74]. Its biochemical properties were found to be similar to those of the TCDD-receptor in rat liver cytosol. The dissociation constant of the [^{3}H]TCDD-receptor complex in rat intestinal mucosa was 0.7-3.1 nM, and it was present at a concentration of 70-80 fmol/mg protein. Starvation did not significantly increase the receptor level. The affinities of some potential dietary ligands for the TCDD receptor in rat intestinal mucosa were also studied [74]. Indole-3-carbinol had 1/2,600 of the affinity of TCDD for the receptor protein. Butylated hydroxyanisol (BHA), transstilbene oxide and quercetinpentamethylether competed even more weakly with [^{3}H]TCDD for binding to the receptor. The biological significance of the occurrence of low-affinity ligands of dietary origin for the TCDD receptor is uncertain at the present time.

Compounds in Urban Air Competing with [^{3}H]TCDD for Binding to the Receptor

Filter collected urban particulate matter contains a complex mixture or organic compounds. The mutagenicity of extracts of such particulate matter in the Ames' Salmonella assay in the absence of rat liver metabolic activation [75, 76] indicates the presence of compounds different from conventional polycyclic aromatic hydrocarbons. Presently much interest is focused on nitroaromatics [77-85]. Several compounds belonging to this class, e.g., nitropyrenes and dinitropyrenes, have been shown to be extremely powerful direct-acting mutagens [81, 86, 87] and are believed to represent a major fraction of the direct-acting mutagens in particulate emissions from diesel engines [81, 82, 88].

Acetone extracts of filter-collected urban atmospheric matter were found to contain compounds which can competetively inhibit [^{3}H]TCDD-binding to the rat liver TCDD receptor protein [89]. The concentration of conventional polycyclic aromatic hydrocarbons (PAH) or chlorinated dioxins and dibenzofurans could not account for more than 1-30% of the observed displacement of TCDD from the receptor protein. The difference in potency between samples collected in urbanized areas during different periods of the year and a background samples was 25-400-fold [89].

When reference nitroarenes were tested with respect to their affinity for the TCDD receptor, it was found that some of them, especially 3-nitroperylene ($EC_{50} \sim 10$ nM) and 4-nitrobenzo(ghi)perylene ($EC_{50} \sim 20$ nM), possess a high affinity for the receptor protein [90]. The nitro substitution is essential since the parent compounds have a very much lower affinity. Furthermore, the position of the nitro group is critical, as shown by the difference between 1- and 3-nitrobenzo(e)pyrene (EC_{50}'s > 300 and $\sim$180 nm, respectively) and between 4- and 7-nitrobenzo(ghi)perylene (EC_{50}'s $\sim$ 20 and >300 nM, respectively) [90]. A high stereospecificity has earlier been demonstrated for chlorinated analogs to TCDD and halogenated biphenyls with respect to receptor binding [91]. From these studies it seems that in order to have a high receptor affinity a compound should be planar or coplanar and fit into a rectangle of 3 × 10 A.u. with halogen atoms in the four corners. In addition to planarity, net polarizability has been suggested to be important [92]. The molecular structure of 3-nitrobenzo(e)pyrene, 3-nitroperylene and 4-nitrobenzo(ghi)perylene roughly fits this model and the nitro group confers the polarizability. Also the reduced binding affinity of 6-nitrochrysene ($EC_{50} \sim 120$ nM) and 6-nitrobenzo(a)pyrene ($EC_{50} \sim 170$ nM) compared to their respective unsubstituted parent compounds (EC_{50}'s $\sim$ 60 and $\sim$20 nM, respectively) is in accordance with these structural requirements [90].

Strains of mice with high AHH inducibility are more susceptible to pulmonary cancers caused by 3-methylcholanthrene than strains with low inducibility indicating a link between AHH activity and appearance of pulmonary tumors [93]. The rat lung has a high content of receptor protein [9] and the capability of the human lung to metabolize B(a)P [94] indicates the presence of the receptor protein also in this tissue. Animal experiments suggest that also other responses such as thymic atrophy and incidence of cleft palate correlate to receptor affinity [95]. Although TCDD has been shown to be a potent carcinogen in chronic feeding studies [96] it is apparent that TCDD shows no or very little mutagenic activity in *in vitro* bacterial test

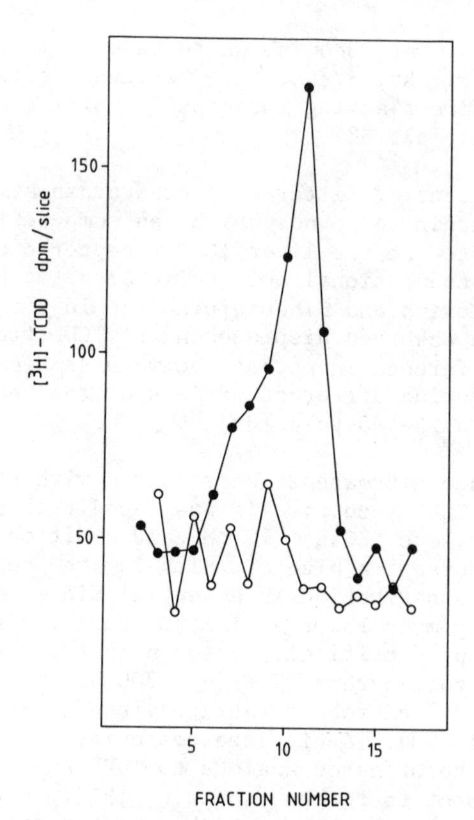

Fig. 2. Presence of specific TCDD-binding in cytosol from human leukemic cells. Cytosol from leukemic cells from a patient with acute lymphatic leukemia was incubated with 1.5 nM [^{3}H]TCDD for 2 h at 4°C in the presence (○–○) or absence (●–●) of a 100-fold excess of TCDBF. Following removal of free ligand with dextran-coated charcoal the incubation mixtures were analyzed with isoelectric focusing in polyacrylamide gel.

systems such as the Ames' test [97] and a very low covalent binding to DNA in vivo [98]. In a recent study, however, it was shown that TCDD is a potent promoting agent for hepatocarcinogenesis initiated by diethylnitrosamine [99]. If this effect of TCDD is mediated via binding to the receptor is not yet known. TCDD and possibly also other compounds with an affinity for the same receptor may thus better be described as cocarcinogens and tumor promoters rather than carcinogens.

The presence of compounds with affinity for the receptor in urban particulate matter may be of importance with regard to the health implications of urban air pollution. This type of compounds

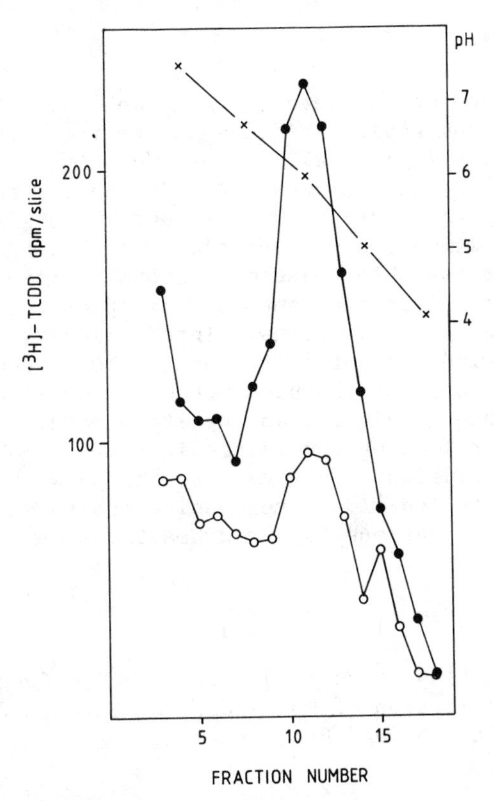

Fig. 3. Presence of specific TCDD-binding in cytosol from a human B lymphocyte cell line. Cytosol from Epstein-Barr virus-transformed human B lymphocytes was incubated with 1.5 nM [^{3}H]TCDD for 2 h at 4°C in the presence (○—○) or absence (●—●) of a 100-fold excess of TCDBF. Following removal of free ligand with dextran-coated charcoal, the incubation mixtures were analyzed with isoelectric focusing in polyacrylamide gel.

may or may not be similar to the components that are mutagenic in the absence of mammalian metabolic activation in the Ames' Salmonella assay.

Presence of the TCDD Receptor in Human Lymphocytes

Efforts are presently being directed in our laboratory towards developing assay methods for the TCDD receptor in human tissues. Specific binding of [^{3}H]TCDD has been found both in lymphocytes from a patient with acute lymphatic leukemia (Fig. 2) and in Epstein-Barr virus-transformed human B lympocytes (Fig. 3). This binding is now being characterized further.

Concluding Remarks

The TCDD receptor has been shown to be present in both cytosol and cell nuclei from several tissues in the rat using isoelectric focusing in polyacrylamide gel, sucrose density gradient centrifugation as well as other techniques. A TCDD-binding species with properties similar to those of the TCDD receptor has also been demonstrated in human lymphocytes. The biochemical characteristics of the TCDD receptor resemble those of steroid hormone receptors although it is quite clear that the TCDD receptor represents a species distinct from the latter proteins. Interestingly, urban air contains compounds with high affinity for the TCDD receptor and it remains to be shown whether the presence of these ligands in city atmosphere constitutes a significant health hazard. The demonstration of a TCDD receptor-like macromolecule in human lymphocytes encourages attempts to develop a TCDD receptor assay with the aim to monitor differences in individual responsiveness to environmental chemicals potentially hazardous to human health.

ACKNOWLEDGMENTS

This work was supported by grants from The Swedish Council for Planning and Coordination of Research, The Swedish Cancer Society and The Swedish Environment Protection Agency.

REFERENCES

1. E. Ward, B. Paigen, K. Steenland, R. Vincent, J. Minowada, H. L. Gurtoo, P. Sartoti, and M. B. Havens, Aryl hydrocarbon hydroxylase in persons with lung or laryngeal cancer, Int. J. Cancer, 22:384-389 (1978).
2. T. L. McLemore, R. R. Martin, N. P. Wray, E. T. Cantrell, and D. L. Busbee, Dissociation between aryl hydrocarbon hydroxylase activity in cultured pulmonary macrophages and blood lymphocytes from lung cancer patients, Cancer Res., 38:3805-3811 (1978).
3. R. E. Kouri, J. Oberdorf, D. J. Slomiany, and C. E. McKinney, A method for detecting aryl hydrocarbon hydroxylase activities in cryopreserved human lymphocytes, Cancer Letters, 14:29-40 (1981).
4. B. Paigen, E. Ward, A. Reilly, L. Houten, H. L. Gurtoo, J. Minowada, K. Steenland, M. B. Havens, and P. Sartori, Seasonal variation of aryl hydrocarbon hydroxylase activity in human lymphocytes, Cancer Res., 41:2757-2761 (1981).
5. T. L. McLemore, R. R. Martin, N. P. Wray, E. T. Cantrell, and D. L. Busbee, Reassessment of the relationship between aryl hydrocarbon hydroxylase and lung cancer, Cancer, 48:1438-1443 (1981).

6. A. Poland, E. Glover, and A. Kende, Stereospecific, high affinity binding of 2,3,7,8-tetrachlorodibenzo-p-dioxin by hepatic cytosol. Evidence that the binding species is receptor for induction of aryl hydrocarbon hydroxylase, J. Biol. Chem., 251: 4936-4946 (1976).
7. J. Carlstedt-Duke, G. Elfström, M. Snochowski, B. Högberg, and J.-Å. Gustafsson, Detection of the 2,3,7,8-tetrachlorodibenzo-p-dioxin (TCDD) receptor in rat liver by isoelectric focusing in polyacrylamide gels, Toxicol. Lett., 2:365-373 (1978).
8. J. M. B. Carlstedt-Duke, G. Elfström, B. Högberg, and J.-Å. Gustafsson, Ontogeny of the rat hepatic receptor for 2,3,7,8-tetrachlorodibenzo-p-dioxin and its endocrine indpendence, Cancer Res., 39:4653-4656 (1979).
9. J. M. B. Carlstedt-Duke, Tissue distribution of the receptor for 2,3,7,8-tetrachlorodibenzo-p-dioxin in the rat, Cancer Res., 39:3172-3176 (1979).
10. J. Carlstedt-Duke, M. Gillner, L.-A. Hansson, R. Toftgård, S. Gustafsson, B. Högberg, and J.-Å. Gustafsson, The molecular basis for the induction of aryl hydrocarbon hydroxylase: characteristics of the receptor protein for 2,3,7,8-tetrachlorodibenzo-p-dioxin (TCDD), in: Biochemistry, Biophysics, and Regulation of Cytochrome P-450 (J.-Å. Gustafsson et al., eds.), pp. 147, Elsevier/North-Holland Biomedical Press (1980).
11. J. M. B. Carlstedt-Duke, U.-B. Harnemo, B. Högberg, and J.-Å. Gustafsson, Interaction of the hepatic receptor protein for 2,3,7,8-tetrachlorodibenzo-p-dioxin with DNA, Biochim. Biophys. Acta, 672:131-141 (1981).
12. L. Poellinger, R. N. Kurl, J. Lund, M. Gillner, J. Carlstedt-Duke, B. Högberg, and J.-Å. Gustafsson, High-affinity binding of 2,3,7,8-tetrachlorodibenzo-p-dioxin in cell nuclei from rat liver, Biochim. Biophys. Acta, 714:516-523 (1982).
13. J. Lund, R. N. Kurl, L. Poellinger, and J.-Å. Gustafsson, Cytosolic and nuclear binding proteins for 2,3,7,8-tetrachlorodibenzo-p-dioxin in the rat thymus, Biochim. Biophys. Acta, 716:16-23 (1982).
14. E. V. Jensen, B. E. Block, S. Smith, K. Kyser, and E. R. DeSombre, Estrogen receptors and breast cancer response to adrenalectomy, NCI Monograph-Prediction of responses in cancer therapy, 34:55-61 (1971).
15. W. L. McGuire and G. C. Chamness, Studies on the estrogen receptor in breast cancer, in: Advances in Experimental Medicine and Biology (B. W. O'Malley and A. R. Means, eds.), Plenum Press, New York, 36:113136 (1973).
16. Ö. Wrange, B. Nordenskjöld, C. Silfverswärd, P. O. Granberg, and J.-Å. Gustafsson, Isoelectric focusing of estradiol receptor protein from human mammary carcinoma - a comparison to sucrose gradient analysis, Europ. J. Cancer, 12:695-700 (1976).
17. Ö. Wrange, B. Nordenskjöld, and J.-Å. Gustafsson, Cytosol estradiol receptor in human mammary carcinoma: an assay based on isoelectric focusing in polyacrylamide gel, Analytical Biochemistry, 85:461-475 (1978).

18. J.-Å. Gustafsson, Å. Pousette, and Ö. Wrange, Predictive tests in treatment of breast and prostatic carcinoma based on steroid receptor assays, Trends in Pharmacological Sciences, 1980, 279-281.
19. Ö. Wrange, J. Carlstedt-Duke, and J.-Å. Gustafsson, Purification of the glucocorticoid receptor from rat liver cytosol, J. Biol. Chem., 254:9284-9290 (1979).
20. G. M. Ringold, Glucocorticoid regulation of mouse mammary tumor virus gene expression, Biochim. Biophys. Acta, 560:487-508 (1979).
21. J. Carlstedt-Duke, J.-Å. Gustafsson, and Ö. Wrange, Formation and characteristics of hepatic dexamethasone-receptor complexes of different molecular weight, Biochim. Biophys. Acta, 497:507-524 (1977).
22. Ö. Wrange and J.-Å. Gustafsson, Separation of the hormone- and DNA-binding sites of the hepatic glucocorticoid receptor by means of proteolysis, J. Biol. Chem., 253:856-865 (1978).
23. J. Carlstedt-Duke, Ö. Wrange, E. Dahlberg, J.-Å. Gustafsson, and B. Högberg, Transformation of the glucocorticoid receptor in rat liver cytosol by lysosomal enzymes, J. Biol. Chem., 254:1537-1539 (1979).
24. S. Okret, J. Carlstedt-Duke, Ö. Wrange, K. Carlström, and J.-Å. Gustafsson, Characterization of an antiserum against the glucocorticoid receptor, Biochim. Biophys. Acta, 677:205-219 (1981).
25. J. Carlstedt-Duke, S. Okret, Ö. Wrange, and J.-Å. Gustafsson, Immunochemical analysis of the glucocorticoid receptor: identification of a third domain separate from the steroid-binding and DNA-binding domains, Proc. Natl. Acad. Sci. USA, 79:4260-4264 (1982).
26. S. Okret, Y.-W. Stevens, J. Carlstedt-Duke, Ö. Wrange, J.-Å. Gustafsson, and J. Stevens, Absence of a glucocorticoid receptor domain responsible for biological effects in glucocorticoid-resistant mouse lymphoma P1798, Cancer Res., 43:3127-3131 (1983).
27. F. Payvar, Ö. Wrange, J. Carlstedt-Duke, S. Okret, J.-Å. Gustafsson, and K. R. Yamamoto, Purified glucocorticoid receptors bind selectively _in vitro_ to a cloned DNA fragment whose transcription is regulated by glucocorticoids _in vivo_, Proc. Natl. Acad. Sci. USA, 78:6628-6632 (1981).
28. J.-Å. Gustafsson, S. A. Gustafsson, B. Nordenskjöld, S. Okret, C. Silfvarswärd, and Ö. Wrange, Estradiol receptor analysis in human breast cancer tissue by isoelectric focusing in polyacrylamide gel, Cancer Res., 38:4225-4228 (1978).
29. S. Okret, Ö. Wrange, B. Nordenskjöld, C. Silfverswärd, and J.-Å. Gustafsson, Estrogen receptor assay in human mammary carcinoma with the synthetic estrogen 11β-methoxy-17α-ethinyl-1,3,5(10)-estratriene-3,17β-diol (R 2858), Cancer Res., 38:3904-3909 (1978).

30. N.-O. Theve, K. Carlström, J.-Å. Gustafsson, S. Gustafsson, B. Nordenskjöld, H. Sköldefors, and Ö. Wrange, Oestrogen receptors and peripheral serum levels of oestradiol-17β in patients with mammary carcinoma, Europ. J. Cancer, 14(1337-1340 (1978).
31. H. Westerberg, B. Nordenskjöld, Ö. Wrange, J.-Å. Gustafsson, S. Humla, N. O. Theve, C. Silfverswärd, and P.-O. Granberg, Effect of antiestrogen therapy on human mammary carcinomas with different estrogen receptor contents, Europ. J. Cancer, 14: 619-622 (1978).
32. C. Silfverswärd, J.-Å. Gustafsson, S. A. Gustafsson, B. Nordenskjöld, A. Wallgren, and Ö. Wrange, Estrogen receptor analysis on fine needle aspirates and on histologic biopsies from human breast cancer, Europ. J. Cancer, 16:1351-1357 (1980).
33. C. Silfverswärd, J.-Å. Gustafsson, S. A. Gustafsson, S. Humla, B. Nordenskjöld, A. Wallgren, and Ö. Wrange, Estrogen receptor concentrations in 269 cases of histologically classified human breast cancer, Cancer, 45:2001-2005 (1980).
34. Ö. Wrange, S. Humla, I. Ramberg, S. A. Gustafsson, L. Skoog, B. Nordenskjöld, and J.-Å. Gustafsson, Progestin-receptor analysis in human breast cancer cytosol by isoelectric focusing in slabs of polyacrylamide gel, J. Steroid Biochem., 14:141-148 (1981).
35. G. L. Greene, F. W. Fitch, and E. V. Jensen, Monoclonal antibodies to estrophilin: Probes for the study of estrogen receptors, Proc. Natl. Acad. Sci. USA, 77:157-161 (1980).
36. G. L. Greene, C. Nolan, J. P. Engler, and E. V. Jensen, Monoclonal antibodies to human estrogen receptor, Proc. Natl. Acad. Sci. USA, 77:5115-5119 (1980).
37. J.-Å. Gustafsson, P. Ekman, M. Snochowski, A. Zetterberg, Å. Pousette, and B. Högberg, Correlation between clinical response to hormone therapy and steroid receptor content in prostatic cancer, Cancer Res., 38:4345-4348 (1978).
38. J.-Å. Gustafsson, P. Ekman, Å. Pousette, M. Snochowski, and B. Högberg, Demonstration of a progestin receptor in human benign prostatic hyperplasia and prostatic carcinoma, Investigative Urology, 15:361-366 (1978).
39. P. Ekman, M. Snochowski, E. Dahlberg, and J.-Å. Gustafsson, Steroid receptors in metastatic carcinoma of the human prostate, Europ. J. Cancer, 15:257-262 (1979).
40. P. Ekman, M. Snochowski, A. Zetterberg, B. Högberg, and J.-Å. Gustafsson, Steroid receptor content in human prostatic carcinoma and response to endocrine therapy, Ancer, 44:1173-1181 (1979).
41. B. R. Rao and W. G. Wiest, Receptors for progesterone, Gynecol. Oncol., 2:239-248 (1974).
42. K. Pollow, M. Schmidt-Gollwitzer, and J. Nevinny-Stickel, Progesterone receptors in normal human endometrium and endometrial carcinoma, in: Progesterone receptors in normal and neoplastic tissues (W. L. McGuire, J. P. Raynaud, and E. E. Baulieu, eds.), Raven Press, New York, 1977, 313-338.

43. M. E. Lippman, R. H. Halterman, B. G. Leventhal, S. Perry, and E. B. Thompson, Glucocorticoid-binding proteins in human acute lymphoblastic leukemic blast cells, J. Clin. Invest., 52:1715-1725 (1973).
44. J. Stevens, Y. W. Stevens, and R. L. Rosenthal, Characterization of cytosolic and nuclear glucocorticoid-binding components in human leukemic lymphocytes, Cancer Res., 39:4939-4948 (1979).
45. L.-A. Hansson, S. A. Gustafsson, J. Carlstedt-Duke, G. Gahrton, B. Högberg, and J.-Å. Gustafsson, Quantitation of the cytsolic glucocorticoid receptor in human normal and neoplastic leukocytes using isoelectric focusing in polyacrylamide gel, J. Steroid Biochem., 14:757-764 (1981).
46. H. Bojar, K. Maar, and W. Staib, The endocrine background of human renal cell carcinoma, I. Binding of the hihgly potent progestin R 5020 by tumor cytosol., Urol. Int., 34:302-311 (1979).
47. G. Concolino, F. DiSilverio, A. Marocchi, and U. Bracci, Renal cancer steroid receptors: Biochemical basis for endocrine therapy, Eur. Urol., 5:90-93 (1979).
48. A. Poland and E. Glover, Comparison of 2,3,7,8-tetrachlorodibenzo-p-dioxin, a potent inducer of aryl hydrocarbon hydroxylase, with 3-methylcholanthrene, Mol. Pharmacol., 10:349-359 (1974).
49. A. Poland and E. Glover, Chlorinated dibenzo-p-dioxins: potent inducers of δ-aminolevulinic acid synthetase and aryl hydrocarbon hydroxylase II. A study of the structure-activity relationship, Mol. Pharmacol., 9:736-747 (1973).
50. A. P. Poland, E. Glover, J. R. Robinson, and D. W. Nebert, Genetic expression of aryl hydrocarbon hydroxylase activity, J. Biol. Chem., 249:5599-5606 (1974).
51. G. W. Lucier, O. S. McDaniel, and G. E. R. Hook, Nature of the enhancement of hepatic uridine diphosphate glucuronyltransferase activity by 2,3,7,8-tetrachlorodibenzo-p-dioxin in rats, Biochem. Pharmacol., 24:325-334 (1975).
52. R. Kirsch, G. Fleischner, K. Kaminaka, and I. M. Arias, Structural and functional studies of ligandin, a major renal organic anion-binding protein, J. Clin. Invest., 55:1009-1019 (1975).
53. D. W. Nebert and H. V. Gelboin, Substrate-inducible microsomal aryl hydrocarbon hydroxylase in mammalian cell culture, II. Cellular responses during enzyme induction, J. Biol. Chem., 243:6250-6261 (1968).
54. D. W. Nebert and H. V. Gelboin, The role of ribonucleic acid and protein synthesis in microsomal aryl hydrocarbon hydroxylase induction in cell culture. The independence of transcription and translation, J. Biol. Chem., 245:160-168 (1970).
55. K. T. Kitchin and J. S. Woods, 2,3,7,8-tetrachlorodibenzo-p-dioxin induction of aryl hydrocarbon hydroxylase in female rat liver, Evidence for de novo synthesis of cytochrome P-448, Mol. Pharmacol., 14:890-899 (1978).

56. D. A. Haugen, M. J. Coon, and D. W. Nebert, Induction of multiple forms of mouse liver cytochrome P-450. Evidence for genetically controlled de novo protein synthesis in response to treatment with β-naphtoflavone or phenobarbitol, J. Biol. Chem., 251:1817-1827 (1976).
57. P. E. Thomas, R. E. Kouri, and J. J. Hutton, The genetics of aryl hydrocarbon hydroxylase induction in mice: a single gene difference between C57BL/7J and DBA/2J, Biochem. Genet., 6: 157-168 (1972).
58. D. W. Nebert and J. E. Gielen, Genetic regulation of aryl hydrocarbon hydroxylase induction in mice, Fed. Proc., 31:1315-1325 (1972).
59. D. W. Nebert, F. M. Goujon, and J. E. Gielen, Aryl hydrocarbon hydroxylase induction by polycyclic hydrocarbons: simple autosomal dominant trait in the mouse, Nat. New Biol., 236: 107-110 (1972).
60. D. W. Nebert, S. S. Thorgeirsson, and G. H. Lamberg, Genetic aspects of toxicity during development, Environ. Health Perspect., 18:35-45 (1976).
61. A. Poland and E. Glover, Genetic expression of aryl hydrocarbon hydroxylase by 2,3,7,8-tetrachlorodibenzo-p-dioxin: evidence for a receptor mutation in genetically non-responsive mice, Mol. Pharmacol., 11:389-398 (1975).
62. D. W. Nebert and H. V. Gelboin, The in vivo and in vitro induction of aryl hydrodarbon hydroxylase in mammalian cells of different species, tissues, strains, and developmental and hormonal states, Arch. Biochem. Biophys., 134:76-89 (1969).
63. B. A. Fowler, G. E. Hook, and G. W. Lucier, Tetrachlordibenzo-p-dioxin induction of renal microsomal enzyme systems: ultrastructural effects on pars recta (S_3) proximal tubule cells of the rat kidney, J. Pharmacol. Exp. Ther., 203:712-721 (1977).
64. R. E. Faith and J. A. Moore, Impairment of thymus-dependent immune functions by exposure of the developing immune system to 2,3,7,8-tetrachlorodibenzo-p-dioxin (TCDD), J. Toxicol. Environ. Health, 3:451-464 (1977).
65. I. P. Lee and R. L. Dixon, Factors influencing reproduction and genetic toxic effects on male gonads, Environ. Health Perspect., 24:117-127 (1978).
66. B. Forsgren, P. Björk, K. Carlström, J.-Å. Gustafsson, Å. Pousette, and B. Högberg, Purification and distribution of a major protein in rat prostate that binds estramustine, a nitrogen mustard derivative of estradiol-17β, Proc. Natl. Acad. Sci. USA, 76:3149-3153 (1979).
67. B. Forsgren, J.-Å. Gustafsson, Å. Pousette, and B. Högberg, Binding characteristics of a major protein in rat ventral prostate cytosol that interacts with estramustine, a nitrogen mustard derivative of 17β-estradiol, Cancer Res., 39:5155-5164 (1979).

68. P. Björk, B. Forsgren, J.-Å. Gustafsson, Å. Pousette, and B. Högberg, Partial characterization and quantitation of a human prostatic estramustine binding protein, Cancer Res., 42:1935-1942 (1982).
69. J.-Å. Gustafsson, P. Söderkvist, T. Haaparanta, L. Busk, Å. Pousette, H. Glaumann, R. Toftgård, and B. Högberg, Induction of cytochrome P-450 and metabolic activation of mutagens in the rat ventral prostate, in: The Prostatic Cell: Structure and Function, Part B, Alan R. Liss, Inc., New York, 191-205 (1981).
70. J. C. Nunnink, A. H. L. Chuang, and E. Bresnick, The ontogeny of nuclear aryl hydrocarbon hydroxylase, Chem.-Biol. Interact., 22:225-230 (1978).
71. D. W. Nebert, F. M. Goujon, and J. E. Gielen, Aryl hydrocarbon hydroxylase induction by polycyclic hydrocarbons: simple automsomal dominant trait in the mouse, Nat. New. Biol., 236: 107-110 (1972).
72. L. W. Wattenberg, Studies of polycyclic hydrocarbon hydroxylases of the intestine possibly related to cancer: effect of diet on benzpyrene hydroxylase activity, Cancer, 28:99-102 (1971).
73. L. W. Wattenberg, W. D. Loub, K. L. Lam, and J. L. Speier, Dietary constituents altering the responses to chemical carcinogens, Fed. Proc., 35:1327-1331 (1976).
74. G. Johansson, M. Gillner, B. Högberg, and J.-Å. Gustafsson, The TCDD receptor in rat intestinal mucosa and its possible dietary ligands, Nutrition and Cancer, 1982, 134-144.
75. J. N. Pitts, D. Grosjean, T. M. Mischke, V. Simmon, and D. Poule, Mutagenic activity of airborne particulate organic pollutants, Toxicol. Letters, 1:65-70 (1977).
76. R. Talcott and E. Wei, Airborne mutagens bioassayed in Salmonella typhimurium, J. Natl. Cancer Inst., 58:449-451 (1977).
77. G. Löfroth, E. Hefner, I. Alfheim, and M. Møller, Mutagenic activity in photocopies, Science, 209:1037-1039 (1980).
78. H. S. Rosenkranz, E. C. McCoy, D. R. Sanders, M. Butler, D. K. Kiriazides, and R. Mermelstein, Nitropyrenes: Isolation, identification, and reduction of mutagenic impurities in carbon black and toners, Science, 209:1039-1043 (1980).
79. E. Agurell and G. Löfroth, Presence of various types of mutagenic impurities in carbon black detected by the Salmoneaella/ microsome assay, in: Short-term Bioassays in the Analysis of Complex Environmental Mixtures III (M. D. Waters, S. S. Sandhu, J. Lewtas, L. Claxton, and S. Nesnow, eds.), Plenum Press, New York (in press).
80. L. Rudling, B. Ahling, and G. Löfroth, Chemical and biological characterization of emissions from combustion of wood and woodchips in small furnaces and stoves, in: Residential Solid Fuels, Environmental Impacts and Solutions (J. A. Copper and D. Malek, eds.), Beaverton, OR: Oregon Graduate Center, 34-53 (1981).
81. J. N. Pitts, Jr., D. M. Lokensgard, W. P. Harger, T. S. Fisher, V. Mejia, J. J. Schuler, G. M. Scorziell, and Y. A. Katzenstein,

Mutagens in diesel exhaust: Identification and direct activities of 6-nitrobenzo(a)pyrene, 9-nitroanthracene, 1-nitropyrene and 5H-phenanthro[4,5-bcd]pyran-5-one, Mut. Res. (in press).

82. D. Schuetzle, Sampling of vehicle emissions for chemical analysis and biological testing, Environ. Health Persp. (in press).
83. C. Y. Wang, M. S. Lee, C. M. King, and P. O. Warner, Evidence for nitroaromatics as direct-acting mutagens of airborne particles, Chemosphere, 9:83-87 (1980).
84. G. Löfroth, Comparison of the mutagenic activity in carbon particulate matter and in diesel and gasoline engine exhaust, in: Short-term Bioassays in the Analysis of Complex Environmental Mixtures II (M. D. Waters, S. S. Sandhu, J. Lewtas Huisingh, L. Claxton, and S. Newnow, eds.), Plenum Press, New York, 319-336 (1981).
85. J. N. Pitts, Jr., W. Harger, D. M. Lokensgard, D. R. Fitz, G. M. Scorziell, and V. Mejia, Diurnal variation in the mutagenicity of airborne particulate organic matter in California's South Coast air basin, Mut. Res. (in press).
86. R. Mermelstein, D. K. Kiriazides, M. Butler, E. C. McCoy, and H. S. Rosenkranz, The extraordinary mutagenicity of nitropyrenes in bacteria, Mut. Res., 89:187-196 (1981).
87. E. C. McCoy, H. S. Rosenkranz, and R. Mermelstein, Evidence for the existence of a family of bacterial nitroreductases capable of activating nitrated polycyclics to mutagens, Environ. Mut., 3:421-427 (1981).
88. T. C. Pederson and J. S. Siak, The role of nitroaromatic compounds in the direct-acting mutagenicity of diesel particle extracts, J. Appl. Toxicol., 1:54-60 (1981).
89. R. Toftgård, J. Carlstedt-Duke, R. Kurl, G. Löfroth, and J.-Å. Gustafsson, Compounds in urban air compete with [^{3}H]2,3,7,8-tetrachlorodibenzo-p-dioxin for binding to the receptor protein, Chem.-Biol. Interact, 46:335-346 (1983).
90. T. Greibrokk, G. Löfroth, L. Nilsson, R. Toftgård, J. Carlstedt-Duke, and J.-Å. Gustafsson, Nitroarenes: mutagenicity in the Ames Salmonella/microsome assay and affinity to the TCDD-receptor protein, in: Toxicity of Nitroaromatic Compounds (D. E. Rickert, ed.), Hemisphere Publ. Corp., Washington, D.C. (in press).
91. A. Poland and E. Glover, 2,3,7,8-tetrachlorodibenzo-p-dioxin: Studies on the mechanism of action, in: The Scientific Basis of Toxicity Assessment (H. Witschi, ed.), Amsterdam/Elsevier, 223-239 (1980).
92. J. D. McKinney and P. Singh, Structure-activity relationships in halogenated biphenyls: Unifying hypothesis for structural specificity, Chem. Biol. Interact., 33:271-283 (1981).
93. R. E. Kouri, L. H. Billups, T. H. Rude, C. E. Whitmire, B. Sass, and C. J. Henry, Correlation of inducibility of aryl hydrocarbon hydroxylase with susceptibility to 3-methylcholanthrene-induced lung cancers, Cancer Letters, 9:277-284 (1980).

94. R. A. Prough, Z. Sipal, S. W. Jakobsson, Metabolism of benzo-(a)pyrene in human lung microsomal fractions, Life Sci., 21: 1629-1636 (1977).
95. A. Poland and E. Glover, 2,3,7,8-tetrachlorodibenzo-p-dioxin: segregation of toxicity with the Ah-locus, Mol. Pharmacol., 17: 86-94 (1980).
96. R. J. Kociba, P. G. Keyes, J. E. Beyer, R. M. Carreon, C. E. Wade, D. A. Dittenberg, R. P. Kalnis, L. E. Frauson, C. N. Park, S. D. Barnard, R. S. Hummel, and C. G. Humiston, Results of a two-year chronic toxicity and oncogenicity study of 2,3,7,8-tetrachlorodibenzo-p-dioxin in rats, Toxicol. Appl. Pharmacol., 46:279-303 (1978).
97. J. S. Wasson, J. E. Huft, and N. Loprieno, A review on the genetic toxicology of chlorinated dibenzo-p-dioxins, Mut. Res., 47:141-160 (1977-78).
98. A. Poland and E. Glover, An estimate of the maximum _in vivo_ covalent binding of 2,3,7,8-tetrachlorodibenzo-p-dioxin to rat liver protein, ribosomal RNA and DNA, Cancer Res., 39: 3341-3344 (1979).
99. H. C. Pitot, T. Goldsworthy, H. A. Campbell, and A. Poland, Quantitative evaluation of the promotion by 2,3,7,8-tetrachlorodibenzo-p-dioxin of hepatocarcinogenesis from diethylnitrosamine, Cancer Res., 40:3616-3620 (1980).

INTERINDIVIDUAL DIFFERENCES IN MONOOXYGENASE ACTIVITIES OF HUMAN LIVER

G. Clare Kahn*, Alan R. Boobis,
and Donald S. Davies

Department of Clinical Pharmacology
Royal Postgraduate Medical School
London W12 OHS, U.K.

*Present address: University of Pennsylvania
The School of Medicine
Med. Labs/G3, 37th and Hamilton Walk
Philadelphia, Pennsylvania 19104

1. INTRODUCTION

A large number of drugs and environmental chemicals can cause cancer when administered to experimental animals [1] and evidence from a variety of sources suggests that at least some of these compounds can cause cancer in man [2]. It has recently been estimated that as much as 90% of cancer in man is environmental in origin, the vast majority of this due to chemicals [3]. The development of a neoplasm following exposure to a chemical carcinogen is obviously a complex, multi-stage process [4]. A schematic diagram, with some of the steps involved, is shown in Fig. 1. The first step is initiation, in which the chemical carcinogen interacts with the target macromolecule, usually believed to be DNA [5]. The resultant damage to the DNA may or may not be repaired by specific DNA repair processes [6]. If repair does not occur then the damaged DNA may be miss-coded, leading to a somatic mutation [4]. The error is then 'fixed' in some way during the promotion stage of tumorigenesis [7]. Interindividual differences in susceptibility to chemical carcinogenesis can obviously come about through variation in any one of the stages shown in Fig. 1.

Although the structures of known chemical carcinogens show considerable diversity, over the last few decades it has become apparent that a common feature is the metabolic activation of otherwise

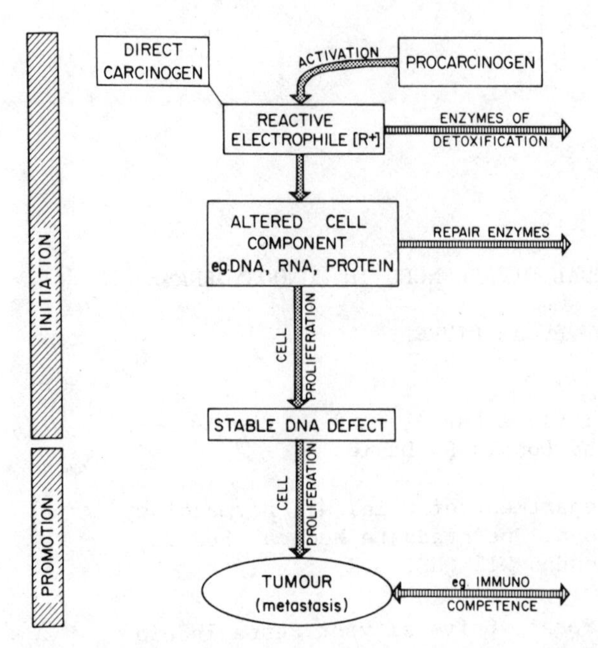

Fig. 1. Stages in chemical carcinogenesis, illustrating the major steps occurring during initiation and promotion in the development of a tumor.

inert procarcinogens to highly reactive, electrophilic intermediates that react with critical macromolecular targets such as DNA [8]. Some of the pathways involved in these activation processes are shown in Fig. 2. Examples of such procarcinogens include polycyclic aromatic hydrocarbons such as benzo(a)pyrene [9], 2-acetylamino-fluorene [10], vinyl chloride [11], dimethylnitrosamine [12], β-naphthylamine [13], and the fungal mold product aflatoxin B_1 [14]. However, not all chemical carcinogens require metabolic activation before they can react with DNA, e.g., nitrogen mustards such as melphalan [15], and nitrosomethylurea [9] are direct acting. Enzyme systems which evolved to inactivate or detoxify both exogenous and endogenous compounds [16, 17] are thus sometimes responsible for converting an otherwise harmless compound into a highly reactive electrophile [18]. Interindividual variation in these enzyme systems may profoundly influence individual susceptibility to chemical carcinogenesis [19]. Although the rate limiting step in the development of cancer is not always known, it is apparent that if the initial exposure does not occur then all other considerations are irrelevant. For example, if certain individuals were genetically incapable of converting benzo(a)pyrene to its ultimate carcinogenic species, the diol-epoxide [20] then such individuals would be completely protected against the development of a bronchogenic carcinoma from cigarette smoking (assuming that benzo(a)pyrene is the causative agent of bronchogenic carcinoma in cigarette smoke [21]). Differences in enzyme activity may explain tissue [22] and species [10] differences in susceptibility to chemical carcinogens and, as stated above, interindividual differences within the human population [19]. A knowledge of the enzymes of activation and of detoxication is thus extremely important in any consideration of the processes of chemical carcinogenesis.

2. THE ROLE OF DRUG METABOLISM* IN CHEMICAL CARCINOGENESIS

2.1. The Enzymes of Drug Metabolism

Metabolic reactions can usefully be divided into two groups [16]; a) the phase one reactions which include oxidations, reductions and hydrolyses. These reactions introduce or expose a functional group in the parent compound. b) The phase two reactions which comprise the conjugation or synthetic reactions. These reactions involve the combination of the parent compound with an endogenous small molecule such as sulfate, glucuronide, or acetate. Thus, the product of a phase one reaction may act as the substrate for a phase two reaction. The net result of these metabolic processes is a considerable increase in the water solubility of what otherwise may be a very lipophilic compound, thereby ensuring its effective elimination by the kidneys [16].

Metabolism most often removes, or at least substantially reduces, any biologic activity of the substrate [23]. Thus, the duration of action of a wide variety of drugs is determined by their rate of metabolism [24]. However, the processes of metabolism can sometimes generate toxic [25] or carcinogenic [26] intermediates, as described above. Metabolism of a compound may also alter, rather than reduce, its biologic effect. For example, phenylbutazone is an anti-inflammatory agent which can be metabolized to an alcohol with uricosuric properties [27]. Finally, the metabolism of otherwise inert compounds may reveal pharmacologic activity, e.g., cyclophosphamide is inactive prior to oxidation [28]. There is there-

*The term "drug metabolism" is widely used to denote the metabolic conversion of foreign compounds, both drugs and environmental chemicals, and its use is not in any way restricted to just drugs. The processes of drug metabolism may also be involved in the metabolism of endogenous compounds. Usually, the same enzyme has to be involved in the metabolism of a foreign compound and an endogenous substrate before the latter is considered the provenance of "drug metabolism." However, the distinction is diffuse and all attempts to change the nomenclature have failed to date and seem destined to do so in the future. In the absence of suitable alternatives, the terms "drug metabolism" and "drug metabolizing enzymes" will be used in the above sense throughout this chapter.

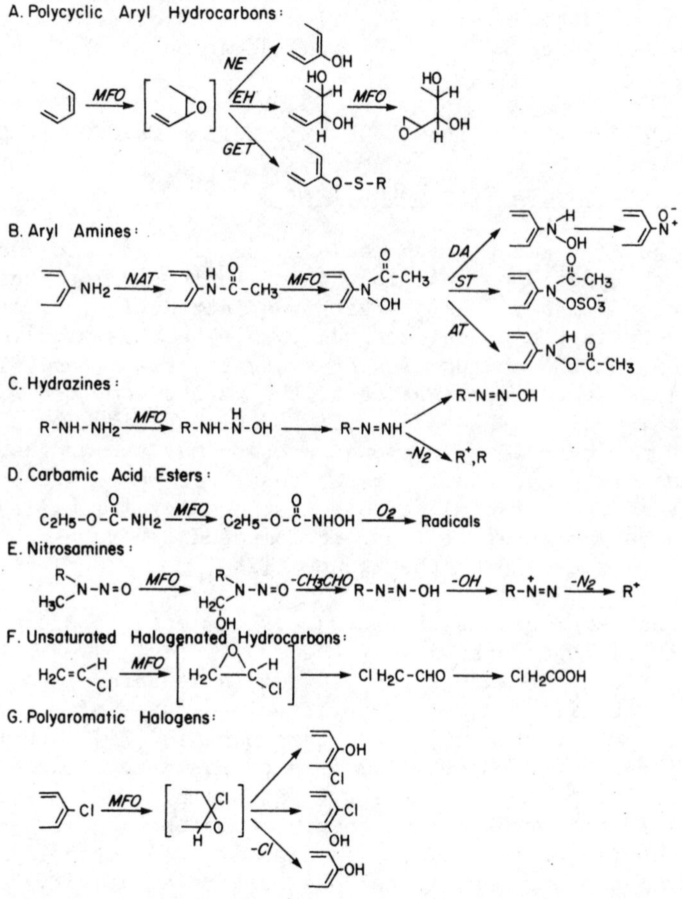

A. Polycyclic Aryl Hydrocarbons:

MFO, NE, EH, MFO, GET

HO, H, OH, H, O-S-R

B. Aryl Amines:

NH2 NAT N-C-CH3 MFO N C-CH3 OH DA ST AT

C. Hydrazines:

R-NH-NH2 MFO R-NH-N(H)-OH → R-N=NH → R-N=N-OH; −N2 → R+, R

D. Carbamic Acid Esters:

C2H5-O-C(=O)-NH2 MFO C2H5-O-C(=O)-NHOH O2 Radicals

E. Nitrosamines:

R(H3C)N-N=O MFO R(H2C(OH))N-N=O −CH3CHO R-N=N-OH −OH R-N+≡N −N2 R+

F. Unsaturated Halogenated Hydrocarbons:

H2C=C(H)Cl MFO [H2C(O)C(H)Cl] → ClH2C-CHO → ClH2COOH

G. Polyaromatic Halogens:

Cl MFO OH Cl OH −Cl OH

Fig. 2. Metabolic pathways involved in the activation of different classes of chemical carcinogens. Abbreviations used are: MFO, mixed function oxidase system; NE, non-enzymic; EH, epoxide hydrolase; GET, glutathione S-epoxide transferase; NAT, N-acetyltransferase; DA, deacetylase; ST, sulfotransferase; At, acyltransferase. (From R. E. Kouri, L. M. Schechtman, and D. W. Nebert, Metabolism of Chemical Carcinogens, in: Genetic Differences in Chemical Carcinogenesis (R. E. Kouri, ed.), pp. 21-66, CRC Press, Boca Raton, Florida (1980), with permission.)

fore a balance between activation and inactivation, with competing pathways determining whether or not the steady state concentration of any reactive metabolite achieved is likely to result in toxicity. This can be illustrated as follows:

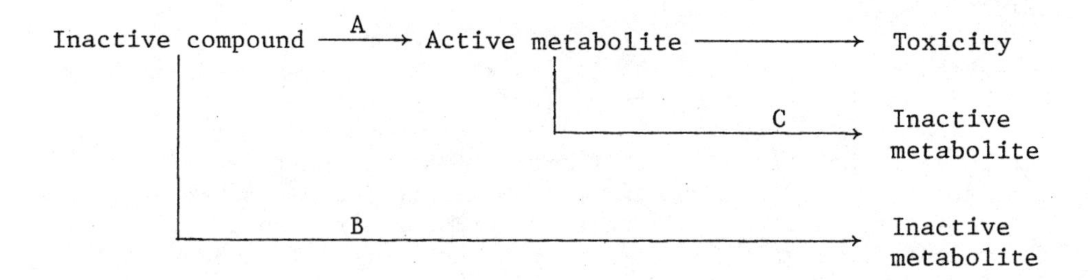

Any increase in the activity of pathway B or C will reduce the toxicity of the compound and conversely any increase in pathway A or any decrease in pathway B or C will lead to an increase in toxicity.

2.2. The Mixed Function Oxidase System

The enzyme system most widely involved in the activation of foreign compounds to chemically reactive intermediates [29] is the cytochrome P-450* dependent mixed function oxidase [30] (also known as monooxygenase [31] and polysubstrate monooxygenase [32]) system. This enzyme system is found predominantly in the liver, kidney, lung, and intestine [33, 34]. It is membrane bound, being located largely in the endoplasmic reticulum [35] although in some tissues there is relatively high activity in the mitochondria [36] and also in the nucleus [37]. The system functions as an electron transport chain [38], with electrons transferred from NADPH, by the flavoprotein

*Cytochrome P-450 is not a single enzyme but a family of isozymes. The term cytochrome P-450 is used in a genetic sense to denote any or all forms, just as the term glucose 6-phosphate dehydrogenase is a blanket name for all the variants of this enzyme. Unfortunately, there is no accepted common terminology for the individual forms of cytochrome P-450 and it is unlikely that there will be one for the next several years.

NADPH-cytochrome P-450 reductase, to cytochrome P-450 [39]. This latter component is a hemoprotein [40] that provides a binding site for the substrate [41] and also catalyzes the activation, and subsequent transfer, of oxygen to the substrate [42]. The monooxygenase system catalyzes a variety of oxidations including side-chain oxidation, aromatic hydroxylation, N- and O-dealkylation, and sulfoxidation [43].

Studies with rat and rabbit liver over the last 10 years have established that there are multiple forms of cytochrome P-450 [44-60]. The forms differ not only in their substrate specificity [61], but also in the metabolite they produce when presented with the same substrate [62]. Although there is often considerable overlap in their specificites [63], this is not always the case [58]. In the rabbit, at least six forms of the hemoprotein have been isolated, purified, and characterized [64, 65]. In the rat, at least eight forms of the enzyme have been purified and reconstituted in active systems [58, 59, 62, 66, 67]. Inter-organ differences in the complement of cytochromes P-450 may explain organ specific toxicity and carcinogenicity of certain chemicals. For example, high concentrations of a specific form of cytochrome P-450 that can demethylate hexamethylphosphoramide, HMPA, have been found in the nasal epithelium of the dog and this may well be the cause of HMPA induced nasal tumors in this species [68].

Levels of the different forms of cytochrome P-450 can be selectively altered by a variety of factors. These include inducers [46], age [69], sex [70], and physiologic [71], and pathologic [72] factors. The specificity of the different forms of cytochrome P-450 gives rise to the possibility of selective competition between substrates for a particular form of the enzyme [73]. For example, in the rat alphanaphthoflavone inhibits the activity of the hydrocarbon induced form of cytochrome P-450 but has little effect of the activity of the phenobarbitone-induced form [74].

Although oxidation by the mixed function oxidase system is often an obligatory first step in the activation of a procarcinogen this may well be followed by one or more metabolic steps leading to further activation [8]. For example, 2-acetylaminofluorene requires oxidation on the nitrogen atom before it is carcinogenic [9]. However, the product of this reaction, N-hydroxy-2-acetylaminofluorene is only a weak mutagen [75], but is further activated by transacetylation or deacetylation [76]. The products of these reactions are potent mutagens and also carcinogens [75, 77, 78]. Thus, the balance between activation and inactivation, maintained by the activities of the different enzymes involved, is further complicated by the interplay that may exist between different metabolic steps in the activation process.

2.3. Extrapolation from Animal Studies to Man

Although studies in animals can provide much information on mechanisms of chemical carcinogenesis [5, 9, 10] ultimately the potential effects of such compounds in man has to be assessed. Activation of procarcinogens by human tissue samples leads to the formation of the same adducts with DNA as are found when activation is catalyzed by animal tissues [79]. Any differences are quantitative rather than qualitative [79]. Examples of compounds for which this has proven to be true are aflatoxin B_1 [80] and benzo(a)pyrene [81]. It thus seems that the initiation step is similar in different species including man. As the consequence of such initiation is tumor formation in both animals and man, one must assume that the processes leading to tumorigenesis are fundamentally similar in animals and man. One might predict that any variation in the activity of those enzymes involved in the metabolism of the procarcinogen will lead to variation in eventual tumorigenesis. This can certainly be demonstrated in laboratory animals [82] and seems a suitable premise from which to start in seeking explanations for some of the differences in susceptibility observed in man.

How then does one go about investigating this problem? One approach, often suggested, is to study the metabolism of each new chemical in animals and in man and then to select the species most like man in its metabolism for long term toxicity and carcinogenicity studies [83]. The difficulty with this approach is that the number of chemicals is almost infinite, four million known compounds at present and increasing continuously [84], and the metabolic pathways are complex [85] so that it is unlikely that any species will resemble man in the formation of all of the metabolites of a complicated organic compound [86]. An alternative approach, possibly more rational, is to study the enzymes of drug metabolism and their control in animals and in man [87]. This has the advantage that although the number of such enzymes is quite large it is finite and their control is unlikely to change in the foreseeable future. To this end the establishment of human tissue banks is of great importance.

3. IN VITRO STUDIES OF THE METABOLISM OF FOREIGN COMPOUNDS IN MAN

3.1. Establishment of a Human Tissue Bank

In order to carry out studies of drug metabolizing enzymes in human tissue a human liver bank has been established at the Royal Postgraduate Medical School. This has now been in existence for 4 years. The source of liver is three-fold.

TABLE 1. Microsomal Mixed Function Oxidase System of Rat and Human Liver

Sample	n	Protein yield[a] (mg/g)	Cytochrome P-450 content (nmol/mg)	NADPH-cytochrome c reductase activity (nmol cyt c reduced/ mg/min)
All human liver samples	57	15.3 ± 0.89[b]	0.41 ± 0.02	159 ± 7
Biopsy samples with normal architecture	30	14.2 ± 1.0	0.49 ± 0.03	168 ± 4
Biopsy samples with abnormal architecture	20	13.3 ± 0.6	0.25 ± 0.03	132 ± 8
Renal donor liver samples	7	18.4 ± 1.2	0.25 ± 0.04	179 ± 23
Rat liver	6	17.5 ± 1.1	0.63 ± 0.04	110 ± 3

[a]Details of the methods used in obtaining the above data are provided in ref. 88.

[b]Values are mean ± SEM.

(The data shown are adapted from ref. 88.)

1. Needle biopsy samples weighing 10-100 mg.
2. Surgical biopsy samples weighing 100-1000 mg.
3. Organ donor samples weighing 100-1000 g.

Material from sources 1 and 2 is excess to histological requirements and is made available with local Research Ethics Committee permission, for research purposes. The third source represents subjects who have met traumatic deaths and who are maintained on life support systems until removal of their kidneys for transplantation.

The material is stored at -80°C either as suspensions of subcellular organelles in 0.25 M phosphate buffer, pH 7.25 containing 30% glycerol or as cubes (10 cm^3) of liver. We [88] and others [89] have established that storage under these conditions maintains the levels of cytochrome P-450 and mixed function oxidase activity at their initial values for periods of up to 4 years.

Human liver samples which appear normal histologically have similar protein yield and cytochrome P-450 content to rat liver [88] (Table 1). NADPH-cytochrome c reductase activity of human liver is, if anything, greater than that of rat liver. Histological evidence of liver disease is associated with decreased cytochrome P-450 content [88]. This is also apparent in liver samples from organ donors [90]. Such samples show evidence of histologic abnormalities. Why this occurs is not known but may be related to the short period of hypoxia that occurs between clamping the aorta and removing the liver sample [91]. The disturbance that has occurred in the pituitary-hypothalamic axis may also contribute to the decrease [92].

3.2. Comparison of the Activities of Human Drug Metabolizing Enzymes with Those of the Rat

A comparison of monooxygenase activities in rat and man (Table 2) reveals that, in general, laboratory animals oxidize foreign compounds more rapidly than man, confirming findings in vivo [93]. However, there are some important exceptions to this. For example, the N-hydroxylation of 2-acetylaminofluorene is catalyzed more effectively by human liver than by rat liver microsomal preparations [94] as is the 3-hydroxylation of biphenyl [95].

There is also wide inter-species variation in the activities of other drug metabolizing enzymes. Human liver is three times more active at catalyzing the hydration of styrene oxide than rat liver, but is less than one third as effective at catalyzing its conjugation with glutathione [96] (Table 2).

3.3. The Activation of 2-Acetylaminofluorene to a Mutagen by Human Liver

3.3.1. Metabolism and Mutagenicity Studies

Is it possible to use such comparisons of in vitro metabolic data to help in extrapolating the results of toxicity tests in experimental animals to man? 2-Acetylaminofluorene (2-AAF) is a synthetic N-acetylarylamine [97] that is hepatocarcinogenic in many laboratory species [10]. Studies in animals have established that 2-AAF requires N-hydroxylation [9], catalyzed by the cytochrome P-450 dependent mixed function oxidase system [98], followed by either deacetylation or transacetylation, before it is carcinogenic [76]. Such studies have shown that N-hydroxylation appears to be the rate

TABLE 2. Comparison of Activities of Drug Metabolizing Enzymes in Rat and Man

Activity	Human[a]	Rat	Human as % of rat
AHH[b]	31.9 ± 2.6[c]	260 ± 20	12.3
EMD	3100 ± 250	8500 ± 450	36.5
ECOD (total)	770 ± 60	3800 ± 840	20.3
PhenOD (total)	1510 ± 300	1740 ± 100	86.8
AP3MH	490 ± 100	3540 ± 290	13.8
AP4H	460 ± 50	1630 ± 120	28.2
APND	310 ± 130	840 ± 80	36.9
2-AAFNH	145 ± 22	38 ± 5	381.6
BP2H	7 ± 1	250 ± 170	2.8
BP3H	138 ± 22	82 ± 50	168.3
BP4H	970 ± 250	780 ± 260	124.4
D4H	69.9 ± 14.3	515 ± 38	13.6
EH[d]	16.4 ± 1.4	5.2 ± 0.6	315.4
GSH T[d]	25.7 ± 2.0	87.3 ± 9.5	29.4
1 Naphthol GT[d]	4.1 ± 1.1	12.0 ± 1.3	34.2

[a]All enzyme activities are expressed in units of pmol product formed/mg protein/min except those indicated in footnote d.

[b]Abbreviations are as follows: AHH, aryl hydrocarbon (benzo[a]pyrene) hydroxylase; EMD, ethylmorphine N-demethylase; ECOD, 7-ethoxycoumarin O-deethylase; PhenOD, phenacetin O-deethylase; AP3MH, AP4H, and APND, antipyrine 3-methylhydroxylase, 4-hydroxylase, and N-demethylase respectively; 2-AAFNH, 2-acetylaminoflorene N-hydroxylase; BP2H, BP3H, and BP4H, biphenyl 2-hydroxylase, 3-hydroxylase, and 4-hydroxylase respectively; D4H, debrisoquine 4-hydroxylase; EH, epoxide (styrene oxide) hydrolase; GSH T, glutathione S-epoxide (styrene oxide) transerase; 1 Naphthol GT, 1 naphthol UDP-glucuronyl transferase.

[c]Values are mean ± SEM.

[d]Activities expressed in units of nmol product formed/mg protein/min.

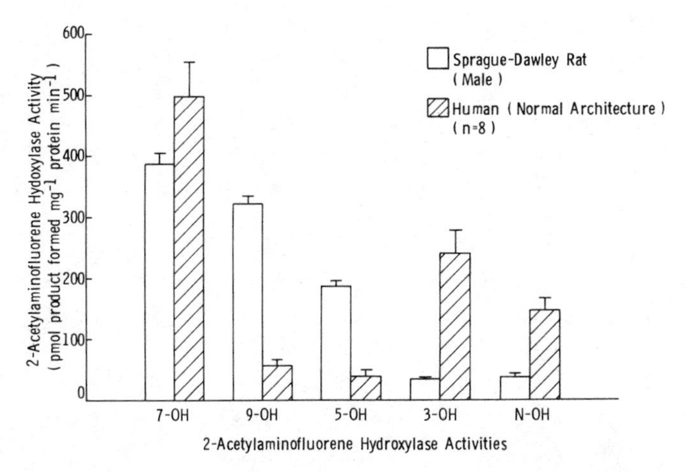

Fig. 3. Hydroxylation of 2-acetylaminofluorene by microsomal fractions from rat and human liver. Values are mean ±SEM.

limiting step in the activation process [99, 100]. For example, the guinea pig is very poor at N-hydroxylating 2-AAF [99, 101] and this species is very resistant to the tumorigenic effects of the compound [99]. In marked contrast, however, the rhesus monkey can N-hydroxylate 2-AAF at similar rates to the rat, yet it is also completely resistant to 2-AAF carcinogenesis [102]. Thus, factors additional to the activation of 2-AAF must also be involved in determining species differences in susceptibility to 2-AAF carcinogenesis. It is not yet known whether 2-AAF is carcinogenic in man [10].

Can any useful information be obtained from in vitro studies? Comparison of 2-AAF oxidation in vitro by microsomal fractions of rat liver with that of human liver (Fig. 3) reveals that man is more active at the N- and 3-hydroxylation of 2-AAF than the rat, but is less active at 9- and 5-hydroxylation [94]. In fact, N-hydroxylase activity of human liver is 4 times that of rat liver. One might therefore predict that man is more susceptible to 2-AAF carcinogenesis than rat. However, it is important to consider the competing, detoxifying, pathways of oxidation. Ring hydroxylation of 2-AAF, catalyzed by different forms of cytochrome P-450 to those catalyzing the N-hydroxylation [103], leads to the formation of non-toxic products [99, 104] and thus diverts substrate from the toxic N-hydroxylation pathway. Johnson et al. [103] have developed a vector model to illustrate this, the direction of the vector representing the balance between activation and detoxication. Such vector analysis reveals that this balance is more towards the detoxication pathways in the rhesus monkey than in the rat. This might explain, in part, the resistance of the former species to 2-AAF tumorigenesis and as would

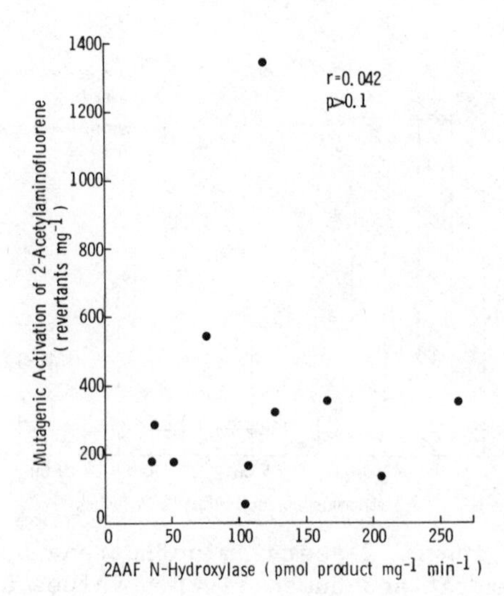

Fig. 4. Comparison of in vitro activation of 2-acetylaminofluorene (2-AAF) to a mutagen with its N-hydroxylation by microsomal fractions of human liver.

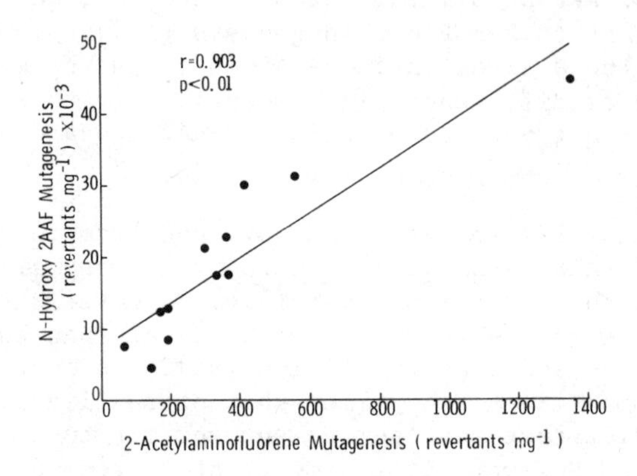

Fig. 5. Correlation between the in vitro activation of 2-acetylaminofluorene (2-AAF) and of N-hydroxy-2-AAF to mutagens by microsomal fractions of human liver. (From A. R. Boobis, M. J. Brodie, M. E. McManus, N. Staiano, S. S. Thorgeirsson, and D. S. Davies, in: Biological Reactive Intermediates. II. Chemical Mechanisms and Biological Effects (R. Snyder, D. V. Parke, J. J. Kocsis, D. J. Jollow, C. G. Gibson, and C. W. Witmer, eds.), Part B, pp. 1193-1201, Plenum Press, New York (1982), with permission.)

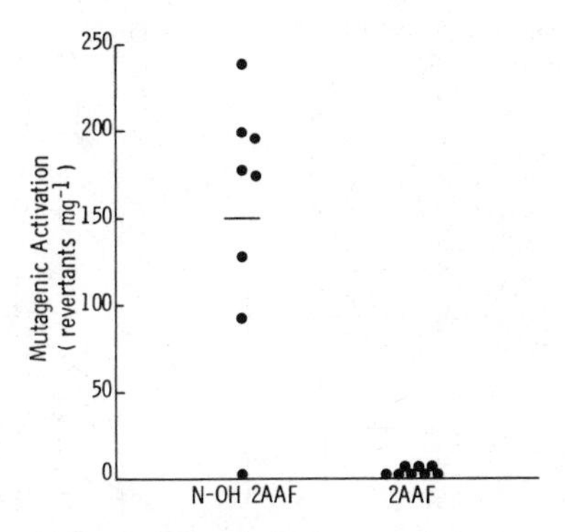

Fig. 6. Activation of 2-acetylaminofluorene (2-AAF) and N-hydroxy-2-AAF to mutagens by S-9 fractions of human lung.

be predicted, subcellular fractions from the livers of rhesus monkeys are much less efficient at catalyzing the activation of 2-AAF to a mutagen than subcellular fractions from rat liver [102]. Similar analysis in man reveals that the balance is less towards detoxication than in the rat, thus favoring activation.

Microsomal fractions from human liver can activate 2-AAF to a mutagen [94], confirming the work of others [105]. In contrast to rodent liver [100], there is no correlation between the ability of human liver to N-hydroxylate 2-AAF and to activate it to a mutagen (Fig. 4). There is no doubt that N-hydroxylation is the initial step in the activation of 2-AAF [10] and in laboratory animals it is rate limiting [10, 100]. Why not in man? As discussed above, N-hydroxy-2-AAF is further activated by deacetylase or transacetylase activity [76]. Microsomal fractions of human liver are very effective at catalyzing the further activation of N-hydroxy-2-AAF to a mutagen [94], due presumably to the activity of a deactylase enzyme. If the ability of liver samples to activate 2-AAF to a mutagen is compared with their ability to activate N-hydroxy-2-AAF an excellent correlation is found (Fig. 5). It is tempting to conclude that deacetylation is the rate limiting step in the activation of 2-AAF to a mutagen by microsomal fractions of human liver.

Without N-hydroxylation, however, 2-AAF is not activated to a mutagen. This is illustrated by results obtained with human lung samples. Human lung has very low monooxygenase activity [106] and does not activate 2-AAF to a mutagen (Fig. 6). However, lung samples do activate N-hydroxy-2-AAF to a mutagen, demonstrating the presence of those enzymes necessary for the subsequent activation of the N-hydroxy product [94]. The inability of lung to activate 2-AAF itself must therefore be due to a very low rate of conversion to the N-hydroxy derivative.

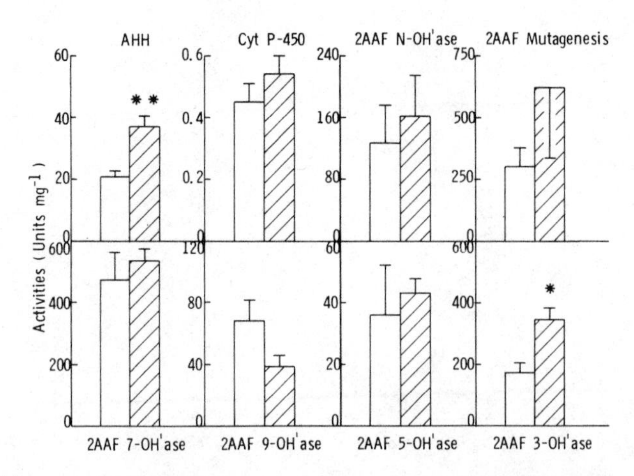

Fig. 7. Effect of cigarette smoking on aryl hydrocarbon (benzo(a)-pyrene) hydroxylase (AHH) and 2-acetylaminofluorene (2-AAF) hydroxylase activities of microsomal fractions of human liver. Values are mean ±SEM from nonsmokers (□) and smokers (▨). *p < 0.05, **p < 0.01 - smokers compared to nonsmokers by Student's t-test. Further details can be found in Ref. 94.

3.3.2. Multiplicity of Human Cytochromes P-450 Oxidizing 2-AAF

In rodents the N-hydroxylation of 2-AAF [107-109] and the hydroxylation of benzo(a)pyrene (AHH) [109-111] are inducible by polycyclic aromatic hydrocarbons. In the mouse, this inducibility is controlled by a pair of alleles at the Ah locus [110, 112]. Induction of these pathways results in increased mutagenesis of the parent compounds [100, 113]. In the rabbit the pattern of induction is somewhat different [114]. 2-AAF N-hydroxylase is inducible only in the liver of adult animals, whereas AHH is inducible in the liver of neonatal rabbits and in kidney at all ages. In man, a common source of exposure to polycyclic aromatic hydrocarbons is from cigarette smoking [115]. This causes induction of hepatic AAH activity of microsomal fractions of human liver [88, 116] but not of 2-AFF N-hydroxylase activity [94] (Fig. 7). Only the 3-hydroxylation of 2-AAF is induced in liver samples from cigarette smokers. As might be predicted from the lack of effect on N-hydroxylation, cigarette smoking does not significantly alter the mutagenic activation of 2-AAF in man [94]. Thus, the inducibility of 2-AAF N-hydroxylase and AHH activities in man by polycyclic aromatic hydrocarbons resembles the situation in the rabbit more closely than that in rat or mouse.

TABLE 3. Correlation Matrix[a] for 2-Acetylamino-fluorene Hydroxylase Activities of Human Liver

	Cyt P-450	AHH	7-OH	9-OH	5-OH	3-OH
AHH[a]	0.790[c]					
7-OH	0.863[d]	0.668[e]				
9-OH	0.495	0.063	0.721[e]			
5-OH	0.743[c]	0.439	0.814[c]	0.649[e]		
3-OH	0.703	0.778[c]	0.575	-0.015	0.534	
N-OH	0.810[e]	0.554	0.739[c]	0.478	0.888[d]	0.698[e]

[a]Abbreviations used are: AHH, aryl hydrocarbon (benzo[a]pyrene) hydroxylase; 7-OH, 9-OH, 5-OH, 3-OH, and N-OH, 2-acetylamino-fluorene 7-hydroxylase, 9-hydroxylase, 5-hydroxylase, 3-hydroxylase, and N-hydroxylase respectively.

[b]Values shown are correlation coefficients (r) obtained with 11 biopsy samples.

[c]$p < 0.01$

[d]$p < 0.001$

[e]$p < 0.05$

(From A. R. Boobis, M. J. Brodie, M. E. McManus, N. Staiano, S. S. Thorgeirsson, and D. S. Davies in: Biological Reactive Intermediates - II. Chemical Mechanisms and Biological Effects (R. Snyder, D. V. Parke, J. J. Kocsis, D. J. Jollow, C. G. Gibson, and C. W. Witmer, eds.), Part B, pp. 1193-1201, Plenum Press, New York (1982), with permission).

Although the N-hydroxylation of 2-AAF is not inducible by cigarette smoking in man, whereas AHH activity is, this does not necessarily preclude a single form of cytochrome P-450 from catalyzing both reactions in the constitutive state. For example, a constitutive form of cytochrome P-450 in the rabbit (form 3) is very effective at catalyzing the oxidation of both aminopyrine and biphenyl whereas the hydrocarbon inducible form (form 4) catalyzes the hydroxylation only of biphenyl [56]. Correlation analysis of the pathways of 2-AAF metabolism with AHH was performed [94] (Table 3). This suggests the involvement of at least three different forms of cytochrome P-450 in the metabolism of 2-AAF by human liver. One form would be involved primarily in the 5-, 7-, and N-hydroxylation of 2-AAF, whereas a second form catalyzes primarily the 9-hydroxylation. The third form, inducible by cigarette smoking, would participate in the 3-hydroxylation of 2-AAF and in the oxidation of benzo(a)pyrene.

3.4. Characterization of Probe Substrates for Different Forms of Cytochrome P-450 that Could be Used in vivo in Man

3.4.1. The Approach to the Problem

Studies on the metabolism of 2-AAF *in vitro* by microsomal fractions of human liver can therefore provide useful information on the number of different forms of cytochrome P-450 involved and their response to hydrocarbon induction. However, in the search for information to help in the extrapolation of animal data to man, it would be of considerable value to be able to administer non-toxic prototype substrates to volunteers *in vivo*. Thus, interindividual variation in the forms of cytochrome P-450 occurring in the population could be studied together with those factors that might control their activity. Several groups have investigated compounds that might possibly be suitable for this purpose [117]. Substrates studied include antipyrine [118, 119], theophylline [120], and amobarbital [121]. Over the past few years we have attempted to characterize the *in vitro* oxidation of a number of compounds by microsomal fractions of human liver with the aim of eventually using them in volunteer studies. (Several such *in vivo* studies have now been performed but space does not permit their consideration in this manuscript.)

3.4.2. Antipyrine Oxidation in vitro

For almost 15 years [122] antipyrine has been extensively used for studies on drug oxidation in man *in vivo* [119]. However, up to 70% [123, 124] of an oral dose of antipyrine is transformed to the oxidative metabolites 4-hydroxyantipyrine [125] (4-OHAP), 3-hydroxymethylantipyrine [126] (3-OHMeAP), and norphenazone [127, 128] (NP). Evidence is now accumulating that the formation of each of these metabolites is catalyzed primarily by a different form of cytochrome P-450 [123, 129-133]. Thus, studies in which only the overall rate of antipyrine metabolism is determined may well overlook important alterations in the activity of a single form of cytochrome P-450 [134]. An assay for the *in vitro* oxidation of antipyrine to its three metabolites was therefore developed to enable the characterization of these enzymic activities in rat and man. The assay utilizes radiometric-HPLC and has been described in detail elsewhere [135]. Kinetic constants are shown in Table 4 for the three pathways of antipyrine oxidation in rat and man [133, 136]. Rat liver activities are 4-6 fold higher than the corresponding activities with human liver. The affinity constants (K_m) are similar between species and also very similar amongst the three pathways.

Most compounds to which man is exposed achieve plasma concentrations well below the K_m of the enzymes involved in their metabolism. In these circumstances activity varies with K_m as well as

TABLE 4. Michaelis-Menten Parameters for the *in vitro* Oxidation of Antipyrine by Rat and Human Liver

Species	Parameter	Metabolite		
		3-OHMeAP[a]	4-OHAP	NP
Human	V_{max}[b]	0.60 ± 0.23[c]	0.57 ± 0.20	0.34 ± 0.18
	K_m[d]	9.0 ± 1.3	7.3 ± 1.1	5.9 ± 0.6
	V_{max}/K_m[e]	0.07	0.08	0.06
Rat	V_{max}	3.80 ± 0.20	1.71 ± 0.18	1.14 ± 0.13
	K_m	2.20 ± 0.51	4.6 ± 1.07	4.6 ± 0.76
	V_{max}/K_m	1.71	0.37	0.26

[a]Abbreviations used are: 3-OHMeAP, 3-hydroxymethylantipyrine; 4-OHAP, 4-hydroxyantipyrine; NP, norphenazone.

[b]Expressed in units of nmol metabolite formed/mg microsomal protein/min.

[c]Values are mean ± SEM of at least three different liver samples.

[d]Expressed in units of mM.

[e]Expressed in units of ml/mg microsomal protein/min × 10^{-3}.

(Data taken from refs. 133 and 136. Details of the analytical methods can be found in ref. 135.)

V_{max} and an estimate of activity can be obtained from the ratio of V_{max} to K_m [137]. In man the values for this parameter are similar for the three pathways whereas in the rat the value for 3-OHMeAP is 5 times that of the next highest value, for 4-OHAP (Table 4). One might therefore predict that *in vivo* values for clearance to the three metabolites of antipyrine in man will be similar whereas in rat clearance to 3-OHMeAP will be much higher than that to the other two metabolites. These predictions have been confirmed, both in man [123, 132] and in rat [130].

However, it is still necessary to prove that metabolite excretion rates *in vivo* are a true reflection of the activity of those enzymes producing them. We have therefore compared the *in vivo* and *in vitro* rates of formation of the three metabolites of antipyrine in a group of patients requiring liver biopsy for diagnostic purposes [136]. There are no significant correlations between the rates of formation of any pair of metabolites, either *in vivo* (Fig.

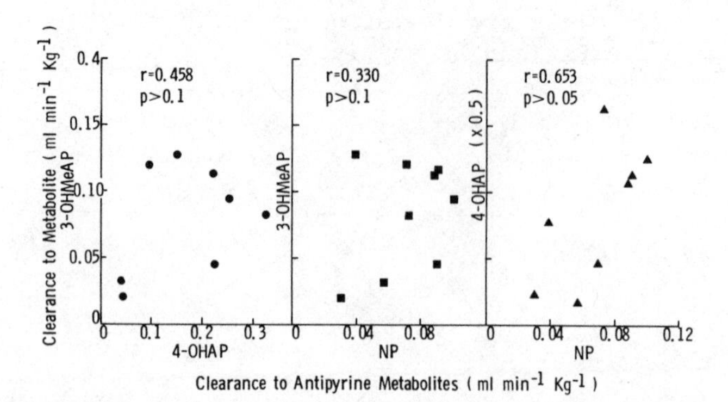

Fig. 8. Correlation between in vivo rates of formation of antipyrine metabolites in 9 patients with suspected liver disease. Clearance to a metabolite was calculated as described in Ref. 136. (From A.R. Boobis, M. J. Brodie, G. C. Kahn, E.-L. Toverud, I. A. Blair, S. Murray, and D. S. Davies, Br. J. Clin. Pharmacol., 12-771-777 (1981), with permission.)

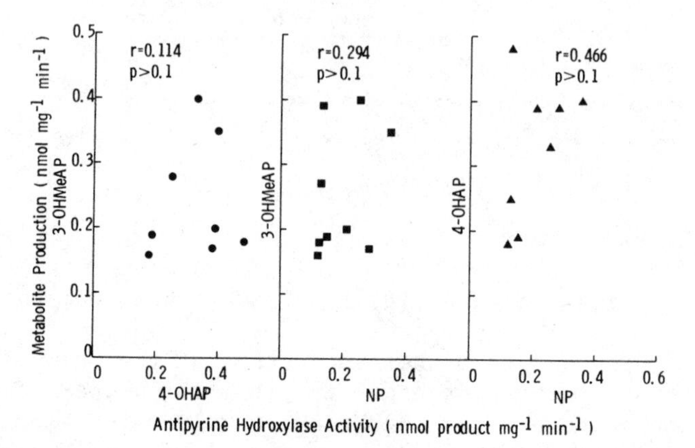

Fig. 9. Correlations between in vitro rates of formation of antipyrine metabolites in 9 patients with suspected liver disease. Enzyme activities were determined in microsomal fractions of hepatic needle biopsy samples with saturating concentrations of antipyrine. (From A. R. Boobis, M. J. Brodie, G. C. Kahn, E.-L. Toverud, I. A. Blair, S. Murray, and D. S. Davies, Br. J. Clin. Pharmacol., 12:771-777 (1981), with permission.)

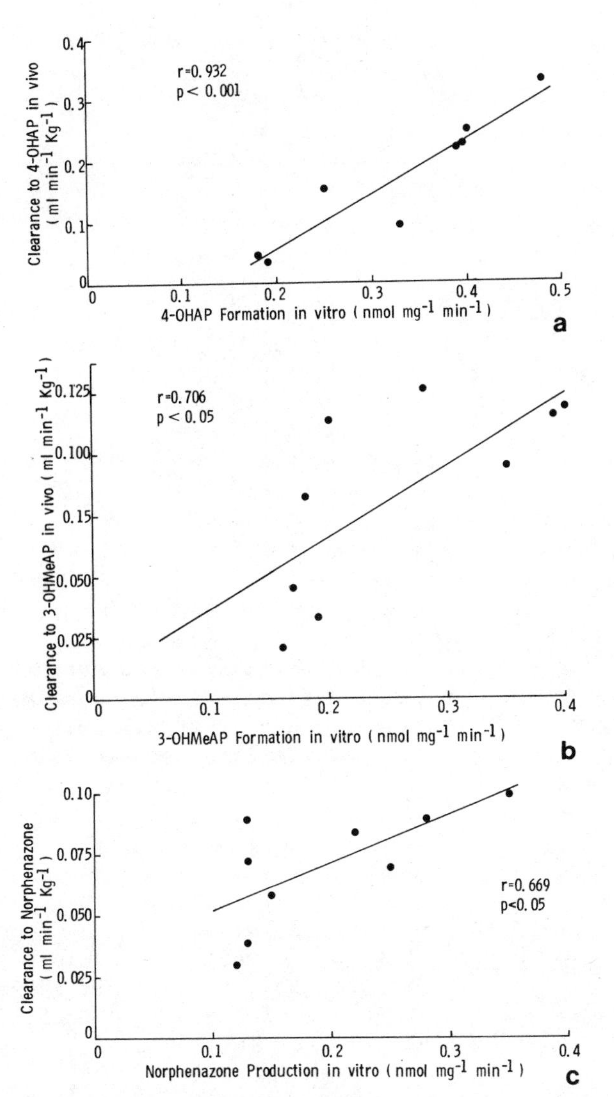

Fig. 10. Correlations between in vivo and in vitro rates for formation of a) 4-hydroxyantipyrine, b) 3-hydroxymethylantipyrine, and c) norphenazone in 9 patients with suspected liver disease. (From A. R. Boobis, M. J. Brodie, G. C. Kahn, E.-L. Toverud, I. A. Blair, S. Murray, and D. S. Davies, Br. J. Clin. Pharmacol., 12:771-777 (1981), with permission.)

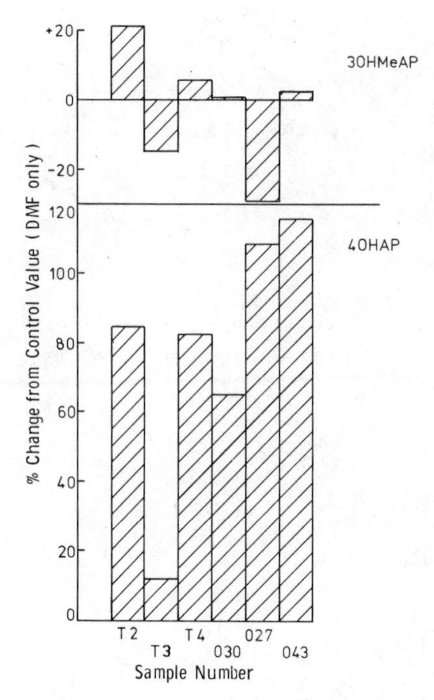

Fig. 11. Effects of betamethasone (10^{-4} M) on antipyrine 4-hydroxylase and 3-methylhydroxylase activities of microsomal fractions of human liver. Each vertical pair of bars shows the results from a single liver sample. DMF, dimethylformamide.

8) or in vitro (Fig. 9). However, for each of the three metabolites there is a significant correlation between its in vivo rate of formation with its in vitro rate of formation (Fig. 10). Thus, it does appear that, for antipyrine at least, the urinary excretion rates of its metabolites provide a good estimate of the activity of the enzymes responsible for their production [136].

The lack of correlation between the rates of formation of each pair of metabolites suggests the involvement of more than one form of cytochrome P-450 in their production [136]. This is supported by observations on the effects of betamethasone on in vitro enzyme activity. This steroid enhances the in vitro production of 2-hydroxybiphenyl from biphenyl by microsomal fractions of both rat [139] and human [95] liver. It causes a differential enhancement of the 4-hydroxylation of antipyrine, the 3-methylhydroxylation being unaffected [95] (Fig. 11). Presumably, therefore, at least 2 forms of cytochrome P-450 are involved in the formation of these two metabolites. The degree of enhancement observed is variable, rang-

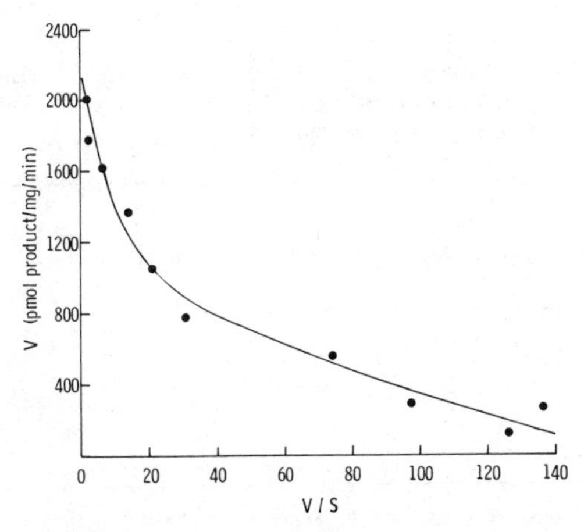

Fig. 12. Eadie-Hofstee plot for phenacetin O-deethylase activity of microsomal fraction of human liver. The points are the experimentally determined values. The solid line is the computer-generated curve of best fit, by iterative non-linear least squares regression analysis, determined as described in Ref. 144.

ing from 10 to 90%, perhaps indicating interindividual variation in the complement of cytochromes P-450 present in these liver samples.

Metabolite production from antipyrine _in vivo_ can be used as an index of the activity of the enzymes involved in individual reactions but there appears to be an overlap in specificity, so that no single form of cytochrome P-450 catalyzes the production of just one metabolite of antipyrine [131, 133]. Thus, although studies with antipyrine may provide useful information on how oxidation pathways vary in man and what factors influence them, it is unlikely to provide definitive information on the activity of a single form of cytochrome P-450. This must await further characterization of the different forms involved.

3.4.3. Multiple Forms of Phenacetin O-Deethylase in Human Liver

For several reasons it seems that the O-deethylation of phenacetin is more likely to provide information on a single form of cytochrome P-450 than antipyrine. Phenacetin O-deethylation is hydrocarbon inducible both in animals [139, 140] and in man [141]. Its metabolism may thus be catalyzed by cytochrome P-448. This is certainly true after induction [142] and hence this reaction could provide an estimate of the activity of a hydrocarbon inducible form of cytochrome P-450 in man. There is also evidence from _in vivo_ studies

TABLE 5. Values for the Four Michaelis-Menten Parameters[a] of Phenacetin O-Deethylase Activity of Rat and Human Liver

Species	$V_{max\ 1}$ (pmol/mg/min)	$K_{m\ 1}$ (μM)	$V_{max\ 2}$ (pmol/mg/min)	$K_{m\ 2}$ (μM)
Human	440 ± 162[b]	6.3 ± 1.5	1060 ± 250	248 ± 75
Rat	145 ± 55	4.9 ± 1.8	1600 ± 90	455 ± 28

[a]The data were analyzed for a 4-parameter, biphasic kinetics, model as described in ref. 144.

[b]Values are mean ± SEM of at least 3 different liver samples.

(From A. R. Boobis, G. C. Kahn, C. Whyte, M. J. Brodie, and D. S. Davies, Biochem. Pharmacol. 30, 2451-2456 (1981), with permission.)

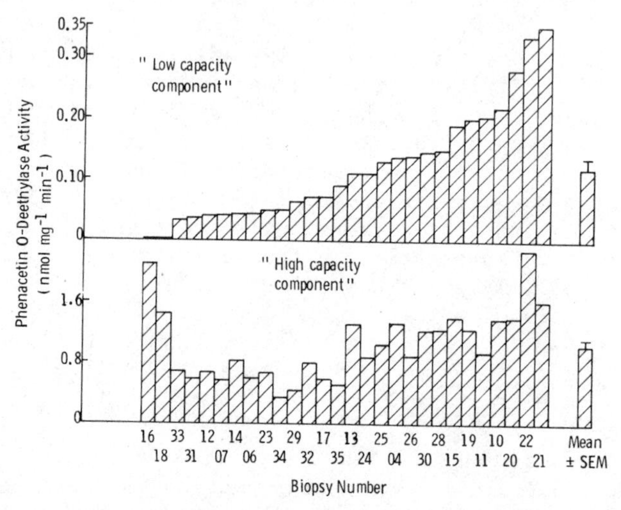

Fig. 13. High affinity, low capacity and low affinity, high capacity components of phenacetin O-deethylase activity of microsomal fractions of human liver. Activities were determined as described in Ref. 144. Each vertical pair of bars represents results from a single liver sample.

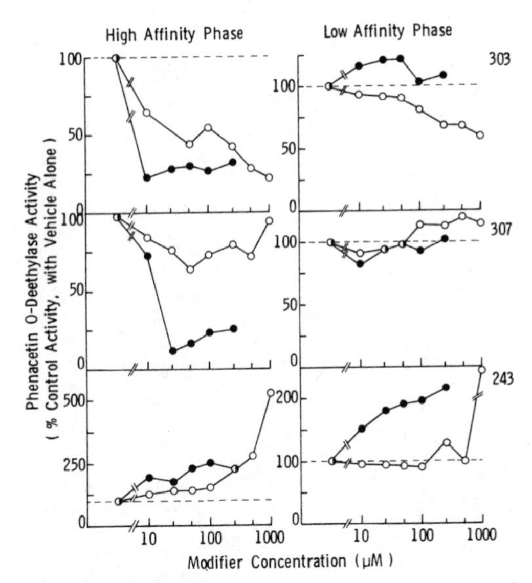

Fig. 14. Effects of alphanaphthoflavone (●) and metyrapone (○) on the two components of microsomal phenacetin O-deethylase activity of human liver. Each horizontal pair of panels shows results with a single liver sample. Further details are provided in Ref. 144.

that a component of phenacetin O-deethylase activity is under monogenic control in man [143], providing evidence for the involvement of a single form of cytochrome P-450 in this reaction.

The in vitro O-deethylation of phenacetin by microsomal fractions from rat and human liver has been studied using a GC-MS assay for paracetamol [144]. In both species the O-deethylation of phenacetin is biphasic, a typical Eadie-Hofstee plot being shown in Fig. 12. The data have been analyzed for a 4-parameter model and the estimates for V_{max} and K_m thus obtained are shown in Table 5. Activity can be divided into a high affinity, low capacity component and a low affinity, high capacity component [144]. A possible explanation for biphasic enzyme kinetics is the involvement of two forms or populations of the enzyme with very different affinities [145]. This appears to be the situation with phenacetin O-deethylation.

In 28 samples of human liver there is only a poor correlation, r = 0.557, between the two components of activity (Fig. 13). In 2 out of 3 liver samples studied, alphanaphthoflavone and metyrapone had selective effects on the two components of phenacetin O-deethylase activity [144] (Fig. 14). In the third liver sample, both compounds enhanced the two components of activity. Further studies on this sample revealed Michaelis-Menten constants of V_{max} ı 150 pmol/

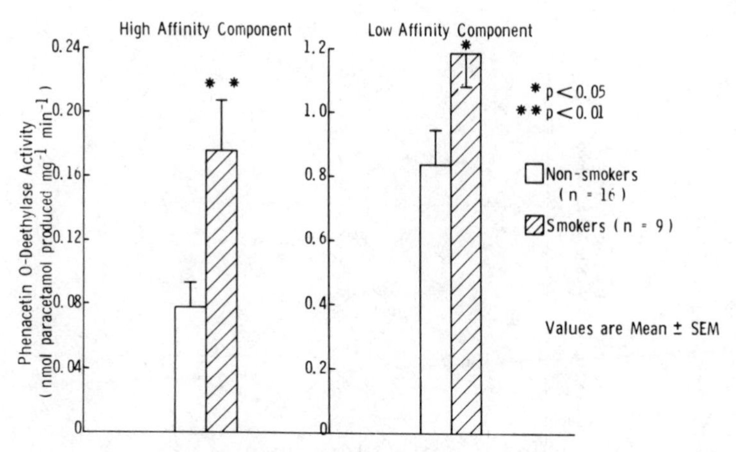

Fig. 15. Effect of cigarette smoking on the two components of microsomal phenacetin O-deethylase activity of human liver.

mg/min, $K_{m\ 1}$ 47 μM, $V_{max\ 2}$ 1700 pmol/mg/min, and $K_{m\ 2}$ 2.6 mM. Thus, despite relatively normal values for the V_{max} of the two components of activity the values for K_m are much higher (7-10 fold) than those normally encountered [144] (Table 5). This result further indicates the heterogeneity in the complement of cytochromes P-450 that occurs in man, a theme to which we will return later.

Cigarette smoking is known to increase the O-deethylation of phenacetin in vivo [141]. Comparison of the in vitro activities for the two components of O-deethylation in smokers and non-smokers shows that both components are significantly increased in liver samples from smokers (Fig. 15). However, the degree of induction is greater with the high affinity component of activity (2-fold) than for the low affinity component (1.4-fold).

All the evidence supports the contention [144] that the two components of phenacetin O-deethylase activity are catalyzed by different forms or populations of cytochrome P-450. The high affinity component of activity may well be catalyzed by a single, hydrocarbon inducible form of cytochrome P-450 so that determination of this activity in control samples may provide an estimate of the constitutive contribution of the hydrocarbon inducible form. However, definitive evidence for this will only become available with the development of specific antibodies to this form of cytochrome P-450.

3.4.4. Debrisoquine 4-Hydroxylase as a Probe for a Single Form of Cytochrome P-450 in Man

Although there is evidence for monogenic control of a proportion of phenacetin O-deethylase activity in vivo, even homozygous

TABLE 6. Cofactor Requirements of Debrisoquine 4-Hydroxylase Activity of Human Liver

Debrisoquine concentration (mM)	NADPH concentration (mM)	Debrisoquine 4-hydroxylase activity[a] (pmol/mg/min)
0	1.2	<2[b]
1.0	0	5.6
1.0	1.2	75.8
Atmosphere		
Air		75.8
80% CO:20% O_2		17.9

[a]Activity was determined as described in ref. 147 using a pool of microsomal fractions from 6 different liver samples.

[b]Values are mean of 2 determinations that varied by less than 10% from each other.

(From G. C. Kahn, A. R. Boobis, S. Murray, M. J. Brodie, and D. S. Davies, Br. J. Clin. Pharmacol. 13, 637-645 (1982), with permission.)

subjects with the impairment have considerable O-deethylase activtiy [143]. Thus, other forms of cytochrome P-450, under different genetic control, can still contribute to the formation of paracetomol. In the absence of any other data this makes the interpretation of results with phenacetin difficult. R. L. Smith's group [146] have shown a polymorphism in debrisoquine 4-hydroxylation *in vivo*, with a very marked difference between phenotypes in the ability to produce 4-hydroxydebrisoquine (4-OHD). It seemed possible that this pathway is catalyzed by a single form of cytochrome P-450, if indeed the defect is due to an enzymic abnormality. To investigate this we developed a GC-MS assay for the *in vitro* production of 4-OHD by microsomal fractions of human liver [147].

The 4-hydroxylation of debrisoquine by microsomal fractions of human liver has all of the characteristics of a cytochrome P-450 mediated reaction [147] (Table 6). At concentrations of debrisoquine up to 1 mM the 4-hydroxylation reaction is monophasic, with mean values for V_{max} of 70.5 ± 8.7 pmol/mg/min (n = 3) and for K_m of 130 ± 24 μM. A small number of patients undergoing diagnostic liver biopsy were phenotyped *in vivo* for their debrisoquine oxidation status [146]. One of the subjects was a poor metabolizer (PM pheno-

TABLE 7. Mixed Function Oxidase Activity of Microsomal Fractions from Liver Samples of Extensive (EM) and Poor (PM) Metabolizers of Debrisquine

Patient no.	Phenotype[a]	Debrisoquine 4-hydroxylase activity[b]	Antipyrine metabolite formation			Phenacetin O-deethylase	
			3-OHMeAP	4-OHAP	NP	high affinity component	low affinity component
Non-phenotyped			370[c] (170-560)[d]	390 (130-620)	170 (110-350)	130 (30-350)	1060 (450-2290)
02019	EM	49.6	ND[e]	ND	ND	200	1250
02035	EM	25.7	200	500	360	90	510
02037	EM	39.3	ND	ND	ND	ND	ND
02047	EM	29.9	ND	ND	ND	ND	ND
02031	PM	0	270	330	200	30	580

[a]Debrisoquine 4-hydroxylation phenotype was assessed from the ratio of debrisoquine/4-hydroxydebrisoquine in the 4 h urine sample following a single oral dose of 10 mg debrisoquine. Subjects were classified as EM if the ratio was <12.(148)

[b]Enzyme activities are expressed in units of pmol product formed/mg microsomal protein/min.

[c]Abbreviations used are as follows: 3-OHMeAP, 3-hydroxymethylantipyrine; 4-OHAP, 4-hydroxyantipyrine; NP, norphenazone.

[d]Values are means from 9 or more different samples from patients that were not phenotyped for their debrisoquine 4-hydroxylation status.

[e]Figures in parentheses are the range of activities encountered in the non-phenotyped samples.

[f]ND indicates that the activity was not determined in this sample.

(Data are taken from ref. 148.)

TABLE 8. Inhibitors of Debrisoquine 4-Hydroxylase Activity of Human Liver

Compound	Type of inhibition[a]	K_i (μM)
Phenformin	Competitive	205
Guanoxan	Competitive	30
Sparteine	Competitive	85
Antipyrine	Non-competitive	19,300
Acetanilide	Non-competitive	1230
Amobarbital	Enhancement[b]	-

[a]Type of inhibition was determined by analysis of Eadie-Hofstee plots of debrisoquine 4-hydroxylase activity determined in the presence of several different concentrations of the inhibitor.

[b]Amobarbital did not inhibit debrisoquine 4-hydroxylase activity but rather caused a slight stimulation on addition in vitro.

type). The in vitro production of 4-OHD was then determined in microsomal fractions isolated from the liver biopsy samples. The PM subject had no detectable 4-hydroxylase activity, whereas the other subjects all had readily detectable activity [148] (Table 7). Despite an inability to produce 4-OHD the PM liver biopsy was still capable of metabolizing antipyrine at normal rates. Previous studies in vivo have shown that PM subjects have normal rates of antipyrine metabolism [149]. Thus, it can be demonstrated that the polymorphism in debrisoquine 4-hydroxylation is due to the deficiency or absence of a specific form of cytochrome P-450 that does not appear to be involved in the formation of any of the metabolites of antipyrine [148].

As the 4-hydroxylation of debrisoquine appears to be catalyzed by a single form of cytochrome P-450 this represented a unique opportunity by which to investigate the specificity of a form of cytochrome P-450 in man. Since the first reports on the debrisoquine oxidation polymorphism [146] the metabolism of a number of drugs has been associated with the defect, from in vivo studies. Such drugs include sparteine [149, 150], phenformin [151], guanoxan [143], and phenacetin [143]. Certain drugs have been excluded from any association with the defect and these include antipyrine [149], amobarbital [149], and acetanilide [152]. Although such in vivo studies can demonstrate whether or not there is an association in the impairment of the metabolism of two drugs, they cannot determine whether or not the compounds share a form of cytochrome P-450. Any asso-

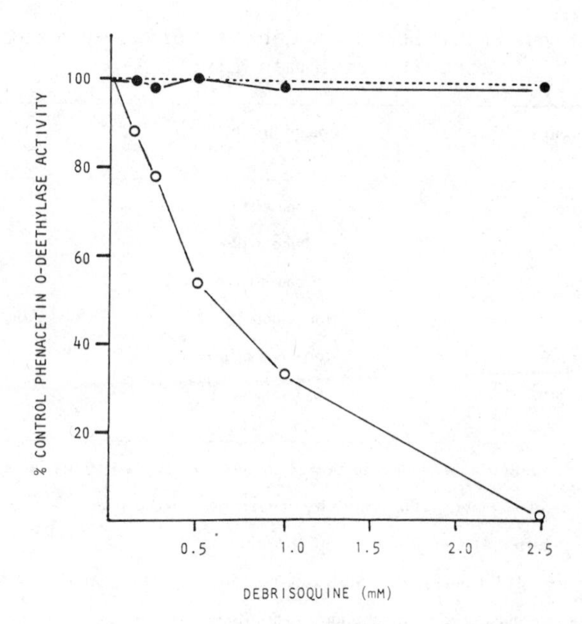

Fig. 16. Effect of in vitro addition of debrisoquine on the high affinity (○) and low affinity (●) components of microsomal phenacetin O-deethylase activity of human liver. Values were obtained using pooled microsomal fractions from 4 different liver samples.

ciation could be due to the involvement of separate, but closely linked loci. If the substrates share a form of cytochrome P-450 they should act as competitive inhibitors of each other's metabolism and the K_i for inhibition should be the same as the K_m for metabolism [73].

The effects of several drugs on debrisoquine 4-hydroxylase activity in vitro has been investigated. The results are shown in Table 8. Guanoxan, phenformin, and sparteine are all competitive inhibitors with low K_i values. Antipyrine and acetanilide are non-competitive inhibitors with high K_i values. Amobarbital enhances 4-hydroxylase activity at concentrations well above its K_m (0.5 mM) [153]. There is thus a perfect correlation between an in vivo association with the debrisoquine oxidation polymorphism and with the ability of a drug to competitively inhibit this activity in vitro. Kalow and his co-workers [154] have investigated sparteine oxidation by human liver. They found that debrisoquine is a competitive inhibitor of this reaction, whereas antipyrine has no effect. A comparison of the values obtained by this group for sparteine with our own values for debrisoquine reveals a very close similarity between K_i for inhibition and K_m for metabolism of each substrate.

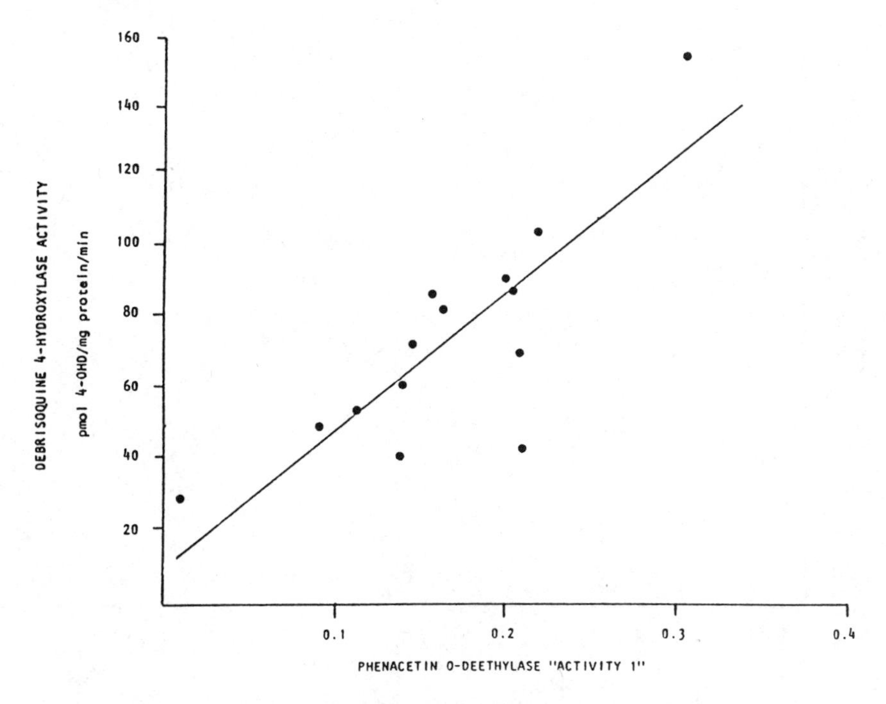

Fig. 17. Correlation between debrisoquine 4-hydroxylase and high affinity (activity 1) phenacetin O-deethylase activities of microsomal fractions of human liver. Results for 14 different liver samples are shown.

There is evidence from in vivo studies that phenacetin O-dedethylase is, in part, controlled by the same, or a closely linked, locus to that controlling debrisoquine 4-hydroxylation [143]. In the PM liver sample previously studied [148] it was found that high affinity phenacetin O-deethylase activity was virtually absent in this sample despite a normal activity for the low affinity component (Table 7). Debrisoquine is a selective, competitive inhibitor for the high affinity component of phenacetin O-deethylase activity [155] (Fig. 16). In addition, the high affinity component of phenacetin O-deethylase activity correlates well with debrisoquine 4-hydroxylase activity (Fig. 17) whereas there is no such correlation with the low affinity component of activity (Fig. 18). Thus, the high affinity component of phenacetin O-deethylase activity appears to be catalyzed by the same form of cytochrome P-450 as that catalyzing the 4-hydroxylation of debrisoquine, whereas a different form (or forms) of cytochrome P-450 is involved in the low affinity component of phenacetin O-deethylase activity.

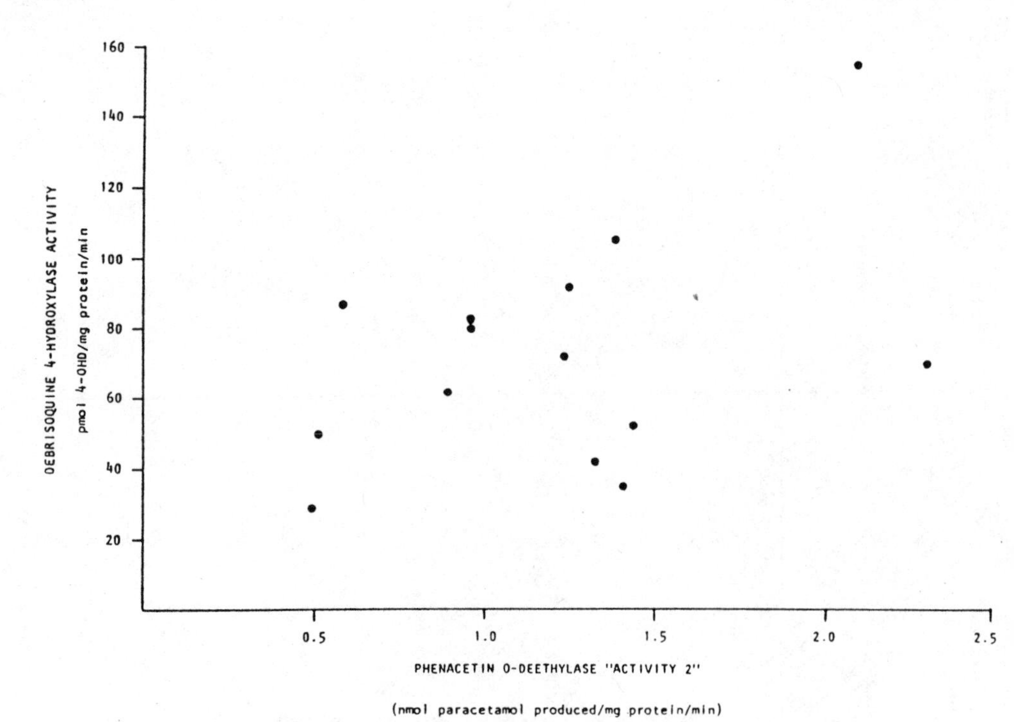

Fig. 18. Comparison between debrisoquine 4-hydroxylase and low affinity (activity 2) phenacetin O-deethylase activities of microsomal fractions of human liver. Data are from the same 14 liver samples as shown in Fig. 17.

The consequences of a genetically determined impairment in the oxidation of debrisoquine, and several other compounds, for the toxicity and carcinogenicity of such compounds are obviously profound. The rate of formation of an active metabolite or the rate of detoxication of the parent compound may differ substantially between phenotypes [156, 157]. A more subtle consequence of the polymorphism occurs when a major, non-toxic, route of metabolism is impaired, permitting more substrate to be diverted to a minor, potentially toxic, pathway that is not impaired. This occurs with phenacetin [158]. The relatively non-toxic pathway of O-deethylation is impaired in PM subjects, whereas the 2-hydroxylation pathway, which leads ultimately to the formation of the potent methemoglobinemic agent 2-hydroxyphenetidine, is not impaired [156]. As a consequence, PM subjects manifest methemoglobinemia at doses of phenacetin that are without effect in EM subjects [158]. The implications of this in the metabolism of carcinogenic chemicals are obvious. However, much more work is necessary to establish the specificity of the debrisoquine form of cytochrome P-450 towards such compounds. These studies can only be performed in vitro, as it is not ethically possible to administer such compounds in vivo.

TABLE 9. Variability in Mixed Function Oxidase Activities of Human Liver

Enzyme activity	Range of V_{max} (pmol product formed/mg microsomal protein/min)
Aryl hydocarbon hydroxylase	5-68
Debrisoquine 4-hydroxylase	0-157
Antipyrine 3-methylhydroxylase	190-800
" 4-hydroxylase	150-560
" N-demethylase	80-690
Ethoxycoumarin O-deethylase	
High affinity component	22-209
Low affinity component	154-1240
Phenacetin O-deethylase	
High affinity component	30-350
Low affinity component	347-2290
2-Acetylaminofluorene 7-hydroxylase	178-754
" 9-hydroxylase	28-112
" 5-hydroxylase	7-97
" 3-hydroxylase	70-418
" 1-hydroxylase	2-11
" N-hydroxylase	35-263

3.5. Interindividual Variability in Monooxygenase Activities in Man

Some of the factors contributing to interindividual variability in the activity of the cytochromes P-450 have been discussed above. Table 9 shows a compilation of the range of activities for a number of oxidation reactions that we have measured in human liver samples with normal histology over the last couple of years [88, 94, 95, 136, 144, 147]. Single gene differences contribute to the wide variation observed in desbrisoquine 4-hydroxylation and high affinity phenacetin O-deethylation [148]. Induction by polycyclic hydrocarbons will contribute to the variability in AHH [88] and phenacetin

O-deethylase activities. In fact, an analysis of the variation in AHH activity occurring in a group of 70 biopsy specimens enabled only 24.5% of the variability to be accounted for in terms of environmental factors [159]. In addition to this constitutive variability, there is variability in the response of the samples to in vitro inhibitors such as alphanaphthoflavone and enhancers such as betamethasone [160].

4. EXTRAPOLATION FROM ANIMALS TO MAN: WHICH WAY NOW?

When one seeks to extrapolate from animals to man, the question that must now be asked is "to what man?" That there are multiple forms of cytochrome P-450 in human liver, as in animal liver, must now be beyond dispute [148, 161]. The complement of these forms obviously varies considerably amongst individuals. The rational way forward must be to determine the number of forms, characterize the specificity of each form and to quantitate the amount of each form present in different individuals. In this way, some idea of the factors affecting individual forms can be obtained and perhaps the reason why some individuals are more susceptible to chemical carcinogens than others will be revealed. One of the techniques likely to receive wide application is that of monoclonal antibodies [162]. Available data from this approach are very preliminary, but are extremely encouraging [163-166].

It would be remiss to close this paper without reference to recent developments in the isolation and characterization of different forms of cytochrome P-450 from animals. These reports could be somewhat disheartening. It has now been demonstrated that different forms of cytochrome P-450 can exist not only amongst different species, or even amongst different strains of a single species, but also amongst different populations of the same strain [67]. These different forms of the enzyme have now been shown to arise from differences in structural genes, and not from post-translational modification of the proteins [167]. Thus, the possibility of allelic variation at multiple loci for the different forms of cytochrome P-450 has now been demonstrated. The extent to which the different loci can show allelic variation has yet to be determined. However, the extent to which such variation can occur with, for example, the hemoglobin system [168] and the glucose-6-phosphate dehydrogenase system [169] is somewhat depressing. Indeed, Nebert [170] has even suggested that there may be hundreds, if not thousands of different forms of cytochrome P-450 in a single species. Monoclonal antibodies to a number of determinants on individual forms of cytochrome P-450 should help in determining the extent of the problem. The application of the techniques of molecular biology to the more traditional area of drug metabolism promises a dramatic increase in our understanding of the mechanisms involved over the next few years.

ACKNOWLEDGMENTS

We wish to thank the Professor and staff of the Department of Surgery, Royal Postgraduate Medical School, for their kind cooperation in making human liver samples available to us for use in many of the studies described here.

We extend our thanks and appreciation to the many collaborators without whom this paper would not be possible. We are particularly grateful to M. J. Brodie, S. Murray, and S. S. Thorgeirsson. We also wish to thank the Medical Research Council for their support of many of these studies.

REFERENCES

1. IARC, Chemicals with Sufficient Evidence of Carcinogenicity in Experimental Animals: IARC Monographs 1-17. IARC Intern. Tech. Rep. No. 70/003, International Agency for Research on Cancer, Lyon (1978).
2. L. Tomatis, C. Agthe, H. Bartsch, J. Huff, R. Montesano, R. Saracci, E. Walker, and J. Willbourn, Evaluation of the carcinogenicity of chemicals: A review of the monograph program of the International Agency for Research on Cancer (1971-1977), Cancer Res., 38:877-885 (1978).
3. L. Tomatis, The predictive value of rodent carcinogenicity tests in the evaluation of human risk, Ann. Rev. Pharmacol. Toxicol., 19:511-530 (1979).
4. H. C. Pitot and A. E. Sirica, The stages of initiation and promotion in hepatocarcinogenesis, Biochim. Biophys. Acta, 605: 191-215 (1980).
5. E. C. Miller and J. A. Miller, Mechanisms of chemical carcinogenesis: Nature of proximate carcinogens and interactions with macromolecules, Pharmacol. Rev., 18:805-838 (1966).
6. B. A. Bridges, in: Environmental Chemicals, Enzyme Function and Human Disease, pp. 67-81, Excerpta Medica, Amsterdam (1980).
7. I. Berenblum, in: Mechanisms of Tumor Promotion and Co-carcinogenesis (T. J. Slaga, A. Sivak, and R. K. Boutwell, eds.), pp. 1-39, Raven Press, New York (1978).
8. E. C. Miller and J. A. Miller, in: Molecular Biology of Cancer (H. Busch, ed.), pp. 377-402, Academic Press, New York (1974).
9. J. A. Miller, Carcinogenesis by chemicals: An overview - G. H. A. Clowes Memorial Lecture, Cancer Res., 30:559-576 (1970).
10. J. H. Weisburger and E. K. Weisburger, Biochemical formation and pharmacological, toxicological, and pathological properties of hydroxylamines and hydroxamic acids, Pharmacol. Rev., 25: 1-66 (1973).
11. B. C. van Duuren, On the possible mechanism of carcinogenic action of vinyl chloride, Ann. N. Y. Acad. Sci., 246:258-267 (1975).

12. P. N. Magee and J. M. Barnes, Carcinogenic nitroso compounds, Adv. Cancer Res., 10:163-246 (1967).
13. E. Kriek, Carcinogenesis by aromatic amines, Biochim. Biophys. Acta, 355:177-203 (1974).
14. R. Schoental, Hepatotoxic activity of retrorsine, senkirkine and hydroxysenkirkine in newborn rats, and the role of epoxides in the carcinogenesis by pyrrolizidine alkaloids and aflatoxins, Nature, 227:401-402 (1970).
15. W. F. Benedict, M. S. Baker, L. Haroun, E. Choi, and B. N. Ames, Mutagenicity of cancer chemotherapeutic agents in the _Salmonella_/microsome test, Cancer Res., 37:2209-2213 (1977).
16. R. T. Williams, Detoxication Mechanisms, Chapman and Hall, London (1973).
17. J. W. Bridges, in: Environmental Chemicals, Enzyme Function, and Human Disease, pp. 5-24, Excerpta Medica, Amsterdam (1980).
18. E. C. Miller and J. A. Miller, in: Biological Reactive Intermediates - II. Chemical Mechanisms and Biological Effects (R. Snyder, D. V. Parke, J. J. Kocsis, D. J. Jollow, C. G. Gibson, and C. W. Witmer, eds.), Part A, pp. 1-21, Plenum Press, New York (1982).
19. D. W. Nebert, Genetic control of carcinogen metabolism leading to cancer risk, Biochimie, 60:1019-1029 (1978).
20. W. Levin, A. W. Wood, P. G. Wislocki, R. L. Chang, J. Kapitulnik, H. D. Mah, H. Yagi, D. M. Jerina, and A. H. Conney, in: Polycyclic Hydrocarbons and Cancer (H. V. Gelboin and P. O. P. Ts'o, eds.), Vol. 1, pp. 189-202, Academic Press, New York (1978).
21. D. Hoffmann and E. L. Wynder, A study of tobacco carcinogenesis, XI. Tumor Initiators, tumor accelerators and tumor promoting activity of condensate fractions, Cancer, 27:848-864 (1971).
22. M. R. Boyd, in: Organ-directed Toxicity, Chemical Indices and Mechanisms (S. S. Brown and D. S. Davies, eds.), pp. 267-272, Pergamon Press, Oxford (1981).
23. H. Remmer, The role of the liver in drug metabolism, Am. J. Med., 49:617-629 (1970).
24. T. C. Butler, Termination of drug action by elimination of unchanged drug, Fed. Proc., 17:1158-1162 (1958).
25. J. R. Mitchell, D. J. Jollow, J. R. Gillette, and B. B. Brodie, Drug metabolism as a cause of drug toxicity, Drug Metab. Dispos., 1:418-423 (1973).
26. C. Heidelberger, Chemical carcinogenesis, Ann. Rev. Biochem., 44:79-121 (1975).
27. J. J. Burns, T. F. Yu, P. G. Dayton, A. B. Gutman, and B. B. Brodie, Biochemical pharmacological considerations of phenylbutazone and its analogues, Ann. N. Y. Acad. Sci., 86:253-262 (1960).
28. N. Brock and J. H. Hohorst, Metabolism of cyclophosphamide, Cancer, 29:900-904 (1967).
29. Biological Reactive Intermediates (D. J. Jollow, J. J. Kocsis, R. Snyder, and H. Vainio, eds.), Plenum Press, New York (1977).

30. H. S. Mason, Oxidases, Ann. Rev. Biochem., 34:595-634 (1965).
31. O. Hayaishi, in: International Congress of Biochemistry, 6th, New York, Proceedings of the Plenary Sessions and the Program, Vol. 33, pp. 31-43, Washington, D.C. (1964).
32. D. W. Nebert, H. J. Eisen, M. Negishi, M. A. Lang, and L. M. Hjelmeland, Genetic mechanisms controlling the induction of polysubstrate monooxygenase (P-450) activities, Ann. Rev. Pharmacol. Toxicol., 21:431-462 (1981).
33. C. L. Litterst, E. G. Mimnaugh, R. L. Reagan, and T. E. Gram, Comparison of in vitro drug metabolism by lung, liver, and kidney of several common laboratory species, Drug Metab. Dispos., 3:259-265 (1975).
34. R. S. Chhabra and J. R. Fouts, Biochemical properties of some microsomal xenobiotic-metabolising enzymes in rabbit small intestine, Drug Metab. Dispos., 4:208-214 (1976).
35. B. B. Brodie, J. Axelrod, J. R. Cooper, L. Gaudette, B. N. La Du, C. Mitoma, and S. Udenfriend, Detoxication of drugs and other foreign compounds by liver microsomes, Science, 121:603-604 (1955).
36. T. M. Guenthner, D. W. Nebert, and R. H. Menard, Microsomal aryl hydrocarbon hydroxylase in rat adrenal: Regulation by ACTH but not by polycyclic hydrocarbons, Mol. Pharmacol., 15:719-728 (1979).
37. C. B. Kasper, Biochemical distinctions between the nuclear and microsomal membranes from rat hepatocytes, J. Biol. Chem., 246:577-581 (1971).
38. J. R. Gillette, Metabolism of drugs and other foreign compounds by enzymatic mechanisms, Prog. Drug. Res., 6:13-73 (1963).
39. R. W. Estabrook and B. Cohen, in: Microsomes and Drug Oxidations (J. R. Gillette, A. H. Conney, G. J. Cosmides, R. W. Estabrook, J. R. Fouts, and G. J. Mannering, eds.), pp. 95-109, Academic Press, New York (1969).
40. T. Omura and R. Sato, The carbon monoxide-binding pigment of liver microsomes. I. Evidence for its hemoprotein nature, J. Biol. Chem., 239:2370-2378 (1964).
41. H. Remmer, J. Schenkman, R. W. Estabrook, H. Sasame, J. Gillette, S. Narasimhulu, D. Y. Cooper, and O. Rosenthal, Drug interaction with hepatic microsomal cytochrome, Mol. Pharmacol., 2:187-190 (1966).
42. R. W. Estabrook, J. Baron, J. Peterson, and Y. Ishimura, in: Biological Hydroxylation Mechanisms (G. S. Boyd and R. M. S. Smellie, eds.), pp. 159-182, Academic Press, London (1972).
43. J. R. Gillette, Biochemistry of drug oxidation and reduction by enzymes in hepatic endoplasmic reticulum, Adv. Pharmacol., 4:219-261 (1966).
44. A. Y. H. Lu and W. Levin, Partial purification of cytochromes P-450 and P-448 from rat liver microsomes, Biochem. Biophys. Res. Commun., 46:1334-1339 (1972).
45. A. P. Alvares and P. Siekevitz, Gel electrophoresis of partially purified cytochromes P-450 from liver microsomes of variously-treated rats, Biochem. Biophys. Res. Commun., 54:923-929 (1973).

46. K. Comai and J. L. Gaylor, Existence and separation of three forms of cytochrome P-450 from rat liver microsomes, J. Biol. Chem., 248:4947-4955 (1973).
47. T. Fujita, D. W. Shoeman, and G. J. Mannering, Differences in P-450 cytochromes from livers of rats treated with phenobarbital and with 3-methylcholanthrene, J. Biol. Chem., 248: 2192-2201 (1973).
48. A. F. Welton and S. D. Aust, Multiplicity of cytochrome P-450 hemoproteins in rat liver microsomes, Biochem. Biophys. Res. Commun., 56:898-906 (1974).
49. D. A. Haugen, T. A. van der Hoeven, and M. J. Coon, Purified liver microsomal cytochrome P-450. Separation and characterization of multiple forms, J. Biol. Chem., 250:3537-3570 (1975).
50. D. Ryan, A. Y. H. Lu, J. Kawalek, S. B. West, and W. Levin, Highly purified cytochrome P-448 and P-450 from rat liver microsomes, Biochem. Biophys. Res. Commun., 64:1134-1141 (1975).
51. P. E. Thomas, A. Y. H. Lu, D. Ryan, S. B. West, J. Kawalek, and W. Levin, Immunochemical evidence for six forms of rat liver cytochrome P-450 obtained using antibodies against purified rat liver cytochromes P-450 and P-448, Mol. Pharmacol., 12:746-758 (1976).
52. R. M. Philpot and E. Arinc, Separation and purification of two forms of hepatic cytochrome P-450 from untreated rabbits, Mol. Pharmacol., 12:483-493 (1976).
53. E. F. Johnson, G. E. Schwab, and U. Muller-Eberhard, Multiple forms of cytochrome P-450: Catalytic differences exhibited by two homogeneous forms of rabbit cytochrome P-450, Mol. Pharmacol., 15:708-718 (1979).
54. K. T. Shiverick and A. H. Neims, Multiplicity of testosterone hydroxylases in a reconstituted hepatic cytochrome P-450 system from uninduced male rats, Drug Metab. Dispos., 7:290-295 (1979).
55. D. E. Ryan, P. E. Thomas, D. Korzeniowski, and W. Levin, Separation and characterization of highly purified forms of liver microsomal cytochrome P-450 from rats treated with polychlorinated biphenyls, phenobarbital, and 3-methylcholanthrene, J. Biol. Chem., 254:1365-1374 (1979).
56. E. F. Johnson, Isolation and characterization of a consitutive form of rabbit liver microsomal cytochrome P-450, J. Biol. Chem., 255:304-309 (1980).
57. H. H. Liem, U. Muller-Eberhard, and E. F. Johnson, Differential induction by 2,3,7,8-tetrachlorodibenzo-p-dioxin of multiple forms of rabbit microsomal cytochrome P-450: Evidence for tissue specificity, Mol. Pharmacol., 18:565-570 (1980).
58. N. A. Elshourbagy and P. S. Guzelian, Separation, purification, and characterization of a novel form of hepatic cytochrome P-450 from rats treated with pregnanolone-16α-carbonitrile, J. Biol. Chem., 255:1279-1285 (1980).
59. D. E. Ryan, P. E. Thomas, and W. Levin, Hepatic microsomal cytochrome P-450 from rats treated with isosafrole. Purification and characterization of four enzymic forms, J. Biol. Chem., 255:7941-7955 (1980).

60. P. P. Lau and H. W. Strobel, Multiple forms of cytochrome P-450 in liver microsomes from β-naphthoflavone-pretreated rats. Separation, purification, and characterization of five forms, J. Biol. Chem., 257:5257-5262 (1982).
61. F. P. Guengerich, Separation and purification of multiple forms of microsomal cytochrome P-450. Activities of different forms of cytochrome P-450 towards several compounds of environmental interest, J. Biol. Chem., 252:3970-3979 (1977).
62. S. S. Lau and V. G. Zannoni, Bromobenzene metabolism in the rabbit. Specific forms of cytochrome P-450 involved in 2,3- and 3,4-epoxidation, Mol. Pharmacol., 20:234-235 (1981).
63. D. A. Haugen and M. J. Coon, Properties of electrophoretically homogeneous phenobarbital-inducible and β-naphthoflavone-inducible forms of liver microsomal cytochrome P-450, J. Biol. Chem., 251:7929-7939 (1976).
64. R. L. Norman, E. F. Johnson, and U. Muller-Eberhard, Identification of the major cytochrome P-450 form transplacentally induced in neonatal rabbits by 2,3,7,8-tetrachlorodibenzo-p-dioxin, J. Biol. Chem., 253:8640-8647 (1978).
65. M. J. Coon, The 1980 Bernard B. Brodie Award Lecture, Drug metabolism by cytochrome P-450: Progess and perspectives, Drug Metab. Dispos., 9:1-4 (1981).
66. D. Ryan and W. Levin, Comparisons between a major and minor form of liver microsomal cytochrome P-450 from rats treated with Aroclor 1545, Fed. Proc., 40:1640 (1981).
67. G. P. Vlasuk, J. Ghrayeb, D. E. Ryan, L. Reik, P. E. Thomas, W. Levin, and F. G. Walz, Jr., Multiplicity, strain differences, and topology of phenobarbital-induced cytochromes P-450 in rat liver microsomes, Biochemistry, 21:789-798 (1982).
68. A. R. Dahl, W. M. Hadley, F. F. Hahn, J. M. Benson, and R. O. McClellan, Cytochrome P-450-dependent monooxygenases in olefactory epithelium of dogs: Possible role in tumorigenicity, Science, 216:57-59 (1982).
69. W. Levin, D. Ryan, R. Kuntzman, and A. H. Conney, Neonatal imprinting and the turnover of microsomal cytochrome P-450 in rat liver, Mol. Pharmacol., 11:190-200 (1975).
70. H. Chao and L. W. K. Chung, Neonatal imprinting and hepatic cytochrome P-450. Immunochemical evidence for the presence of a sex-dependent and neonatally imprinted form(s) of hepatic cytochrome P-450, Mol. Pharmacol., 21:744-752 (1982).
71. B. G. Lake, J. M. Tredger, M. D. Burke, J. Chakraborty, and J. W. Bridges, The circadian variation of hepatic microsomal and steroid metabolism in the golden hamster, Chem.-Biol. Interact., 12:81-90 (1976).
72. G. C. Farrell, W. G. E. Cooksley, and L. W. Powell, Drug metabolism in liver disease: Activity of hepatic microsomal metabolizing enzymes, Clin. Pharmacol. Ther., 26:483-492 (1979).
73. A. Rubin, T. R. Tephly, and G. J. Mannering, Kinetics of drug metabolism by hepatic microsomes, Biochem. Pharmacol., 13:1007-1016 (1964).

74. F. J. Wiebel, J. C. Leutz, L. Diamond, and H. V. Gelboin, Aryl hydrocarbon (benzo[a]pyrene) hydroxylase in microsomes from rat tissues: Differential inhibition and stimulation by benzoflavone and organic solvents, Arch. Biochem. Biophys., 144: 78-86 (1971).
75. B. N. Ames, E. G. Gurney, J. A. Miller, and H. Bartsch, Carcinogens as frameshift mutagens: Metabolites and derivatives of 2-acetylaminofluorene and other aromatic amine carcinogens, Proc. Nat. Acad. Sci. USA, 69:3128-3132 (1972).
76. S. S. Thorgeirsson, P. J. Wirth, N. Staiano, and C. L. Smith, in: Biological Reactive Intermediates - II. Chemical Mechanisms and Biological Effects (R. Snyder, D. V. Park, J. J. Kocsis, D. J. Jollow, C. G. Gibson, and C. W. Witmer, eds.), Part B, pp. 897-919, Plenum Press, New York (1982).
77. H. Bartsch, C. Malaveille, H. F. Stich, E. C. Miller, and J. A. Miller, Comparative electrophilicity, mutagenicity, DNA repair induction activity, and carcinogenicity of some N- and O-acyl derivatives of N-hydroxy-2-aminofluorene, Cancer Res., 37: 1461-1467 (1977).
78. J. A. Miller and E. C. Miller, in: Progress in Experimental Tumor Research (F. Hamburger, ed.), Vol. 11, pp. 273-301, S. Karger, Basel/New York (1969).
79. C. C. Harris, Individual differences in cancer susceptibility, Ann. Intern. Med., 92:809-825 (1980).
80. H. Autrup, J. M. Essigmann, R. G. Croy, B. F. Trump, G. N. Wogan, and C. C. Harris, Metabolism of aflatoxin B_1 and identification of the major aflatoxin B_1-DNA adducts formed in cultured human bronchus and colon, Cancer Res., 39:394-398 (1979).
81. H. Autrup, F. C. Wefald, A. M. Jeffrey, H. Tate, R. D. Schwartz, B. F. Trump, and C. C. Harris, Metabolism of benzo[a]pyrene by cultured tracheobronchial tissues from mice, rats, hamsters, bovines, and humans, Intl. J. Cancer, 25:293-300 (1980).
82. D. W. Nebert and N. M. Jensen, The Ah locus: Genetic regulation of the metabolism of carcinogens, drugs, and other environmental chemicals by cytochrome P-450-mediated monooxygenases, Crit. Rev. Biochem., 6:401-437 (1979).
83. D. M. Morton, in: Testing for Toxicity (J. W. Gorrod, ed.), pp. 11-19, Taylor and Francis, London (1981).
84. T. H. Maugh, II, Chemicals: How many are there? Science, 199:162 (1978).
85. J. Caldwell, in: Enzymatic Basis of Detoxication (W. B. Jakoby, ed.), Vol. 1, pp. 85-114, Academic Press, New York (1980).
86. R. L. Smith and J. Caldwell, in: Drug Metobolism - From Microbe to Man (D. V. Parke and R. L. Smith, eds.), pp. 332-356, Taylor and Francis, London (1977).
87. A. E. M. McLean, in: Long-term Hazards from Environmental Chemicals, pp. 179-197, The Royal Society, London (1979).

88. A. R. Boobis, M. J. Brodie, G. C. Kahn, D. R. Fletcher, J. H. Saunders, and D. S. Davies, Monooxygenase activity of human liver in microsomal fractions of needle biopsy specimens, Br. J. Clin. Pharmacol., 9:11-19 (1980).
89. C. von Bahr, C.-G. Groth, H. Jansson, G. Lundgren, M. Lind, and H. Glaumann, Drug metabolism in human liver _in vitro_: Establishment of a human liver bank, Clin. Pharmacol. Ther., 27:711-725 (1980).
90. A. R. Boobis, M. J. Brodie, G. C. Kahn, and D. S. Davies, in: Microsomes, Drug Oxidations, and Chemical Carcinogenesis (M. J. Coon, A. H. Conney, R. W. Estabrook, H. V. Gelboin, J. R. Gillette, and P. J. O'Brien, eds.), Vol. II, pp. 957-960, Academic Press, New York (1980).
91. R. Kato and J. R. Gillette, Sex differences in the effects of abnormal physiological states on the metabolism of drugs by rat liver microsomes, J. Pharmacol. Exp. Ther., 150:285-291 (1965).
92. J. C. A. Knott and E. D. Wills, Effects of whole-body irradiation and hormones on drug metabolism in the liver endoplasmic reticulum, Rad. Res., 53:65-76 (1973).
93. D. S. Davies, in: Drug Metabolism - From Microbe to Man (D. V. Parke and R. L. Smith, eds.), pp. 357-368, Taylor and Francis, London (1977).
94. A. R. Boobis, M. J. Brodie, M. E. McManus, N. Staiano, S. S. Thorgeirsson, and D. S. Davies, in: Biological Reactive Intermediates - II. Chemical Mechanisms and Biological Effects (R. Snyder, D. V. Parke, J. J. Kocsis, D. J. Jollow, C. G. Gibson, and C. W. Witmer, eds.), Part B, pp. 1193-1201, Plenum Press, New York (1982).
95. D. J. Benford, J. W. Bridges, A. R. Boobis, G. C. Kahn, M. J. Brodie, and D. S. Davies, The selective activation of cytochrome P-450 dependent microsomal hydroxylases in human and rat liver microsomes, Biochem. Pharmacol., 30:1702-1703 (1981).
96. G. M. Pacifici, A. R. Boobis, M. J. Brodie, M. E. McManus, and D. S. Davies, Tissue and species differences in enzymes of epoxide metabolism, Xenobiotica, 11:73-79 (1981).
97. J. W. Cramer, J. A. Miller, and E. C. Miller, N-Hydroxylation: A new metabolic reaction observed in the rat with the carcinogen 2-acetylaminofluorene, J. Biol. Chem., 235:885-888 (1960).
98. S. S. Thorgeirsson, D. J. Jollow, H. A. Sasame, I. Green, and J. R. Mitchell, The role of cytochrome P-450 in N-hydroxylation of 2-acetylaminofluorene, Mol. Pharmacol., 9:398-404 (1973).
99. E. C. Miller, J. A. Miller, and M. Enomoto, The comparative carcinogenicities of 2-acetylaminofluorene and its N-hydroxy metabolite in mice, hamsters, and guinea pigs, Cancer Res., 24:2018-2031 (1964).

100. J. S. Felton, D. W. Nebert, and S. S. Thorgeirsson, Genetic differences in 2-acetylaminofluorene mutagenicity _in vitro_ associated with mouse hepatic aryl hydrocarbon hydroxylase induced by polycyclic aromatic compounds, Mol. Pharmacol., 12:225-233 (1976).
101. C. C. Irving, Enzymatic N-hydroxylation of the carcinogen 2-acetylaminofluorene and the metabolism of N-hydroxy-2-acetylaminofluorene-9-C^{14} _in vitro_, J. Biol. Chem., 239: 1589-1596 (1964).
102. S. S. Thorgeirsson, S. Sakai, and R. H. Adamson, Induction of monooxygenase in Rhesus monkeys by 3-methylcholanthrene: Metabolism and mutagenic activation of N-2-acetylaminofluorene and benzo[a]pyrene, J. Nat. Cancer Inst., 60:365-369 (1978).
103. E. F. Johnson, D. S. Levitt, U. Muller-Eberhard, and S. S. Thorgeirsson, Catalysis of divergent pathways of 2-acetyl-amino-fluorene metabolism by multiple forms of cytochrome P-450, Cancer Res., 40:4456-4459 (1980).
104. E. C. Miller, J. A. Miller, and H. A. Hartmann, N-Hydroxy-2-acetylaminofluorene: A metabolite of 2-acetylaminofluorene with increased activity in the rat, Cancer Res., 21:815-824 (1961).
105. E. Dybing, C. von Bahr, T. Aune, H. Glaumann, D. S. Levitt, and S. S. Thorgeirsson, _In vitro_ metabolism and activation of carcinogenic aromatic amines by subcellular fractions of human liver, Cancer Res., 39:4206-4211 (1978).
106. M. E. McManus, A. R. Boobis, G. M. Pacifici, R. Y. Frempong, M. J. Brodie, G. C. Kahn, C. Whyte, and D. S. Davies, Xenobiotic metabolism in the human lung, Life Sci., 26:481-487 (1980).
107. S. S. Thorgeirsson, J. S. Felton, and D. W. Nebert, Genetic differences in the aromatic hydrocarbon-inducible N-hydroxylation of 2-acetylaminofluorene and acetaminophen-produced hepatotoxicity in mice, Mol. Pharmacol., 11:159-165 (1975).
108. P. B. Lotlikar, M. Enomoto, J. A. Miller, and E. C. Miller, Species variations in the N- and ring-hydroxylation of 2-acetyl-aminofluorene and effect of 3-methylcholanthrene pretreatment, Proc. Soc. Exp. Biol. Med., 125:341-346 (1976).
109. S. S. Thorgeirsson, S. A. Atlas, A. R. Boobis, and J. S. Felton, Species differences in the substrate specificity of hepatic cytochrome P-448 from polycyclic hydrocarbon-treated animals, Biochem. Pharmacol., 28:217-226 (1979).
110. J. E. Gielen, F. M. Goujon, and D. W. Nebert, Genetic regulation of aryl hydrocarbon hydroxylase induction, II. Simple Mendelian expression in mouse tissue _in vivo_, J. Biol. Chem., 247:1125-1137 (1972).
111. D. W. Nebert and H. V. Gelboin, The _in vivo_ and _in vitro_ induction of aryl hydrocarbon hydroxylase in mammalian cells of different species, tissues, strains, and developmental and hormonal states, Arch. Biochem. Biophys., 134:76-89 (1969).

112. A. P. Poland, E. Glover, J. R. Robinson, and D. W. Nebert, Genetic expression of aryl hydrocarbon hydroxylase activity. Induction of monooxygenase activities and cytochrome P_1-450 formation by 2,3,7 8-tetrachlorodibenzo-p-dioxin in mice genetically "nonresponsive" to other aromatic hydrocarbons, J. Biol. Chem., 249:5599-5606 (1974).
113. O. Pelkonen, A. R. Boobis, R. C. Levitt, R. E. Kouri, and D. W. Nebert, Genetic differences in the metabolic activation of benzo[a]pyrene in mice. Attempts to correlate tumorigenesis with binding of reactive intermediates to DNA and with mutagenesis in vitro, Pharmacology, 18:281-293 (1979).
114. S. A. Atlas, A. R. Boobis, J. S. Felton, S. S. Thorgeirsson, and D. W. Nebert, Ontogenetic expression of polycyclic aromatic compound-inducible monooxygenase activities and forms of cytochrome P-450 in rabbit, J. Biol. Chem., 252:4712-4721 (1977).
115. D. Hoffmann, I. Schmeltz, S. S. Hecht, and E. L. Wynder, in: Polycyclic Hydrocarbons and Cancer. Environment, Chemistry, and Metabolism (H. V. Gelboin and P. O. P. Ts'o, eds.), Vol. 1, pp. 85-117, Academic Press, New York (1978).
116. O. Pelkonen, P. Jouppila, E. H. Kaltiala, and N. T. Kärki, Aryl hydrocarbon hydroxylase and cytochrome P-450 in the human liver: Fetal development and cigarette smoking, in: Proc. Eur. Soc. Toxicol. Developmental and Genetic Aspects of Drug and Environmental Toxicity (W. A. M. Duncan, D. Julou, and M. Kramer, eds.), Vol. XVI, pp. 181-185, Excerpta Medica, Amsterdam (1975).
117. A. H. Conney and R. Kato, in: Clinical Pharmacology and Therapeutics (P. Turner, ed.), pp. 49-85, MacMillan Publishers, London (1980).
118. M. Danhof, Antipyrine Metabolite Profile as a Tool in the Assessment of the Activity of Different Drug Oxidizing Enzymes in Man, Ph.D. Thesis, University of Leiden, The Netherlands (1980).
119. E. S. Vesell, The antipyrine test in clinical pharmacology: Conceptions and misconceptions, Clin Pharmacol. Ther., 26:275-286 (1979).
120. J. J. Grygiel and D. J. Birkett, Effect of age on patterns of theophylline metabolism, Clin Pharmacol. Ther., 28:456-462 (1980).
121. W. Kalow, B. K. Tang, D. Kadar, L. Endrenyi, and F.-Y. Chan, A method for studying drug metabolism in populations: Racial differences in amobarbital metabolism, Clin. Pharmacol. Ther., 26:766-776 (1979).
122. E. S. Vesell and J. G. Page, Genetic control of drug levels in man: Antipyrine, Science, 161:72-73 (1968).
123. E.-L. Toverud, A. R. Boobis, M. J. Brodie, S. Murray, P. N. Bennett, V. Whitmarsh, and D. S. Davies, Differential induction of antipyrine metabolism by rifampicin, Eur. J. Clin. Pharmacol., 21:155-160 (1981).

124. M. Danhof, A. van Zuilen, J. K. Boeijinga, and D. D. Breimer, Studies of the different metabolic pathways of antipyrine in man. Oral versus i.v. administration and the influence of urinary collection time, Eur. J. Clin. Pharmacol., 21:433-441 (1982).
125. B. B. Brodie and J. Axelrod, The fate of anitpyrine in man, J. Pharmacol. Exp. Ther., 98:97-104 (1950).
126. H. Yoshimura, H. Shimeno, and H. Tsukamoto, Metabolism of drugs. LIX. A new metabolite of antipyrine, Biochem. Pharmacol., 17:1511-1516 (1968).
127. R. Schüppel, Die Stickstoff demethylierung am Phenazon, Naunyn-Schmiedeberg's Arch. Pharmacol., 255:71-72 (1966).
128. J. D. Baty and D. A. Price Evans, Norphenazone, a new metabolite of phenazone in human urine, J. Pharmacol. Exp. Ther., 25:83-84 (1973).
129. D. H. Huffman, D. W. Shoeman, P. Pentikäinen, and D. I. Azarnoff, The effect of spironolactone on antipyrine metabolism in man, Pharmacology, 10:338-344 (1973).
130. M. Danhof, D. P. Krom, and D. D. Breimer, Studies on the different metabolic pathways of antipyrine in rats: Influence of phenobarbital and 3-methylcholanthrene treatment, Xenobiotica, 9:695-702 (1979).
131. T. Inaba, M. Lucassen, and W. Kalow, Antipyrine metabolism in the rat by three hepatic monooxygenases, Life Sci., 26:1977-1983 (1980).
132. M. Danhof, R. M. A. Verbeek, C. J. van Boxtel, J. K. Boeijinga, and D. D. Breimer, Differential effects of enzyme induction on antipyrine metabolite formation, Br. J. Clin. Pharmacol., 13: 379-386 (1982).
133. G. C. Kahn, A. R. Boobis, S. Murray, and D. S. Davies, Differential effects of 3-methylcholanthrene and phenobarbitone treatment on the oxidative metabolism of antipyrine _in vitro_ by microsomal fractions of rat liver, Xenobiotica (in press).
134. G. H. Kellermann and M. Luyten-Kellermann, Antipyrine metabolism in man, Life Sci., 23:2485-2490 (1978).
135. G. C. Kahn, A. R. Boobis, I. A. Blair, M. J. Brodie, and D. S. Davies, A radiometric high-pressure liquid chromatography assay for the simultaneous determination of the three main oxidative metabolites of antipyrine in studies _in vitro_, Anal. Biochem., 113:292-300 (1981).
136. A. R. Boobis, M. J. Brodie, G. C. Kahn, E.-L. Toverud, I. A. Blair, S. Murray, and D. S. Davies, Comparison of the _in vivo_ and _in vitro_ rates of formation of the three main oxidative metabolites of antipyrine in man, Br. J. Clin. Pharmacol., 12:771-777 (1981).
137. A. Rane, G. R. Wilkinson, and D. G. Shand, Prediction of hepatic extraction ratio from _in vitro_ measurement of intrinsic clearance, J. Pharmacol. Exp. Ther., 200:420-424 (1977).

138. J. M. Tredger, F. J. McPherson, J. Chakraborty, J. W. Bridges, and D. V. Parke, The effect of some glucocorticoids on hepatic microsomal hydroxylation in the rat and hamster, Naunyn-Schmiedeberg's Arch. Pharmacol., 292:267-270 (1976).
139. A. H. Conney, M. Sansur, F. Soroko, R. Koster, and J. J. Burns, Enzyme induction and inhibition in studies on the pharmacological actions of acetophenetidin, J. Pharmacol. Exp. Ther., 151:133-138 (1966).
140. P. J. Poppers, W. Levin, and A. H. Conney, Effects of 3-methylcholanthrene treatment on phenacetin O-dealkylation in several inbred mouse strains, Drug Metab. Dispos., 3:502-506 (1975).
141. E. J. Pantuck, R. Kuntzman, and A. H. Conney, Decreased concentration of phenacetin in plasma of cigarette smokers, Science, 175:1248-1250 (1972).
142. S. S. Thorgeirsson, P. J. Wirth, W. L. Nelson, and G. H. Lambert, in: Origins of Human Cancer (H. H. Hiatt, J. D. Watson, and J. A. Winsten, eds.), pp. 869-886, Cold Spring Harbor Laboratory, New York (1977).
143. T. P. Sloan, A Mahgoub, R. Lancaster, J. R. Idle, and R. L. Smith, Polymorphism of carbon oxidation of drugs and clinical implications, Br. Med. J., 2:655-657 (1978).
144. A. R. Boobis, G. C. Kahn, C. Whyte, M. J. Brodie, and D. S. Davies, Biphasic O-deethylation of phenacetin and 7-ethoxycoumarin by human and rat liver microsomal fractions, Biochem. Pharmacol., 30:2451-2456 (1981).
145. F. P. Guengerich, Separation and purification of multiple forms of microsomal cytochrome P-450. Partial characterization of three apparently homogeneous cytochromes P-450 prepared from livers of phenobarbital- and 3-methylcholanthrene-treated rats, J. Biol. Chem., 253:7931-7939 (1978).
146. A. Mahgoub, J. R. Idle, L. G. Dring, R. Lancaster, and R. L. Smith, Polymorphic hydroxylation of debrisoquine in man, Lancet, ii:584-586 (1977).
147. G. C. Kahn, A. R. Boobis, S. Murray, M. J. Brodie, and D. S. Davies, Assay and characterization of debrisoquine 4-hydroxylase activity of microsomal fractions of human liver, Br. J. Clin. Pharmacol., 13:637-645 (1982).
148. D. S. Davies, G. C. Kahn, S. Murray, M. J. Brodie, and A. R. Boobis, Evidence for an enzymatic defect in the 4-hydroxylation of debrisoquine by human liver, Br. J. Clin. Pharmacol., 11:89-91 (1981).
149. T. Inaba, S. V. Otton, and W. Kalow, Deficient metabolism of debrisoquine and sparteine, Clin. Pharmacol. Ther., 27:547-549 (1980).
150. L. Bertilsson, H. J. Dengler, M. Eichelbaum, and H.-U. Schulz, Pharmacogenetic covariation of defective N-oxidation of sparteine and 4-hydroxylation of debrisoquine, Eur. J. Clin. Pharmacol., 17:153-155 (1980).

151. R. R. Shah, N. S. Oates, J. R. Idle, and R. L. Smith, Genetic impairment of phenformin metabolism, Lancet, i:1147 (1980).
152. L. A. Wakile, T. P. Sloan, J. R. Idle, and R. L. Smith, Genetic evidence for the involvement of different oxidative mechanisms in drug oxidation, J. Pharm. Pharmacol., 31:350-352 (1979).
153. H. S. Fraser, F. M. Williams, D. L. Davies, G. H. Draffan, and D. S. Davies, Amylobarbitone hydroxylation kinetics in small samples of rat and human liver, Xenobiotica, 6:465-472 (1976).
154. S. V. Otton, T. Inaba, W. A. Mahon, and W. Kalow, In vitro metabolism of sparteine by human liver: Competitive inhibition by debrisoquine, Can. J. Physiol. Pharmacol., 60:102-105 (1982).
155. D. S. Davies, G. C. Kahn, S. Murray, M. J. Brodie, and A. R. Boobis, Polymorphic oxidation of debrisoquine by human liver, in: Abstracts of the Eighth International Congress of Pharmacology, Tokyo, Japan (July 1981), p. 275.
156. J. C. Ritchie, T. P. Sloan, J. R. Idle, and R. L. Smith, in: Environmental Chemicals, Enzyme Function and Human Disease, pp. 219-244, Excerpta Medica, Amsterdam (1980).
157. M. Eichelbaum, Defective oxidation of drugs: Pharmacokinetic and therapeutic implications, Clin. Pharmacokinet., 7:1-22 (1982).
158. N. T. Shahidi, Acetophenetidin-induced methemoglobinemia, Ann. N.Y. Acad. Sci., 151:822-832 (1968).
159. M. J. Brodie, A. R. Boobis, C. J. Bulpitt, and D. S. Davies, Influence of liver disease and environmental factors on hepatic monooxygenase activity in vitro, Eur. J. Clin. Pharmacol., 20:39-46 (1981).
160. A. R. Boobis, M. J. Brodie, G. C. Kahn, C. Whyte, and D. S. Davies, Interindividual differences in enhancement and inhibition of human hepatic monooxygenase activity in vitro, in: Abstracts of the Seventh European Workshop on Drug Metabolism, Zurich, Switzerland (September 1980), Abstr. No. 422.
161. R. Kato and T. Kamataki, in: Clinical Pharmacology and Therapeutics (P. Turner, ed.), pp. 80-85, MacMillan Publishers, London (1980).
162. G. Kohler and C. Milstein, Continuous cultures of fused cells secreting antibody of predefined specificity, Nature, 256:495-497 (1975).
163. A. R. Boobis, B. Slade, C. Stern, K. M. Lewis, and D. S. Davies, Monoclonal antibodies to rabbit liver cytochrome P-448, Life Sci., 29:1443-1448 (1981).
164. S. S. Park, S.-J. Cha, H. Miller, A. V. Persson, M. J. Coon, and H. V. Gelboin, Monoclonal antibodies to rabbit liver cytochrome P-450_{LM_2} and cytochrome P-450_{LM_4}, Mol. Pharmacol., 21:248-258 (1982).

165. K. M. Lewis, A. R. Boobis, M. B. Slade, and D. S. Davies, Immunopurification of cytochrome P-448 from microsomal fractions of rabbit liver with retention of metabolic activity, Biochem. Pharmacol., 31:1815-1817 (1982).
166. T. Fujino, S. S. Park, D. West, and H. V. Gelboin, Phenotyping of cytochromes P-450 in human tissues with monoclonal antibodies, Proc. Nat. Acad. Sci. USA, 79:3682-3686 (1982).
167. F. G. Walz, G. P. Vlasuk, C. J. Omiecinski, E. Bresnick, P. E. Thomas, D. E. Ryan, and W. Levin, Multiple, immunoidentical forms of phenobarbital-induced rat liver cytochromes P-450 are encoded by different mRNAs, J. Biol. Chem., 257: 4023-4026 (1982).
168. R. M. Winslow and W. F. Anderson, in: The Metabolic Basis of Inherited Disease (J. B. Stanbury, J. B. Wyngaarden, and D. S. Fredrickson, eds.), 4th edn., pp. 1465-1507, McGraw-Hill, New York (1978).
169. E. Beutler, in: The Metabolic Basis of Inherited Disease (J. B. Stanbury, J. B. Wyngaarden, and D. S. Fredrickson, eds.), 4th edn., pp. 1430-1451, McGraw-Hill, New York (1978).
170. D. W. Nebert, Multiple forms of inducible drug-metabolizing enzymes. A reasonable mechanism by which any organism can cope with adversity, Mol. Cell Biochem., 27:27-46 (1979).

INVOLVEMENT OF PROSTAGLANDIN SYNTHETASE IN THE METABOLIC ACTIVATION OF CHEMICAL CARCINOGENS - PHENACETIN AS AN EXAMPLE

Peter Moldéus*, Bo Andersson*,
Roger Larsson*, and Magnus Nordenskjöld†

*Departments of Forensic Medicine and
†Clinical Genetics
Karolinska Institutet
S-104 01 Stockholm, Sweden

INTRODUCTION

Prostaglandin synthetase (PGS) catalyzes the oxygenation of polyunsaturated fatty acids to hydroxy endoperoxides (PGH) [1]. The most important substrate *in vivo* is arachidonic acid (AA). The PGS contains two activities. The fatty acid cyclooxygenase activity, catalyzes the oxygenation of arachidonic acid to a hydroperoxy endoperoxide (PGG_2), while the hydroperoxidase activity catalyzes the reduction of PGG_2 to a hydroxy endoperoxide (PGH_2) [2-3]. PGH_2 represents a branch in the metabolism of arachidonic acid, and can undergo tissue specific metabolism, to different biological active substrates. PGS activity is present in different tissues. The highest activities have been found in seminal vesicles, platelets, lung and kidney while the activity in liver is quite low. The enzyme is associated with the endoplasmic reticulum and nuclear membranes, and is thus present in microsomal preparations.

The cyclooxygenase activity of PGS is inhibited by non-steroid anti-inflammatory agents, such as indomethacin and acetylsalicylic acid, while there are no known specific inhibitors of the hydroperoxidase activity.

The interest in PGS mediated metabolic activation, has been focused on the possible activation of chemical carcinogens in specific target organs. Thus for instance, the formation of the ultimate carcinogenic benzo(a)pyrene-7,8-dihydrodiol-9,10-epoxide from benzo(a)pyrene-7,8-dihydrodiol is to a significant extent catalyzed by PGS in microsomes from different organs including human lung [4].

There are also evidence suggesting, that the induction of skin cancer in benzo(a)pyrene treated rats, is to a great extent PGS dependent [1].

Another group of chemicals, for which PGS may be involved in the formation of ultimate DNA damaging species, are some inducers of urinary tract tumors, e.g., benzidine N-|4-(5-nitro-2-furyl)-2-thiazolyl| formamide and 3-hydroxymethyl-1-{|3-(5-nitro-2-furyl)-allydidene|amino}hydantoin which undergo PGS catalyzed activation in kidney medulla microsomes and in vivo [5-7].

These examples suggest that PGS mediated cooxidation, could be an important mechanism for the formation of DNA binding metabolites of certain chemical carcinogens in specific organs.

Phenacetin, 4-ethoxyacetanilide, is a component in several analgesic and antipyretic formulations. There are several reports of different kinds of serious kidney damage in man, following prolonged treatment with these drugs, and there are also indications of an increased rate of cancer in the kidney and urinary tract, after long term phenacetin exposure [8-9]. Phenacetin has also been shown to induce tumors in the kidney and urinary tract in rat [10-11].

Since the target organ for the phenacetin toxicity is the kidney, an organ with relatively high PGS activity, we decided to investigate the involvement of this enzyme system in the metabolic activation of phenacetin, and two of its major primary metabolites, paracetamol and p-phenetidine.

RESULTS AND DISCUSSION

The two major metabolites of phenacetin in vivo are paracetamol (70-80%) and p-phenetidine (4-10%) [12]. As shown in Table 1, both of these metabolites could be further metabolized to reactive products, binding irreversible to protein, in the presence of ram seminal vesicle (RSV) microsomes and arachidonic acid. Phenacetin itself is not activated by this system [13]. The activation of paracetamol and p-phenetidine is enzymatic in nature, dependent on arachidonic acid and inhibited in the presence of the cyclooxygenase inhibitor indomethacin (Table 1).

It is evidently the peroxidase activity of PGS, catalyzing the reduction of PGG_2 to PGH_2, which is responsible for the metabolic activation of paracetamol and p-phenetidine. This conclusion is supported by the observation, that AA can be replaced by linolenic acid hydroperoxide (Table 1). This is also in accordance with previous findings, using substrates other than these phenacetin metabolites [5, 14-15].

TABLE 1. Irreversible Binding of Phenacetin Metabolites to RSV Microsomes

Incubation conditions	nmol/mg protein/min 200 μM paracetamol	nmol/mg protein/30 sec 50 μM p-phenetidine
Complete	1.5	15.2
- Arachidonic acid	0.1	0
Boiled microsomes	0	0
+ Indomethacin, 100 μM	0.2	0.5
- Arachidonic acid + Linolenic acid hydroperoxide	1.6	13.3
- Arachidonic acid + Linolenic acid hydroperoxide + Indomethacin, 100 μM	1.7	11.5

Incubations with |^{3}H|-paracetamol were performed at 25°C in 0.1 M phosphate buffer, pH 8.0, at a protein concentration of 1.0 mg/ml for 1 min. Concentrations of arachidonic acid and linolenic acid hydroperoxide were 300 μM. In the experiments with |^{14}C|-p-phenetidine 1 mM EDTA was included in the buffer and 100 μM arachidonic acid or linolenic acid hydroperoxide was used. Incubations were for 30 seconds. Irreversible binding was determined according to Jollow et al.(22). Data from refs(13) and(23).

In order to investigate whether the reactive products of paracetamol and p-phenetidine also had genotoxic potential, their ability to induce DNA single strand breaks in cultured human fibroblasts was investigated. The human fibroblasts were thus incubated with RSV-microsomes, AA and either phenacetin, paracetamol or p-phenetidine. After 30 minutes of incubation the amount of DNA single strand breaks was analyzed (Table 2). As shown in this table, only p-phenetidine produces any significant increase in the amount of single stranded DNA. Both phenacetin and paracetamol are inactive. Thus, only the activation of p-phenetidine seem to result in the formation of a genotoxic product.

The mechanism of activation of paracetamol and p-phenetidine by PGS is not yet established, nor is the nature of the reactive metabolites. However, the reaction which is catalyzed by the peroxidase function of PGS, probably involves one electron oxidation, which could lead to hydrogen abstraction to yield the phenoxy radical of paracetamol and the nitreniumradical of p-phenetidine [13,

TABLE 2. Effect of Phenacetin, Paracetamol and p-Phenetidine on the Induction of DNA Strand Breaks in Cultured Human Skin Fibroblasts Coincubated with RSV Microsomes and Arachidonic Acid

Additions		100 μM arachidonic acid	+100 μM indomethacin
None		24.3±7.7	18.7±3.7
Phenacetin,	200 μM	18.7±5.9	-
Paracetamol,	200 μM	23.0±6.3	-
p-Phenetidine,	50 μM	41.5±9.8	20.8±2.8

Incubations were performed in Dulbeccos phosphate-buffered saline, pH 7.6, at 37°C in the presence of 100 μM AraC and 1.0 mg/ml RSV microsomes.

DNA strand breaks were quantitated by the rate of DNA unwinding in alkali[24]. The values given, represents the fraction of radioactive DNA eluted in single strand form.

A more detailed description of the methodology is given in ref.[13].

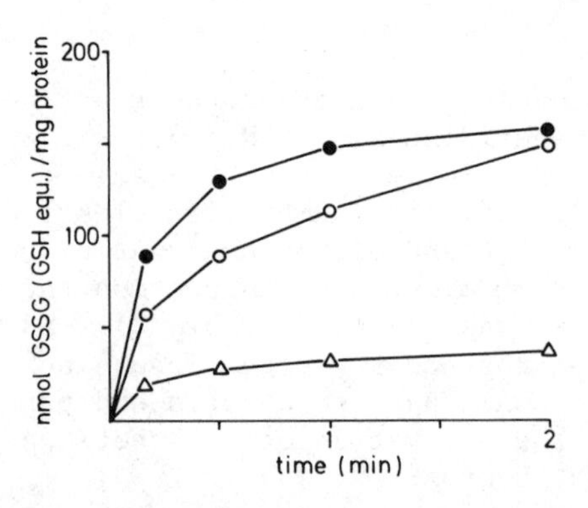

Fig. 1. Oxidation of GSH by RSV microsomes in the presence of 300 μM arachidonic acid (Δ): 300 M arachidonic acid + 200 μM paracetamol (●) or 300 μM arachidonic acid + 50 μM p-phenetidine (○). Incubations were performed with 1.0 mg RSV microsomes/ml incubate in 100 mM phosphate buffer, pH 8.0, at 25°C. The initial GSH concentration was 1 mM.

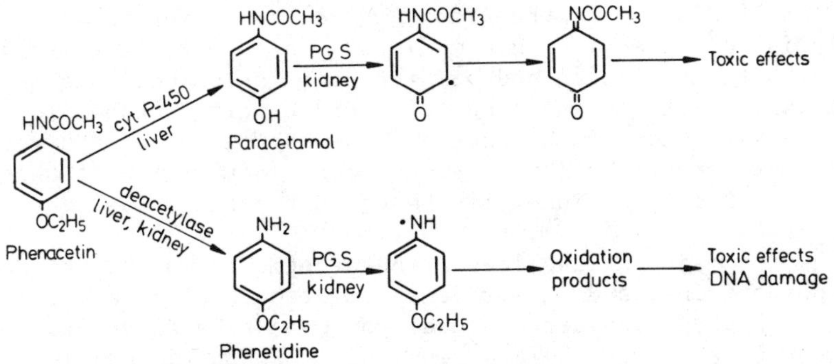

Fig. 2. Possible mechanism for phenacetin induced nephrotoxicity.

16]. Support is lent to the formation and existence of such radicals by the extremely rapid oxidation of GSH during the course of the reaction (Fig. 1). GSH could thus reduce this radical metabolite back to paracetamol and p-phenetidine. The positive identification of the existence of a phenoxy radical of paracetamol during a horseradish catalyzed reaction has also recently been made using EPR spectroscopy [17].

It is unclear whether the proposed radicals themselves are responsible for the DNA damage and/or protein binding. Oxidation and condensation products from these radicals have recently been shown to be formed from both radicals (unpublished) and these products may be responsible for the toxic effects (see Fig. 2).

There is clearly a difference in the nature of the reactive products formed from p-phenetidine, compared to that from paracetamol. Even though it does react with protein and GSH, the paracetamol metabolite does not cause any detectable damage to the fibroblast DNA. This may be explained in terms of the reactivity of the metabolite formed. The paracetamol intermediate may be too reactive to reach the DNA, or interacts with DNA without inducing strand breaks.

The relevance *in vivo*, as well as the toxicological implications of the PGS catalyzed reaction with the phenacetin metabolites, remains to be established. The PGS catalyzed metabolic activations is of particular interest regarding the nephropathy of phenacetin. Phenacetin has, in experimental animals, been shown to be both toxic, causing papillary necrosis [18] and carcinogenic, giving rise to tumors of the renal pelvis and the lower urinary tract after long term exposure [10-11]. Phenacetin has also been suggested to be a kidney carcinogen in man [19]. The nephrotoxic activity of phenacetin has generally been assumed to occur via N-hydroxylation, followed by conjugation with sulfate, glucuronate or acetate in the liver, and transport to the kidney where the conjugates undergo hydrolysis to a reactive intermediate. In support of this is the finding that acetylation and sulfate conjugation activate N-hydroxy-phenacetin to mutagenic and nucleic acid binding metabolites [20-21]. Deacetylation of phenacetin to p-phenetidine or its dealkylation to paracetamol followed by subsequent activation of these compounds by PGS in the kidney may be another alternative.

The two major metabolites of phenacetin in man are paracetamol and p-phenetidine, and it may be speculated that these metabolites in the kidney and bladder can undergo PGS catalyzed activation to acute toxic (paracetamol) and genotoxic (p-phenetidine) intermediates which may, in turn, be responsible for the observed phenacetin induced lesions (see Fig. 2).

In conclusion, the results presented, support the hypothesis that PGS mediated cooxidation may be of importance in the activation of extrahepatic, e.g., urinary tract, toxic and carcinogenic chemicals. Furthermore, this mechanism ought to be taken into account when evaluating chemicals in *in vitro* screening tests. These tests usually utilize activating systems derived from rodent liver, with low PGS activity.

ACKNOWLEDGMENTS

Supported by grants from the Swedish Medical Research Council, Swedish Council for Planning and Coordination of Research, and Funds from the Karolinska Institute.

REFERENCES

1. L. J. Marnett, Polycyclic aromatic hydrocarbon oxidation during prostaglandin biosynthesis, Life Science, 29:531-546 (1981).
2. T. Miyamoto, N. Ogino, S. Yamamoto, and O. Hayaishi, Purification of prostaglandin endoperoxide synthetase from bovine vesicular gland microsomes, J. Biol. Chem., 251:2629-2636 (1976).
3. P. J. O'Brien and A. Rahimtula, The possible involvement of a peroxidase in prostaglandin biosynthesis, Biochem. Biophys. Res. Commun., 70:832-838 (1976).
4. K. Sivarajah, J. M. Lasker, and T. E. Eling, Prostaglandin synthetase-dependent cooxidation of (±) benzo(a)pyrene 7,8-dihydrodiol by human lung and other mammalian tissues, Cancer Res., 41:1834-1839 (1981).
5. T. V. Zenser, M. B. Mattammal, H. J. Armbrecht, and B. B. Davis, Benzidine binding to nucleic acids mediated by the peroxidative activity of prostaglandin endoperoxide synthetase, Cancer Res., 40:2839-2845 (1980).

6. T. V. Zenser, T. M. Balasubramanian, M. B. Mattammal, and B. B. Mattammal, and B. B. Davis, Transport of the renal carcinogen 3-hydroxymethyl-1-{|3-(5-nitro-2-furyl) allydidene|amino}-hydantoin by renal cortex and cooxidative metabolism by prostaglandin endoperoxide synthetase, Cancer Res., 41:2032-2037 (1981).
7. S. M. Cohen, T. V. Zenser, G. Murasaki, S. Fukushima, M. B. Mattammal, N. S. Rapp, and B. B. Davis, Aspirin inhibition of N-|4-(5-nitro-2-furyl)-2-thiazolyl|formamid-induced lesions of the urinary bladder correlated with inhibition of metabolism by bladder prostaglandin endoperoxide synthetase, Cancer Res., 41:3355-3359 (1981).
8. U. Bengtsson, S. Johansson, and L. Angervall, Malignancies of the urinary tract and their relation to analgesic abuse, Kidney Internat., 13:107-113 (1978).
9. G. Carro-Ciampi, Phenacetin abuse: a review, Tox., 10:311-339 (1978).
10. H. Isaka, H. Yoshii, A. Otsuji, M. Koike, Y. Nagai, M. Koura, K. Sugiyasu, and T. Kanabayashi, Tumors of Sprague-Dawley rats induced by long-term feeding of phenacetin, Gann, 70:29-36 (1979).
11. S. L. Johansson, Carcinogenecity of analgesics: long-term treatment of Sprague-Dawley rats with phenacetin, phenazone caffeine and paracetamol (acetamidophen), Int. J. Cancer, 27: 521-529 (1981).
12. R. L. Smith and J. A. Timbrell, Factors affecting the metabolism of phenacetin. 1. Influence of dose, chronic dosage, route of administration and species of the metabolism of |1-^{14}C-acetyl|-phenactin, Zenob., 4:489-501 (1974).
13. B. Andersson, M. Nordenskjöld, A. Rahimtula, and P. Moldéus, Prostaglandin synthetase catalyzed activation of phenacetin metabolites to genotoxic products, Mol. Pharmacol. (in press).
14. L. J. Marnett and G. A. Reed, Peroxidatic oxidation of benzo-(a)pyrene and prostaglandin biosynthesis, Biochem., 18:2923-2929 (1979).
15. R. W. Egan, P. H. Gale, W. J. A. van den Heuvel, E. M. Baptista, and F. A. Kuchl, Jr., Mechanism of oxygen transfer by prostaglandin hydroperoxidase, J. Biol. Chem., 255:323-326 (1980).
16. P. Moldeus, B. Andersson, A. Rahimtula, and M. Berggren, Prostaglandin synthetase catalyzed activation of paracetamol, Biochem. Pharmacol., 31:1363-1368 (1982).
17. S. D. Nelson, D. C. Dahlin, E. J. Rauckman, and G. M. Rosen, Peroxidase-mediated formation of reactive metabolites of acetaminophen, Mol. Pharmacol., 20:195-199 (1981).
18. E. Clausen, Histological changes in rabbit kidneys induced by phenacetin and acetaylsalicyclic acid, Lancet, II, 123-124 (1964).
19. N. Hultengren, C. Lagergren, and A. Ljungqvist, Carcinoma of the renal pelvis in renal papillary necrosis, Acta Chir. Scand., 130:314-320 (1965).

20. J. B. Vaught, P. B. McGarvey, M.-S.Lee, C. D. Garner, C. Y. Wang, E. M. Linsmaier-Bednar, and C. M. King, Activation of N-hydroxyphenacetin to mutagenic and nucleic acid-binding metabolites by acyltransfer, dealkylation and sulfate conjugation, Cancer Res., 41:3424-3429 (1981).
21. P. J. Wirth, E. Dybing, C. von Bahr, and S. S. Thorgeirsson, Mechanism of N-hydroxyacetyl aryl amine mutagenicity in the salmonella test-system: Metabolic activation of N-hydroxyphencetin by liver and kidney fractions from rat, mouse, hamster, and man, Mol. Pharmacol., 18:117-127 (1980).
22. D. J. Jollow, J. R. Mitchell, W. Z. Potter, D. C. Davis, J. R. Gillette, and B. B. Brodie, Acetaminophen-induced hepatic necrosis. II. Role of covalent binding _in vivo_, J. Pharmacol. Exp. Ther., 187:195-202 (1973).
23. B. Andersson, R. Larson, A. Rahimtula, and P. Moldéus, Hydroperoxide-dependent activation of p-phenetidine catalyzed by prostaglandin synthetase and other peroxidases (submitted for publication).
24. M. Nordenskjöld, S. Söderhäll, and P. Moldéus, Studies of DNA-strand breaks induced in human fibroblasts by chemical mutagens/carcinogens, Mutation Res., 63:393-400 (1979).

THE USE OF IMMUNOLOGICAL AND MOLECULAR BIOLOGICAL TECHNIQUES IN THE ASSESSMENT OF OCCUPATIONAL AND ENVIRONMENTAL DISEASE

Alf Fischbein*, J. George Bekesi*,
Irving J. Selikoff*, and Ernest Borek†

*The Mount Sinai Medical Center
New York, New York 10029

and

†AMC Cancer Research Center and Hospital
Lakewood, Colorado 80214

1. INTRODUCTION

Environmental factors are playing an increasing etiologic role in the development of disease [1]. This includes both non-malignant and malignant illnesses. The occupational diseases do not differ biologically from other, non-occupationally-induced diseases, but are unique in their relation to the occupational-environmental conditions under which they have been induced. Recognition of risk to a certain population and the identification of causative environmental factors is, therefore, of profound importance; in fact, this forms the basis upon which the success of controlling environmental health hazards ultimately will depend.

2. OCCUPATIONAL CANCER - CLINICAL LATENCY

One important characteristic of occupational and environmental cancer is the long period of clinical latency between first onset of exposure and the development of disease [2]. The period of clinical latency implies that there is a "silent" period between first onset of exposure and the time when the disease becomes clinically observable. For one of the asbestos-associated cancers, i.e., mesothelioma, this latency is commonly 30-40 years. The difficulty and delay in recognizing the cancer risk associated with previous asbestos exposure is now evidence in a large number of clinical cases, which have developed into a major public health problem in the United

States. It is estimated that about 4,000,000 people experienced significant direct or indirect exposure to asbestos-containing insulation material during World War II, particularly in shipbuilding and maintenance [3]. In the United States the total number of individuals with significant asbestos exposure is estimated to be around 27,000,000. The long delay in disease development after exposure has also been recognized for other occupationally-induced cancers such as arsenic and vinyl chloride-related neoplasms.

Although a multifactorial origin has been suggested for most environmental cancers, little information is available concerning the biological events occurring during the latency period. Thus a multidiscriplinary approach seems necessary to further elucidate the pathogenesis of environmental cancer. The clarification of "subclinical" events is essential and may have important implications for the prevention of these diseases.

The studies presented here were designed to evaluate the usefulness of new approaches such as immunological and molecular biological techniques in detecting environmentally-induced cancer in particular. The immunodiagnostic tests were applied to a population exposed to polybrominated biphenyls (PBB) while the molecular biological technique was used on individuals who had been exposed to asbestos.

3. IMMUNOTOXICOLOGY

Immunotoxicology, which deals with the effects of chemicals and environmental agents on the immune system, has emerged as a subdiscipline of toxicology [4]. For several decades immunology was firmly related to bacteriology. Only in the early 1960s was the structure of antibodies described and the role of the lymphocyte in cellular immunity discovered. The revelation that certain chemicals could produce alterations in either humoral-mediated or cell-mediated immunity at levels that were considered "nontoxic" by traditional methods of toxicological evaluation, made it essential to evaluate the immune system in the assessment of environmentally-induced health effects [5]. It has been suggested that the evaluation of the immune system may provide an additional sensitive, diagnostic tool for the demonstration of toxicity at an early stage; and when dealing with human toxicology, this stage might be characterized as "pre" - or "subclinical." Moreover, an association between carcinogenesis and immunodeficiency has been suggested, and both depressed antibody response to sheep erythrocytes [6] and reduced homograft rejection and stimulation of the growth of transplantable tumors have been reported [7]. In addition, the thymus dependent immune system may also play a role in determining the degree of tumor proliferation. An increased incidence of tumors has been reported in patients treated with immunosuppressive therapy following an organ transplant.

Consequently, it would seem important to evaluate the role of chemicals in the induction of immunodeficiency states which, per se, may cause malignant disease or render individuals more sensitive to the effect of carcinogens. It should be emphasized, however, that overstimulation of the immune system, which may be caused by certain chemicals, is also considered abnormal and could lead both to allergic conditions and autoimmune responses.

It is evident from the foregoing that the immune response is a complex phenomenon. This is also reflected in the great number of techniques available for the assessment of immunotoxicity. Different chemicals may affect different facets of the immune response mechanism, and it has been necessary to employ a wide spectrum of immunodiagnostic tests [5]. Some typical test batteries and their applications will be discussed elsewhere in this paper.

4. MOLECULAR BIOLOGY

As mentioned earlier, one of the characteristics of environmental cancer is the long period of clinical latency between onset of exposure and the time when the disease becomes clinically observable. In the majority of cases, the basic biologic mechanisms operating during this period are still unknown.

Interference with nucleic acid metabolism and cell proliferation has been suggested as an important aspect of carcinogenesis. As evidence of this, it has been demonstrated that patients with certain malignant diseases excrete increased amounts of modified purines and pyrimidines in their urine [8, 9]; nucleosides in urine originate from high turnover rate of tRNA in tumor tissues [10]. Transfer RNA is a most complex biomacromolecule capable of undergoing various modifications. Methylation of the bases is the most frequent modification, and the excretion of some methylated tRNA breakdown products in the urine is of particular interest. Aberrant tRNA methylase activity occurs in tumor tissue and breakdown products resulting from such enzymatic activity may therefore serve as an index of malignant disease. Since different patterns of excretion of these compounds have been manifested, it is possible that specific patterns of excretion may be characteristic for a specific tumor. Furthermore, since the excretion of modified nucleosides is low and very constant in normal individuals [8], it would be of interest to study the excretion patterns of such nucleosides in individuals known to be at very high neoplastic risk. This might have the potential for clarifying biological events at a "precancerous" stage and could serve as an early diagnostic test which may predict the development of malignant diseases.

5. POLYBROMINATED BIPHENYLS

The production of flame retardant chemicals increased greatly biodegradation, there is potential for environmental accumulation of PBBs similar to that found for the chemically-related polychlorinated biphenyls (PCBs).
This was first used in 1970 and produced exclusively by the Michigan Chemical Corporation, St. Louis, Michigan. Hexabromobiphenyl, the major component of Firemaster PB-6, is insoluble in water but highly soluble in organic solvents and fat. Because of their low degree of during the 1970s, primarily in response to requirements of fire safety legislation. A mixture of polybrominated biphenyls (PBB), called Firemaster BP-6, was widely used as a flame retardant for plastics.

Some time during the summer of 1973, 500 to 1000 pounds of PBB were inadvertently mixed into animal feed, which was widely distributed among Michigan dairy farmers [11]. The PBBs were accidentally substituted for a magnesium oxide feed supplement. The consequences of this accidental contamination became one of the most serious agricultural disasters ever to have occurred in the United States. Widespread contamination of the livestock occurred, and disease and death resulted in the cattle that had consumed the PBB-containing feed [12].

The toxic effects observed in the cattle included decreased milk production, hyperkeratosis, swelling and inflammation of the joints, mastitis, and cutaneous and other soft tissue infections often associated with abscss formation [13]. In the most severe cases, progressive emaciation occurred with ultimate death. The clinical syndrome observed in animals, especially the severe infections, suggested a possible effect on the immune response. Subsequent animal studies demonstrated thymus atrophy in animals with PBB toxicosis [14], and severely depressed cell-mediated immunity in both mice and rats [15]; the latter findings indicated impaired responsiveness of splenic lymphocytes to mitogenic stimulation by polyclonal T-cell activators. Humoral immunity was also found to be impaired as reflected in abnormalities in serum immunoglobulin concentrations.

More than 500 farms were quarantined and approximately 23,000 dairy cattle, 1.6 million chickens, and 5 million eggs were destroyed. Before the extent of the problem was fully evaluated, contaminated meat and dairy products had been distributed throughout the State of Michigan for approximately nine months. PBB contaminated items appeared on the market until 1978, and there was thus a potential for widespread exposure to individuals consuming food products. Although it was demonstrated that dairy farmers and their families, accustomed to eating their own products, were at the greatest risk for heavy exposure, it was subsequently shown that wide dissemination of PBB had also occurred among the general population of the State of Michigan [16].

Prior to this accident, the toxic effects of PBBs on humans were unknown, and concern for potentially serious public health consequences led to the initiation of clinical investigations. One major clinical and laboratory examination of PBB-exposed Michigan farm residents was conducted in 1976 by the Environmental Sciences Laboratory, The Mount Sinai Medical Center, New York, following an invitation from Michigan State authorities to conduct a health hazard evaluation.

This initial investigation demonstrated _inter alia_ that 96% of the Michigan population had detectable levels of PBB in their serum. As expected, individuals living on or consuming products from quarantined farms had a higher PBB level than those living in non-quarantined farms [17]. Follow-up studies in 1977 showed that both serum and fat PBB levels were unchanged when compared with levels found at the initial examination in 1976 [18]. With regard to clinical findings, high prevalence of neurological, muscular and arthritis-like symptoms were found [17].

Pursuant to the clinical field survey of dairy farmers in 1976, a group of 45 farmers and members of their families were examined with regard to their immunological status. For comparison, a group of 46 dairy farm residents from the State of Wisconsin and 79 healthy individuals from the New York Metropolitan area were examined with identical techniques. A detailed description of laboratory techniques has been published elsewhere [19].

5.1. Results

The results of these studies, summarized in Table 1, demonstrate that in 18 of the 45 PBB-exposed individuals there was a significant reduction in peripheral blood lymphocyte response to PHA and PWM as well as a reduced reaction in mixed leukocyte cultures. Maximal stimulation values as measured by the uptake of (^{3}H) thymidine were approximately one-third of values found in normal controls. In addition, 27 of the Michigan dairy farm residents showed a decrease in the percentage of E [T-lymphocytes] and EAC [B-lymphocytes] rosettes, and an increase in the number of lymphocytes without detectable service markers, so-called "null cells." These findings were demonstrated in individuals whose mitogenic response was within normal range and who had normal values for absolute numbers of E [T-lymphocytes] and EAC [B-lymphocytes] rosette-forming lymphocytes. In 18 of the PBB-exposed individuals, a significantly reduced number of E [T-lymphocytes] and EAC [B-lymphocytes] rosette-forming lymphocytes was found. A consistent dose-response relationship between plasma concentration of PBBs and lymphocyte function could not be established. However, measurement of the concentration of the most abundant isomer of PBB (i.e., 2, 2', 4, 4', 5', 5'-hexabromobiphenyl) in various compartments of the blood demonstrated that there was a higher concentration of this isomer in leukocytes than in plasma and erythrocytes which suggested an assoication between the accumulation

TABLE 1. Lymphocyte Function in PBB Exposed Michigan Dairy Farmers and Michigan Chemical Workers

Population	N	Phytohemagglutinin		Pokeweed Mitogen	
		Maximum stimulation c.p.m.	SI	Maximum stimulation c.p.m.	SI
Wisconsin Dairy Farm Residents	46	97,662±2,693*	292	95,130±6,369	194
PBB-Exposed Michigan Farm Residents					
With normal lymphocyte function	37	89,438±3,749	239	86,782±5,263	169
With decreased lymphocyte function	18	28,457±3,406**	71	39,159±3,527**	75
Michigan Chemical Workers					
With < 1 ppb plasma level (not involved in PBB production)	6	91,652±4,256	241	93,574±4,926	186
With 40-1200 ppb plasma level (directly exposed to PBB)	4	54,251±6,228**	133	72,750±8,7705	146

The Stimulation index (SI) is the number of lymphocytes undergoing maximum stimulation divided by the number of unstimulated lymphocytes.

* Mean ± standard error.

** Statistical significance (Student's t-Test) between blastogenesis of Wisconsin farm residents and that of Michigan farm residents or Michigan Chemical Workers ($P<0.001$).

of PBB on the target organ (leukocyte) and an impaired immunological response [20].

The results of the *in vivo* immunological assessment showed delayed cutaneous hypersensitivity response to antigens. There was also a significantly higher prevalence of complete anergism among the Michigan dairy farmers when compared with the control group. Increase in sensitivity and heightened response to mumps and varidase antigens was accompanied by a reduction of T cells and impaired function in the peripheral blood lymphocytes [21].

In order to assess the degree to which the observed immunological abnormalities were chronic, the examined populations have been followed prospectively and re-examined 4-5 years after the initial examination, i.e., in 1980-1981. Initial analysis of these studies, which are in progress, indicates that the immune dysfunction noted in 1976 was not a transient phenomenon, but that most of the effects are persistent. This is so despite the fact that the possibility for renewed exposure to PBB through the food chain was very slight after 1977 when most of the contaminated meat and dairy products had been eliminated. It should be recognized, however, that the clearance of PBB from the body, and from fatty tissues in particular, is very slow, and that there is still a biologically active internal dose which could be associated with the long-term immunological dysfunction now being observed.

Continual observation of this population in the future is essential for the clarification of morbidity and mortality patterns. The incidence of malignant disease is now being monitored, and initial observations raise some concern in that 12 cases of various neoplasms appear in the followup of a population of 345 Michigan farm residence. Epidemiological studies are necessary to further clarify the association between morbidity-mortality patterns and PBB exposures. They may also shed light on the predictive value of environmentally-induced immune dysfunction and cancer development.

6. ASBESTOS

Asbestos-related disease is an important public health problem in many countries. Although used since antiquity, it was only at the beginning of this century that asbestos production and use reached significant levels; subsequently, asbestos materials have been distributed throughout the entire spectrum of the human environment and are now present in some 3000 industrial and consumer products. Its unique and useful physical properties include heat resistance, tensile strength, flexibility and resistance to chemical action. The most common uses are in thermal insulation, reinforced cement, heat resistant textiles and other fireproofing applications, pipe insulation and brake linings. A significant use of asbestos in the United

States was for ship insulation material, particularly during World War II [2]]. Because of the known period of clinical latency, this exposure source is now related to an unprecedented increase in the number of cases of asbestos-associated diseases.

Inhalation of asbestos fibers is also associated with a wide spectrum of signs and symptoms. Asbestosis is the designation of the pulmonary interstitial fibrosis caused by asbestos. Roentgenographic abnormalities are characterized by a diffuse interstital pattern, often more prominent in the lower parts of the lungs. Pleural thickening, plaques and calcifications (sometimes diaphragmatic) are all radiologic signs, often diagnostic or even pathognomonic, for asbestos-induced disease. Although asbestosis is the most common disease associated with exposure to asbestos, it is the asbestos-induced malignant diseases which are now the most common cause of asbestos-related mortality.

In the mid-1930s, attention was called to the possibility that asbestos might be implicated in the development of lung cancer. A most definitive study was reported in 1955 [23] indicating that lung cancer had occurred ten times more often than expected among asbestos workers employed for twenty or more years.

Subsequent to the initial observations of the association between lung cancer and previous asbestos exposure, the gloomy spectrum of the mortality experience of asbestos workers has been confirmed in several mortality studies [24, 25, 26]. The excess mortality in lung cancer is extraordinary, especially among asbestos workers who smoke cigarettes. A synergism was demonstrated between cigarette smoking and lung cancer [25].

Moreover, further evidence indicated that previous asbestos exposure was also related to the development of malignant pleural and peritoneal mesothelioma [24]. When the first observation of this relationship was established [27], this disease was rare in the general population and has remained so until the present. In contrast, it is a common neoplastic disease among asbestos workers with almost 10% of all deaths among this group being due to malignant mesothelioma.

This disease, in the form of a highly malignant tumor, is resistant to most modalities of therapy [28]; the prognosis is poor even after diagnosis has been obtained as early as possible with currently available diagnostic means, including chest X-ray, CT-scan, gallium-scum, biopsy, and histochemical analysis.

There is an obvious need for the development of new diagnostic methods which would enable us to identify pathophysiological changes at a stage when early therapeutic intervention might be of great advantage.

Consequently we have initiated studies to assess the urinary excretion patterns of modified nucleosides in patients with malignant mesothelioma, and to evaluate this diagnostic test as a new approach to the early preclinical detection of this disease. For the latter purpose, individuals with a history of long term exposure to asbestos, but without any evidence of malignant disease, were also studied. It was assumed that information obtained on such individuals, who are approaching the end of their clinical latency, could be of great predictive value.

The populations consisted of eleven individuals with malignant mesothelioma and a group of thirteen asbestos workers who had worked in a factory producing asbestos insulation material during the 1940s and 1950s.

The methods for determination of methylated purines and pyrimidines in the urine have been described in detail [29, 30, 31]. The following eight compounds were determined: pseudouridine (ψ), 1-methyladenosine (m'A), 2-pyridone-5-carboxamide-N'-ribofuranoside (PCNR), 1-methylinosine (m'I), 1-methyl-guanosine (m'G), N^2-methyl-guanosine (m^2G), N^2N^2-dimethyl-guanosine (m_2^2G), and β-aminoisobutyric acid (BAIB). Creatinine was measured in three different dilutions of the urine by a dedicated creatinine analyzer. All samples were coded and analyzed in a double blind manner, and neither technicians nor investigators were able to identify the origin of the samples until after the analysis was completed and the results discussed.

6.1. Results

Several nucleosides were elevated in the urine of all eleven patients and the excretion was characteristic of malignant disease. The elevation of pseudouridine was ubiquitous. For comparison, the results obtained on the asbestos workers without patent malignancy showed that the majority of these also had elevated levels of nucleosides in their urine, but the degree of elevation was less than among the patients with mesothelioma. A detailed description of these observations will be reported elsewhere [39].

The significance of these initial findings should be interpreted in the light of currently available epidemiologic evidence suggesting that the incidence rate of asbestos-related neoplastic diseases will increase during the next few decades. The effect of current safety standards will not be reflected in any sharp reduction of disease until the next century [33]. At the present time, diagnosis of mesothelioma is almost exclusively made at the stage when therapeutic intervention is of limited value in terms of long-term survival. The data described above demonstrate that mesothelioma produces elevated excretion of nucleosides, and that this may be an additional tool for early diagnosis. The findings in two patients give further credence to this assumption. In one instance, the initial nucleoside analysis was interpreted as "high normal." At that time, the patient was clinically well, following a decortication procedure of the pleura. A follow-up sample six months later was "positive" despite clinical remission. However, some six weeks later, clinical progress of the disease was noticed. This patient is still being followed.

The second case, of particular interest in terms of early diagnosis, concerns an asbestos worker participating in a clinical field survey who was found to have radiological evidence of pleural thickening, but the nature of this was not such as to warrant any immediate clinical investigation. Because of some discomfort in his chest, however, it was felt that further workup was indicated. Despite the absence of typical radiological evidence of mesothelioma, we found elevated levels of nucleosides in the urine suggesting that diagnosis. This was subsequently verified by pleural biopsy, and intervention at a very early stage seems feasible in this case.

The potentially useful predictive value of nucleoside excretion patterns is also reflected in the data presented on several individuals whose asbestos exposure preceded the examination by 30-40 years but who had elevated levels of nucleosides. This is of interest because of the high neoplastic risk of this population. It should be emphasized that followup information on three workers who had elevated nucleosides gave evidence of malignancy. Although a limited number of asbestos-exposed individuals were examined in this study, there is indication that measurement of urinary nucleosides may have a predictive value in addition to facilitating the diagnosis. More investigations of this nature, coupled with immunological evaluation of asbestos workers, may shed further light on the metabolic events occurring during the latency period, and add to our understanding of these mechanisms, which could assist in developing improved medical surveillance procedures for prevention of occupational and environmental cancer.

REFERENCES

1. I. J. Selikoff, in: Scientific Basis for Control of Environmental Health Hazards, in: Public Health and Preventive Medicine (Maxcy-Rosenau and John M. Last, eds.), 11th edition, pp. 529-548, Appleton-Century-Crofts (1980).
2. I. J. Selikoff, E. C. Hammond, and H. Seidman, Latency of asbestos disease among insulation workers in the United States and Canada, Cancer, 46:2736-2740 (1980).
3. I. J. Selikoff and E. C. Hammond, Asbestos-associated disease in the United States Shipyards, CA (A Journal for Clinicians), 28:87-99 (1978).
4. Immunologic Considerations in Toxicology (Raghubir P. Sharma, ed.), CRC Press, Inc., Boca Raton, Florida (ISBN 0-8493-5271-1) (1981).

5. J. H. Dean, M. I. Luster, G. A. Boorman, K. Chae, L. D. Lauer, R. W. Luebke, L. D. Lawson, and R. E. Wilson, Assessment of Immunotoxicity Induced by Environmental Chemicals 2,3,7,8-Tetrachlorodibenzo-p-dioxin, Diethylstilbestrol and Benzo(a)pyrene in 'Proceedings of the First International Conference on Immunopharmacology', July, 1980, Brighton, England, Adv. Immunopharmacology (J. Hadden, L. Chedid, P. Kullen, and F. Sprefacio, eds.), Pergamon Press (1981).
6. R. A. Malmgren, B. E. Bennison, T. W. McKinley, Jr., Reduced antibody titers in mice treated with carcinogenic and cancer chemotherpeutic agents, Proc. Soc. Exp. Biol. Med., 79:484-488 (1952).
7. R. T. Phren, Function of depressed immunologic reactivity during carcinogenesis, J. Nat. Cancer Inst., 31:791-805 (1963).
8. C. W. Gherke, K. C. Kuo, T. P. Waalkes, and E. Borek, Patterns of urinary excretion of modified nucleosides, Cancer Res., 39: 1150-1153 (1978).
9. J. Speer, C. W. Gehrke, K. C. Kuo, T. P. Waalkes, and E. Borek, tRNA breakdown products as markers for cancer, Cancer, 44: 2120-2123 (1979).
10. E. Borek, B. S. Baliga, C. W. Gehrk, C. W. Kuo, S. Belman, W. Troll, and T. P. Waalkes, High turnover rate of transfer RNA in tumor tissue, Cancer Res., 37:3362-3366 (1977).
11. K. Kay, Polybrominated biphenyls (PBB) environmental contamination in Michigan, 1973-1976, Environ. Rev., 13:74-93 (1977).
12. T. F. Jackson and F. L. Halbert, Toxic syndrome associated with feeding of polybrominated biphenyl-contaminated protein concentrate to dairy cattle, J. Amer. Vet. Med. Assoc., 165:437-439 (1974).
13. M. W. Wastell, D. L. Moody, and J. F. Plog, Jr., Effects of polybrominated biphenyl on milk production, reproduction, and health problems in Holstein Cows, Env. Hlth. Persp-ct., 23:99-103 (1978).
14. P. D. Moorhead, Pathology of experimentally-induced polybrominated biphenyl toxicosis in pregnant heifers, J. Amer. Vet. Med. Assoc., 170:307-313 (1977).
15. M. I. Luster, R. E. Faith, and J. A. Moore, Effects of PBB on immune response in rodents, Env. Hlth. Perspect., 23:227-232 (1978).
16. M. S. Wolff, H. A. Anderson, and I. J. Selikoff, Human tissue burdens of halogenated aromatic chemicals in Michigan, J. Amer. Med. Assoc., 247:2112-2116 (1982).
17. H. A. Anderson, M. S. Wolff, R. Lilis, E. C. Holstein, J. A. Valciukas, K. E. Anderson, M. Petrocci, L. Sarkozi, and I. J. Selikoff, Symptoms and clinical abnormalities following ingestion of polybrominated biphenyl-contaminated food products, Ann. N.Y. Acad. Sci., 320:684-702 (1979).
18. M. S. Wolff, H. A. Anderson, K. D. Rosenman, and I. J. Selikoff, Equilibrium of polybrominated biphenyl (PBB) residues in serum and fat of Michigan residents, Bull. Env. Contam. Toxicol., 21: 775-781 (1979).

19. J. G. Bekesi, J. F. Holland, H. A. Anderson, A. S. Fischbein, W. Rom, M. S. Wolff, and I. J. Selikoff, Lymphocyte function of Michigan dairy farmers exposed to polybrominated biphenyls, Science, 199:1207-1209 (1978).
20. J. Roboz, R. K. Suzuki, J. G. Bekesi, J. F. Holland, and I. J. Selikoff, Mass spectrometric identification and quantification of polybrominated biphenyls in blood compartments of Michigan chemical workers, J. Env. Pathol. Toxicol., 3:363-378 (1979).
21. J. G. Bekesi, H. Anderson, J. Roboz, and I. J. Selikoff, Investigation of the immunobiological effects of polybrominated biphenyls in Michigan farmers, in: "Biological Relevance of Immune Suppression as induced by Genetic Therapeutic and Environmental Factors" (J. Dean and M. Padarathsingh, eds.), Van Nostrand Reinhold Company, pp. 119-135, New York (1981).
22. I. J. Selikoff, Occupational respiratory disease: Asbestos-associated disease, in: Public Health and Preventive Medicine, (Maxcy-Rosenau and John M. Last, eds.), 11th edition, pp. 568-598 and 641-644, Appleton-Century-Crofts (1980).
23. R. Doll, Mortality from lung cancer in asbestos workers, Brit. J. Ind. Med., 12:81-86 (1955).
24. I.J. Selikoff, Lung cancer and mesothelioma during prospective surveillance of 1249 asbestos insulation workers, 1963-1974; Ann. N.Y. Acad. Sci., 271:488-456 (1976).
25. I.J. Selikoff, H. Seidman, and E. C. Hammond, Mortality effects of cigarette smoking among amosite asbestos factory workers, J. Natl. Cancer Inst., 65:507-513 (1980).
26. I. J. Selikoff, E. C. Hammond, and H. Seidman, Mortality experience of insulation workers in the United States and Canada 1943-1976, Ann. N.Y. Acad. Sci., 330:91-116 (1979).
27. J. C. Wagner, C. A. Sleggs, and P. Marchand, Diffuse pleural mesothelioma and asbestos exposure in North Western Cape Province; Br. J. Ind. Med., 17:260-271 (1960).
28. K. H. Antman, R. H. Blum, J. S. Greenberger, G. Flowerdew, A. T. Skarin, and G. P. Canellos, Multimodality therapy for malignant mesothelioma based on a study of natural history, Amer. J. Med., 68:356-362 (1980).
29. K. C. Kuo, C. W. Gehrke, R. A. McCune, T. P. Waalkes, and E. Borek, Rapid, quantitative, high-performance liquid column chromatography of Pseudouridine, J. Chromatogr., 145:383-392 (1978).
30. C. W. Gehrke, K. C. Kuo, G. E. Davis, R. D. Suits, T. P. Waalkes, and E. Borek, Quantitative high performance liquid chromatography of nucleosides in biological materials, J. Chromatogr., 150:455-476 (1978).
31. K. C. Kuo, T. F. Cole, C. W. Gehrke, T. P. Waalkes, and E. Borek, Dual-column, cation-exchange chromatographic method for β-aminoisobutyric acid and β-alanine in biological samples, Clin. Chem., 24:1373-1380 (1978).
32. A. Fischbein, O. K. Sharma, I. J. Selikoff, and E. Borek, Urinary excretion of nucleosides in patients with malignant mesothelioma, Cancer Research, 43:2971-2974 (1983).

33. W. J. Nicholson, G. Perkel, and I. J. Selikoff, Occupational exposure to asbestos: population at risk and projected mortality-1980-2030, Am. J. Ind. Mad., 3:259-312 (1982).

INTERINDIVIDUAL VARIATION IN THE DNA BINDING OF CHEMICAL GENOTOXINS FOLLOWING METABOLISM BY HUMAN BLADDER AND BRONCHUS EXPLANTS

F. B. Daniel*, G. D. Stoner† and H. A. J. Schut†

*Health Effects Research Laboratory
U.S. Environmental Protection Agency
26 W. St. Clair Street
Cincinnati, Ohio 45268

†Department of Pathology
Medical College of Ohio
3000 Arlington Avenue
Toledo, Ohio 43614

ABSTRACT

The DNA-binding of three chemical carcinogens, benzo(a)pyrene (BP), aflatoxin B_1 (AFB) and 2-acetylaminofluorene (AAF) subsequent to their metabolic activation by explant cultures of human bladder and bronchus tissues has been studied. The tissue specimens, obtained at autopsy, were cultured in serum free medium in a rocking culture system under an atmosphere of 50% O_2 - 47% N_2 - 3% CO_2. The epithelial cell DNA binding levels were determined 24 hours following the addition of 1 μM of the tritium labeled hydrocarbon and isolation of the DNA by hydroxylapatite chromatography. Following hydrolysis of the purified DNA, the carcinogen-DNA adducts were analyzed via high pressure liquid chromatography (HPLC). The BP-DNA adducts obtained from both tissues co-chromatographed with adduct standards formed by the reactions of 7,8-dihydroxy-7,8,9,10-tetrahydrobenzo(a)pyrene at the exocyclic amino nitrogen (N^2) of guanine. The AFB-DNA adducts co-chromatographed with adducts found in rat liver *in vivo* and formed by the attack of 8,9-dihydro-8,9-oxyaflatoxin B_1 at the N7 position of guanine. The range of DNA binding interindividual variation was higher in the bladder than in the bronchus for all three chemicals. Further, weak but significant, correlations between the carcinogen-DNA binding level in explants of the two tissues from the same individual was observed for all three carcinogens. For both human bladder and bronchus there was a significant correlation between BP and AAF DNA binding levels in tissues from the same individual whereas no such significant correlation existed between BP and AFB DNA binding levels. The utility of the explant culture method for measuring interindividual variation with respect to genotoxin metabolism is discussed.

INTRODUCTION

Quantitative and qualitative aspects of an individual's ability to metabolize xenobiotic chemicals are important parameters in an individuals susceptibility to various toxic effects by these agents. Consequently, it is important to have some understanding of the range of human interindividual variability with respect to both metabolic activation and detoxification processes for these chemicals. It is axiomatic that such information is critical if adequate regulatory safeguards for human exposure to toxic chemicals are to be developed. This is particularly true with respect to genotoxins (e.g., mutagens, carcinogens) since it has become clear that a large number of these compounds are active only subsequent to cellular metabolism. Numerous pharmacological studies have shown a wide interindividual variation with respect to the metabolism of therapeutic drugs by human cells [1, 2]. Likewise, wide range variations (20 to 76-fold) have been observed in the levels of various oxidative enzyme systems in human liver, bronchus, and lung [3-5]. There are obvious legal and moral impediments to the study of genotoxic chemical metabolism in humans and thus, until recently most of the data (carcinogenesis/mutagenesis) has come from studies with human fibroblast cell lines or human cell homogenates. However, developments in the techniques for culturing and maintaining human tissue via explant culture has greatly facilitated studies of carcinogen metabolism by human cells. Recently, Harris, et al., have reviewed the literature and noted that at least eleven human tissues (including breast, bronchus, trachea, colon, prostate, pancreas, bladder, esophagus, stomach, and others) have been successfully maintained in explant culture [6]. Thus, this methodology not only permits the generation of a data bank on human interindividual variation with respect to metabolism of genotoxins but it also allows direct comparison of human tissue metabolism with the analogous tissues (cultured in an identical manner) of various test animals used in bioassay systems.

In these studies we compare the metabolism-induced DNA binding of three genotoxic compounds: benzo(a)pyrene, (BP); aflatoxin B_1, (AFB); and 2-acetylaminofluorene, (AAF) (Fig. 1) in cultured explants of human bladder and bronchus.

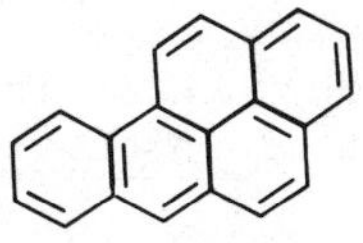

BENZO(A)PYRENE

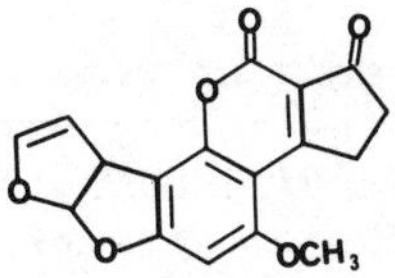

AFLATOXIN B 1

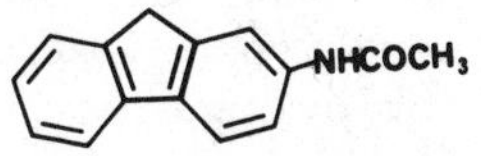

2-ACETYLAMINO-FLOURENE

Fig. 1. Line structures of the three carcinogens used in this study.

MATERIALS AND METHODS

Chemicals

Generally tritiated BP (20-30 Ci/mmole, Amersham Corp., Il.), catalytically tritiated AFB, (10-15 Ci/mmole, Moravek Biochemicals, Brea, Ca.) and [7-^{3}H]AAF (15 Ci/mmole, Moravek) were diluted with the respective non-radiolabeled chemicals to a specific activity of 3-5 Ci/mmole. As necessary HPLC was employed to assure at least 96% radiochemical and chemical purity. The specific activity of each preparation was checked by UV spectroscopy and liquid scintillation counting. The radiolabeled chemicals were dissolved in dimethylsulfoxide: methanol (4:1) at 0.1 mM and stored at -20°C until use. Hydroxylapatite (HTP-DNA grade) was purchased from Bio-Rad (Richmond, Ca.) and stored at 4°C until usage. All solvents were HPLC grade and were obtained from either Fisher Scientific (Pittsburgh, Pa.) or from Burdick-Jackson (Muskegon, Mi.).

Tissue Specimens

Grossly normal bladder and bronchial tissues were taken from 16 patients at autopsy within 2-6 hours from the time of death. Tissues from patients with sepsis, metastatic cancer, serum hepatitis, or a history of tuberculosis were not used in these studies. The tissues were removed and transported to the tissue culture laboratory in ice-cold L-15 medium. Under aseptic conditions, bronchial specimens were trimmed of adherent lung tissue. Bladder specimens were trimmed of muscle and connective tissue while leaving the epithelial layer intact. Samples of each specimen were fixed in phosphate buffered 4% formaldehyde-1% glutaraldehyde and embedded in Epon for assessment by high-resolution light microscopy after staining with toluidine blue.

Explant Culture Conditions

Bronchial and bladder specimens were cut into approximately 0.3-0.5 cm^2 pieces and placed on the etched surface of 60-mm plastic tissue culture dishes (1 explant/dish) with the epithelium oriented towards the gas-liquid interface. Attempts were made to culture approximately the same amount of epithelial tissue per dish (see Discussion). To each dish was added 2.5 ml of PFMR-4 medium supplemented with crystalline bovine insulin (1 μg/ml, Eli Lilly, Indianapolis, In.), hydrocortisone hemisuccinate (35 ng/ml, Upjohn Co., Kalamazoo, Mi.), B-retinyl acetate (0.1 μg/ml, Hoffman LaRoche, Nutley, N. J.), gentamicin (50 μg/ml), penicillin G (100 U/ml), kanamycin (100 μg/ml), and amphotericin B (2.5 μg/ml). The cultures were maintained in an air-tight chamber at 36.5°C in an atmosphere of 50% O_2-47% N_2-3% CO_2 at a pressure of 3 psi. The chamber was placed on a rocker platform (Bellco Glass Co., Vineland, N. J.) and rocked in the dark at approximately 5 cycles/min so that the tissues were alternately exposed to the medium and to the atmosphere. Every 2 days the medium was replaced and the atmosphere of the cultures was restored by flushing the chamber with the gas mixture at 5 liter/min for 5 min.

Incubation with Carcinogens

On the third day of explant culture the radiolabeled carcinogens (3-5 Ci/mmole) were added to the culture medium at a concentration of 1 μM (approximately 10 μCi/dish). After incubation with the chemical for 24 hours, the medium and tissues were separated and the media was stored at -20°C. The explants were rinsed with phosphate buffered saline containing 0.1% glucose after which the epithelial cells were scraped from the stroma and were pooled and stored at -70°C.

Isolation of the Carcinogen Labeled DNA

DNA was isolated from the cells as described previously in detail [7]. Briefly, the epithelial preparations were homogenized in 0.05 M sodium phosphate buffer (pH 6.5) containing 0.01 M EDTA and 0.1 M EGTA, adjusted to 1.5% (v/v) in sodium dodecylsulfate and incubated with pronase B (Calbiochem, San Diego, Ca.) for one hour. After incubation the samples were adjusted to 6% 4-aminosalicylate and 1% NaCl (w/v) and then extracted with 1 volume of Kirby's phenol solution. The DNA was then isolated from the aqueous phase via hydroxylapatite chromatography and quantitated by fluorimetry following reaction with 3,5-diaminobenzoic acid. The radioactivity was determined by standard liquid scintillation techniques.

DNA-Carcinogen Adduct Analysis

The DNA preparation eluted from the hydroxylapatite was dialyzed against water and redissolved in 0.1 M Tris buffer, pH 7.0, containing 10 mM $MgCl_2$ and 0.1 mM $ZnCl_2$. For the BP-DNA adducts the DNA was enzymatically digested to the deoxyribonucleoside level (DNase, snake venom phosphodiesterase, and alkaline phosphatase, Worthington, Freehold, N. J.) and the BP-DNA adducts were separated from the unmodified deoxyribonucleosides by extraction in 1 volume of n-butanol (water-saturated). The butanol was evaporated under a stream of N_2 at 37°C and the adduct residue was dissolved in 1 ml of 50% methanol. The radioactive adducts were compared to BP-DNA adduct standards via HPLC (Varian, Model 5000, Palo Alto, Ca.) using a Zorbax-ODS (4.6 × 250 mm) column (Dupont, Wilmington, De.) and eluted at room temperature with a 30 min, 50 to 80% methanol in water linear gradient and a flow rate of 1 ml/min. The effluent was monitored at 330 nm for UV (Varian Vari-Chrom) and for fluorescence (Excitation = 330 nm, Emission = 389 nm) (Model GM970, Schoeffel, Westwood, N. J.). Fractions of 0.5 ml were collected and analyzed for radioactivity in an LKB model 1216 liquid scintillation counter equipped with a DPM-Plot package. For the AFB adducts, the dialyzed DNA preparation was evaporated to dryness (flash evaporator) and redissolved in 0.5 ml of 80% formic acid [15]. After 1 hour, 1 ml of water was added. The solution was then dried, redissolved in 10% methanol in water and allowed to stand at -20°C overnight. The precipitate was removed by centrifugation (20 min at 12,000 xg) and the AFB-DNA adducts in the supernatant were analyzed by HPLC on the same system as described for the BP adduct analysis except for the following modifications. A linear, 60 min, 10 to 70% methanol in water gradient was used at a flow rate of 1 ml and 60 fractions of 1 ml each were collected while monitoring the effluent for UV (365 nm) and fluorescence (Excitation = 365 nm, Emission = 418 nm).

Calculations

When data from other investigators is quoted it is expressed in binding units of μmoles compound/mole deoxyribonucleotide. For this calculation the average molecular weight of a deoxyribonucleotide is assumed to be 325 g/mole.

RESULTS

Previously, we demonstrated that both human bronchus and bladder tissues could be successfully maintained in rocking organ culture for up to twenty days if removed from the donor within 6 hours after death [7]. High-resolution light microscopy of bladder and bronchus explants prepared both at the time of culturing and after 7 days in culture revealed that the morphology was well maintained [40]. The

TABLE 1. Patient Data and DNA-Binding Levels of Three Carcinogens in Human Bladder and Bronchus Explants[a]

Case	Time[c]	DNA- Binding Levels[b]					
		BP		AFB		2AAF	
(Sex, Age)	(Hr)	Bladder	Bronchus	Bladder	Bronchus	Bladder	Bronchus
5 (M,15)	6	10.4	6.0	1.6	2.9	5.1	8.1
9 (F,22)	2.5	4.4	5.8	1.1	2.1	2.2	3.4
14 (F,55)	2.5	8.4	5.2	8.9	5.3	3.0	3.4
15 (M,10)	3	8.7	1.9	8.7	1.9	7.2	0.5
17 (F,64)	5	15.9	12.0	4.5	-	16.0	4.2
19 (F,50)	2.5	13.1	3.8	2.6	6.2	2.6	6.2
20 (M,57)	6	-	10.8	-	-	-	
21 (M,54)	3	4.8	-	-	-	-	-
22 (M,33)	4	8.3	2.1	2.1	7.2	2.9	0.8
23 (M,29)	6	0.5	4.1	0.13	0.5	3.2	4.2
24 (M,55)	3	17.1	5.7	0.42	0.5	8.7	4.5
31 (M,55)	5	1.7	1.1	0.07	0.1	0.62	0.42
33 (M,75)	2.5	2.9	1.5	0.23	0.29	0.50	0.45
34 (F,42)	3.5	1.6	2.5	0.21	1.3	0.31	1.32
35 (M,37)	5	1.0	0.6	0.27	0.31	0.80	0.55
36 (M,80)	-	1.0	-	-	-	-	-
72 (F,42)	3.5	-	1.0	-	0.5	-	0.55
73 (M,76)	5	2.7	0.6	0.09	0.55	0.14	0.48
77 (M,63)	4.5	2.0	2.1	3.8	4.5	0.67	4.6
93 (F,42)	5	0.8	-	0.28	-	0.28	-
98 (M,58)	5.3	2.0	-	-	-	-	-
100 (M,25)	3	10.9	2.4	-	-	-	-
103 (M,67)	3	8.8	5.0	-	-	-	-
Mean ± S.D.		6.2 ± 5.0	3.8 ± 3.1	2.1 ± 2.8	2.3 ± 2.3	3.7 ± 4.1	2.7 ± 2.4

[a]Binding level measured 24 hours after administration of 1 uM carcinogen.

[b]Units umoles compound/mole deoxyribonucleotide.

[c]Hours elapsed between patient death and explant culturing.

viability of all tissues used in this study were evaluated by high-resolution light microscopy following treatment with the test chemicals. Regression analysis of the carcinogen-DNA binding levels of the three carcinogens versus time elapsed after death until culture of the tissue revealed no significant correlations (bladder: r_{BP} = 0.34, r_{AFB} = -0.36, r_{AAF} = 0.20; bronchus: r_{BP} = .19, r_{AFB} = -0.28, r_{AAF} = 0.12). Table 1 lists patient data, time after death before removal of the tissue, and the carcinogen-DNA binding levels for the three carcinogens used.

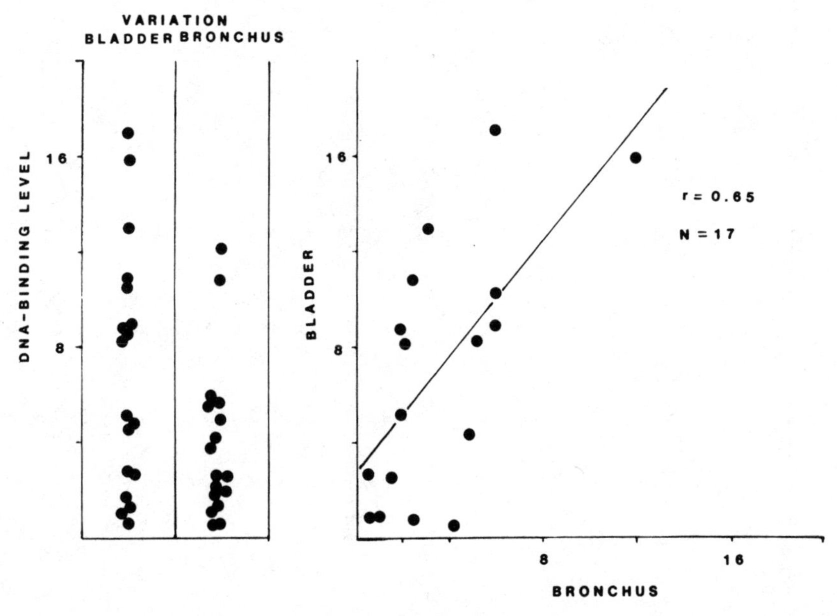

Fig. 2. Scatter diagram and correlation plots of BP-DNA binding in human bladder and bronchus from the same individual. All binding units are µmole BP/mole deoxyribonucleotide.

TABLE 2. DNA Binding and Distribution of BP-Deoxynucleoside Adducts Formed in Human Bronchus and Bladder Explants

TISSUE	COVALENT BINDING[b]	(N)	Percentage of Total [^{3}H] BP Covalently Bound to the Purified DNA.[a]				
			UNKNOWN I	UNKNOWN II	7S-BPDE I-dG	7R-BPDE I-dG	7(R/S)-BPDE II-dG
BLADDER	6.2 ± 5.0	(21)	5.9 - 6.2	8.9 - 10.1	12.2 - 12.8	58.8 - 60.8	N/O
BRONCHUS	3.8 ± 3.1	(19)	3.2 - 5.7	10.2 - 11.4	12.4 - 13.1	60.0 - 62.5	N/O

a. Values represent range of two or more determinations.

b. Units = moles BP/mole deoxyribonucleotide, measured after 24 hr. exposure to 1 µM, BP.

c. N/O = not observed.

The mean level of DNA-binding for BP was significantly higher ($p = 0.027$) in the human bladder explants ($N = 21$) than in the bronchus explants ($N = 19$) (Table 1). Figure 2 shows a scatter diagram for BP-DNA binding levels obtained in bladder and bronchus explants from 17 patients. There was a weak correlation ($r = 0.65$, $p = 0.01$) between the level of BP-DNA binding in the bladder with

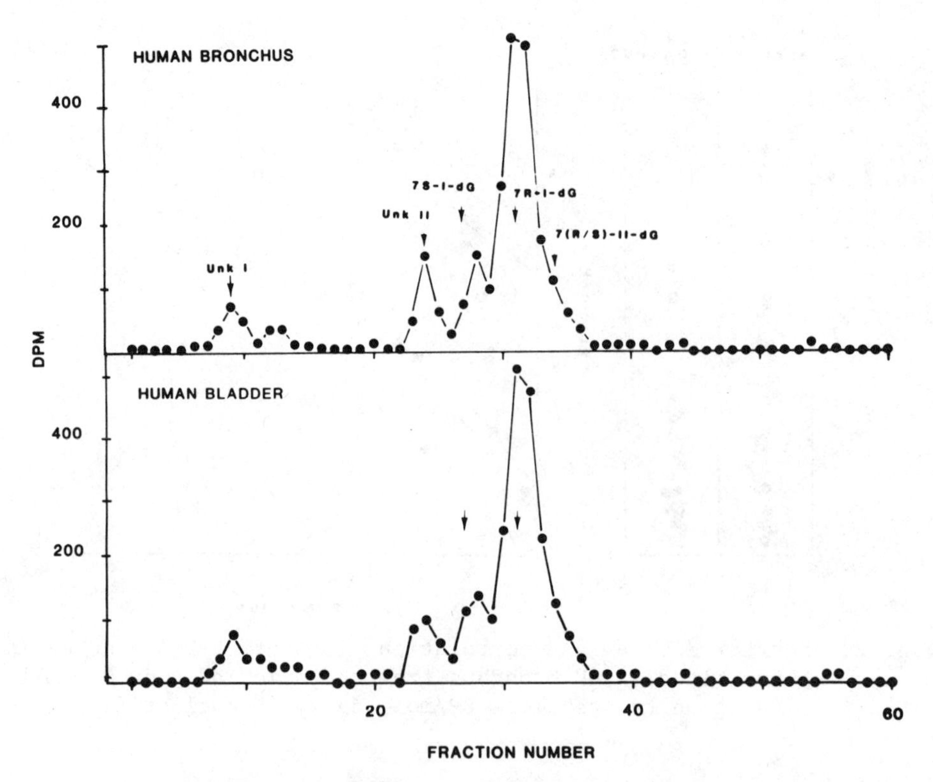

Fig. 3. HPLC traces of BP-deoxyribonucleotide adducts isolated from cultured human bladder and bronchus tissues 24 hour after exposure to 1 μM BP.

the corresponding level in the bronchus of the same patient (Fig. 2). The interindividual variation was somewhat higher in the bladder explants with a range of 0.5-17.1 μmoles BP/mole deoxyribonucleotide (34-fold) versus 0.6-12.0 μmoles BP/mole deoxyribonucleotide (20-fold) for the bronchus. Figure 3 depicts typical HPLC chromatograms for the BP-deoxyribonucleoside adducts obtained from human bladder and bronchus explants. Table 2 details the various relative amounts of the chromatographically resolved BP-DNA adducts. In both human tissues 58-62% of the adducts co-chromatographed with the 7R-BPDE I-dG adduct standard with smaller amounts of the enantiomeric 7S-BPDE I-dG also present. Both of these adducts result from the attack of the (+)7β,8α-dihydroxy-9α,10α-epoxy-7,8,9,10-tetrahydrobenzo(a)pyrene (BPDE I) at the exocyclic N^2-amino function of guanine [8]. In our chromatograms there was no evidence for the existence of adducts of the (-)7β,8α-dihydroxy-9β,10β-epoxy-7,8,9,10-tetrahydrobenzo(a)pyrene (BPDE II) although minor amounts of BPDE II-dG adducts have been reported by Autrup et al., for human bronchus [9, 10, 12] colon [9, 10, 12] and esophagus [9, 10]. Both our laboratory [11] and others

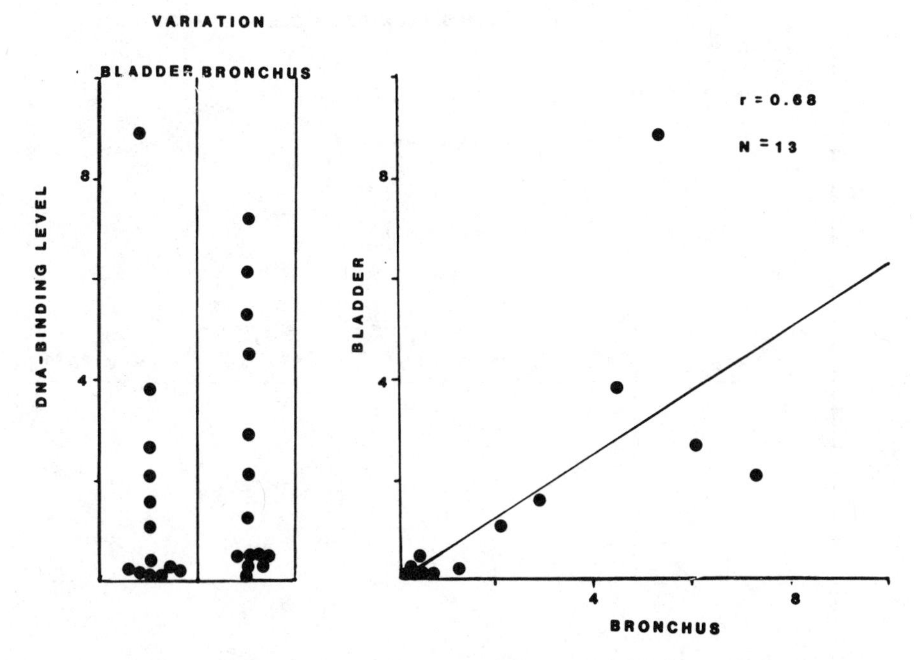

Fig. 4. Scatter diagrams and correlation plots of AFB-DNA binding in human bladder and bronchus from the same individual. All binding units are μmole AFB/mole deoxyribonucleotide.

[12] observed BPDE II-dG adducts in the DNA hydrolysates from BP-treated explants of animal species.

The DNA-binding level of AFB was not significantly different between the two tissues employed in this study although the mean level of AFB-DNA binding was slightly higher in the bronchus (Table 1). The range of interindividual variation in AFB-DNA binding was higher in both tissues than for BP although the absolute level of binding was lower. However, as for BP, the range of variation was higher in the bladder (0.07-0.89 μmole AFB/mole deoxyribonucleotide, 127-fold) than in the bronchus (0.1-7.2 μmole AFB/mole deoxyribonucleotide, 72-fold) (Fig. 4). A significant but weak correlation ($r = 0.68$, $p = 0.01$) between AFB-DNA binding levels between the bladder and bronchus from the same individual was found (Fig. 4).

When the purified AFB-DNA preparations were subjected to formic acid hydrolysis and HPLC analysis two major peaks of radioactivity were observed (Fig. 5). These materials co-chromatographed with the major adducts found in rat liver 4 hours after administration of AFB. Recently, Croy et al. [14] and Essigmann et al. [15] have shown these adducts are formed from the reaction of the 8,9-dihydro-8,9-epoxy-

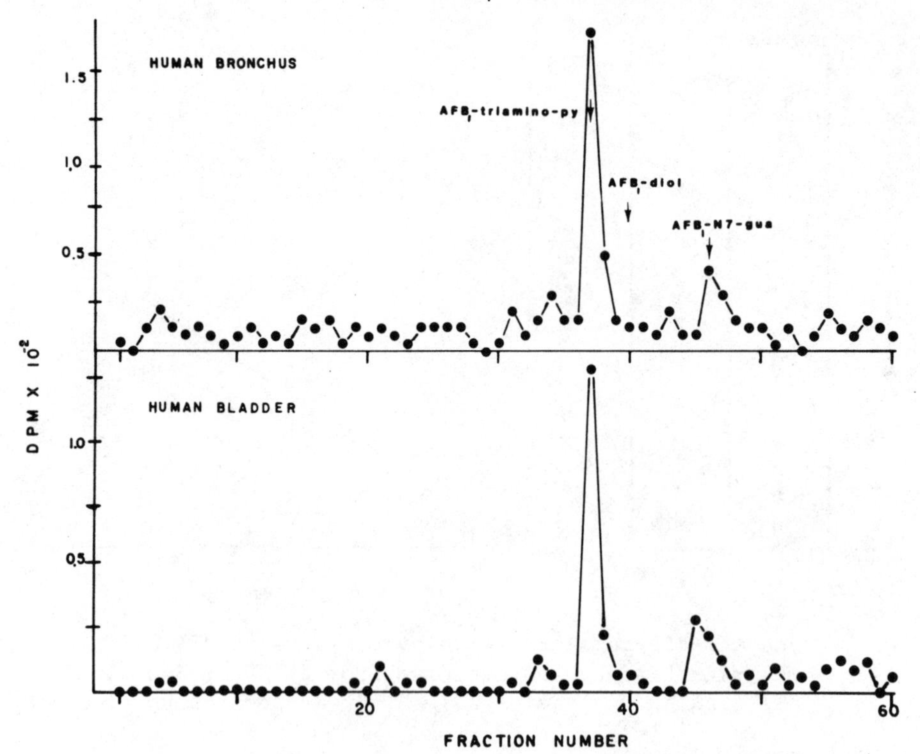

Fig. 5. HPLC traces of AFB-base adducts isolated from cultured human bladder and bronchus tissues 24 hours after exposure to 1 μM AFB.

TABLE 3. DNA Binding and Distribution of AFB_1-DNA Adducts Formed in Cultured Explants of Human Bronchus and Bladder[a]

			Percentage of Total $[^3H]AFB_1$ Covalent Bound to the Purified DNA[a]		
TISSUE	COVALENT BINDING[b]	(N)	POLAR ADDUCT	AFB_1-Triamino-py[c]	AFB_1-N7-gua
BLADDER	2.1 ± 2.8	(16)	2-5	57-70	14-18
BRONCHUS	2.3 ± 2.3	(15)	0-2	61-67	7-18

a. Represents the results of three determinations

b. Units = umoles AFB_1/mole deoxyribonucleotide, measured after 24 hr exposure to 1 μM AFB.

c. In previous work (40) adduct referred to as AFB_1-FAPyr.

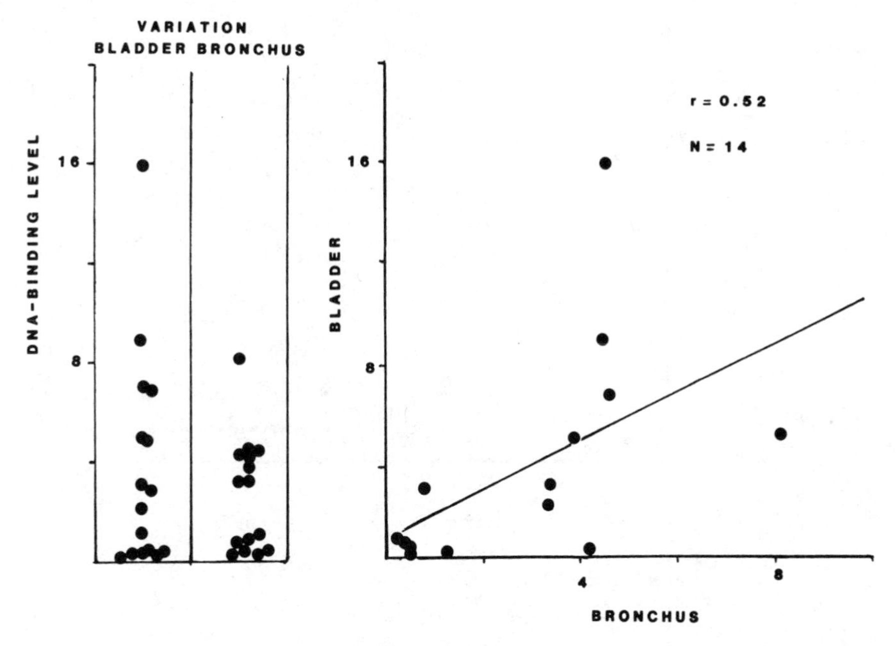

Fig. 6. Scatter diagrams and correlation plots of AAF-DNA binding in human bladder and bronchus tissues from the same individual. All binding units are µmole AAF/mole deoxyribonucleotide.

aflatoxin B_1 (AFB-8,9-epoxide) with the N7-position of guanine. The primary adduct, 8,9-dihydro-2(N7-guanyl)-9-hydroxyaflatoxin B_1 (AFB-N7-gua) is unstable due to the formal positive charge in the imidazole ring and undergoes transformation to a more stable, ring-opened, triaminopyrimidine adduct, 8-(N^7-formyl-2',5',6'-triamino-4'-oxo-N^5-pyrimidyl)-9-hydroxyaflatoxin B_1 (AFB-triamino-py) [16]. Earlier Autrup et al., reported that similar DNA adducts were detected from the incubation of AFB with human bronchus [17] and colon [17, 18] explants. The relative amounts of these two adducts in bladder and bronchial explant DNA after 24 hours exposure to AFB was similar (Table 3). These data are in good agreement with those of Autrup et al. [17, 18] and are very comparable to the ratios observed 24 hours after AFB treatment in rat liver *in vivo* [19], and in cultured human lung cells [20, 21].

For AAF, as for the other two carcinogens employed in this study, the interindividual variation in DNA binding was higher in bladder explants, (0.28-16.0 µmoles AAF/mole deoxyribonucleotide, 57-fold) than in bronchial explants (0.42-8.1 µmoles AAF/mole deoxyribonucleotide, 19-fold) (Table 1, Fig. 6). Similar to the situation with the other two carcinogens a weak correlation (r = 0.52,

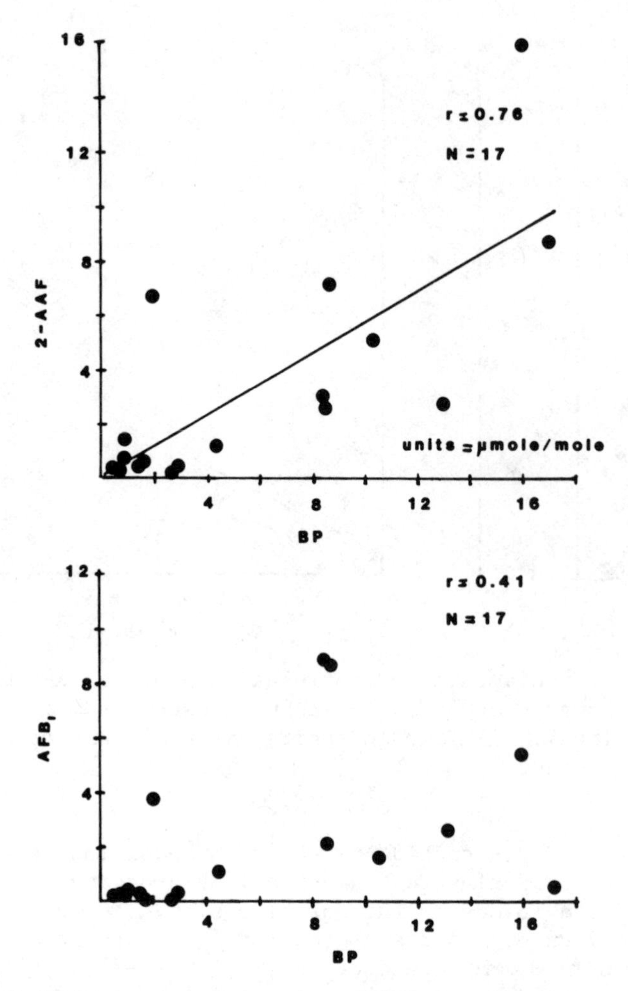

Fig. 7. Correlation plots of BP v. AAF and BP v. AFB DNA binding between bladder explants from the same individual. All binding units are μmole carcinogen/mole deoxyribonucleotide and all data are collected 24 hours after exposure to 1 μM carcinogen.

$p = 0.05$) existed between the DNA binding levels of AAF in the bladder and those in bronchus explants from the same individual (Fig. 6).

For some patients the DNA binding of all three carcinogens was studied in one or both tissues from the same individual allowing correlations to be made between the binding levels of the three chemicals in tissues from the same patient. Figure 7 correlates the level of BP-DNA binding with the corresponding parameter for AFB and AAF in the human bladder explants. Figure 8 shows the analogous

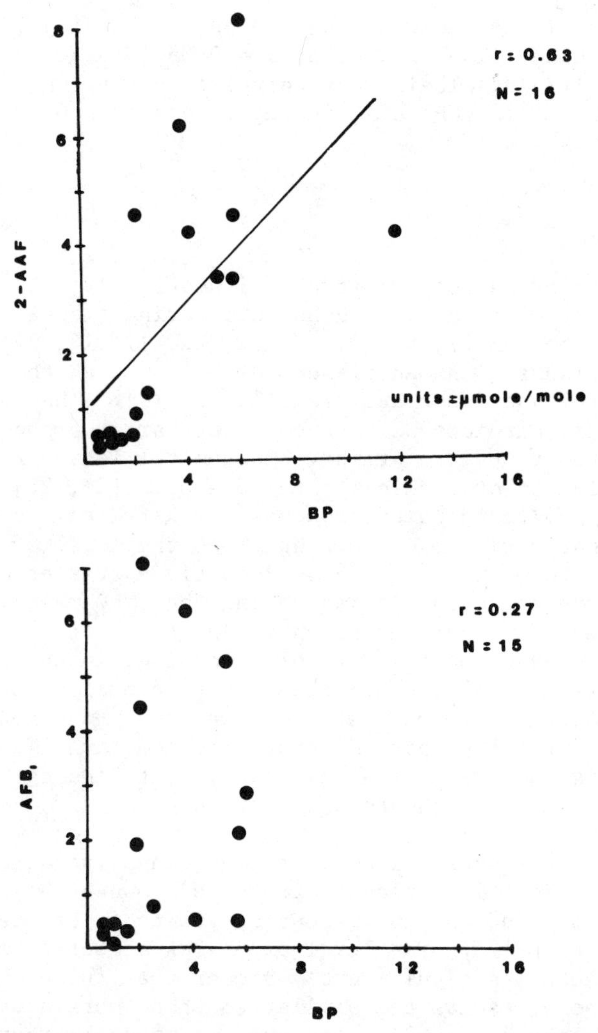

Fig. 8. Correlation plots of BP v. AAF and BP v. AFB DNA binding between bronchus explants from the same individual. All binding units are µmole carbinogen/mole deoxyribonucleotide and all data are collected 24 hours after exposure to 1 µM carcinogen.

plots for the bronchial specimens. For both tissues there was a significant correlation between the binding levels observed for BP and AAF in explants from the same individual (bladder: $r = 0.76$; $p = 0.01$; bronchus: $r = 0.63$; $p = 0.01$). However, there was a much less significant correlation between the levels of BP-DNA binding and AFB-DNA binding (bladder: $r = 0.41$, $p = 0.05$; bronchus: $r = 0.27$; $p = 0.05$). In a previous study Autrup et al. [18] had also observed a lack of correlation between the AFB and BP binding levels observed in cultured human colon ($r = 0.23$, $p = 0.05$, $N = 20$), but there was a highly significant albeit weak correlation between BP and 1,2-dimethylhydrazine (1,2-DMH) DNA binding levels ($r = 0.63$, $p = 0.001$, $N = 42$).

DISCUSSION

The human organ culture (explant) system [22, 23] has permitted the development of data on both human interindividual variation of genotoxin metabolism [7, 9, 10, 11, 12, 12, 17, 18, 24] as well as direct comparisons of human tissue metabolism with the analogous tissue from relevant test animals [11, 12, 18]. The covalent binding of chemical carcinogens to the DNA of target organs has often been postulated to be the primary biochemical lesion responsible for the initiation of chemical carcinogenesis [25, 26]. Early studies using cultured human bronchus demonstrated a 60 to 75-fold interindividual variation with respect to the level of BP-DNA covalent adduct formation [27, 28]. Recently Harris et al. [5, 6] have summarized the evidence indicating that the majority of this observed interindividual variation is due to genetic differences rather than to experimental variables involved in the explant culturing process. Similar conclusions have been made in other investigations [7, 10]. Thus, in this study, we have employed the human tissue explant system for the purpose of providing additional information on the interindividual variation in the covalent binding to DNA of three chemical genotoxins which are also carcinogens.

Both of the tissues, bladder and bronchus, used in this study are well maintained in explant culture as shown by high-resolution light microscopy and the consistent outgrowth of epithelial cells from the explants [7]. The level of BP-DNA adduct formation in the bladder explants was significantly higher than those observed in the bronchus. These results are similar to those obtained in an earlier study in our laboratory [7]. In a recent study by Moore et al. [29] of BP-DNA binding in human bladder explants derived from tissue obtained from 14 patients who underwent voluntary cytoscopical examination, both the levels of BP-DNA binding (5.9 ± 3.9 μmol BP/mol deoxyribonucleotide) and the range of interindividual variation observed (50-fold) are in good agreement with the data from this study of bladder tissue obtained at autopsy (Table 1). As noted above both the level of BP-adduct formation and the range of interindividual variation was smaller in the bronchus than in the bladder (Table 1). In earlier studies of similar design, employing human bronchus, Harris, and co-workers [10, 27, 28] observed slightly lower BP-DNA binding levels (1.0 ± 0.6 to 1.2 ± 1.2 μmoles BP/mole deoxyribonucleotide) than those found in this study. The interindividual variations in BP-DNA adduct levels observed in the studies of Harris et al. [10, 27, 28] varied from 16 to 75-fold. However, there was

TABLE 4. Benzo(a)pyrene-DNA Binding Levels in Cultured Human Tissues

TISSUE	BINDING LEVEL[a,b] (N)	VARIATION (FOLD)	STUDY
BRONCHUS	1.1 ± 1.8 (37)	75	Harris et al, 1976 (27)
	1.2 ± 1.2 (28)	60	Harris et al, 1977 (28)
	0.5 + 0.1 (33)	33	Autrup et al, 1980 (12)
	3.8 ± 3.1 (20)	21	Stoner et al, 1982 (7)
	1.0 ± 0.6 (15)	16	Autrup et al, 1982 (10)
TRACHEA	0.3 ± 0.2 (5)	6	Autrup et al, 1980 (12)
PERIPHERAL LUNG	0.6 ± 0.5 (4)	3	Stoner et al, 1978 (48)
ESOPHAGUS	0.3 ± 0.3 (13)	99	Harris et al, 1979 (24)
	0.6 ± 0.6 (15)	31	Autrup et al, 1982 (10)
DUODENUM	0.5 ± 0.5 (15)	31	Autrup et al, 1982 (10)
COLON	0.2 ± 0.2 (103)	125	Autrup et al, 1980 (9)
	0.3 ± 0.2 (15)	31	Autrup et al, 1982 (10)
BLADDER	6.2 ± 5.0 (21)	34	Stoner et al, 1982 (7)
	5.9 ± 3.9 (14)	50	Moore et al, 1982 (29)
MACROPHAGES	0.75 ± 0.8 (20)	9	Autrup et al, 1978 (49)

a. Mean ± S.D.

b. Binding units all in umole BP / mole deoxyribonucleotide

good agreement between the two laboratories with respect to maximal BP-DNA binding levels to human bronchus explants. Autrup et al. [9, 10] have summarized the BP-DNA binding levels obtained from studies in cultured explants of several human tissues. Based on these studies as well as those reported here we may rank human tissues with respect to the extent of BP-DNA binding as bladder > bronchus > trachea ∿ esophagus ∿ duodenum ∿ peripheral lung > colon (Table 4).

The BP-DNA adduct distribution in bladder and bronchial explants, as determined by HPLC on Zorbax-ODS (Table 2), are essentially identical with those reported previously by Autrup et al. [9, 10] with the major adduct (52-62%) co-chromatographing with the 7R-BPDEI-dG adduct standard. Likewise the relative amount of the minor adduct observed in this study, 7S-BPDEI-dG, is similar to that found by others [10]. Recently Autrup et al. [10] presented tentative evi-

TABLE 5. DNA-Binding of Genotoxic Agents (Carcinogens) in Cultured Human Tissues

TISSUE	COMPOUND	BINDING LEVEL [a,b] (N)	VARIATION (FOLD)	STUDY
BRONCHUS	AFB_1	2.3 ± 2.3 (15)	72	This study, 1982
	AFB_1	0.3 ± 0.4 (7)	23	Autrup et al, 1979 (17)
	2-AAF	2.5 ± 2.3 (13)	18	This study, 1982
	1,2-DMH	124.4 ± 72 (3)	5	Harris et al, 1977 (52)
	DMN	16.7 ± 5.3 (3)	2	Harris et al, 1977 (52)
	DMBA	3.5 ± 2.5 (28)	15	Harris et al, 1977 (28)
ESOPHAGUS	DMN	N.A. (8)	10	Harris et al, 1979 (24)
COLON	AFB_1	0.05 ± 0.06 (24)	70	Autrup et al, 1980 (18)
	AFB_1	0.08 ± 0.06 (6)	10	Autrup et al, 1979 (17)
	1,2-DMH	18.6 ± 19.3 (66)	80	Autrup et al, 1980 (18)
BLADDER	AFB_1	2.1 ± 2.8 (17)	127	This study, 1982
	2-AAF	3.7 ± 4.1 (16)	114	This study, 1982
LIVER	AFB_1	3.6 ± 2.7 (6)	12	Booth et al, 1981 (39)
LYMPHOCYTES	DMBA	0.6 to 4.4 (25)	7	Pero et al, 1976 (53)

a. Mean ± S.D.

b. Binding units all in μmole/mole deoxyribonucleotide

c. N.A. = not available

dence for the formation of small amounts of BP-deoxyadenosine adducts from cultured human duodenum and colon. The observation of the relatively high levels of adduct formation in the human bladder and bronchus may be of significance since increased cancer incidences in both tissues have been linked to cigarette smoking [30, 31] and tobacco smoke condensate contains, BP as well as other genotoxins.

The DNA-binding of AFB, a mold toxin, was similar for the two human explant tissues studied. AFB has been epidemiologically linked to human liver cancer incidence in studies on two continents [34, 35]. Likewise it is a potent hepatocarcinogen for the rat [33, 36], Rhesus monkey [37, 38], duck [34] and numerous other animal species [34]. Thus, it was included in this study since it is a human carcinogen and the failure to observe carcinogenic activity in other human tissues may merely indicate a lack of power in the epidemiological studies. Table 5 lists results from other DNA-binding studies with AFB in human tissues. It is particularly noteworthy that the levels of AFB-DNA adducts formed by the incubation of AFB with fresh human liver slices [39] are very similar to those obtained in this study using human bronchus and bladder explants. The interindividual variation in the formation of AFB-DNA adducts was larger than for BP in both tissues and somewhat larger than the interindividual variation in AFB-DNA binding previously reported for human liver [39], bronchus [17] and colon [17, 18].

The AFB-DNA adducts were resolved by HPLC following acid hydrolysis to hydrolyze the N-glycosidic linkages of the AFB-deoxyriboguanosine adducts [14, 15]. The ratio of the primary adduct, AFB-N7-gua, to the secondary adduct, the ring-opened, AFB-triamino-py was similar for both tissues and in good agreement with previous studies which employed cultured human bronchus [17] and colon [17, 18].

The aromatic amine, AAF, is a liver and bladder carcinogen for mice, rats and dogs [41, 43, 44], and has been used as a model compound for studying aromatic amine metabolism and DNA-binding. Exposure to aromatic amines, in particular 4-aminobiphenyl, 2-naphthylamine and benzidine [45, 46] has been strongly linked to human bladder cancer and thus metabolism of AAF by human tissues is particularly relevant. It has been postulated that conjugates of aromatic amines formed in other tissues (e.g., liver, kidney), which are excreted via the urine, can be reconverted to reactive intermediates by enzymes (e.g., β-glucuronidase) in the urine or epithelial cells of the bladder. Subsequently, these regenerated reactive intermediates could form cancer-initiating DNA-adducts. However, in a recent study Moore et al. [29] demonstrated that human bladder explants can metabolize AAF to a variety of intermediates, including the proximate carcinogen N-hydroxy-2-acetylaminofluorene (N-OH-AAF), to an extent similar to that of bladder explants from the Fisher 344 rat, a tissue susceptible to AAF carcinogenesis. Likewise in these studies we show that human bladder and bronchus are fully capable of converting AAF to DNA-binding forms. The significance of these results however must await comparative DNA-adduct analysis by HPLC. In other work in our laboratory [47] we have shown that bladder explants from the dog and the CD1 rat, both of which are susceptible to AAF-induced bladder cancer, convert AAF to DNA-binding metabolites less effectively than the human bladder explants.

We observed a significant, albeit weak, correlation between the levels of BP-DNA binding in bronchus and bladder explants from the same individual. Autrup et al. [10] observed a strong correlation between BP-DNA binding levels in bronchus, colon and duodenum explants from the same individual, but a weak correlation between BP-DNA binding levels in bronchial and esophageal explants also from the same individual. In addition, in this study, there was also a significant correlation between the BP-DNA and AAF-DNA binding levels both in the bladder (Fig. 7) and bronchus (Fig. 8) tissues. The existance of a correlation in binding levels of various carcinogens between different tissues of the same individual may raise the possibility that an easily accessible tissue may be used for identifying individuals at high risk.

On the other hand, the lack of correlation between BP and AFB DNA-binding and the weak correlation between BP and AAF binding in the same individual's (Figs. 7 and 8) indicates that estimates of a particular individual's ability to metabolize (and/or susceptibility to the effects of) one compound may not be predictive of that individual's ability to metabolize another chemical. That no correlation exists between BP and AFB DNA-binding in these explant tissues is analogous to the observations of Gurtoo et al. [51] who showed that the hepatic microsomal mixed function oxidase complexes that are involved in the generation of aflatoxin reactive intermediates are different than those responsible for polycyclic aromatic hydrocarbon oxidation. However, in a previous study, with two polycyclic aromatic hydrocarbons, on a weak ($r = 0.42$, $p = 0.001$) correlation was observed between the lvels of DNA binding of two polycyclic aromatic hydrocarbons, BP and 7,12-dimethylbenz(a)anthracene (DMBA) in cultured human bronchus of 28 patients [28]. Further, in this same study [28] the range of interindividual variation in DNA binding of BP was 60-fold while that for DMBA, in the same explants, was only 15 fold. Thus, considerable caution must be exercised in attempting to make estimates of either a particular individual's susceptibility or the range of interindividual variability within the human population on the basis of studies with a single compound.

The results obtained in this study are in good agreement with those of others who have employed human explant tissues to study the activation of chemical genotoxins in that we find a wide range of interindividual variation in humans, which ranged from 20-fold (bronchus-BP) to 127-fold (bladder-AFB). Tables 4 and 5, respectively, list the range of variation found for BP and other genotoxic chemicals in various human explant tissues by other investigators. For BP in most tissues the range of variation is from 20 to 100-fold except in trachea [12], peripheral lung [48] and alveolar macrophages [49], where smaller variations were observed. It should be noted however that both the trachea [12] and peripheral lung [48] studies are based on an extremely small number of samples. Further Cohen et al. [50] found a 44-fold variation in the propensity of short-term cultures of human peripheral lung to metabolize BP to organo-soluble metabolites. Likewise, with other genotoxic chemicals wide ranges of interindividual variation have been observed in several laboratories (Table 5). It is possible that some of the interindividual variation observed in our studies and those of others is due to variation in explant viability and to differences in the amount of tissue cultured per unit volume of media. However, we believe that such influences are minor rather than major since: 1) the variation in the total amount of DNA isolated from the explants used in these studies was small (bladder: 40 to 250 mg, 6.3-fold; bronchus 20-125 mg, 6.6-fold) and, more importantly, there was no correlation between the amount of DNA isolated from the various explants and the carcinogen-DNA binding levels observed in those explants for any of the three compounds employed in this study, 2) no correlation was observed between binding level and the time elapsed between patient death and the culturing of the tissue, 3) high-resolution light microscopy indicated that the explants had normal morphology, and 4) individual tissues with low DNA-binding for one compound were often high with respect to binding for another compound.

In conclusion, the human explant culture system can provide valuable data on the range of interindividual variation in ability to metabolize chemical genotoxins. Studies are continuing in our laboratories to expand both the types of human tissue that can be successfully cultured as well as expanding the list of chemical genotoxins studied in human tissues.

Finally, the human explant culture system may be used to study other aspects of target tissue metabolism that may be related to individual variation in susceptibility to the genotoxic effects of chemicals (e.g., DNA repair).

REFERENCES

1. J. R. Idle and R. L. Smith, Polymorphisms of Carbon Centers of Drugs and Their Clinical Significance, Drug Metabol. Rev., 9: 301-317 (1979).
2. E. S. Vesell, Pharmacogenetics, Biochem. Pharmacol., 24:445-450 (1975).
3. O. Pelkonen, E. Sotaniemi, and R. Mokka, The In vitro Oxidative Metabolism of Benzo(a)pyrene in Human Liver Measured by Different Assays, Chem. Biol. Interactions, 16:13-21 (1977).
4. N. Sabadie, C. Malaveille, A. M. Campus, and H. Bartsch, Comparison of the Hydroxylation of Benzo(a)pyrene with the Metabolism of Vinyl Chloride, N-Nitrosomorphine and N-Nitroso-N'-Methylpiperazine to Mutagens by Human and Rat Liver Microsomal Fractions, Cancer Res., 40:119-26 (1980).
5. M. W. Kahng, M. W. Smith, and B. F. Trump, Aryl Hydrocarbon Hydroxylase in Human Bronchial Epithelium and Blood Monocyte, J. Natl. Cancer Inst., 66:222-232 (1981).
6. C. C. Harris, B. F. Trump, R. Grafstrom, and H. Autrup, Differences in Metabolism of Chemical Carcinogens in Cultured Human Epithelial Tissues and Cells, in: Mechanisms in Chemical Carcinogenesis (C. C. Harris and P. Cerutti, eds.), A. L. Liss Publishing Co., Inc., New YOrk, pp. 289-298 (1983).
7. G. D. Stoner, F. B. Daniel, K. M. Schenck, H. A. J. Schut, P. J. Goldblatt, and D. W. Sandwisch, Metabolism and DNA Binding of Benzo(a)pyrene in Cultured Human Bladder and Bronchus, Carcinogenesis, 3:195-201 (1982).

8. J. Remsen, D. Jerina, H. Yagi, and P. Cerutti, In vitro Reaction of Radioactive 7β,8β-Dihydroxy-9α,10α-epoxy-7,8,9-tetrahydrobenzo(a)pyrene and 7β,8α-Dihydroxy-9β,10β-epoxy-7,8,9,10-tetrahydrobenzo(a)pyrene with DNA, Biochem. Biophys. Res. Comm., 74:934-940 (1977).
9. H. Autrup, A. M. Jeffrey, and C. C. Harris, Metabolism of Benzo(a)pyrene in Cultured Human Bronchus, Trachea, Colon, and Esophagus, in: Polynuclear ARomatic Hydrocarbons: Chemistry and Biological Effects (A. Bjorseth and A. J. Dennis, eds.), pp. 89-105, Battelle Press, Columbus, Ohio (1980).
10. H. Autrup, R. C. Grafstrom, M. Brugh, J. F. Lechner, A. Haugen, B. F. Trump, and C. C. Harris, Comparison of Benzo(a)pyrene Metabolism in Bronchus, Esophagus, Colon, and Duodenum from the Same Individual, Cancer Res., 42:934-1938 (1982).
11. F. B. Daniel, H. A. J. Schut, G. D. Stoner, D. W. Sandwisch, K. M. Schenck, C. A. Hoffman, and J. R. Patrick, Interspecies Comparisons of Benzo(a)pyrene Metabolism and DNA-Adduct Formation in Cultured Human and Animal Model Bladder and Tracheobronchial Tissues, Cancer Res., 43:4723-4729 (1983).
12. H. Autrup, F. C. Wefald, A. M. Jeffrey, H. Tate, R. D. Schwartz, B. F. Trump, and C. C. Harris, Metabolism of Benzo(a)pyrene by Cultured Tracheobronchial Tissues from Mice, Rats, Hamsters, Bovines, and Humans, Int. J. Cancer, 25:293-300 (1980).
13. H. Autrup, C. C. Harris, B. F. Trump, and A. M. Jeffrey, Metabolism of Benzo(a)pyrene and Identification of the Major Benzo(a)pyrene-DNA Adducts in Cultured Human Colon, Cancer Res., 38: 3689-3696 (1978).
14. R. G. Croy, J. M. Essigmann, V. N. Reinhold, and G. N. Wogan, Identification of the Principal Aflatoxin B_1-DNA Adduct Formed In vitro in Rat Liver, Proc. Natl. Acad. Sci., U.S.A., 75: 1745-1749 (1978).
15. J. M. Essigmann, R. G. Croy, A. M. Nadzan, W. F. Busby, Jr., V. N. Reinhold, G. Buchi, and G. N. Wogan, Structural Identification of the Major DNA Adduct Formed by Aflatoxin B_1 In vitro, Proc. Natl. Acad. Sci., U.S.A., 74:1870-1874 (1977).
16. T. V. Wang and P. Cerutti, Spontaneous Reaction of Aflatoxin B_1 Modified Deoxyribonucleic Acid In vitro, Biochemistry, 19:1692-1698 (1980).
17. H. Autrup, J. M. Essigmann, R. G. Croy, B. F. Trump, G. N. Wogan, and C. C. Harris, Metabolism of Aflatoxin B_1 and Identification of the Major Aflatoxin B_1-DNA Adducts Formed in Cultured Human Bronchus and Colon, Cancer Res., 39:694-698 (1979).
18. H. Autrup, R. D. Schwartz, J. M. Essigmann, L. Smith, B. F. Trump, and C. C. Harris, Metabolism of Aflatoxin B_1, Benzo(a)-pyrene, and 1,2-Dimethylhydrazine by Cultured Rat and Human Colon, Tetato. Carcino. Muta., 1:3-13 (1980).
19. P. J. Hertzog, J. R. Lindsay-Smith, and R. C. Garner, A High Pressure Liquid Chromatography Study on the Removal of DNA-Bound Aflatoxin B_1 in Rat Liver and In vitro, Carcinogenesis, 1:787-793 (1980).

20. T. V. Wang and P. A. Cerutti, Formation and Removal of Aflatoxin B_1-Induced DNA Lesions in Epithelioid Human Lung Cells, Cancer Res., 39:5165-5170 (1979).
21. S. A. Leadon, R. M. Tyrrell, and P. A. Cerutti, Excision Repair of Aflatoxin B_1-DNA Adducts in Human Fibroblasts, Cancer Res., 41:5125-5129 (1981).
22. L. A. Barrett, E. M. McDowell, A. L. Frank, C. C. Harris, and B. F. Trump, Long-Term Organ Culture of Human Bronchial Epithelium, Cancer Res., 36:1003-1006 (1976).
23. G. D. Stoner, Y. Katoh, J. M. Foidart, G. A. Meyers, and C. C. Harris, Identification and Culture of Human Bronchial Epithelial Cells, in: Methods in Cell Biology XXI A (C. C. Harris, B. F. Trump, G. D. Stoner, eds.), Academic Press, New York, pp. 15-35 (1980).
24. C. C. Harris, H. Autrup, G. D. Stoner, B. F. Trump, E. Hillman, P. W. Schafer, and A. M. Jeffrey, Metabolism of Benzo(a)pyrene, N-Nitrosodimethylamine and N-Nitrosopyrrolidine and Identification of the Major Carcinogen-DNA Adducts Formed in Cultured Human Esophagus, Cancer Res., 39:4401-4406 (1979).
25. P. Brookes, Covalent Interaction of Carcinogens with DNA, Life Sciences, 16:331-344 (1975).
26. W. K. Lutz, _In vivo_ Covalent Binding of Organic Chemicals to DNA as a Quantitative Indicator in the Process of Chemical Carcinogenesis, Mutation Res., 65:289-356 (1979).
27. C. C. Harris, H. Autrup, R. Connor, L. A. Barrett, E. M. McDowell, and B. F. Trump, Interindividual Variation in Binding of Benzo(a)pyrene to DNA in Cultured Human Bronchi, Science, 194:1067-1069 (1976).
28. C. C. Harris, H. Autrup, G. D. Stoner, S. K. Yang, J. C. Leutz, H. V. Gelboin, J. K. Selkirk, R. J. Connor, L. A. Barrett, R. T. Jones, E. McDowell, and B. F. Trump, Metabolism of Benzo(a)pyrene and 7,12-Dimethylbenz(a)anthracene in Cultured Human Bronchus and Pancreatic Duct, Cancer Res., 37:3349-3355 (1977).
29. B. P. Moore, R. M. Hicks, M. A. Knowles, and S. Redgrave, Metabolism and Binding of Benzo(a)pyrene and 2-Acetylaminofluorene by Short-Term Organ Cultures of Human and Rat Bladder, Cancer Res., 42:642-648 (1982).
30. U.S. Public Health Service, The Consequences of Smoking, U.S. Department of Health, Education, and Welfare, Public Health Service, Washington, D.C., DHEW Publ. No. (HSM) 71-7513 (1971); (HSM) 73-8704 (1974); and (CDS) 76-8704 (1976).
31. P. Cole, R. R. Morrison, H. Haring, and G. M. Friedel, Smoking and Cancer of the Lower Urinary Tract, N. Engl. J. Med., 284: 129-134 (1971).
32. E. Wynder and D. Hoffman, Experimental Tobacco Carcinogenesis, Science, 162:862-871 (1968).
33. G. N. Wogan, Aflatoxin Carcinogenesis, in: Methods in Cancer Res. (H. Busch, ed.), 7:309-444 (1974).
34. F. G. Peers and C. A. Linsell, Dietary Aflatoxins and Liver Cancer - A Population Based Study in Kenya, Br. J. Cancer, 27: 473-484 (1973).

35. R. C. Shank, N. Bharmaropravoti, J. E. Gordon, and G. N. Wogan, Dietary Aflatoxin and Human Liver Cancer in Two Municipal Populations of Thailand, Fd. Cosmet. Toxicol., 10:171-179 (1972).
36. G. N. Wogan, S. Paglvalunga, and P. M. Newberne, Carcinogenic Effect of Low Dietary Levels of Aflatoxin B_1 in Rats, Fd. Cosmet. Toxicol., 12(681-687 (1971).
37. C. Gopalon, P. G. Tupule, and D. Krishnamuthi, Induction of Hepatic Carcinoma and Aflatoxin in the Rhesus Monkey, Fd. Cosmet. Toxicol., 10:519-521 (1972).
38. R. H. Adamson, P. Correa, S. Sieber, K. R. McIntire, and D. W. Dalgard, Carcinogenicity of Aflatoxin B_1 in Rhesus Monkey: Two Cases of Primary Liver Cancer, J. Natl. Cancer Inst., 57: 67-78 (1976).
39. S. C. Booth, H. Bosenberg, R. G. Garner, P. J. Hertzog, and K. Norpoth, The Activation of Aflatoxin B_1 in Liver Slices and in Bacterial Mutagenicity Assays Using Livers from Different Species Including Man, Carcinogenesis, 2:1063-1068 (1981).
40. G. D. Stoner, F. B. Daniel, K. M. Schenck, H. A. J. Schut, D. W. Sandwisch, A. F. Gohara, DNA Binding and Adduct Formation of Aflatoxin B_1 in Cultured Human and Animal Tracheobronchial and Bladder Tissues, Carcinogenesis, 3:1345-1348 (1982).
41. D. B. Clayson, in: The Biology and Clinical Management of Bladder Cancer (E. H. Cooper and R. E. Williams, eds.), Blackwell Scientific Publications, London, pp. 1-35 (1975).
42. W. F. Dunning, M. R. Curtis, and M. E. Maun, The Effect of Dietary Tryptophan and the Occurrence of 2-Acetylaminofluorene-Induced Liver and Bladder Cancer in Rats, Cancer Res., 10:454-459 (1950).
43. M. Wood, Factors Influencing the Induction of Tumors of the Urinary Bladder and Liver by 2-Acetylaminofluorene in the Mouse, Europ. J. Cancer, 5:41-47 (1969).
44. D. B. Clayson and R. C. Garner, in: Chemical Carcinogens (C. E. Searle, ed.), American Chemical Society, Washington, D.C. (1976).
45. R. A. M. Case, M. E. Hosker, D. B. McDonald, and T. J. Pearson, Tumor of the Urinary Bladder in Workmen Engaged in the Manufacture and Use of Certain Dyestuff. Intermediates in the British Chemical Industry. The role of Aniline, Benzidine, Alpha-Naphtylamine and Beta-Naphtylamine, Br. J. Ind. Med., 11:75-104 (1954).
46. IARC Monograph on the Evaluation of the Carcinogenic Risk of Chemicals to Man, Vol. 4, International Agency for Research on Cancer, Lyon, pp. 3-127 (1974).
47. F. B. Daniel, G. D. Stoner, and H. A. J. Schut, Unpublished Observations (1982).
48. G. D. Stoner, C. C. Harris, H. Autrup, B. F. Trump, E. W. Kingsbury, and G. A. Meyers, Explant Culture of Human Peripheral Lung, I. Metabolism of Benzo(a)pyrene, Lab. Invest., 38:685-692 (1978).

49. H. Autrup, C. C. Harris, G. D. Stoner, J. K. Selkirk, P. W. Schafer, and B. F. Trump, Metabolism of [^{3}H]Benzo(a)pyrene by Cultured Human Bronchus and Cultured Pulmonary Alveolar Macrophages, Lab. Invest., 38:217-224 (1978).
50. G. M. Cohen, R. Mehta, and M. Merdith-Brown, Large Interindividual Variations in Metabolism of Benzo(a)pyrene by Peripheral Lung Tissue from Lung Cancer Patients, Int. J. Cancer, 24:129-133 (1979).
51. H. L. Gurtoo and L. V. Dave, In vitro Metabolic Conversion of Aflatoxins and Benzo(a)pyrene to Nucleic Acid-Binding Metabolites, Cancer Res., 35:382-389 (1975).
52. C. C. Harris, H. Autrup, G. D. Stoner, E. M. McDowell, B. F. Trump, and P. Schafer, Metabolism of Dimethylnitrosamine and 1,2-Dimethylhydrazine in Cultured Human Bronchi, Cancer Res., 37:2309-2311 (1977).
53. R. W. Pero, C. Bryngelsson, F. Mitelman, T. Thulin, and A. Nordon, High Blood Pressure Related to Carcinogen-Induced Unscheduled DNA Synthesis, DNA Carcinogen Binding and Chromosomal Aberrations in Human Lymphocytes, Proc. Natl. Acad. Sci., U.S.A., 73:2496-2500 (1976).

INTRINSIC FACTORS THAT CAN AFFECT SENSITIVITY TO CHROMOSOME ABERRATION INDUCTION*

R. Julian Preston

Biology Division
Oak Ridge National Laboratory
Oak Ridge, Tennessee 37830

1. Introduction

The question to be addressed in this paper, namely are there individuals who are hypersensitive, or who are more likely to be hypersensitive, to the induction of chromosome aberrations by radiation and chemicals, is, at this time, difficult to answer. Furthermore, if such individuals can be identified there are social and legal implications concerning what action, if any, should be taken as far as their occupational or medical status is concerned. This will not be discussed here, as it is felt that such a topic is outside the domain of the author, a cytogeneticist. While the emphasis will be on factors that could contribute to an increased sensitivity, it is also appropriate to consider if there are also factors that could contribute to a decreased sensitivity to chromosome aberration induction. Some of the discussion regarding DNA repair will be abbreviated, because much more detail can be found in the paper given by Dr. James Cleaver at this same Workshop.

2. Identifiable Syndromes Conferring Increased Sensitivity to Chromosome Aberration Induction or Cell Killing

There are a large number of published studies of the increased sensitivity to chromosome aberration induction, cell killing, or mutation induction of cells from persons having recognizable syn-

*Research sponsored by the Office of Health and Environmental Research, U.S. Department of Energy, under contract W-7405-eng-26 with the Union Carbide Corporation.

domes, that are associated with DNA repair or replication defects. These syndromes include xeroderma pigmentosum, ataxia telangiectasia, Bloom's syndrome, Fanconi's anemia, Cockayne's syndrome, and Rothmund-Thompson syndrome. These probably represent the extremes of increased sensitivity, and as such have been of value in providing information on DNA repair mechanisms, and the relationship between DNA repair or replication defects and other biological end-points. However, when considering individuals that might have a higher than normal sensitivity to chromosome aberration induction from occupational exposures to chemical or physical agents, they are unlikely to be a contributing factor. All these syndromes are clearly recognizable, generally at an early age, with early death being fairly common. Thus, the presence of such individuals in an occupational group is either unlikely or clearly identifiable.

3. Increases in Sensitivity of Persons Hetrozygous for DNA Repair or Replication Defects

The frequency of persons in the population with identifiable DNA repair or replication defects is very low. However, since the genes responsible for the defects are recessive, it can readily be seen that the frequency of persons who will be heterozygous for these genes can be quite high - of the order of 1% of the population (Swift et al., 1976). It will, therefore, be important to know if these heterozygotes have an increased sensitivity to chromosome aberration induction by chemical or physical agents when compared to the sensitivity of some control group. It is appreciated that any control group could contain either unidentified heterozygotes for the same repair defects, or individuals with increased sensitivities for unsuspected reasons. However, if a control group is fairly large, this problem is to some extent alleviated.

For obvious reasons, the heterozygotes that have been studied to date are obligate ones, i.e., they are the parents of individuals who have identifiable syndromes that are considered to be due to DNA repair or replication defects. Furthermore, the amount of information is small. Paterson et al. (1979) reported an increased sensitivity to cell killing by γ-rays given under hypoxic conditions for three of five presumed ataxia telangiectasia (AT) heterozygotes. The sensitivity was intermediate between that for children with AT and normal individuals. For two of the presumed heterozygotes the sensitivity was the same as that for normal individuals. The three heterozygotes of intermediate sensitivity to cell killing also carried out DNA repair replication at a level intermediate between normal and AT individuals. However, it is noteworthy that there was no difference in sensitivity between the AT heterozygotes and normal individuals when irradiations were given in oxic conditions. These results indicate some increase in sensitivity to cell killing by γ-rays under hypoxic conditions for some AT heterozygotes. Similar results have also been reported by Chen et al. (1978) and Arlett and Lehmann (1978).

It has also been reported that the spontaneous chromosome aberration frequency in AT heterozygotes is not significantly higher than for normal individuals, whereas it is often somewhat elevated in persons having AT. In addition, the increase in aberration frequency following radiation exposure is similar to that in normal individuals, whereas AT individuals are markedly more sensitive to x-ray induced chromosome aberration induction (Paterson and Smith, 1978; Harnden, 1978).

Ivanova et al. (1980) analyzed chromosome aberrations in lymphocytes of AT heterozygotes and normal individuals following treatment with thiophosphamide. The results indicated that there was no overall difference in aberration frequency between the two groups. The authors suggested that there was a significant increase in the frequency of aberrant cells in the AT heterozygotes (not manifest as an increase in aberration frequency on a per cell basis), but it would appear that further analysis is needed before this can be considered conclusive.

Auerbach et al. (1980) measured the frequency of chromosome aberrations in the lymphocytes from individuals with Fanconi anemia (FA), heterozygotes for FA, and normal individuals following *in vitro* treatment with diepoxybutane. The spontaneous frequency of chromosome aberrations in FA heterozygotes and normal individuals was identical, and considerably lower than that for FA individuals. The frequencies of aberrations induced by diepoxybutane were significantly elevated in some of the eleven FA heterozygotes compared to those in normal individuals, but the mean frequency was some 50 times lower than that for FA homozygotes. This was a fairly small study, and only sensitivity to diepoxybutane was measured. Thus more data are needed on FA heterozygotes, and heterozygotes for other disorders in order for any general conclusion to be drawn.

It seems to be clear that even in those few cases where an increased sensitivity of a heterozygous condition is indicated, it is not consistent, i.e., only some heterozygotes for a specific condition appear to have an increased sensitivity. Clearly more work in this area is needed. It is also particularly important to note that the studies to date on specific heterozygotes are for obligate heterozygotes, and where the homozygous condition is known to be especially sensitive. The question of particular importance is whether or not it is possible to identify persons who might be only very slightly more sensitive to chromosome aberration or mutation induction by specific agents, and for whom there is no available genetic information that would indicate such an increased sensitivity. If such individuals are identifiable, then with a sufficient quantity of information it might be possible to determine what intrinsic factors would be responsible for an increased sensitivity to aberration or mutation induction. In parallel with such studies it would also be essential to determine what extrinsic factors could also influence the sensitivity of individuals to chemical or physical agents.

4. The Mechanism of Induction of Chromosome Aberrations and Differences in Sensitivity

It is very difficult to determine the best, or even a suitable approach for identifying persons who might show an increased sensitivity to the induction of chromosome aberrations by chemical or physical agents, and subsequently to determine if there are intrinsic factors that can influence this sensitivity. The approach we have taken as a starting point is to try and better understand the mechanism of induction of chromosome aberrations, and then begin to determine what steps in the induction process are subject to alteration, or variation and if such alterations or variations could result in increases or decreases in sensitivity to aberration induction.

Our initial results on the use of DNA repair inhibitors for studying the mechanism of aberration induction by radiation and chemicals have been published (Preston, 1980; Preston, 1981; Preston and Gooch, 1981; Preston, 1982a; Bender and Preston, 1982; Preston, 1982b) and will not be presented in detail here. The conclusion from these studies are, however, important for the subsequent discussion.

It was shown that cytosine arabinoside (ara-C) can inhibit excision repair following UV, X-ray, or chemical treatment. This inhibition probably occurs at the polymerase step, and results in single strand gaps in the DNA (Hiss and Preston, 1977; Dunn and Regan, 1979). The inhibition can be reversed by the addition of deoxycytidine. If the single strand gaps accumulated by ara-C incubation, following X-irradiation for example, can interact to form aberrations, on reversing the inhibition with deoxycytidine, then the aberration frequency would be expected to be greatly enhanced for cells X-irradiated and incubated with ara-C, compared to that in cells X-irradiated but with no post-treatment with ara-C. This was the rationale behind the experiments. If the time required to repair all the induced DNA damage that can result in aberrations is long, of the order of an hour or more, then the aberration frequency should increase with increasing duration of post-treatment ara-C incubations.

All the chromosome aberration studies reported here utilized human lymphocytes, largely because they represent a rather synchronous population before and after mitogenic stimulation, at least for the first *in vitro* cell cycle, that was all that was necessary for these experiments. The first *in vitro* cell cycle is also long (about 44-48 h) and so it is possible to carry out post-treatment incubations in ara-C for several hours without the complication of cells progressing into subsequent stages of the cell cycle (particularly G_1 to S).

Unstimulated (G_0) lymphocytes were irradiated with 200 rad of X-rays and incubated with ara-C for 0, 1, 2, or 3 hours. The ara-C inhibitory action was reversed by the addition of deoxycytidine, and the cultures were incubated for 48 hours, following mitogenic stimulation with phytohemagglutinin (pha). The frequency of chromosome-type aberrations was considerably enhanced when cells were incubated with ara-C after X-irradiation, and the increase was approximately linear with increasing ara-C incubation times (Preston, 1980). These results show that ara-C inhibits the repair of DNA damage that is converted into aberrations, and that the damage resulting in aberrations takes 3 hours or more to be completely repaired. Since ara-C inhibits excision repair, and not the repair of directly induced strand breaks (Hiss and Preston, 1977), and because the aberration frequency increases with ara-C incubation times up to 3 hours, it was concluded that X-ray-induced base damage was the DNA lesion that was converted into aberrations during the repair process. In order to obtain an aberration a single strand induced event has to be converted into a double strand gap during repair and this can be accomplished either by a single strand nuclease (Natarajan and Obe, 1978) or by having coincident base damages on the two strands of the DNA double helix (Ahnstrom and Bryant, 1982).

These results led us to propose a general hypothesis for the induction of chromosome aberrations (Preston, 1980). An aberration can be formed by the interaction (or misrepair) of two repairing lesions. The repair of these two lesions must be coincident in time and they must be close together spatially (within the rejoining distance) for an aberration to be formed. The probability of obtaining an aberration will be dependent upon the likelihood of having coincident lesions. This in turn will be dependent upon the amount of the specific DNA damage that can be converted into aberrations, and upon the rate of repair of this damage - the more rapid the repair the greater the probability of coincident repairing lesions.

It can be seen that this hypothesis will explain the induction of aberrations by chemical agents, and also the aberration spectrum and cell cycle variations observed with most chemical agents. If cells are treated in G_1 with the majority of chemical agents, the aberrations observed at the first post-treatment metaphase are of the chromatid-type, and are formed during or after DNA synthesis.

If the reapir of chemically-induced damage that can be converted into aberrations is slower than that for X-ray-induced damage, then the probability of obtaining coincident lesions will be lower. Thus, the probability of producing aberrations in G_1 cells is low. As a consequence of the replication of damaged DNA, the probability of two lesions being coincident during or after DNA synthesis is much higher than in G_1 cells, and aberrations can be formed - these will be of the chromatid-type.

Pha-stimulated lymphocytes were treated with 4-nitroquinoline-N-oxide (4NQO) and methyl methanesulfonate (MMS) in G_1 (9 hours after stimulation), and no aberrations were observed. If cells were treated with 4NQO or MMS in G_1 and then incubated with ara-C for 6 hours aberrations were observed. These aberrations were unequivocally of the chromosome type, and had to have been formed during G_1, i.e., prior to DNA synthesis (Preston and Gooch, 1981). Thus it can be concluded that ara-C inhibits the repair of 4NQO or MMS-induced damage that can be converted into aberrations, accumulating single-strand gaps that can interact to form aberrations on reversing the ara-C inhibition with deoxycytidine. The probability of obtaining coincidentally repairing regions is greatly increased by essentially accumulating 6 hours of repairing regions by ara-C inhibition. If this probability is increased, then aberrations can be produced in G_1 following chemical treatment as evidenced by the observation of chromosome-type aberrations. There is supportive evidence for this hypothesis, since it has been shown that the repair of 4NQO- and MMS-induced DNA damage is inhibitible by ara-C, that this repair is of the "long-patch" type, and that the duration of repair of this chemically induced damage is considerably longer than that for X-rays (Waters et al., 1982; Snyder and Regan, 1982).

If chromosome aberrations are indeed produced by the interaction of two coincidentally repairing regions, then it would be predicted that if the rate of repair of the damage leading to aberrations is more rapid in a particular cell type then the aberration frequency would be higher in that cell type at any dose level. As a possible example of such a situation we utilized lymphocytes of Down individuals. It was reported by Sasaki and Tonomura (1969) and Evans and Adams (1970) that the frequencies of X-ray-induced chromosome-type aberrations in G_0 lymphocytes of Down individuals were about 1.6 times as high as the frequencies in normal lymphocytes. Since the amount of X-ray-induced DNA damage is expected to be the same because the DNA content of the Down cells is only about 1% higher than in normal cells, it seemed plausible that some differences in rate of repair of the damage leading to aberrations could account for this increased sensitivity of Down cells.

Unstimulated lymphocytes from Down individuals were X-irradiated and incubated with ara-C for 0, 1/2, 1, or 2 hours. It was found that the increase in frequency of chromosome-type aberrations as a function of ara-C incubation time was more rapid in Down cells than normal cells (Preston, 1981). Thus, for example, the frequency of dicentrics in Down cells for 1 hour post-treatment in ara-C is about the same as the frequency after 2 hours ara-C incubation for normal cells. Thus, the repair of X-ray-induced DNA damage, that can be inhibited by ara-C, and that can be converted into chromosome aberrations, appears to be more rapid in Down G_0 lymphocytes than normal lymphocytes. This more rapid repair will increase the probability of having coincidentally repairing regions that are able to interact to form aberrations. A possible explanation for differences in the rate of repair of X-ray-induced DNA damage, that is converted into aberrations, in Down and normal lymphocytes has been offered by Leonard and Merz (1983). They have shown that the G_0 Down lymphocytes behave as though partially stimulated, i.e., the time of entry into the S-phase, following mitogenic stimulation, and the first cell cycle in Down cells are shorter than in normal cells. It is also important to note that Down cells are equally sensitive to aberration induction when irradiated in G_0 or G_1, following stimulation, whereas normal lymphocytes are more sensitive when irradiated in G_1 compared to irradiation in G_0, i.e., a G_1-irradiated normal lymphocyte has the same sensitivity to aberration induction as Down lymphocytes irradiated in G_1 or G_0 (Lambert et al., 1976; Leonard and Merz, 1983).

These results indicate that aberration frequency can be influenced by the rate of repair of the specific DNA damage that is converted into chromosome aberrations - the more rapid the rate of repair, the higher the probability of having coincidentally repairing regions, and the higher the probability of producing an aberration. It is possible, therefore, to argue that differences in sensitivity to chromosome aberration induction by chemical and physical agents can be reflective of differences in the rate of repair of the specific DNA damage that results in aberrations.

The preceding discussion relates rather specifically to how differences in the rate of repair of DNA damage induced in G_1 cells could influence the frequency of chromosome-type aberrations. However, it can also be argued that an increase in this rate of repair could decrease the frequency of aberrations, specifically chromatid-type aberrations formed during or after DNA replication. If, for example, the DNA damage induced in cells prior to replication is repaired more rapidly than normal (or average), less of this damage will be present at the time of replication, and thus the probability of producing a chromatid-type aberration will be lower. The converse would be expected for a slower rate of repair than normal. Therefore, for a more rapid repair of damage than normal, the probability of producing chromosome-type aberrations is increased, but the probability of producing chromatid-type aberrations will be decreased, and vice versa for a slower than normal rate of repair. The transmission and possible consequences of the different aberration types may be quite different, and so increases or decreases of different types could be very important.

As discussed in Sections 2 and 3, increased sensitivity to aberration induction can be the result of the presence of defective

DNA repair or replication processes. It was considered that such defects represent the extreme of sensitivity. The influences of perhaps small changes in the rate of repair of DNA damage in apparently normal individuals would account for much smaller differences in sensitivity. Furthermore, it is the consideration of small differences in sensitivity that is probably the most important for estimating potential consequences of occupational or environmental exposures, because there is unlikely to be a phenotypic characteristic that will identify such slightly sensitive individuals. Clearly the hypothesis presented needs further study, particularly with regard to attempting to relate differences in rate of repair of specific DNA damage with sensitivity to aberration induction for a range of chemical and physical agents. It is perhaps important to add here that differences in rate of repair could also account, in part, for differences in sensitivity between the same cell type in different species, or between different cell types in the same species. Such considerations could well be valuable when attempting to assess potential health effects of exposure to chemical or physical agents, where extrapolation generally have to be made from data obtained with laboratory animals to man, or from somatic to germ cells.

It is appreciated that the preceding discussion deals with only one possible intrinsic factor that could influence sensitivity. However, it is proposed simply as a first step towards trying to understand what is clearly a most complex problem. It is hoped that by determining the mechanism of induction of chromosome aberrations, that it will be possible to learn where to start looking for factors that could influence this process.

REFERENCES

G. Ahnstrom and P. E. Bryant, DNA double-strand breaks generated by the repair of X-ray damage in Chinese hamster cells, Int. J. Radiat. Biol., 41:671 (1982).

C. F. Arlett and A. R. Lehmann, Human disorders showing increased sensitivity to the induction of genetic damage, Annu. Rev. Genet., 12:95 (1978).

A. J. Auerbach, B. Adler, and R. S. K. Chaganti, Fanconi anemia: Pre- and post-natal diagnosis and carrier detection by a cytogenetic method, Am. J. Hum. Genet., 32, Abstract 187, 61A (1980).

M. A. Bender and R. J. Preston, Role of base damage in aberration formation: interaction of aphidicolin and X-rays, in: "Progress in Mutation Research, Vol. 4" (A. T. Natarajan, G. Obe, and H. Altmann, eds.), Elsevier Biomedical Press, Amsterdam, pp. 37-46 (1982).

P. C. Chen, M. F. Lavin, and C. Kidson, Identification of ataxia telangiectasia heterozygotes, a cancer prone population, Nature (London), 274:484 (1978).

W. C. Dunn and J. D. Regan, Inhibition of DNA excision repair in human cells by arabinofuranosyl cytosine: effect on normal and xeroderma pigementosum cells, Molec. Pharmacol., 15:367 (1979).

H. J. Evans and A. Adams, X-ray-induced chromosome aberrations in human lymphocytes irradiated in vitro: the influence of exposure conditions, genotype and age on aberration yields, in: "Advances in Radiation Research, Vol. 1" (J. F. DuPlan and A. Chapiro, eds.), Gordon and Breach, London, pp. 335-348 (1970).

D. G. Harnden, Ataxia telangiectasia syndrome: cytogenetic and cancer aspects, in: "Chromosomes and Cancer" (J. German, ed.), John Wiley and Sons, Inc., New York, pp. 619-636 (1974).

E. A. Hiss and R. J. Preston, The effect of cytosine arabinoside on the frequency of single-strand breaks in mammalian cells following irradiation or chemical treatment, Biochim. Biophys. Acta, 478:1 (1977).

Y. E. Ivanova, N. P. Bochkov, and K. N. Yakovenko, Effect of thiophosphamide on the frequency of chromosomal aberrations in ataxia telangiectasia heterozygotes, Bull. Exptl. Biol. and Med., 89:977 (1980).

B. Lambert, K. Hansson, T. H. Bui, F. Funes-Cravioto, J. Lindsten, M. Holmberg, and B. Strausmanis, DNA repair and frequency of X-ray and UV-light induced chromosome aberrations in leukocytes from patients with Down's syndrome, Ann. Hum. Genet., 39:293 (1976).

J. C. Leonard and T. Merz, The relationship of cell kinetics to chromosome radiosensitivity in trisomy 21, Mutat. Res., 109: 111 (1983).

A. T. Natarajan and G. Obe, Molecular mechanisms involved in the production of chromosomal aberrations, I. Utilization of Neurospora endonuclease for the study of aberration production in G_2 stage of the cell cycle, Mutat. Res., 52:137 (1978).

M. C. Paterson, A. K. Anderson, B. P. Smith, and P. J. Smith, Enhanced radiosensitivity of cultured fibroblasts from ataxia telangiectasia heterozygotes manifested by defective colony-forming ability and reduced DNA repair replication after hypoxic γ-irradiation, Cancer Res., 39:3725 (1979).

M. C. Paterson and P. J. Smith, Ataxia telangiectasia: An inherited human disorder involving hypersensitivity to ionizing radiation and related DNA-damaging chemicals, Annu. Rev. Genet., 13: 291 (1979).

R. J. Preston, The effect of cytosine arabinoside on the frequency of X-ray-induced chromosome aberrations in normal human leukocytes, Mutat. Res., 60:71 (1980).

R. J. Preston, X-ray-induced chromosome aberrations in Down lymphocytes: an explanation of their increased sensitivity, Environ. Mutagen., 3:85 (1981).

R. J. Preston, DNA repair and chromosome aberrations: Interactive effects of radiation and chemicals, in: "Progress in Mutation Research, Vol. 4" (A. T. Natarajan, G. Obe, and H. Altmann, eds.), Elsevier Biomedical Press, Amsterdam, pp. 25-35 (1982a).

R. J. Preston, The use of inhibitors of DNA repair in the study of the mechanisms of induction of chromosome aberrations, Cytogenet. Cell Genet., 33:20 (1982b).

R. J. Preston and P. C. Gooch, The induction of chromosome-type aberrations in G_1 by methyl methanesulfonate and 4-nitroquinoline-N-oxide, and the non-requirement of an S-phase for their production, Mutat. Res., 83:395 (1981).

M. S. Sasaki and A. Tonomura, Chromosomal radiosensitivity in Down's syndrome, Jap. J. Hum. Genet., 14:81 (1969).

R. D. Snyder and J. D. Regan, DNA repair in normal human and xeroderma pigmentosum Group A fibroblasts following treatment with various methanesulfonates and the demonstration of a long patch (U.V.-like) repair component, Carcinogenesis, 3:7 (1982).

M. Swift, L. Sholman, M. Perry, and C. Chase, Malignant neoplasms in the families of patients with ataxia telangiectasia, Cancer Res., 36:209.

R. Waters, R. Mirzayans, J. Meredith, G. Mallalah, N. Danford, and J. M. Parry, Correlations in mammalian cells between types of DNA damage, rates of DNA repair, and the biological consequences, in: "Progress in Mutation Research, Vol. 4" (A. T. Natarajan, G. Obe, and H. Altmann, eds.), Elsevier Biomedical Press, Amsterdam, pp. 247-259 (1982).

INDIVIDUAL VARIABILITY IN THE FREQUENCY OF SISTER CHROMATID EXCHANGE IN HUMAN LYMPHOCYTES*

Bo Lambert

Department of Clinical Genetics
Karolinska Hospital
104 01 Stockholm, Sweden

and

Robert Olin

Royal Institute of Technology
100 44 Stockholm and ASTRA Inc.
151 85 Södertälje, Sweden

INTRODUCTION

Sister chromatid exchange (SCE) can be demonstrated in metaphase cells by autoradiographic techniques [1] or by more recently developed staining techniques [2]. Although the mechanism(s) and the nature of the alterations leading to SCE are largely unknown, there are a number of observations suggesting a relationship between mutagenic and/or carcinogenic damage and an increased rate of SCEs [3]. Firstly there is a linear correlation between the induction of SCE and single gene mutation for many chemical agents, and both of these events show a similar spontaneous rate of occurrence over the entire genome [4]. Secondly, a great number of mutagens and carcinogens have been shown to induce SCEs, in some cases at very low concentrations, in various mammalian cells *in vitro* as well as *in vivo* [5, 6]. Finally, the induction of SCE is S-phase dependent, and appears to be closely related to DNA replication [3].

Alkylating agents and a variety of DNA-binding compounds are particularly effective in inducing SCE and do so at concentrations

*Work supported by the Swedish Work Environmental Fund, the Swedish Medical Research Council and the Swedish Council for Planning and Coordination of Research.

much lower than required to detect chromosome aberrations, whereas other clastogenic agents producing DNA strand breaks, e.g., X-rays and bleomycin, induce very few, if any, detectable SCEs [5, 6]. These and other observations suggest that the nature of the lesions and the mechanism(s) responsible for the formation of conventional chromosome aberrations are different from those leading to SCE [3].

Several comprehensive reviews dealing with various aspects on the formation and detection of SCE have appeared recently [5, 7-11]. The SCE assay is generally considered to be an easy and rapid method for the study of chemically induced genetic toxicity *in vitro* [5], and it can be used for *in vivo* studies in a number of animal cells and tissues, including germ cells [12]. The assay is also useful for the monitoring of human populations exposed to genotoxic factors in the environment by the analysis of SCE in short term cultured human lymphocytes [13]. The present paper deals mainly with differences between human subjects in the individual levels of SCE in peripheral lymphocytes, and with the various methodological, biological, and environmental factors that may be responsible for these differences.

METHODOLOGICAL FACTORS INFLUENCING THE SCE LEVELS IN CULTURED HUMAN LYMPHOCYTES

Freshly collected blood lymphocytes from healthy donors are readily available for the study of SCE induction by chemical agents *in vivo* as well as *in vitro*. In order to produce differentially stained sister chromatids, the cells are grown for two replication cycles in the presence of bromodeoxyuridine (BrdUrd) [2, 14]. Only SCEs originating during this short-term *in vitro* cultivation are detected, whereas SCEs formed *in vivo* in replication cells will not be visible. Accordingly, an increase of SCEs in human lymphocytes exposed *in vivo* must be due to either remaining activity of the exposing agent in the cell fraction or to persistent lesions in the cellular DNA.

A number of conditions related to the culture techniques have been shown to influence the SCE levels. BrdUrd is in itself an SCE-promoting agent, and may cause a twofold increase of SCEs over the practically used concentration range between 10 to 100 μM [13, 14]. At higher concentrations the dose-dependent increase of SCEs levels off [15]. Most workers therefore choose to work in the low concentration range, or at the apparent plateau level.

Both the growth medium [16] and the type of serum [13, 17] used in the culture have been shown to influence the SCE level. This may partly reflect differences in the cellular ability to utilize BrdUrd under these various culture conditions. Also the rate of cell proliferation has been shown to be of importance, as significantly lower

SCE levels were found in rapidly replicating cultures than in slowly replicating ones [13, 18].

The most frequently used staining techniques to visualize SCEs are the direct Hoechst 33258 fluorescence method [2] and the FPG-(fluorescence plus Giemsa)-method [19], which appear to yield very similar results [14]. Other available staining methods have either given the same [13] or a slightly lower [16] incidence of SCE than the abovementioned techniques. Thus, the scoring of SCEs appears to be relatively reproducible and safe, provided that the concentration of BrdUrd is kept constant and suitable stained preparations are produced. In contrast, a number of less discrete parameters related to the culture conditions have a considerable influence and may cause an approximately two-fold variation in the incidence of SCEs in human lymphocyte cultures. It is therefore important to standardize the culture conditions as much as possible, and, where appropriate to use concurrent as well as historical controls.

INDIVIDUAL VARIATIONS IN THE BASELINE LEVELS OF SCE

Even though methodological factors such as those mentioned above are carefully controlled, several workers have been able to demonstrate significant differences in the baseline levels of SCEs in short-term cultured human lymphocytes from different individuals [13, 15, 16, 20]. Usually, the individual SCE level is determined from the average number of SCEs in 20-40 metaphases in a single cell culture from each donor. The range of individual levels of SCE, as well as the group mean levels, may differ greatly between laboratories, mainly because of the differences in culturing techniques as discussed above. However, also within laboratories a two-fold difference in the individual SCE levels is often encountered. As shown in Fig. 1a, the difference in SCE levels between 2-3 replicate cultures from the same donor is usually small and insignificant, giving account for the standardized methodological procedure. When the SCE analysis is repeated one or several times in the same donor at monthly intervals or longer, the differences tend to increase, but the differences between subjects are nevertheless conserved to a great extent (Fig. 1b).

Attempts to explain these individual differences by the influence of sex or age have not been successful so far (Fig. 2). Although females in many studies show slightly higher SCE levels than males [13, 20, 21] these differences have not been statistically significant, and have not been confirmed in other studies [22]. Similarly, age in adults has not proven to influence the individual SCE levels [20, 23, 24], except in one study which showed a small but statistically significant increase of SCEs in older adults as compared to younger ones [22]. In contrast, children [21, 23] and newborn [25] have shown SCE levels which are considerably lower than

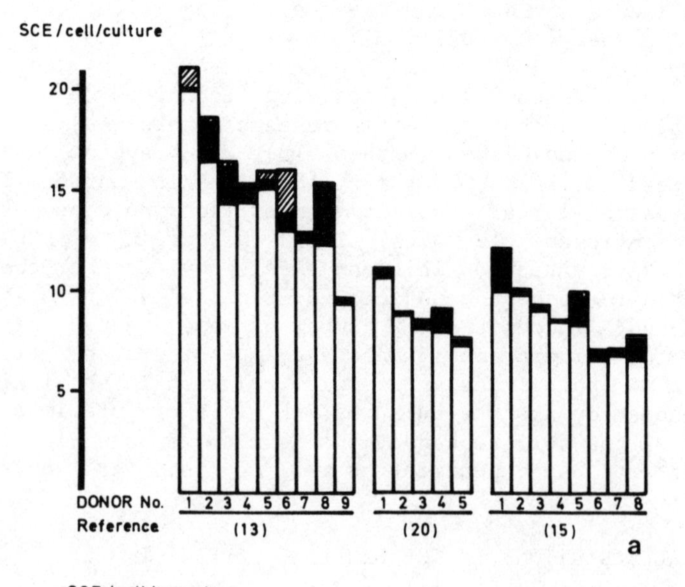

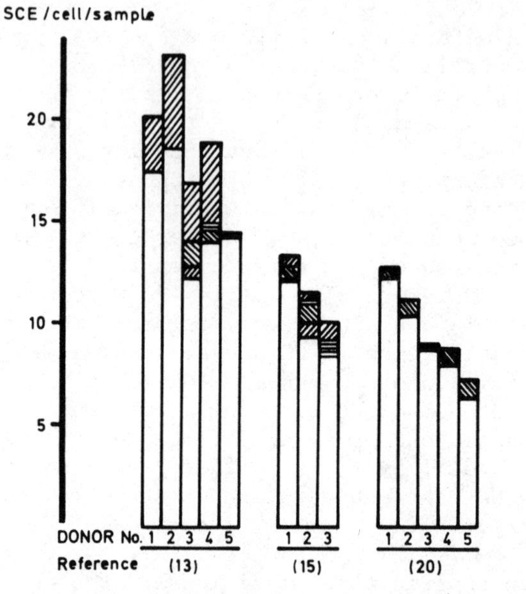

Fig. 1. Differences in SCE levels in replicate (a) and repeat (b) cultures from single donors. Replicate cultures were run in duplicate or triplicate and the repeat cultures were run 2-6 times at monthly intervals or longer. Horizontal bars, separated by open, filled, or hatched areas, indicate the SCE level in the various cultures from each donor. The figures in parenthesis refer to the original work in the reference list.

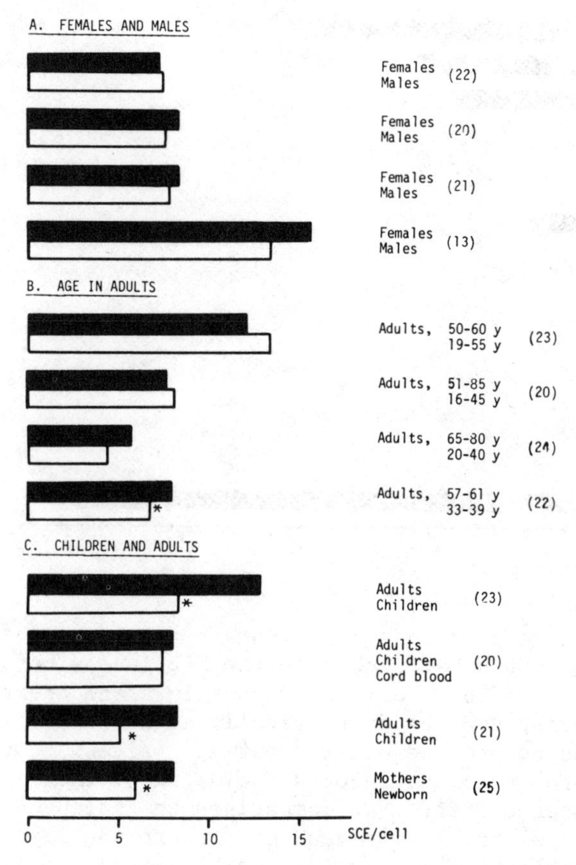

Fig. 2. SCE levels in groups of healthy males and females and in newborn, children, and adults at different ages. The figures in parenthesis refer to the original work in the reference list.

in adults, although one study failed to show a statistically significant difference [20].

The difference in SCE levels between children and adults (Fig. 2) may be due to a continuous accumulation of SCE forming lesions during life, but it is also possible that the lower SCE levels in young children are related to their different composition of lymphocyte subpopulations. Human B lymphocytes have signficantly lower SCE levels than T lymphocytes [18, 26]. However, this difference between T and B cells has no detectable influence on the SCE levels in conventional lymphocyte cultures from adult donors [18].

Thus, the differences in SCE levels between adult subjects seem to be relatively indpendent of age and sex. Moreover, twin studies

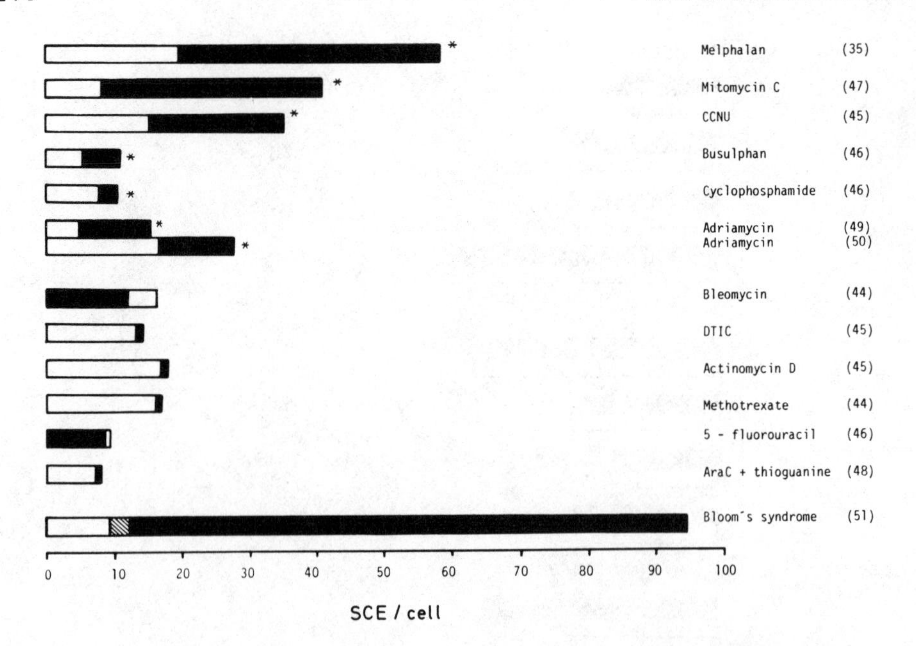

Fig. 3. SCE levels in cancer patients receiving cytostatic treatment. Open bars indicate the SCE levels before treatment, and closed bars during or shortly after treatment. An asterisk denotes statistically significant difference between before and after treatment values. For details confer the original reports indicated by their numbers in the reference list. For comparison are included the SCE levels observed in Bloom's syndrome, where the majority of cells show high SCE levels (closed bar) and a smaller proportion of the cells (hatched bar) show SCE levels close to that found in normal donors (open bar).

have not been able to disclose any influence of genetic factors on the individual SCE level [13], 22, 27]. Other biological, habitual, and environmental factors have therefore to be explored.

THE EFFECTS OF SMOKING, CYTOSTATIC TREATMENT AND OCCUPATIONAL EXPOSURE ON THE SCE LEVELS

The original observation of significantly higher SCE levels in smokers than in non-smokers [28] has recently been confirmed in a number of reports [21, 29, 32]. The SCE levels seem to be increased by only about 15-20% in moderate smokers and by 20-40% in heavy smokers [13, 28]. Accordingly, there is a considerable overlap in the range of individual SCE levels between groups of smokers and non-smokers, which may explain why some laboratories still do not observe the difference [22, 33, 34]. Yet, the SCE-promoting effect of smoking has been observed in healthy subjects with various occupations [21, 29, 30, 32] as well as in patients with various disorders [28, 35]. Thus, smoking seems to be one important factor causing individual differences in the SCE levels.

Studies on SCE levels in various inherited human disorders and acquired disease states have recently been reviewed by Evans [36]. Apart from the very significant increase of SCEs in Bloom's syndrome (Fig. 3) these studies have not given much information about the cause or origin of SCEs and they will not be commented upon further here. More relevant in this context are reports on the effect of different medical treatments on the SCE levels, which may provide more detailed information about SCE-inducing agents and the kinetics of induction and disappearance of SCEs *in vivo*.

The treatment of psoriasis with PUVA, a combination of oral 8-methoxypsoralen and whole body irradiation with long wave ultraviolet light (UVA), has been studied by several workers. In general, no PUVA-induced alterations in the SCE levels were found, neither after intensive daily treatments for several weeks, nor after sustained maintenance therapy during several months [37-39]. However, PUVA is indeed a very potent inducer of SCEs *in vitro*, and when blood from patients who have taken psoralen tablets is irradiated *in vitro* with the clinical UVA dose there is a significant increase of SCEs [37, 40, 41]. Thus, the serum concentration of psoralen *in vivo* is sufficiently high to induce SCEs if only the cells are reached by the UVA irradiation. The single, clinical UVA dose is apparently not large enough to induce a detectable increase of SCEs. Nevertheless, it was found that PUVA-treated psoriasis patients when studied 1-6 years after initiation of the therapy, showed a small but significant increase of SCEs in the peripheral lymphocytes [42]. This finding suggests that PUVA-induced DNA damage is accumulated in the circulating and non-dividing lymphocytes, and may eventually lead to a detectable increase of SCE also *in vivo*.

More pronounced increases in the SCE levels of individual patients have been found in cancer chemotherapy (Fig. 3). The malignant disorders *per se* appear not to be associated with any change in the levels of SCE, as these patients before therapy show SCE levels in the normal range [35, 43]. After single, clinical doses of directly acting alkylating cytostatics such as melphalan, mitomycin C, CCNU and busulphan there is a pronounced and highly significant increase of SCEs in the peripheral lymphocytes [33, 35, 43-47]. Similar, although less pronounced increases have also been demonstrated after treatment with cyclophosphamide, which undergoes metabolic activation [46, 48] and the intercalating and DNA-damaging agent adriamycin [46, 49, 50]. In contrast, bleomycin, which is not S-dependent in its mode of action, and other potent clastogenic and/or DNA-binding agents such as DTIC and actinomycin D, did not show any

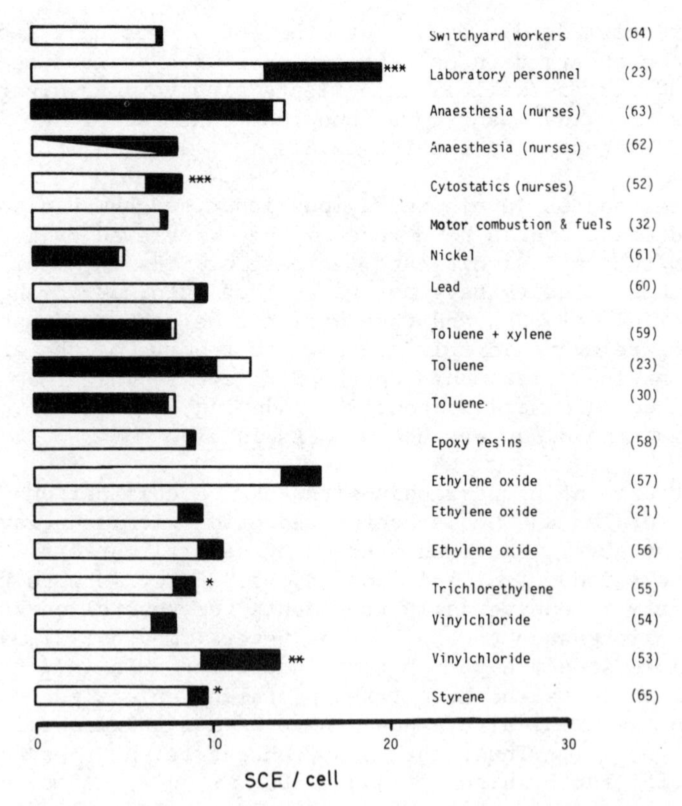

Fig. 4. SCE levels in groups of occupationally exposed subjects (closed vars) and the corresponding controls (open bars). Figures in parenthesis refer to the original publication in the reference list.

SCE-inducing effects, neither in vivo [35, 44, 45] nor in vitro [50]. Thus, the induction of SCEs by DNA-active cytostatic drugs in vivo seems to depend greatly on the type of DNA damage induced, and in fact, all of the drugs which so far have been shown to induce SCEs in vivo are potentially crosslinking agents (Fig. 3).

The increased SCE levels in patients treated with adriamycin [49] or alkylating cytostatics [35, 46, 48] usually reach a maximum within a few hours or days after drug administration, and then decline gradually during 1-8 weeks after a single administration. The kinetics of the disappearance of SCEs from the exposed cell population indicate that a relatively slow repair process may be active in removing the SCE-forming DNA lesions [35]. This notion is supported by findings after repeated administrations of alkylating cytostatics at 4 to 6 week intervals, when there is a gradual increase in the SCE levels, suggesting the accumulation of SCE-forming damage in the cells [35, 43].

Cytostatic agents which act by metabolic inhibition of DNA synthesis have not been found to induce SCEs _in vivo_. Thus neither methotrexate [44] and 5-fluorouracil [46], nor the combination of araC and thioguanine [48] give rise to a detectable increase of the SCE levels (Fig. 3).

In view of the pronounced increase of SCE caused by some cytostatic agents it was not unexpected to find a significant, although much smaller increase of the SCE levels in personnel working in oncology clinics [52]. A number of other occupationally exposed groups have also been studied (Fig. 4). Increased SCE levels were found in laboratory personnel exposed to a variety of chemicals in research and hospital institutions [23], and in factory workers exposed to vinyl chloride [53, 54], styrene [65] and trichloroethylene [55]. In addition, small, but statistically insignificant increases in the SCE levels have been reported in personnel exposed to ethylene oxide [21, 56, 57], whereas no significant changes were found in workers exposed to epoxy resins [58], in switchyard workers [64], nurses practicing inhalation anaesthesia [62, 63], truck and lorry drivers [32], nickel [61] and lead [60] exposed workers as well as in subjects in the printing and painting industry exposed mainly to toluene and xylene [23, 30, 59]. Obviously further studies are warranted in order to evaluate the possible effect of occupational exposure on the SCE levels.

THE DISTRIBUTION OF SCE IN A POPULATION OF EMPLOYEES IN A PHARMACEUTICAL INDUSTRY

The Population under Study

The subjects under study were employed in a pharmaceutical industry located in a community with about 75,000 inhabitants and situated 40 km south of Stockholm. More than 70 different medical drugs are produced at this plant, including antibiotics, local anaesthetics and analgesic drugs, tranquilizers, and drugs used in the treatment of cardiovascular disorders. The total number of employees exceed 2000, including factory workers, research and laboratory personnel, and administrative officials.

The SCE studies were coordinated with the concurrent health surveillance program of employees carried out by the occupational health services. A total number of 303 subjects were studied during two years. Blood samples were taken on the morning after at least one day of regular work, i.e., usually on Tuesdays, and at least one week of full work was allowed to pass after vacations before sampling was resumed. At the time of blood sampling each subject was asked to answer a written questionnaire relating to previous medical history and drug consumption, personal habits and working conditions. The SCE analyses were carried out as described previously [13]. The re-

TABLE 1. The Numbers of Subjects and the SCE Levels in Groups of Employees Occupied with Different Types of Work and in Different Departments

OCCUPATION GROUPS	NUMBER OF SUBJECTS							SCE/ cell group mean ± S.D.
	WORKING DEPARTMENTS						All units	
	FDV	IAC	AF	VIK	MF	HS		
Administrative officials	11	0	2	0	2	1	16	14,4± 2,5
Work supervisors	0	2	5	0	3	14	24	14,4± 3,0
Laboratory personnel	0	45	41	0	0	3	99	13,6± 2,8
Factory workers	0	0	0	27	0	68	95	14,3± 2,8
Construction workers	0	0	0	1	28	4	33	14,7± 2,5
Others*	0	1	5	5	0	25	36	13,3± 3,0
Total no. of subjects	11	48	53	33	33	115	293	
SCE/cell group mean ± S.D.	14,2 ± 2,0	13,7 ± 3,6	13,8 ± 2,6	15,2 ± 4,0	14,6 ± 2,8	14,0 ± 3,5		14,2 ± 3,0

*See Tables 3 and 5.

sults are still under evaluation, and only some preliminary data related to individual and occupational factors will be presented.

Employees in six different departments were studied (Table 1). HS is the organic synthetic manufacturing plant. VIK is the department for manufacture of pharmaceutical products intended for injections. This highly clean working area contains packing lines where the use of silicon oil and ethyl acetate has been under suspicion of causing eye irritation. MF is the department responsible for general construction and maintenance works within the facilities. IAC is the chemical quality control laboratory where a variety of chemicals and analytical processes are handled. AF is a pharmacological research laboratory involved in the development of new penicillins and other synthetic and semisynthetic antibiotics. FDV is the scientific library and documentation unit of the company.

On the basis of statements in the questionnaires about work assignment and working conditions, the subjects were divided into six major occupation groups. The number of subjects and the mean levels of SCE in these groups are shown in Table 1. In ten cases the questionnaires were not complete or the cultures failed and were not repeated. The SCE frequency in the remaining population of 293 sub-

TABLE 2. Sister Chromatid Exchange (SCE) in the Peripheral Lymphocytes from Non-Smoking and Smoking Males and Females

Groups of subjects	SCE/cell Mean ± S.D.*	Age (y) Mean + S.D.	No of subjects
MALES			
Non-smokers	12.9 ± 2.8	42.1 ± 12.8	106
Smokers	15.2 ± 3.4	40.8 ± 11.9	78
FEMALES			
Non-smokers	13.7 ± 2.9	39.8 ± 10.7	75
Smokers	16.6 ± 3.1	35.8 ± 12.0	35

*The figures shown were calculated from individual means based on the numbers of SCE in 20 cells. Statistical analysis (t-test) showed the difference between non-smokers and smokers in each group of males and females to be significant ($P < 0.001$). The difference between smoking males and females was significant at $P < 0.05$, whereas the difference between non-smoking males and females was not significant ($t = 1,91$; $0.05 < P < 0.1$).

jects studied so far was found to be 14, 2 ± 3.0 SCE/cell, which is slightly lower but not significantly different from that observed in a group of 82 blood donors studied previously [13]. The group mean levels of SCE in the occupation groups ranged from 13.6-14.7 SCE/cell, and in the different department groups from 13.7-15.2 SCE/cell (Table 1).

The Influence of Sex, Age, and Smoking Habits on the SCE Levels

The SCE levels in the subpopulations of non-smoking and smoking males and females are shown in Table 2. In the non-smoking groups, males showed a slightly lower SCE-frequency than females, but the difference was not statistically significant. Smokers showed a highly significant increase of SCE compared to non-smokers in both groups of males and females ($P < 0.001$), and the SCE frequency in female smokers was significantly increased over that in male smokers ($P < 0.05$).

The mean age ranged between 35.8 and 42.1 years in the four groups (Table 2). Female smokers were found to have a significantly lower mean age than male smokers ($P < 0.05$), but there was no statistical differences in the age distribution between smokers and non-smokers within the groups of males and females respectively.

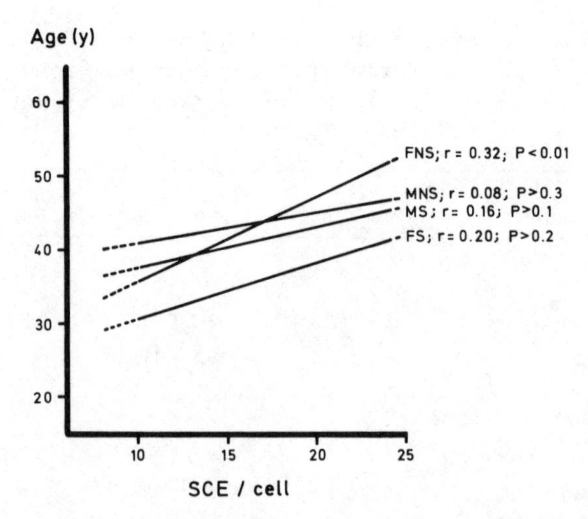

Fig. 5. Correlation between the SCE frequency and age. FNS and FS are female non-smokers and smokers respectively, and MNS and MS are the non-smoking and smoking males. The numbers of individuals in the groups were: FNS = 74; FS = 35; MNS = 106, and MS = 78. The lines represent the best fit of data by linear regression analysis (r = correlation coefficient).

The correlation between age and SCE-frequency was studied separately in the four groups of non-smoking and smoking males and females. A weak positive correlation was found in all four groups but statistical significance was obtained only in the group of non-smoking females (Fig. 5).

These results show that smoking habits must be taken into account in the further evaluation of individual SCE levels, and that sex as well as age may influence the individual SCE frequency in this population, albeit to a smaller extent.

The Frequency Distribution of Individual SCE Levels and of SCE Numbers in Single Cells

The frequency distributions of individual SCE levels in non-smoking males and females are shown in Fig. 6. There was no significant deviation from the normal distribution in any of the groups, although there seemed to be a slight under representation of subjects with very low SCE-levels, and a corresponding over representation of high SCE levels.

The numbers of SCE in single cells from the entire population of non-smokers was found to range more than tenfold, from 3 to 33.

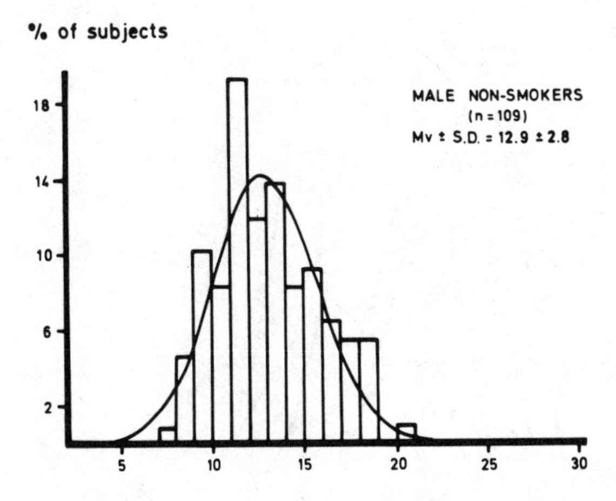

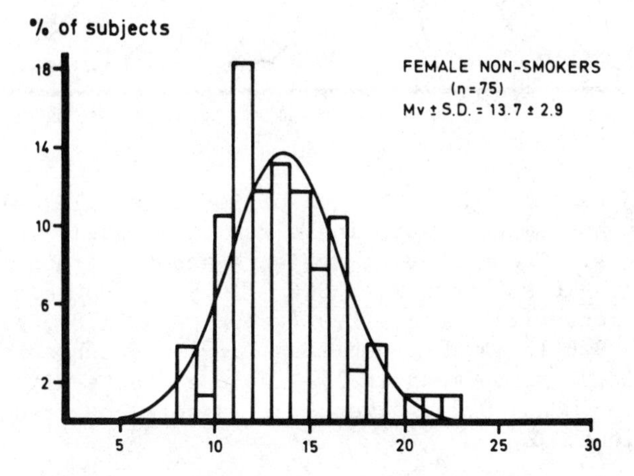

Fig. 6. The frequency distribution of the individual SCE frequency in peripheral lymphocytes from male non-smokers and female non-smokers. The solid line represents the theoretical normal distribution.

In groups of subjects with similar individual SCE levels, the numbers of SCE/cell showed a more narrow distribution. As the group mean SCE levels increased, cells with high SCE counts became relatively more overrepresented (Fig. 7). However, no clear biphasic distribution was observed, and consequently no evidence was obtained for the presence of discrete subpopulations of cells with markedly different SCE-rates in these lymphocyte cultures.

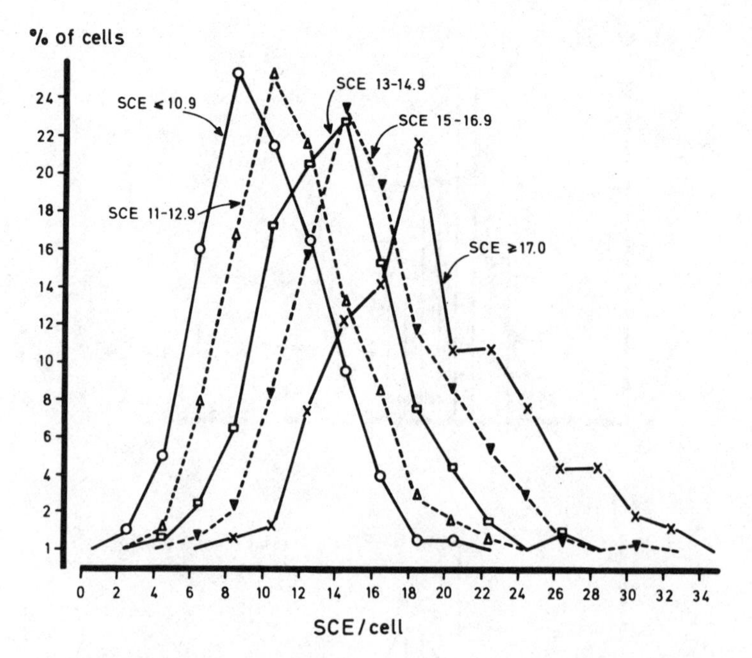

Fig. 7. The frequency distribution of SCE in single cells from female non-smokers with individual SCE levels in the indicated ranges. The number of subjects and the group mean levels were: 10 subjects with SCE $\leqslant$ 10.9 (group mean 10.1), 25 subjects with SCE 11-12.9 (group mean 11.8), 17 subjects with SCE 13-14.9 (group mean 13.9), 15 subjects with SCE 15-16.9 (group mean 16.1), and 8 subjects with SCE $\geqslant$ 17.0 (group mean 19.2). Twenty cells were analyzed from each subject.

The SCE Levels in Groups of Subjects with Different Occupations and Working Conditions

Because of the obvious influence of smoking on the individual SCE-levels, further analysis of the relationship between SCE-levels and occupational factors was carried out in subgroups of smokers and non-smokers. Males and females were also studied separately since there was a possible sex difference in the baseline SCE levels, and, in addition, the professional activities as well as working conditions often differ between males and females with similar education or work assignment.

The group mean levels of SCE in male employees with different occupations are shown in Table 3. In general, the differences between the various occupation groups were found to be of the same order of size as that between smokers and non-smokers. The factory

TABLE 3. Group Mean SCE Levels in Non-smoking and Smoking Males with Various Occupations

OCCUPATION	SCE/cell: Group mean ± S.D. (No of subjects)	
	NONSMOKERS	SMOKERS
Administrative officials	14.2 ± 1.8 (6)	-
Work supervisors	13.2 ± 2.8 (14)	15.8 ± 3.2 (11)
Laboratory personnel	13.6 ± 2.8 (27)	14.3 ± 5.0 (6)
Factory workers	12.0 ± 2.6 (29)	15.2 ± 3.6 (40)
Construction workers	13.5 ± 2.4 (18)	16.1 ± 2.6 (15)
Others*	11.4 ± 3.1 (12)	12.6 ± 2.2 (6)
All subjects	12.9 ± 2.8 (106)	15.2 ± 3.4 (78)

*Others were mainly engineers, technicians and assistents involved in the planning, instruction and operation of production, working part time in the office and the factory

TABLE 4. Group Mean SCE Levels in Male Factory Workers Occupied in Different Production Units

WORKING UNIT	SCE/cell: Group mean ± S.D. (No of subjects)	
	NONSMOKERS	SMOKERS
HSO	11.4 ± 2.5 (9)	16.1 ± 4.0 (12)
HSH	11.2 ± 3.4 (5)	13.5 ± 1.6 (5)
HSA	11.2 ± 1.2 (8)	13.6 ± 3.2 (11)
HSL	14.8 ± 2.4*(6)	17.5 ± 5.5 (5)
HSK	-	15.7 ± 1.2 (6)

*Significantly different from each of the other groups of nonsmokers by t-test and rank-test ($P < 0.05$).

TABLE 5. Group Mean SCE Levels in Non-smoking and Smoking Females with Various Occupations

OCCUPATION	SCE/cell: Group mean ± S.D. (No of subjects)	
	NONSMOKERS	SMOKERS
Administrative officials	14.7 ± 3.0 (8)	13.6 ± 2.3 (2)
Laboratory personnel	12.9 ± 2.4 (44)	15.5 ± 2.9 (12)
Factory workers	14.4 ± 3.4 (16)	17.2 ± 3.5 (10)
Others*	15.4 ± 3.6 (7)	17.8 ± 2.8 (11)
All subjects	13.7 ± 2.9 (75)	16.6 ± 3.1 (35)

*Others were mainly personnel occupied with cleaning, dishing and various maintenance works in the laboratory and factory facilities.

workers had significantly lower SCE levels than laboratory workers and construction workers ($P < 0.05$, t-test). However, within the group of male factory workers, the employees at one unit, HSL in the organic synthetic manufacturing plant, were found to have significantly higher SCE levels than employees at other working units in this plant ($P < 0.05$, t-test). The significance of this observation was further supported by the finding of high SCE levels in smokers from the same unit (Table 4). At HSL the activities take place in two separate buildings which are designed for big scale operations. Most syntheses are closed processes in reactors, but in one of the buildings the processes were mostly open at the time of investigation, giving the possibility of exposure to various organic solvents. Further studies are carried out at present in order to explore factors or conditions in the working environment that may be responsible for the increased SCE levels in this group of workers.

Within the group of non-smoking, male laboratory personnel, high SCE levels were found in one working unit of seven employees involved in the development of new penicillins. The subjects in this group were exposed to various organic solvents, and showed an SCE level of 14.6 ± 2.5, as compared to three other working units with 5-8 employees each, in which the group mean levels ranged from 12.3 to 13.8 SCE/cell.

In the females, the laboratory personnel were found to have the lowest SCE levels, and the group with mixed occupation ("others") the highest (Table 5). The difference between these two groups was statistically significant ($P < 0.05$, t-test). The group with mixed occupations, which also showed the highest SCE levels among the smokers, included mainly personnel involved in cleaning and dishing services in the various factory premises as well as in the laboratories.

The SCE levels in the female, non-smoking factory workers were significantly increased ($P < 0.02$) over that in the corresponding group of males factory workers. These two groups of female and male factory workers were occupied in different departments and with different types of production. Chemical exposure is probably more continuous among the presently studied female factory workers as they have assembly line activities, whereas most male jobs are discontinuous and consist of the supervising of processes combined with moments of manual close activities. Thus, the higher SCE levels in the females may well be related to the working environment.

Although the females in the total population of employees showed higher SCE levels than males, this was not the case in laboratory personnel. The increased SCE levels in the group of non-smoking, male laboratory personnel was, however, mainly due to the high SCE levels noticed among employees in a discrete working unit, as mentioned above.

CONCLUDING REMARKS

The numbers of SCE in lymphocytes form a single individual may show a more than four-fold range, and the individual SCE levels, based on the average number of SCE in 20 cells, may differ by a factor of three. The causes of these great variations on the cellular and individual levels are still largely unknown. Even when factors related to culture conditions and the scoring of SCE are carefully standardized, and the SCE inducing effect of smoking is eliminated by the exclusion of smokers, the difference between individuals may still cover a 2.5-fold range.

Age in adults and sex differences have a very small influence, if any at all, on the individual SCE levels. No other human disorders than Bloom's syndrome has so far been reported to show a significant increase of SCE. Twin studies have not been able to show that genetic factors play a major role in the determination of the individual SCE level. Thus, by exclusion it appears likely that environment factors may be responsible for a great part of the observed variation.

Several different types of _in vivo_ exposure have been shown to increase the SCE levels in human lymphocytes. Most convincing are the highly significant increases of SCE in studies of patients receiving single drug treatment with alkylating and potentially DNA cross-linking agents, whereas smoking and some types of occupational exposure have been associated with a small and significant increase of SCE in some, but not all reports.

Taken together, the results from studies on smoking, PUVA therapy, cytostatic treatment and occupational exposure indicate that the SCE assay is potentially useful for the study of _in vivo_ genotoxic damage in human cells. However, there are also certin important limitations in the use of this method. The great individual variation in the levels of SCE prevents meaningful comparisons between single subjects, and relatively large groups of individuals have to be studied in order to obtain reliable data related to for example occupational exposure. A possibly very useful approach would therefore be to undertake repeated SCE analyses in single subjects and try to cover variations in exposure over time, e.g., before and after working periods and vacations, etc.

The finding that some potent genotoxic agents fail to induce a detectable increase of SCEs _in vivo_, even at cancer chemotherapeutic doses, may indeed cause problems in the interpretation of negative observations. Moreover, as certain SCE forming lesions seem to be removed from the circulating lymphocytes _in vivo_, it is possible that an increased SCE level may be of short duration and therefore difficult to pick up. On the other hand, there are indications that some SCE forming lesions may accumulate over time in the non-dividing, peripheral lymphocytes, suggesting that repeated exposure even to small doses may eventually be detectable.

Finally, the causes of the individual variation in the SCE levels are still largely unknown and call for explanations. It would be of interest to explore other habitual conditions than smoking, for example factors related to the diet, the conditions of living and the site of residence in urban vs rural communities. The present investigation of employees in a pharmaceutical industry shows that significant differences in the SCE levels may exist between groups of subjects having very similar working conditions but being involved in different chemical production work. Until further information is available, various specific or mixed types of occupational exposure must be suspected as one important cause of the individual variation in SCE levels, and for that reason further studies of groups of chemically exposed subjects are warranted.

REFERENCES

1. J. H. Taylor, Sister chromatid exchanges in tritium-labelled chromosome, Genetics, 43:515-529 (1958).
2. S. A. Latt, Microfluorometric detection of deoxyribonucleic acid replication in human metaphase chromosomes, Proc. Natl. Acad. Sci., U.S.A., 70:3395-3399 (1973).
3. S. Wolff, Chromosome aberrations, sister chromatid exchanges, and the lesions that produce them, in: Sister chromatid exchange (S. Wolff, ed.), pp. 41-57, John Wiley and Sons, New York (1982).
4. A. V. Carrano and L. H. Thompson, Sister chromatid exchange and single-gene mutation, in: Sister chromatid exchange (S. Wolff, ed.), pp. 59-86, John Wiley and Sons, New York (1982).
5. S. A. Latt, J. Allen, S. E. Bloom, A. Carrano, E. Falke, D. Kram, E. Schneider, R. Schreck, R. Tice, B. Whitfield, and S. Wolff, Sister chromatid exchanges: A report of the genetox program, Mutation Research, 87:17-62 (1981).
6. S. Takehisa, Induction of sister chromatid exchanges by chemical agents, in: Sister chromatid exchange (S. Wolff, ed.), pp. 87-147, John Wiley and Sons, New York (1982).
7. S. Wolff (ed.), Sister chromatid exchange, John Wiley and Sons, New York (1982).
8. M. S. Sasaki, Chromosome aberration formation and sister chromatid exchange in relation to DNA repair in human cells, in: DNA repair and mutagenesis in eukaryotes (W. M. Generoso, M. D. Shelby, and F. J. de Serres, eds.), pp. 285-313, Plenum Press, New York and London (1980).
9. S. A. Latt and R. R. Schreck, Sister chromatid exchange analysis, Am. J. Hum. Genet., 32:297-313 (1980).
10. S. A. Latt, R. R. Schreck, K. S. Loveday, C. P. Dougherty, and C. F. Shuler, Sister chromatid exchanges, in: Advances in human genetics (H. Harris and K. Hirschhorn, eds.), Vol. 10, pp. 267-331 (1980).
11. P. E. Perry, Chemical mutagens and sister chromatid exchange, in: Chemical mutagens: Principles and methods for their detection (F. J. de Serres and A. Hollaender, eds.), pp. 1-39, Plenum Press, New York and London (1980).
12. E. L. Schneider, In vivo methods for detecting sister chromatid exchange, in: Sister chromatid exchange (S. Wolff, ed.), pp. 229-242, John Wiley and Sons, New York (1982).
13. B. Lambert, A. Lindblad, K. Holmberg, and D. Francesconi, The use of sister chromatid exchange to monitor human populations for exposure to toxicologically harmful agents, in: Sister chromatid exchange (S. Wolff, ed.), pp. 149-182, John Wiley and Sons, New York (1982).
14. S. A. Latt, Sister chromatid exchange: New methods for detection, in: Sister chromatid exchange (S. Wolff, ed.), pp. 17-40, John Wiley and Sons, New York (1982).

15. A. V. Carrano, J. L. Minkler, D. G. Stetka, and D. H. Moore, II, Variation in the baseline sister chromatid exchange frequency in human lymphocytes, in: Environmental Mutagenesis 2, 325-337, Alan Liss, Inc., New York (1980).
16. W. F. Morgan and P. E. Crossen, Factors influencing sister chromatid exchange rate in cultured human lymphocytes, Mutation Res., 81:395-402 (1981).
17. H. Kato and A. A. Sandberg, The effects of sera on sister chromatid exchangs in vitro, Exp. Cell Res., 109:445-448 (1977).
18. A. Lindblad and B. Lambert, Relation between sister chromatid exchange, cell proliferation and proportion of B and T cells in human lymphocyte cultures, Hum. Henet., 57:31-34 (1981).
19. P. Perry and S. Wolff, New Giemsa method for the differential staining of sister chromatids, Nature, 251:156-158 (1974).
20. W. F. Morgan and P. E. Crossen, The incidence of sister chromatid exchanges in cultured human lymphocytes, Mutation Res., 42:305-312 (1977).
21. K. Husgafvel-P-rsiainen, J. Mäki-Paakanen, H. Norpa, and M. Sorsa, Smoking and sister chromatid exchange, Hereditas, 92: 247-250 (1980).
22. H. Waksvik, P. Magnus, and K. Berg, Effects of age, sex, and genes on sister chromatid exchange, Clin. Genet., 20:449-454 (1981).
23. F. Funes-Cravioto, C. Zapata-Gayon, B. Kolmodin-Hedman, B. Lambert, J. Lindsten, E. Norberg, M. Nordenskjöld, R. Olin, and Å. Swensson, Chromosome aberrations and sister chromatid exchange in workers in chemical laboratories and a rotoprinting factory and in children of women laboratory workers, Lancet, 2:332-325 (1977).
24. F. Zanzoni, J. W. A. Baumann, and E. G. Jung, Sister chromatid exchange (SCE) in human lymphocytes. Effect of UV C irradiation and age, Arch. Dermatol. Res., 265:281-287 (1979).
25. G. Ardito, L. Lamberti, E. Ansaldi, and P. Ponzetto, Sister chromatid exchanges in cigarette-smoking human females and their newborns, Mutation Res., 78:209-212 (1980).
26. B. Santesson, K. Lindahl-Kiessling, and A. Mattsson, SCE in B and T lymphocytes. Possible implications for Bloom's syndrome, Clin. Genet., 16:133-135 (1979).
27. C. Pedersen, E. Oláh, and U. Merrild, Sister chromatid exchanges in cultured peripheral lymphocytes from twins, Hum. Genet., 52: 281-294 (1979).
28. B. Lambert, A. Lindblad, M. Nordenskjöld, and B. Werelius, Increased frequency of sister chromatid exchanges in cigarette smokers, Hereditas, 88:147-149 (1978).
29. P. Bala Krishna Murthy, Frequency of sister chromatid exchanges in cigarette smokers, Hum. Genet., 52:343-345 (1979).
30. J. Mäki-Paakanen, K. Husgafvel-Pursiainen, P.-L. Kalliomäki, J. Tuominen, and M. Sorsa, Toluene-exposed workers and chromosome aberrations, J. Toxicol. Environ. Health, 6:775-781 (1980).

31. J. M. Hopkin and H. J. Evans, Cigarette smoke-induced DNA damage and lung cancer risk, Nature, 283:388-390 (1980).
32. K. Fredga, personal communication.
33. D. H. Hollander, M. S. Tockman, Y. W. Liang, D. S. Borgoankar, and J. K. Frost, Sister chromatid exchanges in the peripheral blood of cigarette smokers and in lung cancer patients; and the effect of chemotherapy, Hum. Genet., 44:165-171 (1978).
34. P. E. Crossen and W. F. Morgan, Sister chromatid exchange in cigarette smokers, Hum. Genet., 53:425-426 (1980).
35. B. Lambert, U. Ringborg, A. Lindblad, and M. Sten, The effects of DTIC, melphalan, actinomycin D and CCNU on the frequency of sister chromatid exchanges in peripheral lymphocytes of melanoma patients, in: Adjuvant Therapy of Cancer II (S. E. Jones and S. E. Salmon, eds.), pp. 56-61, Grune and Stratton, New York (1979).
36. H. J. Evans, Sister chromatid exchanges and disease states in man, in: Sister chromatid exchange (S. Wolff, ed.), pp. 183-228, John Wiley and Sons, New York (1982).
37. A. Brøgger, H. Waksvik, and P. Thune, Psoralen/UVA treatment and chromosomes. II. Analyses of psoriasis patients, Arch. Dermatol. Res., 261:287-294 (1978).
38. B. Lambert, M. Morad, A. Bredberg, G. Swanbeck, and M. Thyresson-Hök, Sister chromatid exchanges in lymphocytes from psoriasis patients treated with 8-methoxypsoralen and longwave ultraviolet light, Acta Dermatovenereol., 58:13-16 (1978).
39. E. Wolff-Schreiner, M. Carter, H. G. Schwarzacher, and K. Wolff, Sister chromatid exchanges in photochemotherapy, J. Invest. Dermatol., 69:387-391 (1977).
40. M. Faed and D. Mourelatos, Sister chromatid exchanges in lymphocytes treated with 8-methoxypsoralen and exposed to long-wave ultraviolet light, in: Mutagen-induced Chromosome Damage in Man (H. J. Evans and D. C. Lloyd, eds.), Edinburgh University Press, pp. 216-220 (1978).
41. D. M. Carter, K. Wolff, and W. Schnedl, 8-Methoxypsoralen and UVA promote sister chromatid exchanges, J. Invest. Dermatol., 67:548-551 (1976).
42. A. Bredberg, B. Lambert, A. Lindblad, G. Swanbeck, and G. Wennersten, Studies of DNA and chromosome damage in skin fibroblasts and blood lymphocytes from psoriasis patients treated with 8-methoxypsoralen and UVA irradiation, J. Invest. Dermatol., 81:93-97 (1983).
43. E. Gebhart, B. Windolph, and F. Wopfner, Chromosome studies on lymphocytes of patients under cytostatic therapy. II. Studies using the BUDR-labelling technique in cytostatic interval therapy, Hum. Genet., 56:157-167 (1980).
44. B. Lambert, U. Ringborg, E. Harper, and A. Lindblad, Sister chromatid exchanges in lymphocyte cultures of patients receiving chemotheraphy for malignant disorders, Cancer Treat. Rep., 62:1413-1419 (1978).

45. B. Lambert, U. Ringborg, and A. Lindblad, Prolonged increase of sister chromatid exchanges in lymphocytes of melanoma patients after CCNU treatment, Mutat. Res., 59:295-300 (1979).
46. J. Musilová, K. Michalová, and J. Urban, Sister chromatid exchanges and chromosomal breakage in patients treated with cytostatics, Mutat. Res., 67:289-294 (1979).
47. L. G. Littlefield, S. P. Coyler, and R. J. DuFrain, Comparison of sister chromatid exchanges in human lymphocytes after G_0 exposure to mitomycin in vivo vs. in vitro, Mutat. Res., 69: 191-197 (1980).
48. T. Raposa, Sister chromatid exchange studies for monitoring DNA damage and repair capacity after cytostatics in vitro and in lymphocytes of leukaemic patients under cytostatic therapy, Mutat. Res., 57:241-251 (1978).
49. N. P. Nevstad, Sister chromatid exchanges and chromosomal aberrations induced in human lymphocytes by the cytostatic drug adriamycin in vivo and in vitro, Mutat. Res., 57:253-258 (1978).
50. M. Sten and B. Lambert, unpublished observations.
51. J. German, S. Schonberg, E. Louie, and R. S. K. Chaganti, Bloom's syndrome. IV. Sister chromatid exchanges in lymphocytes, Am. J. Hum. Genet., 29:248-255 (1977).
52. A. Brøgger, O. Klepp, and H. Waksvik, Chromosome analyses of nurses handling cytostatic agents, Mutat. Res., 85:254 (1981).
53. M. Kucerová, Z. Polívková, and J. Bátora, Comparative evaluation of chromosomal aberrations and the SCE numbers in peripheral lymphocytes of workers occupationally exposed to vinyl chloride monomer, Mutat. Res., 67:97-100 (1979).
54. D. Anderson, C. R. Richardson, I. F. H. Purchase, H. J. Evans, and M. L. O'Riordan, Chromosomal analysis in vinyl chloride exposed workers: Comparison of the standard technique with the sister chromatide exchange technique, Mutat. Res., 83:137-144 (1981).
55. Zu Wei Gu, B. Sele, P. Jalbert, M. Vincent, F. Vincent, C. Marka, D. Chmara, and J. Faure, Induction of sister chromatid exchange by trichloroethylene and its metabolites, Toxicological European Research Volume II, No. 2/Maars (1981).
56. B. Högstedt, personal communication.
57. B. Lambert and A. Lindblad, Sister chromatid exchange and chromosome aberrations in lymphocytes of laboratory personnel, J. Toxicol. Env. Health, 6:1237-1242 (1980).
58. F. Mitelman, S. Fregert, K. Hedner, and K. Hillbertz-Nilsson, Occupational exposure to epoxy resins has no cytogenetic effect, Mutat. Res., 77:345-348 (1980).
59. U. Haglund, I. Lundberg, and L. Zech, Chromosome aberrations and sister chromatid exchanges in Swedish paint industry workers, Scand. J. Env. Health, 6:291-298 (1980).
60. J. Mäki-Paakanen, M. Sorsa, and H. Vainio, Chromosome aberrations and sister chromatid exchanges in lead-exposed workers, hereditas, 94:269-275 (1981).

61. H. Waksvik, M. Boysen, A. Brøgger, H. Saxholm, and A. Reitz, In vivo and in vitro studies of mutagenicity and carcinogenicity of nickel compounds in man, Poster presented at the 10th annual EEMS meeting, Athens, September 14-19, 1980.
62. B. Husum and H. C. Wulf, Sister chromatid exchanges in lymphocytes in operating room personnel, Acta Anaesth. Scand., 24: 22-24 (1980).
63. K. Holmberg, B. Lambert, J. Lindsten, and S. Söderhäll, DNA and chromosome alterations in lymphocytes of operating room personnel and in patients before and after inhalation anesthesia, Acta Anaesth. Scand., 26:531-539 (1982).
64. M. Bauchinger, R. Hauff, E. Schmid, and J. Dresp, Analysis of structural chromosome changes and SCE after occupational long-term exposure to electric and magnetic fields from 380 kV-systems, Radiat. Environ. Biophys., 19:235-238 (1981).
65. H. C. Andersson, E. Å. Trauberg, A. H. Uggla, and G. Zetterberg, Chromosomal aberrations and sister chromatid exchanges in lymphocytes of men occupationally exposed to styrene in a plastic boat factory, Mutat. Res., 73:387-341 (1980).

VARIATIONS IN MITOTIC INDEX AND CHROMOSOMAL ABERRATION RATES IN WOMEN

Beryl Hartley-Asp

Pharmacology Department
A B LEO
Box 941
S-25109 Helsingborg
Sweden

SUMMARY

Chromosome analysis was carried out on peripheral blood lymphocytes from 12 women at 0, 1, and 4 weeks. Two women were also consecutively sampled for 2 years. The mean rate for chromatid breaks was 3.0% range 0-13 and for gaps 8.7% range 2-21.

The intra-individual fluctuations were as great as the inter-individual variations. The increases in the aberration yield could not be correlated with the phase of the menstrual cycle, the mitotic index or with any clinically manifest infection.

INTRODUCTION

During the last decennium the analysis of chromosomal aberrations in human peripheral lymphocytes for monitoring the effects of exposure to chemical and physical mutagens [1-2] and in experimental cytogenetics [3-4] has become common. However, in contrast to experimental cytogenetics, where the control level of aberrations can always be assessed, in studies dealing with environmental exposure the control level usually is obtained from an unexposed, sometimes matched, population. Thus, the number of chromosomal aberrations and their fluctuation in control populations is extremely important.

Unfortunately experiments specifically designed to study control populations have been few [5-7]. Lubs and Samuelson found an average of 5.1% breaks/cell for men and 8.3% breaks/cell for women on single sampling of 5 women and 5 men. They also carried out 3 consecutive samplings for one women and found a variation of 6 to 20% breaks/cell. On single sampling of 524 people Ayme [7] also found a mean break rate of around 5%. The most detailed study was that carried out by Littlefield and Gok [8] who on consecutive sampling found a mean break frequency of 5.6% for men and 6.5% for women. They also showed that the intra-individual variation was much larger in women than men when sampling every 4 to 6 weeks. This suggested that the hormonal status could be involved in this variation. Therefore we designed this study to investigate the role of the menstrual cycle in the fluctuations found in aberration yields.

Experimental Design

Samples were taken from twelve women three times, at 0, 1 week and 4 weeks, i.e., throughout a whole menstrual cycle.

Two women were followed over a 2 year period in which sampling was carried out at irregular intervals. For women A February, March, and April were sampled both years for women B February, March, and December reoccurred. Investigation of the distribution into 1st, 2nd, and 3rd metaphase was carried out in conjunction with 4 of these sampling occasions.

Results

Chromosome analysis was carried out on 5200 metaphases from 14 women. The mean frequency for the number of gaps was 8.7% range 2-21 and for breaks 3.0% range 0-13. The break frequencies given include, as they were so few, the 6 iso-chromatid breaks and the 3 chromatid exchanges found in this population.

The number of gaps and breaks varied widely between the individual subjects but this was no greater than the intra-individual variation. In Fig. 1 the subjects in whom the fluctuations in gaps and breaks followed one another are grouped together whereas no such parallelism was found for those in Fig. 2. In women, number 1, 4, and 5 the frequencies of gaps and breaks increased substantially in sample 3, but in none of these women could this be attributed to any clinically manifest illness. Neither did the menstrual cycle appear to play any role in these fluctuations. The mitotic index was very similar for all these women 3-4% and remained stable throughout the month. Thus no correlation between mitotic index and aberration yield was found.

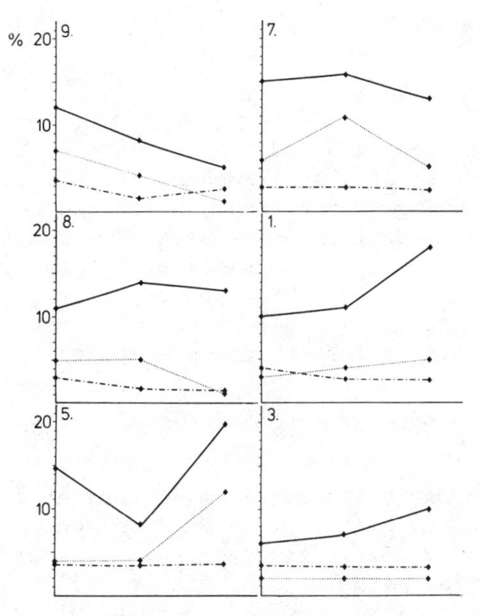

Fig. 1. The mitotic index (-.♦.-) and the frequencies of chromatid breaks (···♦···) and gaps (-♦-) found in 6 women. Sampling at 0 and after 1 and 4 weeks.

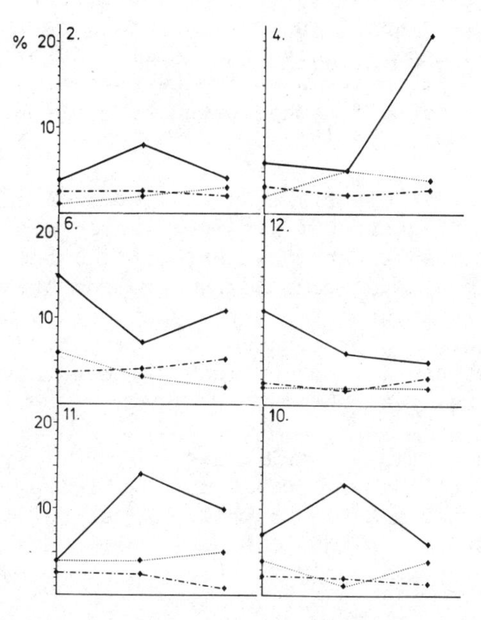

Fig. 2. The mitotic index (-.♦.-) and the frequencies of chromatid breaks (···♦···) and gaps (-♦-) found in 6 women. Sampling at 0 and after 1 and 4 weeks.

MATERIALS AND METHODS

Description of Subjects

Twelve women age 20-30 and two age 46 and 48 years took part in the study. Twelve of these women had slight urogenital infections at the beginning of the study, which did not require medication, but had no clinical symptoms by second sampling. Approximately half of these women were smokers. Of the two women 26 and 28 years old participating in the longer study one is a smoker and one has a bronchial allergy to rodents. Neither of these women had any clinically manifest infections on sampling throughout the two year period.

Culture Method

Venous blood was collected in a vacutainer tube and stored for at least 2 hr at 4°C. Eight drops whole blood were added to McCoys 5A medium containing 25% foetal calf serum, 1% glutamine 3.4% phytohaemagglutin, streptomycin and penicillin. The cultures used for cell division analysis also contained 10 μg/ml BrdU and were incubated at 37°C in 5% CO_2 for 72 hours.* Colcemid, 10 μg/ml, was added 1 hr before harvesting. Slide for chromosome analysis were made according to Evans and O'Riordan [9] and stained in 2% Giemsa. Slides for cell division analysis were immersed for 12 min in 0.5 μg/ml bisbenzimid (Hoechst 33258) mounted directly, illuminated for 2 hr under UVA light and then stained for 5 min in 2% Giemsa. All slides were coded and 100 metaphases analyzed for chromosome aberrations and 3000 for mitotic index by the same person. The criteria for defining the types of damage were those of Evans and O'Riordan [9].

In Fig. 3, it is clearly seen that woman A exhibited a larger variation in the frequency of gaps than woman B. The frequencies of the chromatid breaks were however, not substantially different for these two women. In woman A, the gap and break frequencies followed one another throughout the two year period while this parallelism was not quite so pronounced for woman B. The mitotic index was relatively stable for both women, neither varying more than 1%, throughout this two year period.

Table 1 demonstrates that these women's lymphocytes have very different cell kinetics. Woman B having very few metaphases in 3rd division at 72 hr while almost 50% of woman A's metaphases were in 3rd division.

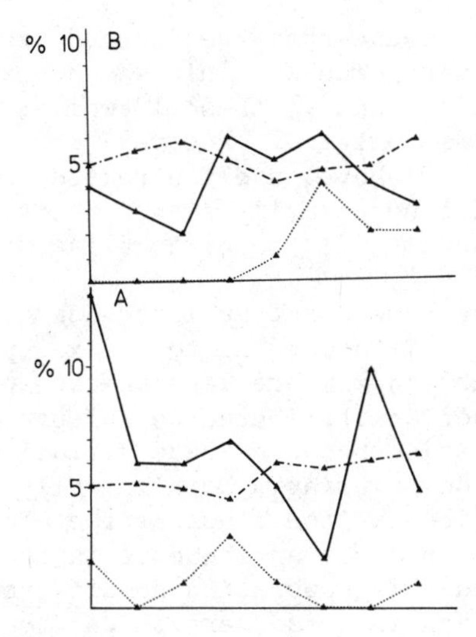

Fig. 3. The mitotic index (-·♦·-) and the frequencies of gaps (···♦···) and chromatid breaks (-♦-) found over a 2 year period for two women.

TABLE 1. The Frequency of 1st, 2nd, and 3rd Division Metaphases in Two Women

DONOR	I	II	III
A	36.5	47.0	16.5
B	9.5	44.0	46.5

DISCUSSION

In the present study considerable variation was found on consecutive sampling, during the short period of one month, for both gaps and breaks. The mean and range for breaks fell however within those previously published [5-7]. The number of gaps was always higher and the variation in frequency was much larger than that found for breaks. This puts the issue of the use of gaps as an indicator of genetic damage to the forefront. Certain authors regard them as toxic events [9] or artefacts of staining and processing open to highly subjective analysis [10]. Others, however, such as Gebhart [11], demonstrated that in human lymphocytes gaps induced by chemicals are dose dependent and have a non-random distribution similar to that of breaks. The dose dependency has also been confirmed in rats by Anderson and Richardson [12] who suggested that gaps could be a useful and sensitive indicator of chemically induced damage *in vivo*. They found that the induced frequency of gaps increased in parallel with breaks. This was not always the case for the spontaneous levels found in these 12 women, only about half of them conforming to this pattern. However, in all subjects the frequency of gaps remained above, and fluctuated more, than that of breaks. This does not necessarily negate their usefulness but demonstrates their sensitivity to unknown factors.

All these cultures were set up in the same medium and incubated and prepared for analysis according to a fixed protocol. Therefore, the incidence of high and low breakage in consecutive cultures from the same person cannot be attributed to culture conditions. Neither was there any correlation between these variations and the phase of the menstrual cycle on sampling. No clinically manifest infection could either be ascribed as the cause of the elevations. Heavy smoking has been shown to produce an increase in the chromosomal aberration rate [13] but in this population no difference could be found between smokers and non-smokers, perhaps due to the fact that the number of cigarettes per day consumed was too low.

The mitotic index in this group showed very little intra- or inter-individual variation (±1%) in contrast to Crossen and Morgon who found a range of 1.1% to 8.3% in 10 normal individuals [14]. Thus, no correlation between the mitotic index and fluctuations in aberration yield was found in this study.

The 2 women followed over a 2 year period were incubated for 72 h. They had very similar mitotic indicies although they were composed of very different fractions of 1st, 2nd, and 3rd metaphases. Woman A had a larger proportion of cells in 1st metaphase than women B and exhibited larger fluctuations in the frequency of gaps. It is interesting to speculate if women A's hypersensitivity to rodents and thus a possible different immunological status plays a role in the delayed reaction to PHA compared with women B. It also indicates that gaps are perhaps more prevalent in 1st division metaphase than in later divisions.

This study shows that the inter-individual variation in aberration yield reflects the normal intra-individual variation found in women. Thus, studies designed to study environmental exposure to chemicals should include sufficient subjects to allow us to distinguish real differences between the exposed and unexposed group. This can prove difficult if gaps are to be used as an indicator, due to their very large fluctuations, but until more data accumulates it would be advantageous if this aberration type was always included in the analysis.

REFERENCES

1. K. E. Buckton, G. E. Hamilton, L. Paton, and A. O. Langlands, Chromosome aberrations in irradiated Ankylosing spondylitis patients, in: Mutagen-Induced Chromosome Damage in Man (H. J. Evans and D. C. Lloyd, eds.), 142-150, Edinburgh Univ. Press, Edinburgh.
2. F. Funes-Cravioto, B. Lambert, J. Lindsten, L. Ehrenberg, A. T. Natarajan, and S. Osterman-Golker, Chromosome aberrations in workers exposed to vinyl chloride, Lancet, 1:459-461 (1975).
3. B. A. Kihlman, K. Hansson, F. Palitti, H. C. Andersson, and B. Hartley-Asp, Potentiation of induced chromatid-type aberrations by hydroxyurea and caffeine in G_2, in: Progress in Mutation Res., Vol. 4 (A. T. Natarajan, G. Obe, and H. Altman, eds.), 11-24, Elsivier North Holland (1982).
4. A. T. Natarajan, G. Obe, and F. N. Dalout, The effect of caffeine post-treatment on X-ray-induced chromosomal aberrations in human blood lymphocytes in vitro, Hum. Genetics, 54:183-189 (1980).
5. H. A. Lubs and J. Samuelson, Chromosome abnormalities in lymphocytes from normal human subjects, Cytogenetics, 6:402-411 (1967).
6. S. Aymé, J. F. Mattei, M. G. Mattei, Y. Aurran, and F. Giraud, Non-random distribution of chromosome breaks in cultured lymphocytes of normal subjects, Hum. Genet., 31:161-175 (1976).
7. L. G. Littlefield and K. O. Gok, Cytogenetic studies in control men and women. I. Variations in aberration frequencies in 29,709 metaphases from 305 cultures obtained over a three-year period, Cytogenet. Cell Genet., 12_17-34 (1973).
8. H. J. Evans and M. L. O'Riordan, Human peripheral blood lymphocytes for the analysis of chromosome aberrations in mutagen tests, Mutation Res., 31:135-148 (1975).
9. J. A. Di Paolo and N. C. Popescu, Relationships of chromosome changes to neoplastic cell transformation, Am. J. Pathol., 85: 709-738 (1976).
10. A. Schinzel and W. Schmid, Lymphocyte chromosome studies in humans exposed to chemical mutagens: The validity of the method in 67 patients under cytostatic therapy, Mutation Res., 40:139-166 (1976).
11. E. Gebhart, Experimentelle Beitrage zum problem der lokalen achromasien (Gaps), Hum. Genet., 13:98-107 (1971).
12. D. Anderson and C. R. Richardson, Issues relevant to the assessment of chemically induced chromosome damage in vivo and and their relationship to chemical mutagenesis, Mutation Res., 90:261-272 (1981).
13. G. Obe and J. Herha, Chromosomal aberrations in heavy smokers, Hum. Genet., 41:259-263 (1978).
14. P. E. Crossen and W. F. Morgan, Lymphocyte proliferation in Down's Syndrome measured by sister chromatid differential staining, Hum. Genet., 53:311-313 (1980).

MICRONUCLEI IN CULTURED LYMPHOCYTES AS AN INDICATOR OF GENOTOXIC EXPOSURE

Benkt Högstedt and Felix Mitelman

Department of Clinical Genetics
University Hospital
Lund, Sweden

1. INTRODUCTION

The study of chromosomal aberrations in lymphocytes is a sensitive but time consuming method for revealing mutagen effects in humans. An interesting alternative and/or complement could be to use micronuclei as an indicator of chromosomal damage. Micronuclei are formed during cell division from acentric fragments or lagging chromosomes and thus reflect both damage of the chromosomes and the spindle apparatus. They are easily detected in interphase cells as free intracytoplasmatic bodies of nuclear origin.

Countryman and Heddle and Heddle et al. [1, 2] have shown that *in vitro*-expose to ionizing radiation gives a dose related increase of micronuclei in cultured lymphocytes. They scored micronuclei in cell preparations where the cytoplasm had been destroyed by hypotonic treatment. Using the same method Norman et al. [3] found an increased level of micronuclei in patients after X-ray examinations with iodine-containing contrast medium. Linnainmaa et al. [4] found a dose related effect in micronuclei of *in vitro* exposure to styrene. We have used this method in a study of persons exposed to ethylene oxide and found no effect of exposure but a statistically significant one of smoking [5].

When scoring micronuclei in preparations after hypotonic treatment it was evident that there were certain disadvantages with this method: it was often difficult to differentiate micronuclei from nuclei/cellular debris from other cells, since the cytoplasm of the cell was lacking; cells with segmentation or fragmentation of the nucleus could easily be taken for lymphocytes with micronuclei; since the cytoplasm of the cells was destroyed, micronuclei might also be torn away from the neighborhood of the original nucleus.

The aims of the present study were to introduce a new method for analyzing micronuclei in cultured lymphocytes with preserved cytoplasm, to compare the results of different culture times, to determine how many lymphocytes that ought to be scored to obtain an optimal result, and to compare this method with the method of Countryman and Heddle in a study of styrene exposed persons.

2. MATERIAL AND METHODS

2.1. Individuals Investigated

The exposed group consisted of 38 males aged 19-63 (median 38.5, mean 38.7) years, who had been working with styrene 1-23 (median 6, mean 7.9) years in a factory producing fiberglassreinforced polyester plastic articles. Sixteen were smokers. The controls were 20 male workers aged 22-59 (median 34, mean 36.4) years from a mechanical industry in the same town. Eight were smokers. Blood specimens from both groups and urine samples from the exposed ones were obtained in two days in May 1980. None of the exposed persons were exposed to levels of styrene in work room air exceeding 40 ppm during the last five years. The mean concentration of styrene was 13 ppm at the moment of the investigation.

2.2. Micronuclei in Lymphocytes with Preserved Cytoplasm

A standard microculture technique was used. Whole blood was added to a medium of 80% Mc Coy 5a and 20% new born calf serum. Two cultures were set up for each person. After incubation with PHA in 37°C for 72 and 96 h, the cells were suspended in the medium. After centrifugation the medium was removed and the cells were suspended in an equal volume of medium, smeared on slides, air-dried and stained with May-Grünwald-Giemsa stain. The accumulated frequency of intracellular micronuclei was noted after scoring 250, 500, 750, 1000 lymphocytes.

The micronuclei were of different sizes but had to be not larger than one third of the main nucleus and had to be clearly divided from it. They were of almost the same color and structure as the main nucleus. The lymphocyte cytoplasm stained blue and these cells had to be differentiated from other nucleated cells with fragmentation or segmentation of the nucleus. The slides were examined with an oil immersion lens with a total magnification of ×1000.

2.3. Micronuclei in Lymphocytes Treated with Hypotonic Saline

The same microculture technique as above was used. The cells were cultured for 96 h and then treated for 1 h with colchicin and

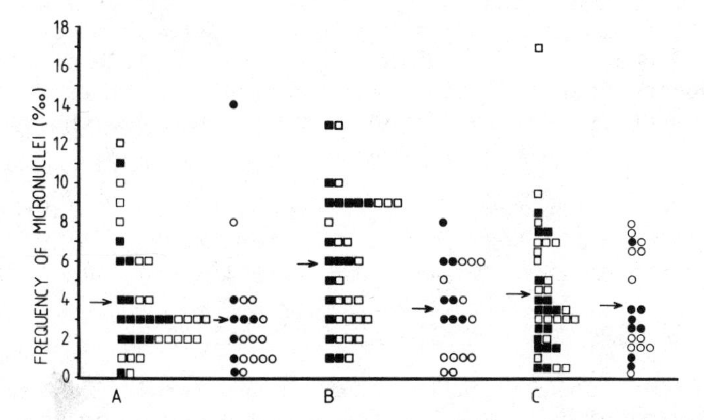

Fig. 1. (A) Micronuclei in lymphocytes with preserved cytoplasm cultured for 72 h. (B) Micronuclei in lymphocytes with preserved cytoplasm cultured for 96 h. (C) Micronucli in hypotonic treated lymphocytes cultured for 96 h. Individuals exposed to styrene (squares) and controls (circles); smokers closed and non-smokers open symbols.

thereafter with 0.75 M KCl and finally fixed in methanolacetic acid (3:1). The cells were spread on slides, air-dried and stained in Giemsa. The frequency of micronuclei was estimated by scoring 2000 lymphocyte nuclei. Almost the same criteria as above were used for micronuclei. However, since the cytoplasm was lacking, the micronuclei had to be situated within two nuclear diameters from the main nucleus. The preparations were examined with the same magnification as above. This is the same method as earlier described by Countryman and Heddle [1].

2.4. Statistical Methods

To evaluate the influences of exposure, smoking and age on the effect parameters a multivariate analysis was performed. Since this type of analysis demands normal distribution of the variables, the values were transformed into logarithms to meet this requirement approximately. In the analysis of correlation the Spearman rank correlation coefficient test was used. Statistical significance denotes $p < 0.05$. All tests were two-sided.

3. RESULTS

3.1. Effect of Styrene Exposure

When the effects of smoking and age were eliminated in the statistical analysis, there was a significant effect of exposure in micronuclei in lymphocytes with preserved cytoplasm, cultured for 96 h (p = 0.005, Fig. 1). This effect was significant already when 500 lymphocytes per individual were scored. After scoring 750 cells the final statistical significance was reached and thus it did not increase when 1000 lymphocytes were analyzed.

Also in lymphocytes with preserved cytoplasm cultured for 72 h and in hypotonic treated lymphocytes there were effects of exposure to styrene, though not statistically significant (Fig. 1).

3.2. Association between Effect Parameters

There were statistically significant correlation coefficients between micronuclei in lymphocytes with preserved cytoplasm cultured for 96 and 72 h (r = 0.30, p = 0.024) and between micronuclei in cells with preserved cytoplasm cultured for 72 h and micronuclei in hypotonic treated lymphocytes (r = 0.26, p = 0.044).

4. DICUSSION

Styrene is used in producing fiberglassreinforced polyester plastic. The compound gains mutagen properties through metabolic activation in somatic cells by forming an epoxide, styrene 7.8 oxide. Styrene/styrene oxide is mutagenic in different test systems [6, 7, 8, 9]. Styrene has in several investigations of exposed workers been associated with an increased level of chromosomal aberration in lymphocytes and also with an increased frequency of leukemias [10, 11, 12, 13].

The modified method presented, which means analyzing micronuclei in lymphocytes with preserved cytoplasm, showed a highly significant effect of styrene exposure. The exposure level was rather low and there was no effect on the frequencies of chromsomal aberrations in lymphocytes (unpublished data). This lack of effect of styrene exposure below 40 ppm on chromosomal aberrations has recently been confirmed by three independent investigators [A. Brögger, I.-L. Hansteen, and I. Nordensen; personal communication]. Nor is there any effect on SCE of these or higher doses of styrene [A. Brögger and I.-L. Hansteen; personal communication]. Therefore the new method for scoring micronuclei in cells with preserved cytoplasm seems to be very sensitive. The method is cytologically simple as the staining of the cells makes the identification of lymphocytes with true intracytoplasmatic micronuclei easy.

Compared to the method with hypotonic treated cells, the new one is a little more time consuming, but a good preparation can be analyzed in 30 minutes or even faster. When using isolated lymphocytes or leucocytes from 'buffy coat,' which easily can be done, the method is extremely fast.

The method is at present used as a possible *in vitro* test in our laboratory.

ACKNOWLEDGMENT

This study has been financially supported by the Swedish Work Environment Fund.

REFERENCES

1. P. J. Countryman and J. A. Heddle, The production of micronuclei from chromosome aberrations in irradiated cultures of human lymphocytes, Mutation Research, 41:321-332 (1976).
2. J. A. Heddle, C. V. Lue, E. F. Saunders, and R. D. Benz, Sensitivity to five mutagens in Fanconi's anemia as measured by the micronucleus method, Cancer Research, 38:2983-2988 (1978).
3. A. Norman, F. H. Adams, and R. F. Riley, Cytogenetic effects of contrast media and triiodobenzoic acid derivatives in human lymphocytes, Radiology, 129:199-203 (1978).
4. K. Linnainmaa, T. Meretoja, M. Sorsa, and H. Vainio, Cytogenetic effects of styrene and styrene oxide on human lymphocytes and Allium cepa, Scand. J. Work Environ. and Health, 4: 156-162 (1978).
5. B. Högstedt, B. Gullberg, K. Hedner, A.-M. Kolnig, F. Mitelman, and S. Skerfving, Chromosome aberrations and micronuclei in bone marrow cells and peripheral blood lymphocytes in humans exposed to ethylene oxide, Hereditas, 98:105-113 (1983).
6. K. Sugiura, T. Kimura, and M. Goto, Mutagenicities of styrene oxide derivatives on Salmonella typhimurium (TA 100), Mutation Research, 58:159-165 (1978).
7. M. Donner, M. Sorsa, and H. Vainio, Recessive lethals induced by styrene and styrene oxide in Dorosophila melanogaster, Mutation Research, 67:373-376 (1979).
8. L. Fabry, A. Leonard, and M. Roberfroid, Mutagenicity tests with styrene oxide in mammals, Mutation Research, 51:377-381 (1978).
9. H. Norypa, The *in vitro* induction of sister chromatid exchanges and chromosome aberrations in human lymphocytes by styrene derivatives, Carcinogenesis, 2:237-242 (1981).
10. T. Meretoja, H. Vainio, M. Sorsa, and H. Härkönen, Occupational styrene exposure and chromosomal aberrations, Mutation Research, 56:193-197 (1977).
11. B. Högstedt, K. Hedner, E. Mark-Vendel, F. Mitelman, A. Schutz, and S. Skerfving, Increased frequency of chromosome aberrations in workers exposed to styrene, Scand. J. Work Environ. and Health, 5:333-335 (1979).

12. H. C. Andersson, E. A. Tranberg, A. H. Uggia, and G. Zetterberg, Chromosomal aberrations and sister chromatid exchanges in lymphocytes of men occupationally exposed to styrene in a plastic boat factory, Mutation Research, 73:387-401 (1980).
13. M. G. Ott, R. C. Kolesar, H. C. Scharnweber, E. J. Schneider, and J. R. Venable, A mortality survey of employees engaged in the development or manufacture of styrene-based products, J. Occupational Med., 22:445-460 (1980).

DETECTION OF MUTATED ERYTHROCYTES IN MAN*

William L. Bigbee, Elbert W. Branscomb,
and Ronald H. Jensen

Biomedical Sciences Division
University of California
Lawrence Livermore National Laboratory
Livermore, California 94550

ABSTRACT

Two assay systems are being developed to measure the level of _in vivo_ somatic mutations in human cells. Both are based on immunologic recognition and fluorescence-activated cell sorter enumeration of cells carrying variant proteins. The first assay is based on the detection of erythrocytes containing the single amino acid-substituted hemoglobins S or C. Frequencies of anti-hemoglobin S- and C-labeled red cells in the blood of normal hemoglobin A individuals were determined. In five samples, the S-frequencies ranged from 1.1×10^{-8} to 1.1×10^{-7}. C-frequencies from 6.7×10^{-8} to 2.6×10^{-7} were observed in three samples. Methods to test the genetic validity of these results and to extend the measurements to additional point- and frameshift-mutant hemoglobins through the use of monoclonal antibodies are discussed. The second assay seeks to detect variant red cells arising as a result of "gene expression loss" and point mutations in the genes for the membrane-associated protein glycophorin A. Monoclonal antibodies, specific for the M and N allelic forms of the protein, have been produced and can be used to screen blood samples from MN heterozygotes for variant red cells which fail to present one of the glycophorin A forms. These antibodies may also be capable of detecting red cells in homozygotes expressing single amino acid-substituted glycophorin A. Such mea-

*Work performed under the auspices of the U.S. Department of Energy by the Lawrence Livermore National Laboratory under contract number W-7405-ENG-48.

surements, when properly validated, will estimate the integrated mutational damage which has occurred in the hemoglobin and glycophorin A genes in the erythroid stem cells and hence reflect the accumulated mutagenic burden incurred by each individual.

INTRODUCTION

We wish to develop methods capable of detecting the results of mutations occurring in single genes in the DNA of somatic cells in people. The great advantage of this approach is that mutant cells, rather than individuals presenting transmitted mutant phenotypes, are identified. The frequency of such cells may represent the integrated rate of mutations occurring in that gene in a particular tissue. Inasmuch as such damage can be taken as representative of the entire genome in every tissue of the body, such a measurement may serve as a monitor of the overall genetic insult, both background and environmentally induced, suffered by that individual. The motivation for such development is both of a practical and basic scientific nature. On the practical side, establishment of such methods would lead to screening procedures to quantify accumulated environmental genotoxic burden to human individuals in the short term. If mutational injury is an initiating mechanism in the development of human cancer, then application of these tests may have important clinical and health monitoring significance. Screening could be used to recognize exposed workers in occupational settings and to identify atypically vulnerable people in the population due to variations in genetic susceptibility and/or lifestyle. On the basic biological side, such methods will measure background mutational frequencies in somatic tissues leading to estimates of mutation rates, correlation with germinal mutation rates and the possible link to carcinogenesis.

To be useful, the assay system must satisfy several genetic and biochemical criteria. Since the frequency of mutant cells is low, the tissue to be sampled must be readily accessible and capable of yielding large numbers of intact single cells. Also, the mutant cells must appear at a measurable frequency. Here two general strategies have been pursued. The first, for example the detection of thioguanine-resistant lymphocytes [1], is based on mutations in the X-linked gene for HGPRT which functionally exists in a single copy per cell. Thus, a single mutational event, resulting in the lack of functional gene product, will produce a HGPRT "null" cell. For an autosomal gene, such a variant will be extremely rare since a "null" cell could only occur as a result of damage to both copies of the gene in a single cell. A second strategy, and the one employed in our assays, attempts to detect heterozygous normal/mutant cells, i.e., cells containing co-dominantly expressed autosomal genes in which one allele is normal and the other is mutant. Such cells should, in the case of neutral mutations, produce both normal and mutant gene products, or in the case of "null" mutations, synthesize protein coded for by only one allele. For accuracy, the mutant cells must not be selected against *in vivo* and the selection method must be powerful and specific; either clonogenic assay of mutant cells or immunologic detection of mutant gene products. Finally, an independent means must exist to validate that the identified cells result from a structural gene mutation. For example, maintenance of the mutant phenotype under non-restrictive growth conditions or the direct biochemical demonstration of the synthesis of a variant protein.

In this paper, I will describe two assay systems we are developing based on high speed cell sorter detection of immunologically identified variant human erythrocytes. Both take advantage of the ability of the flow sorter to rapidly screen large numbers of cells to enumerate rare, presumptively mutant, cells labeled with fluorescent antibodies.

DETECTION OF RARE RED BLOOD CELLS CONTAINING VARIANT HEMOGLOBIN

Nearly 400 variant hemoglobins, the majority characterized by a single amino acid substitution, have been identified in the human population [2]. Individuals who synthesize any one of these mutant hemoglobins do so as a result of inheriting a mutated hemoglobin gene. The presence of such rare individuals carrying such germinal mutations suggests that these same mutational events may also occur somatically in the hemoglobin genes of the blood-forming stem cells. In normal hemoglobin A individuals, such mutations would give rise to rare circulating red blood cells containing variant hemoglobin in addition to the normal form.

To identify variant hemoglobin-containing cells, Dr. George Stamatoyannopoulos at the University of Washington prepared two antibodies, one specific for hemoglobin S [3] and the other for hemoglobin C [4]. These antibodies recognize and bind tightly to hemoglobin S or C but not to normal hemoglobin A. These variant hemoglobins differ from the normal hemoglobin A amino acid sequence by single amino acid substitutions at the sixth position of the β chain:

Hemoglobin	Amino-acid Substitution	Nucleotide Change
A → S	Glu → Val	GAG → GUG
A → C	Glu → Lys	GAG → AAG

As shown above, either one of these amino acid substitutions can occur as a result of a single base change in the triplet codon corresponding to that position in the β globin gene. Hence in normal hemoglobin A individuals, these point mutations in the β globin genes in erythroid stem cells will give rise to circulating red cells containing hemoglobin S or C in addition to hemoglobin A. To identify such cells, the hemoglobin S antibody was conjugated with FITC and incubated with red cells fixed on slides. The preparations were then manually examined under a fluorescence microscope for the presence of antibody-labeled cells. Long and laborious counting effort revealed their presence at a frequency of about one labeled cell in 10^7 unlabeled cells [5]. While these results were encouraging, this assay method was not practical since a single measurement was so time consuming (about one man-month per sample) and it was not possible to demonstrate biochemically that the labeled cells did, in fact, contain hemoglobin S.

It was at this point we established a collaboration with Dr. Stamatoyannapoulos to explore the potential application of flow sorter technology to this problem. Two technical issues needed to be addressed; (1) could a hemoglobin antibody staining method be developed for red cells in suspension and (2) could the sorter process such samples rapidly enough to allow examination of a statistically significant number of cells in a reasonable time. To permit antibody labeling of hemoglobin, a method needed to be devised which would maintain the integrity of individual erythrocytes, fix intracellular hemoglobin in situ and then permit access to antibody. Such a procedure was suggested by the work of Wang and Richards [6], in which membrane permeable cross-linking reagents were used to study the topography of red cell membrane proteins. They showed that, in intact red cells, these reagents cross-linked intracellular hemoglobin to itself and to many of the peripheral and intrinsic membrane proteins (Bands 1, 2, 3, 4.1, 4.3, 5, 6, and 7). We successfully adapted this procedure to cross-link cells and produce ghosts which retain about 10^6 hemoglobin molecules covalently bound to the cell membrane, remain permeable to antibody and retain the native antigenicity of hemoglobin A, S, and C [7].

To test the specificity of antibody labeling, artificial mixtures of variant and normal cells were processed together and examined. Fluorescence microscope observation revealed the approximate number of expected fluorescent ghosts with the normal hemoglobin-containing ghosts nearly invisible. These ghost suspensions were then analyzed quantitatively on the FACS cell sorter. By integrating the number of fluorescence signals of intensity characteristic of antibody-labeled hemoglobin AS or AC ghosts and knowing the total number of ghosts processed by the sorter, the frequency of positive ghosts could be calculated and compared to the known frequency in the prepared mixture. As shown in Fig. 1, artificial mixtures as dilute as one variant in $\sim 10^5$ normal cells could be accurately reconstructed before fluorescent background noise obscured the rare labeled ghosts.

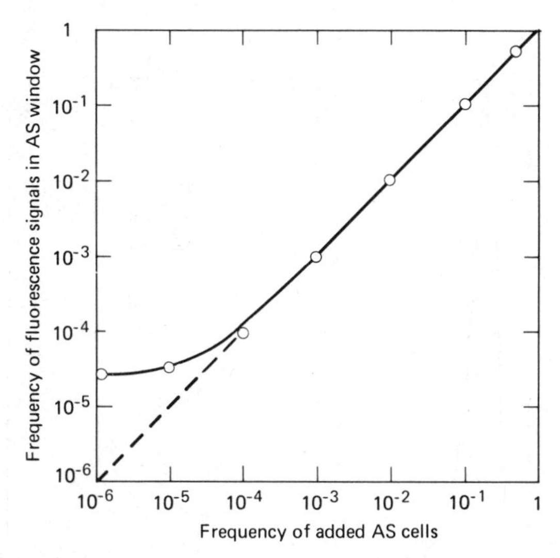

Fig. 1. Reconstruction of artificial mixtures of hemoglobin AS and hemoglobin AA cells. AS and AA cells were mixed, ghosts prepared and the suspension incubated with FITC-anti-hemoglobin S. The mixtures were then analyzed on the cell sorter as described in the text. Similar results were obtained for AC/AA mixtures labeled with FITC-anti-hemoglobin C [7].

Next, the issue of sorter processing speed was examined. Assuming a background frequency of one anti-hemoglobin S positive-cell per 10^7 negative cells to be correct, one would need to analyze at least 10^9 total cells per sample. To do this efficiently requires a sorter throughput rate of $\sim 10^6$ per sec, a rate that is about 300 times higher than is normally used. Such high flow rates require a sample density of about 5×10^8 cells/ml resulting in about 20 cells being in the laser beam simultaneously. Using cross-linked red cell ghosts, we wished to determine if rare fluorescent ghosts could be reliably detected under such conditions. Suspensions of fluorescent ghosts in the absence and the presence of dense suspensions of unlabeled ghosts were analyzed. As illustrated in Fig. 2, such rare ghosts could be detected with no loss of resolution under flow conditions resulting in an overall ghost processing rate of greater than 10^6 per sec.

Since unavoidable fluorescent background noise prevented a direct machine count of the background frequency of rare antibody labeled ghosts from bloods of normal individuals, we designed a two-step method in which the sorter served as a powerful enrichment device. This procedure first involves sorting all objects of fluores-

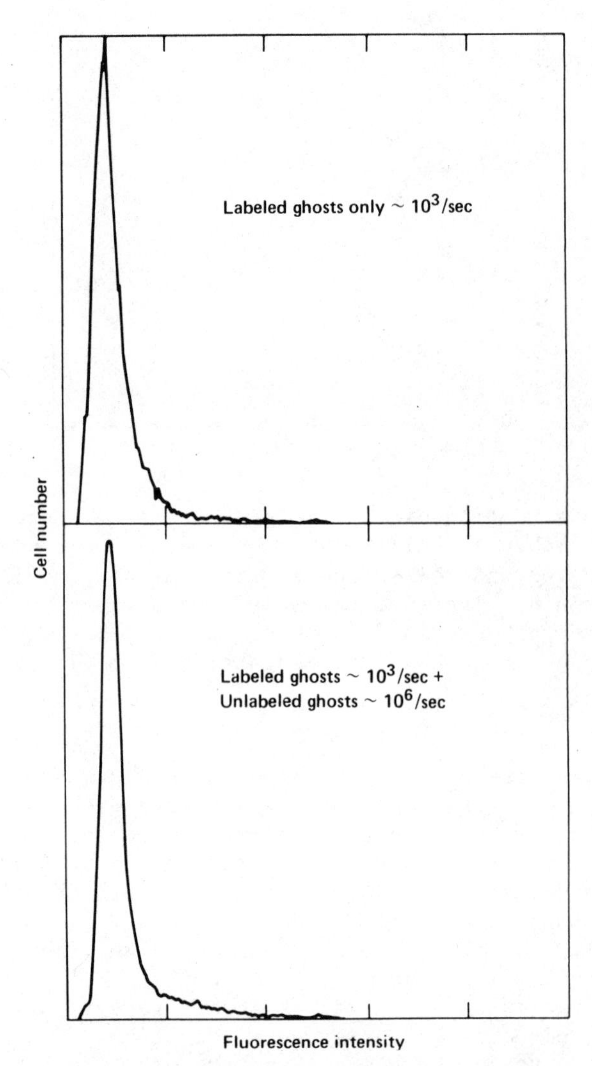

Fig. 2. High-speed detection of fluorescent red cell ghosts in the presence of a high density of non-fluorescent red cell ghosts. Fluorescent red cell ghosts were first analyzed alone at a normal rate of approximately 10^3 per sec (top panel). The fluorescent ghosts were then mixed with a 1000-fold excess of unlabeled ghosts and the mixtures analyzed at the same rate (lower panel). Under these conditions the overall throughput rate of the unlabeled ghosts exceed 10^6 per sec with no degradation in the histogram generated by the labeled ghosts.

TABLE 1

	Hb S	
(LLNL)	$1.1 \times 10^{-8} - 1.1 \times 10^{-7}$	(5)
(U. of Wash.)	$4 \times 10^{-8} - 3 \times 10^{-7}$	(15)
	Hb C	
(LLNL)	$6.7 \times 10^{-8} - 2.6 \times 10^{-7}$	(3)

cence intensities characteristic of antibody-labeled ghosts then immobilizing the sorted sample on microscope slides. The labeled ghosts, which can be discerned from fluorescent debris by their characteristic shape and membrane-associated fluorescence, were then counted manually using a fluorescence microscope equipped with a computer-controlled scanning stage. Typically, a sorted sample was enriched in fluorescent to non-fluorescent ghosts approximately 3000-fold over the initial ghost suspension and could be quantitatively scanned under the microscope in about six hours. Table 1 lists our initial results using this two-step method together with previously reported slide-based results obtained by Dr. Stamatoyannopoulos [5]. Table 1 also includes our results with the anti-hemoglobin C antibody. The frequencies of anti-hemoglobin S labeled cells using the two techniques are consistent and the S and C frequencies comparable. In spite of this initial success, this approach is compromised in several respects. The signals from the rare, dimly fluorescent ghosts cannot be separated from the fluorescent background artifacts, thus the slide preparations are contaminated and difficult to analyze. The ghosts themselves are very fragile structures which we could not immobilize without significant and variable losses. Thus our results are only semi-quantitative; a large part of the order of magnitude variation of the reported variant cell frequencies may be attributable to these technical problems. We believe a more quantitative procedure is required in order to detect an increase in the number of variant cells with subtle environmental mutagen exposures. Also, the ghosts contain only about 1% residual hemoglobin which makes biochemical verification difficult.

"HARD-CELL" LABELING APPROACH

Because of the technical problems encountered with the ghost approach, we are exploring another red cell preparation procedure. A potentially useful strategy was suggested by the work of Aragon et al. [8], in which red cells could be made permeable to substrates of intracellular enzymes without concominant loss of the protein themselves. This was accomplished by heavy cross-linking with the same membrane permeable reagent, dimethyl suberimidate, used in the ghost procedure. The resulting "hard cells," resistant to hypotonic lysis, are then permeabilized by organic solvent/detergent treatment. We have added a third step of protease digestion in order to remove the outer surface of residual membrane proteins to expose more of the immobilized intracellular hemoglobin. These "hard cells" possess a number of useful properties. They are mechanically very stable and do not lose significant amounts of hemoglobin. This physical stability will permit serial sorting of samples. By resorting the initial sorted sample, we expect to obtain significantly increased enrichments of labeled cells. In model experiments, using artificial mixtures of fluorescent and non-fluorescent "hard cells," approximately 90% recovery efficiencies were obtained in two-step serial sorts. Like the ghost suspensions, these cells can be processed at throughput rates exceeding 10^6 per sec. In addition, these cells can be easily and quantitatively immobilized on microscope slides. Secondly, exterior presentation of the hemoglobin should improve the efficiency and specificity of antibody labeling thus lowering the frequency of fluorescent artifacts. Extended incubation time under less than optimal solubility conditions for IgG is necessary for immunologic labeling of ghosts. The resulting microprecipitates of fluoresceinated antibody contribute significantly to the frequency of false-positive signals. With the "hard cell" approach, staining conditions can be modified for optimal stability of the antibody. Lastly, since essentially the full cellular hemoglobin content is retained in these "hard cells," almost 100-fold more protein per cell can be obtained for biochemical analysis. By using a reversible cross-linking analog of dimethyl suberimidate, 3,3'-dithiobispropionimidate [6],and ultra-thin gel [9] or single cell electrophoretic techniques [10], it appears feasible to directly characterize the hemoglobin content of individual antibody labeled cells.

We are presently refining this technique for immunologic labeling; an intial result demonstrating specificity of FITC-anti-hemoglobin S binding to AS "hard cells" is presented in Fig. 3. If accurate reconstructions of artificial mixtures can be obtained, we plan to adopt this procedure as the one of choice for continuing work.

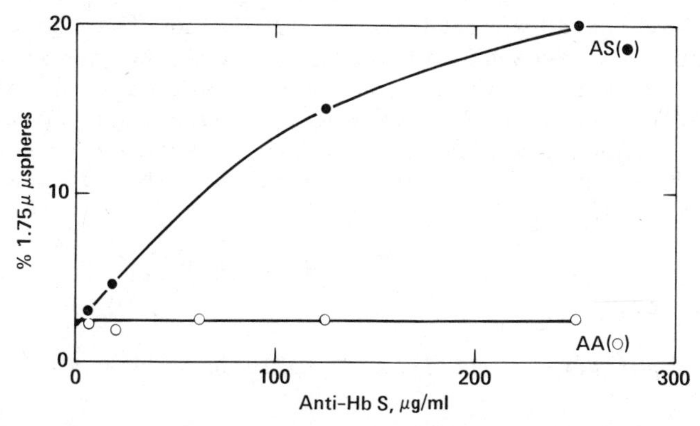

Fig. 3. Specific binding of FITC-anti-hemoglobin S to heterozygous AS "hard cells." Hemoglobin AS and AA red cells were prepared as described in the text; 10^8 "hard cells" were then incubated for 30 min at room temperature with the indicated concentration of FITC-anti-hemoglobin S, washed three times in detergent buffer and analyzed on the flow sorter. The plotted intensities correspond to the peak modal channels of the resulting histograms.

FUTURE OF THE HEMOGLOBIN-BASED ASSAY

Continued progress on the development of this assay approach is dependent of three factors: (1) success of the "hard cell" preparative technique, (2) continued development of automated cytometric analysis methods, and (3) availability of highly specific antibodies to hemoglobin variants. At LLNL there are two ongoing machine development projects relevant to this work. The first is the construction of a high-speed sorter. This device operates at about 20 times the sample stream velocity and droplet formation rate of conventional cell sorters thus allowing an order of magnitude increase in processing speed. Secondly, a slit-laser illuminated microscope system has been constructed. This device has been used in the manual mode for examining microscope slides containing the sorted ghost samples but is also be capable of computer-controlled slide scanning and automated detection of labeled cells. We are also presently addressing the third point above with the development of mouse monoclonal antibodies to a variety of mutant hemoglobins. While whole animal methods have succeeded in producing monospecific antibodies to 11 additional single amino acid-substituted hemoglobins [11], the development of the hybridoma technique offers the most promise for obtaining the desired reagents. This approach has already been successfully applied as mouse monoclonal antibodies to myoglobin and to hemoglobin have been reported [12, 13]. In no other context can the inherent advantages of high purity, exquisite specificity and reliable antibody production of hybridoma-derived antibodies be better exploited than in this system. We have made a comprehensive survey of all known human hemoglobin single amino acid substitution variants and evaluated each for its applicability to this work. The criteria for inclusion were:

- the hemoglobin and/or variant cells be available in useful quantities;
- the variant can result from a single base change in the hemoglobin A gene;
- the variant is a stable hemoglobin, so mutant cells will not be selected against *in vivo*;
- the substituted amino acid be recognizable by the mouse immune system;
- a variety of base changes, particularly those leading to electrophoretic variants, be represented;
- both α and β globin variants be represented.

As shown in Table 2, application of these criteria results in a list of 21 candidate single amino acid-substituted hemoglobins in addition to S and C. We have begun initial immunizations using ten of these variants. Mice immunized with these variants, should respond with a family of monoclonal antibodies, the great majority of which will recognize regions of the hemoglobin molecule common to both the normal and variant forms. However, by screening positive clones against both the variants and hemoglobin A, it will be possible to identify those binding specifically to an altered protein. Ultimately, if such an approach is successful, a library of monoclonal antibodies each specific for a different and defined hemoglobin variant, can be generated. A battery of such antibodies could then be used together to detect a variety of variant hemoglobin-containing cells, thus improving the technical ease of the assay (since the frequency of antibody-labeled cells will be higher) and its generalizability, since damage at many sites in both the α and β globin genes, involving both DNA base transitions and transversions, could then be detected.

In addition to the single amino acid substituted-variants, there exist other variant hemoglobins produced by different mutational mechanisms. As shown in Fig. 4, two such hemoglobins, Cranston and Tak, are characterized by elongated β globin chains as a result of nucleotide insertions having occurred in the vicinity of the carboxyl terminal region of the β globin gene [14, 15]. These are

TABLE 2

Position	Hemoglobin Variant	Amino Acid Substitution	Base Change
α-Chain			
54	Mexico, J, J-Paris-II, Uppsala	Gln → Glu	C → G
	Shimonoseki, Hiroshima	Gln → Arg	A → G
56	Thailand	Lys → Thr	A → C
	Shaare, Zedek	Lys → Glu	A → G
57	L-Persian Gulf	Gly → Arg	G → A
	J-Norfolk, Kagoshima, Nishik-I, II, III	Gly → Asp	G → A
60	Zambia	Lys → Asn	G → T or C
	Dagestan	Lys → Glu	A → G
61	J-Buda	Lys → Asn	G → T or C
85	G-Norfolk	Asp → Asn	G → A
	Atago	Asp → Tyr	G → T
	Inkster	Asp → Val	A → T
90	J-Broussais, Tagawa-I	Lys → Asn	G → T or C
	Rajappen	Lys → Thr	A → C
120	J-Meerut, J-Birmingham	Ala → Glu	C → A
β-Chain			
20	Olympia	Val → Met	G → A
43	G-Galveston, G-Port Arthur, G-Texas	Glu → Ala	A → C
	Hoshida, Chaya	Glu → Gln	G → C
87	D-Ibadan	Thr → Lys	C → A
95	N-Baltimore, Hopkins-I, Jenkins, N-Memphis Kenwood	Lys → Glu	A → G
	Detroit	Lys → Asn	G → T or C

	144 146
β^{A}	. . . -Lys-Tyr-His
	. . . -AAG-TAT-CAC-TAA-GCTCGCTTT CTTGCTGTCCAATTTCTATTAA . . .
	144 157
$\beta^{Cranston}$	. . . -Lys-Ser-Ile-Thr-Lys-Leu-Ala-Phe-Leu -Leu-Ser-Asn-Phe-Tyr
	. . . -AAG-AGT-ATC-ACT-AAG-CTC-GCT -TTC-TTG-CTG-TCC-AAT-TTC-TAT -TAT-TAA-.
	144 157
β^{Tak}	. . . -Lys-Tyr-His-Thr-Lys-Leu-Ala-Phe-Leu -Leu-Ser-Asn-Phe-Tyr
	. . . -AAG-TAT-CAC-ACT-AAG-CTC-GCT -TTC-TTG-CTG-TCC-AAT-TTC-TAT -TAA-.

Fig. 4. Amino acid and nucleotide sequences for normal β chain, $\beta^{Cranston}$ and β^{Tak}. These variant hemoglobins, characterized by 11 additional amino acids at the amino-terminus of the β globin chain, result from the insertion of two nucleotides (underlined) in the normal β globin gene. Any frameshift mutation which produces this same altered reading frame in an otherwise normal β globin gene will cause the synthesis of this same 11 amino acid peptide. Amino acid sequences are from Bunn et al. [14], and Flatz et al. [15]. Nucleotide sequence is from Lawn et al. [29].

stable variants and make up 30-35% of the total hemoglobin produced in heterozygous cells [16, 17]. The 11 amino acid carboxyl-terminal peptide common to these variants is immunologically detectable since specific anti-hemoglobin Cranston has been produced in horses [18]. As discussed by Stamatoyannopoulos et al. [18], this same extended β chain sequence will be produced whenever there is a deletion of one (or 3n + 1) nucleotide or an insertion of two (or 3n + 2) nucleotides in the β globin gene provided that such frameshifts do not lead to an unstable globin chain or create an accidental termination codon within the β globin gene. A mouse monoclonal antibody specific for any part of the extra peptide will thus detect a family of β globin gene frameshift mutations. This monoclonal antibody will be particularly valuable since it will detect a family of variant red cells distinct from the population of rarer variant cells arising from point mutations.

THE GLYCOPHORIN A SYSTEM

In parallel with the variant hemoglobin-based assay we are developing a second independent system based on the biochemically well-studied protein glycophorin A [19]. Glycophorin A is a glycosylated red cell membrane protein present at about $5\text{-}10 \times 10^5$ copies per cell

[20]. Its 131 amino acid residues span the membrane with the amino-terminal portion presented on the red cell surface. The utility of this protein as a basis for a somatic cell mutation marker was suggested by the work of Furthmayer [21] which showed that this protein was responsible for the M and N blood group determinants and that these determinants were defined by a polymorphism in the amino acid sequence of the protein coded for by a pair of co-dominantly expressed alleles. The polymorphic sequence at the amino-terminus is shown below:

Glycophorin A(M)   Ser-Ser(*)-Thr(*)-Thr(*)-Gly-Val-...

Glycophorin A(N)   Leu-Ser(*)-Thr(*)-Thr(*)-Glu-Val-...

(*) indicates a glycosylated amino acid

Except for the amino acid substitutions at positions one and five of the sequence, the two proteins are identical, both in amino acid sequence and sites and structures of glycosylation. Individuals homozygous for the M or N allele synthesize only the A(M) or A(N) sequence respectively, while heterozygotes present equal numbers of the two proteins on their erythrocytes [21].

Two assays can be developed for the glycophorin A system. The first we call our glycophorin A "gene expression loss" or "null mutation" assay. In this approach, we wish to detect rare erythrocytes in the blood of glycophorin A heterozygotes which fail to express one or the other of the two allelic forms of the protein. Such an approach has been described for an in vitro system using human cells heterozygous for the multi-allelic HLA determinants [22, 23]. Using immunologic selection, these researchers have demonstrated that lymphoid cells can lose expression of one or more polymorphic HLA cell surface antigens as a result of spontaneous mutation. Exposure to radiation or chemical mutagens increases the background mutation rate by greater than two orders of magnitude. Analysis of these variants showed the majority to be single HLA gene mutants.

The glycophorin A "gene expression loss" approach has several inherent practical and biological advantages over the hemoglobin-based system. First, this antigen is presented on the surface of the red cell and is firmly anchored in the membrane; thus, cell preparation and antibody labeling procedures are simple and straightforward. Second, since the mutant phenotype detected can result from a variety of mutational lesions, i.e., single nucleotide changes and frameshifts occurring either in the glycophorin A structural gene or its control elements, the frequency of variant cells should be much higher (perhaps 100-1000 times the frequency seen for a single amino acid substitution at a single site). Hence such cells should be easier to detect and the frequency of such cells, representing the

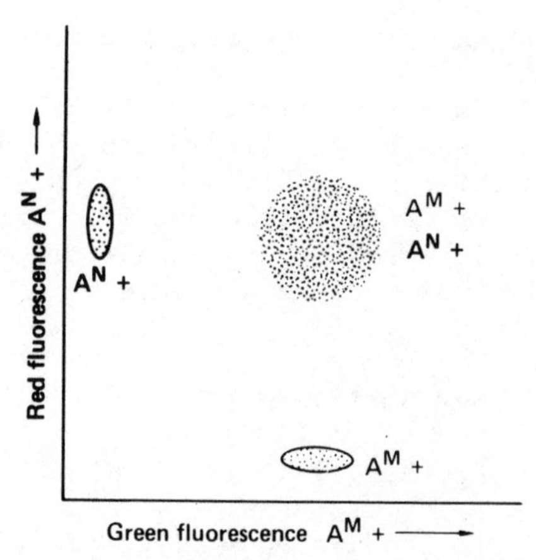

Fig. 5. Dual beam sorter, two color fluorescent detection of glycophorin A "null" variant red cells. The figure depicts a hypothetical two-color fluorescence histogram of heterozygous glycophorin A red cells incubated with monoclonal FITC-anti-glycophorin A(M) and X-RITC-anti-glycophorin A(N). The green and red fluorescence will be detected independently with 488 nm argon laser excitation of FITC and 568 nm krypton laser excitation of X-RITC. The normal cells express both the M and N glycophorin A sequences; they will bind both antibodies and exhibit both green and red fluorescence. The variant cells, lacking the expression of one allele, will fluoresce only green or red.

sum of all of these mutational mechanisms, may more accurately reflect the integrated genetic damage in that individual.

To detect the presence of these functionally hemizygous cells, we are at present generating mouse monoclonal antibodies which differentiate the M and N forms of the protein. Mouse monoclonal antibodies recognizing glycophorin A have been produced by Edwards [24] and we have adopted a variation of his immunization protocol. First, mice were injected with a equal mixture of homozygous MM and NN red cells, then boosted with purified glycophorin A(M) and A(N). The serum was then assayed for the presence of anti-red cell antibodies. The spleens from responding mice were then fused with SP2/0 mouse myeloma cells and anti-red cell producing clones were selected using a red cell enzyme-linked immunosorbant assay (ELISA). Positive clones were then assayed using homozygous MM and NN cells and those showing specificity for either cell type were selected, sub-cloned and expanded. Using this procedure, we have isolated four clones, two of which are specific for glycophorin A(M), one specific for A(N) and one which recognizes a shared determinant. Purified A(M)-

and A(N)-specific monoclonal antibodies will be labeled with green and red fluorophors, e.g., fluorescein isothiocyanate (FITC) and a derivative of rhodamine isothiocyanate (X-RITC) and incubated simultaneously with red cells from MN heterozygotes. Variant cells, defined by binding only one of the antibodies and hence fluorescing only green or red, will be enumerated as shown in Fig. 5 using the LLNL two-color dual-beam flow sorter [25]. The variant frequency will simply be the number of green- or red-only cells divided by the total number of cells processed (the sum of the signals in all three peaks). Because we expect the frequency of these variant cells to be as much as a 1000-fold higher than the frequency of single amino acid substitution-variant cells, direct sorter quantitation should be possible. Also adequate numbers of variant cells should be obtainable by sorting for biochemical analysis.

Since this assay is based on the detection of cells which fail to express a gene product it is critical to insure that the counted variant cells are true glycophorin A structural gene mutants. This is important since there are both genetic and non-genetic mechanisms which could cause the protein to fail to appear on the red cell membrane. For example, mutations outside the glycophorin A locus leading to loss of function of proteins necessary for processing, transporting, glycosylating or inserting glycophorin A into the membrane could produce apparent glycophorin A "null" cells. Non-genetic events include loss of membrane integrity or insufficient levels of substrate sugars for glycosylating enzymes due to metabolic anomalies. This assay is strongly protected against such false positive "phenocopies" since it requires antibody binding to one of the glycophorin A types. The proper cell surface presentation of the glycophorin A product of the unaffected allele insures that the rest of the cell apparatus necessary for the expression of the protein is intact. Finally, we can be assured that the variant cells will not be selected against _in vivo_ since erythrocytes from genetically homozygous glycophorin A "null" individuals, completely lacking expression of the protein, appear to exhibit normal viability [26].

Our second glycophorin A-based assay corresponds exactly to the hemoglobin-based single amino acid substitution approach by taking advantage of the mutational basis underlying the A(M), A(N) polymorphism. Not unexpectedly, the sequence differences at positions one and five can arise as a result of single nucleotide changes in the glycophorin A gene:

	Amino-Acid Difference		Corresponding Triplet Codons	
	A(M)	A(N)	A(M)	A(N)
Position 1	Ser	Leu	TC(A or G)	TT(A or G)
Position 5	Gly	Glu	GGG	GAG

Thus in homozygous MM individuals there should be rare cells containing "N-like" glycophorin A with Leu at position one or, independently, Glu at position five. Likewise in the blood of homozygous NN people should be rare cells with "M-like" glycophorin A with amino acids Ser or Gly at positions one and five respectively. To detect such single substitutions our A(M)- and A(N)-specific monoclonal antibodies must recognize the amino acid differences at positions one and five independently, i.e., the antibody must not depend on the presence of both differences for its binding specificity. It is most likely that our antibodies recognize the amino acid difference at position one. The amino-terminus of the molecule appears to be the immunodominant determinant recognized by animal anti-M and anti-N sera [27], although some sera may be directed at position five [28]. We are presently testing the specificity of our clones with chemically modified glycophorin A to precisely define their target antigenic sites. Given the expected specificity, we will use these antibodies to measure the frequency of these single amino acid-substituted glycophorins in the blood of homozygous MM and NN individuals. These two assays of glycophorin A variant-cells will be particularly useful since it will be possible to compare the frequency of a single nucleotide change with the frequency of the loss of expression of the entire gene in the same structural locus.

SUMMARY

This paper has outlined our approaches for measuring somatic cell mutations in human erythrocytes using high speed sorter technology. These devices are capable of processing large numbers of cells with statistical precision and quantitatively sorting antibody labeled, presumptively mutant, cells for subsequent analysis. All of our methods are based on immunologic detection of variant cells and hence depend on the availability of highly specific antibody reagents. Monoclonal antibodies are ideally suited for this purpose and are now in hand for the glycophorin A-based assays and under development in our laboratory and elsewhere for the hemoglobin-based assay. Armed with a battery of these hybridoma reagents it should be possible to detect variant erythrocytes arising from an ensemble of mutations in the α-, β-globin and glycophorin A genes. Before practical application, these assays must be validated by measurement of reproducible background rates in "unexposed" control populations, dose response in mutagen-exposed individuals and direct biochemical verification of sorted variant cells.

REFERENCES

1. G. H. Strauss and R. J. Albertini, Enumeration of 6-Thioguanine-Resistant Peripheral Blood Lymphocytes in Man as a Potential Test for Somatic Cell Mutations Arising _in vivo_, Mutation Res., 61:353-379 (1979).

2. T. H. J. Huisman and J. H. P. Jonxis, The Hemoglobinopathies: Techniques of Identification, Marcel Dekker, New York (1977).
3. Th. Papayannopoulou, T. C. McGuire, G. Lim, E. Garzel, P. E. Nute, and G. Stamatoyannopoulos, Identification of Haemoglobin S in Red Cells and Normoblasts, Using Fluorescent Anti-Hb S Antibodies, Brit. J. Haemat., 34:25-31 (1976).
4. Th. Papanannouploulou, G. Lim, T. C. McGuire, V. Ahern, P. E. Nute, and G. Stamatoyannopoulos, Use of Specific Fluorescent Antibodies for the Identification of Hemoglobin C in Erythocytes, Amer. J. Hemat., 2:105-112 (1977).
5. G. Stamatoyannopoulos, Possibilities for Demonstrating Point Mutations in Somatic Cells, as Illustrated by Studies of Mutant Hemoglobins, in: Genetic Damage in Man Caused by Environmental Agents (K. Berg, ed.), pp. 49-62, Academic Press, New York (1979).
6. K. Wang and F. M. Richards, Reaction of Dimethyl-3,3-dithiobispropionimidate with Intact Human Erythrocytes, J. Biol. Chem., 250:6622-6626 (1975).
7. W. L. Bigbee, E. W. Branscomb, H. B. Weintraub, Th. Papayannopoulou, and G. Stamatoyannopoulous, Cell Sorter Immunofluorescence Detection of Human Erythrocytes Labeled in Suspension with Antibodies Specific for Hemoglobins S and C, J. Immunol. Meth., 45:117-127 (1981).
8. J. J. Aragón, J. E. Feliu, R. A. Frenkel, and A. Sols, Permeabilization of Animal Cells for Kinetic Studies of Intracellular Enzymes: In situ Behavior of the Glycolytic Enzymes of Erythrocytes, Proc. Natl. Acad. Sci. (U.S.A.), 77:6324-6328 (1980).
9. H. W. Goedde, H.-G. Benkmann, and L. Hirth, Ultrathin-Layer Isoelectric-focusing for Rapid Diagnosis of Protein Variants, Hum. Genet., 57:434-436 (1981).
10. S. I. O. Anyaibe and V. E. Headings, Identification of Inherited Protein Variants in Individual Erythrocytes, Biochem. Genet., 18:455-463 (1980).
11. F. A. Garver, M. B. Baker, C. S. Jones, M. Gravely, G. Altay, and T. H. J. Huisman, Radioimmunoassay for Abnormal Hemoglobins, Science, 196:1334-1336 (1977).
12. J. A. Berzofsky, G. Hicks, J. Fedorko, and J. Minna, Properties of Monoclonal Antibodies Specific for Determinants of a Protein Antigen, Myoglobin, J. Biol. Chem., 255:11188-11191 (1980).
13. G. Stamatoyannopoulos, D. Lindsley, Th. Papayannopoulos, M. Farquhar, M. Brice, P. E. Nute, G. R. Serjeant, and H. Lehmann, Mapping of Antigenic Sites on Human Haemoglobin by Means of Monoclonal Antibodies and Haemoglobin Variants, Lancet ii, 952-954 (1981).
14. H. F. Bunn, G. J. Schmidt, D. N. Haney, and R. G. Dluhy, Hemoglobin Cranston, an Unstable Variant Having an Elongated β Chain due to Nonhomologous Crossover between Two Normal β Chain Genes, Proc. Natl. Acad. Sci. (U.S.A.), 72:3609-3613 (1975).

15. G. Flatz, J. L. Kinderlerer, J. V. Kilmartin, and H. Lehmann, Haemoglobin Tak: A Variant with Additional Residues at the End of the β Chains, Lancet i, 732-733 (1971).
16. J. R. Shaeffer, G. J. Schmidt, R. E. Kingston, and H. F. Bunn, Synthesis of Hemoglobin Cranston, an Elongated β Chain Variant, J. Mol. Biol., 140:377-389 (1980).
17. K. Imai and H. Lehman, The Oxygen Affinity of Haemoglobin Tak, a Variant with an Elongated β Chain, Biochim. Biophys. Acta, 412:288-294 (1975).
18. G. Stamatoyannopoulos, P. E. Nute, Th. Papayannopoulou, T. McGuire, G. Lim, H. F. Bunn, and D. Rucknagel, Development of a Somatic Mutation Screening System Using Hb Mutants, IV. Successful Detection of Red Cells Containing the Human Frameshift Mutants Hb Wayne and Hb Cranston Using Monospecific Fluorescent Antibodies, Am. J. Hum. Genet., 32:482-496 (1980).
19. H. Furthmayer, Structural Analysis of a Membrane Glycoprotein: Glycophorin A, J. Supramol. Struct., 7:121-134 (1977).
20. C. G. Gahmberg, M. Jokinen, and L. C. Andersson, Expression of the Major Red Cell Sialoglycoprotein, Glycophorin A, in the Human Leukemic Cell Line K562, J. Biol. Chem., 254:7442-7448 (1979).
21. H. Furthmayer, Structural Comparison of Glycophorins and Immunochemical Analysis of Genetic Variants, Nature, 271:519-524 (1978).
22. D. Pious and C. Soderland, HLA Variants of Cultured Human Lymphoid Cells: Evidence for Mutational Origin and Estimation of Mutation Rate, Science, 197:769-771 (1977).
23. P. Kavathas, F. H. Bach, and R. DeMars, Gamma Ray-Induced Loss of Expression of HLA and Glyoxalase I Alleles in Lymphoblastoid Cells, Proc. Natl. Acad. Sci. (U.S.A.), 77:4251-4255 (1980).
24. P. A. W. Edwards, Monoclonal Antibodies that Bind to the Human Erythrocyte-Membrane Glycoproteins Glycophorin A and Band 3, Biochem. Soc. Trans., 8:334-335 (1980).
25. P. N. Dean and D. Pinkel, High Resolution Dual Laser Flow Cytometry, J. Histochem. Cytochem., 26:622-627 (1978).
26. M. J. A. Tanner and D. J. Anstee, The Membrane Change in En(a-) Human Erythrocytes, Biochem. J., 153:271-277 (1976).
27. H. Furthmayer, M. N. Metaxas and M. Metaxas-Bühler, M^g and M^c: Mutations within the Amino-Terminal Region of Glycophorin A, Proc. Natl. Acad. Sci. (U.S.A.), 78:631-635 (1981).
28. W. Dahr, M. Kordowicz, K. Beyreuther, and J. Krüger, The Amino-Acid Sequence of the M^c-Specific Major Red Cell Membrane Sialoglyoprotein - An Intermediate of the Blood Group M- and N-Active Molecules, Hoppe-Seyler's Z. Physiol. Chem., 362:363-366 (1981).
29. R. M. Lawn, A. Efstratiadis, C. O'Connell, and T. Maniatis, The Nucleotide Sequence of the Human β-Globin Gene, Cell, 21: 647-651 (1980).

PERIPHERAL BLOOD LYMPHOCYTES AS INDICATOR CELLS FOR *in vivo* MUTATION IN MAN

Gösta Zetterberg

Department of Genetics
University of Uppsala
Uppsala, Sweden

1. INTRODUCTION

For the determination of the mutation rate of a single gene it is necessary to screen a very large number of the individual units chosen as testing objects. Single cell systems have been developed making possible the effective scoring of billions of cells in a short time. The mutants observed and counted are clones, originating from a single mutant cell. Most of the methods having microorganisms as testing objects use selective media in which only the mutants can grow out to colonies large enough to be scored by the naked eye. The same principles have been applied to mammalian cells, transformed to grow continuously, e.g., cell lines of Chinese hamster or humans. Results from such mutation tests are used for the risk evaluation of human exposure to genotoxic agents. However, the information gained in such tests about the genotoxic effects of a chemical in humans has limited value, mainly because human pharmacokinetic and metabolic factors are difficult to mimic. Therefore, there is great potential value in a mutation test performed with human cells constructed so that mutation can occur *in vivo*, the indicator cells can be withdrawn, and the mutant character can be developed *in vitro*.

This approach has been used in a few attempts to develop such test systems. Cells from human blood are the easiest to sample and they have been used in tests including the following genetic markers: (i) loss of agglutinogen A from erythrocytes of AB or A blood groups [1, 2], (ii) persistence of foetal hemoglobin [3, 4], (iii) glucose-6-phosphate dehydrogenase deficiency [3], (iv) mutant hemoglobins [5, 6], and (v) hypoxanthine-guanine-phosphoribosyl transferase (HGPRT) deficiency [7-10].

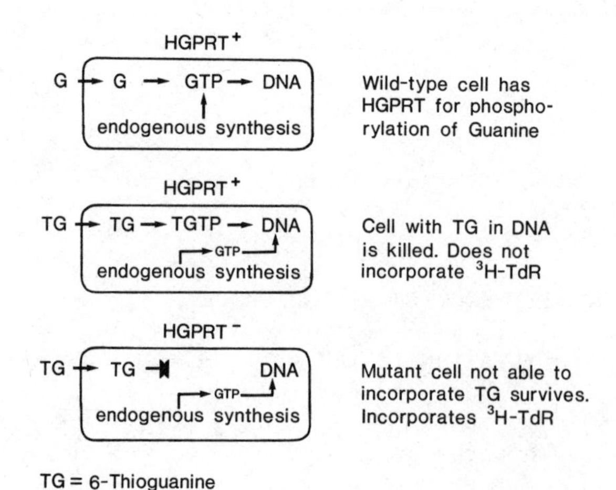

Fig. 1. The toxic effect of 6-thioguanine (TG) on cells with active HGPRT enzyme and the principle for selection of HGPRT defective mutants.

The prototype for the mutation is the cells found in Lesch-Nyhan (LN) syndrome patients [12, 20-22]. This rare syndrome is inherited as an X-chromosome linked disorder and associated with an impaired HGPRT activity. In non-LN persons cells may occur that have lost HGPRT activity and become resistant to AG and TG [8]. Such variant cells may be due to mutation at the HGPRT locus.

2.2. Methods for the Detection of HGPRT Mutants

When peripheral human lymphocytes are cultured in a medium containing AG or TG wild-type cells expressing HGPRT activity are killed while variant cells, defective in HGPRT activity, will survive and go through several cell divisions. In a liquid medium with free cells the colony-forming approach is not applicable so the resistance of the variant cells (AG^r, TG^r, or $HGPRT^-$) is indicated by their ability to synthesize DNA during incubation in medium containing the purine analog. This capacity is shown by giving a 6-12 hr pulse of tritiated thymidine (^{3}H-TdR) and subsequently analyzing autoradiographs of isolated nuclei. Labeled nuclei originate from resistant cells [8-10]. The endpoint of the method is the microscopic counting of rare labeled nuclei on slides with hundreds of thousands of nuclei. This is tedious work and in my laboratory we have developed a method where a flow cytometer is used to enrich for TG^r nuclei. Details of this method have been published elsewhere [10].

2.2.1. The Flow Cytometric Method

The principle on which the use of the flow cytometer is based is that nuclei of cells resistant to TG contain more DNA than sensi-

I shall discuss the restrictions and possibilities of methods using human lymphocytes as testing objects and I shall refer to results obtained in tests using the last-mentioned marker, the HGPRT locus.

2. THE HGPRT SYSTEM

The HGPRT system has been extensively used in studies of induced mutation in fibroblast cell lines of Chinese hamster and man [11]. Strauss and Albertini [8] developed a method for the study of gene mutation occurring *in vivo* in human lymphocytes using the HGPRT system. In my laboratory we have applied flow cytometry to increase the resolution and applicability of this test system in epidemiological investigations [10].

2.1. Rationale for the Use of the HGPRT Marker

Since human somatic cells are diploid a mutant allele can be either dominant, recessive, or anything in between, depending on the criteria used in the method of testing. For example, codominance would mean that both the wild-type gene product and the mutant gene product are synthesized and expressed in the cell. In such a case the method used should allow the identification of the mutant gene product or of the phenotype of the mutant cell.

For genes on the X chromosome the situation is different. In somatic cells from a male the genes on the single X chromosome have no homologous alleles present and are thus expressed. In somatic cells from a female one of the X chromosomes is heterochromatic and silent, thus causing a situation similar to that in a male cell. There seems to be no preference for which of the two X chromosomes becomes heterochromatic and nonfunctional. Theoretically, in a female heterozygous at an X-chromosome linked locus, 50% of the cells express the mutant character and the other 50% the wild-type character. Thus, for calculations of the frequency of mutation of X-chromosome linked genes there is no difference between females and males. The chance of detecting the mutant phenotype is the same since in cells from both sexes there is only one functional X chromosome.

The genetic marker used in the HGPRT system is an X-chromosome linked gene coding for hypothanthine-guanine phosphoribosyltransferase (HGPRT; E.C.2.4.2.8). The enzyme functions in a salvage pathway for free purine phosphorylation converting hypoxanthine and guanine to inosine 5'-monophosphate and guanine 5'-monophosphate, respectively [12-14]. The enzyme also catalyzes the phosphorylation of the purine analogs 8-azaguanine (AG) and 6-thioguanine (TG) which are then incorporated into the nucleic acids and thereby rendered toxic to the cell (Fig. 1). Since the analogs are toxic only after phosphorylation, mutants with impaired HGPRT activity are partially or totally resistant to the analogs [12-19].

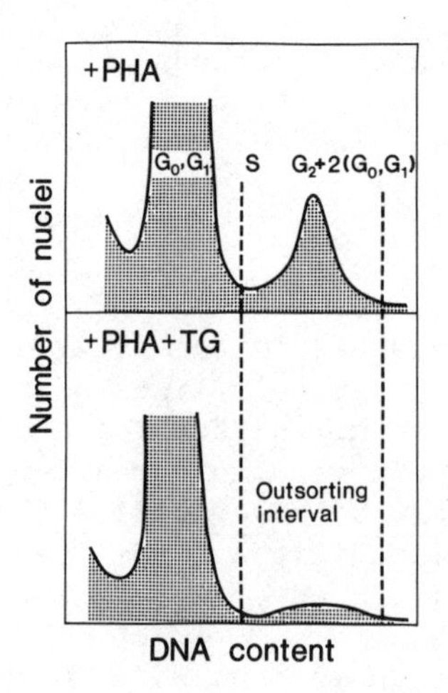

Fig. 2. The distribution of nuclei according to DNA content as registered with the flow cytometer. The lymphocytes were grown for 48 hr in the absence (+ PHA) and in the presence (+ PHA + TG) of 6-thioguanine. Nuclei in the indicated range (outsorting interval) are outsorted which gives an enrichment of nuclei for TG resistant cells.

tive cells do after 48 hr of cultivation in a medium containing TG. At the start of the culture the lymphocytes are in G_0 phase. A mitogen, phytohemagglutinin (PHA), stimulates a certain proportion of the cells to divide fairly synchronously.

The kinetics of the cell culture were studied with the help of flow cytometry. Propidium iodide was used as fluorophor to measure the DNA content of nuclei in samples withdrawn from PHA-stimulated cultures at different times. On the basis of studies of the kinetics of the culture we decided to use cells from cultures incubated for 48 hr. This time was considered optimal with respect to the number of cells (S + G_2) phase of their first cell cycle. At 40 hr of incubation about 30% of the cells were in S-phase, and 8 hr later 12% had reached G_2. The maximum number of cells in (S + G_2) was observed at 54 hr of culturing. It cannot be excluded that a minor fraction of the cells had completed their first cycle at 48 hr, but they should not have progressed to the S-phase of their second cell cycle.

Theoretically, with a flow cytometer that can precisely measure the DNA content of individual cells, the resistant cells can be directly separated from the sensitive cells on the basis of the higher DNA content of the resistant cells allowed to reach the (S + G_2) phase in presence of TG. Thus we selected in a control culture a region of the DNA content distribution starting at the minimum be-

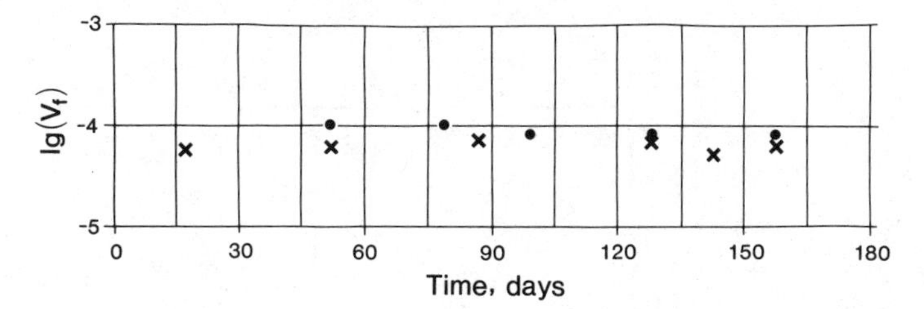

Fig. 3. The variant frequency in blood samples from two donors (X and O) taken at intervals over half a year. [TG] = 2×10^{-4} M.

tween the ($G_0 + G_1$) peak and the G_2 peak and extending beyond the G_2 peak (Fig. 2). When we measured a TG culture for the proportion of nuclei falling within this interval it turned out that it was 1×10^{-4}-5×10^{-3}. This is much higher than one should expect if mutation in the HGPRT locus were reponsible for their TG resistance and increased content of DNA. Autoradiography of outsorted nuclei in the ($S + G_2$) region showed that the majority of the nuclei had not synthesized DNA in the presence of TG. Most of the outsorted nuclei originated from unlabeled sensitive cells and were sorted out as doublets. The flow cytometer cannot effectively discriminate particles according to size. The frequency of these doublets was variable and we could not reduce the frequency of doublets to a sufficiently low level. Thus, an autoradiographic analysis had to be included. Unlike the earlier approaches [8, 9], however, the outsorting of nuclei in the ($S + G_2$) region provided us with an enrichment of nuclei for resistant cells. The enrichment factor was usually about 500-fold, depending on the frequency of doublets in the analyzed suspension of nuclei. This enrichment reduced the time necessary for microscopic counting of ^{3}H-TdR labeled nuclei and increased the accuracy with which the frequency of variant cells could be determined. The variant frequency (Vf) was calculated according to:

$$\text{Vf} = \frac{\text{labeling index in the outsorted region in presence of TG}}{\text{labeling index in the outsorted region in absence of TG}}$$

A study of the methodological variation showed a scatter around the mean of the variant frequency of ±20% for cultures started at the same time, and ±30% for cultures from the same person but started on different occasions. For two persons serving as internal standards the variant frequency was stable within ±20% of the mean during a period of about six months (Fig. 3).

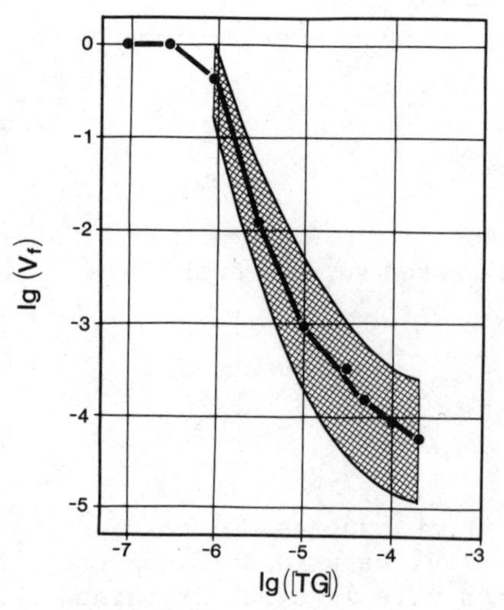

Fig. 4. The variant frequency at different TG concentrations in blood samples from 33 control persons. The scatter around the mean values for each TG concentration is indicated by the shaded area.

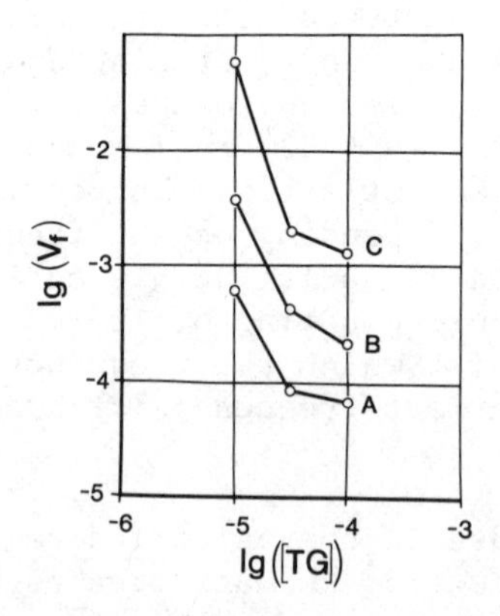

Fig. 5. The variant frequency at three concentrations of TG in blood samples from a cancer patient taken before (A) and after two treatments with cisplatinum (B and C) with 10 days interval.

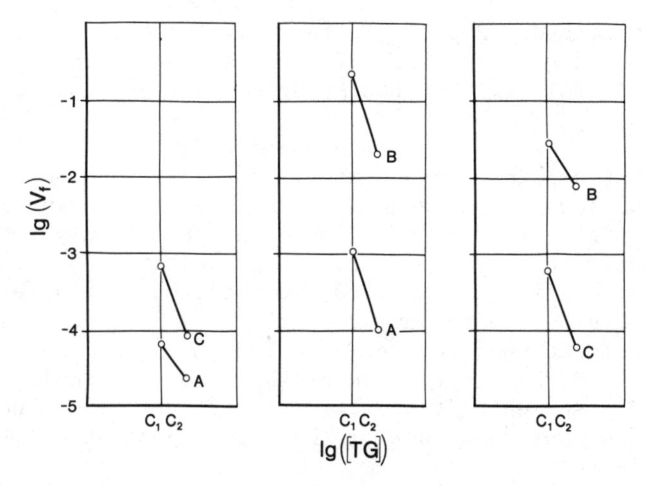

Fig. 6. The variant frequency at 1×10^{-4} M (C_1) and 2×10^{-4} M (C_2) 6-thioguanine in lymphocytes from three patients treated with PUVA. Blood samples were taken before (A), two days after (B), and 10 days after (C) the treatment.

3. RESULTS

Using the methods described above we determined the frequencies of TG resistant cells in blood samples from (i) a group of 33 donors not supposed to have been exposed to mutagenic agents (Fig. 4), (ii) cancer patients treated with cytostatics (Fig. 5), and (iii) patients treated with psoralen and UV light (PUVA) (Fig. 6). For the two latter we found much elevated frequencies of TG-resistant cells. The frequencies of TG^r cells that we found in blood samples from supposedly unexposed persons coincided well with the frequencies reported for control persons by Strauss and Albertini [8] and by Evans and Vijayalaxmi [9], the latter using AG as selective agent.

4. DISCUSSION

4.1. Are the High Frequencies of Variant Cells Found After Treatments with Cytostatics and PUVA Due to Somatic Mutation?

In general, the frequency of mutant cells determined in a test is dependent on the criteria applied in the classification of the mutant phenotype. Usually the mutant character is deduced from a different function at the phenotypic level, e.e.g., an impaired activity of the enzyme coded on the gene for which the mutation rate is to be determined. However, often only the extremes, full function (100%), and total absence of function can be scored in a selective test. It follows that the mutation rate determined in such a system generally will be under-estimated. The system does not take into account many of the possible DNA alterations, extending from base substitutions that do not at all change the amino acid sequence of the enzyme (due to the degenerate nature of the genetic code), to substitutions of amino acids at positions that simply reduce the catalytic activity of the enzyme. Thus, depending on the type and strength of the selection applied the mutation rate determined for a gene may turn out to be very different. As yet no mutation test has been developed that can record all DNA alterations brought about by a mutagen. Such a test would probably be too complicated to be practical in a screening test for suspected mutagens. However, without such a test our knowledge of the "true" gene stability is at the best approximate.

We have discussed elsewhere [10] the background of the highly elevated Vfs in patients treated with cytostatics and PUVA. In summary, we believe it is unlikely that such high frequencies of TG resistant cells can be due to mutation, and we have suggested that the majority are phenocopies, resistant to TG because their HGPRT activity is (temporarily?) too low to make them sensitive. We suggested that new synthesis of the HGPRT enzyme is blocked by an inhibition of transcription.

Evans and Vijayalaxmi [9] also suggested that the high Vf they obtained after *in vitro* treatment with mitomycin C was due to an inhibited transcription of the HGPRT locus. *In vitro* X-irradiation also increased Vf in their experiments and they concluded from data on X-ray induced chromosome breaks in human lymphocytes that the cells became TG resistant because of breaks in the X-chromosome at the HGPRT locus.

4.2. Are the Variant TG^r Lymphocytes Found in Control Persons Due to Somatic Mutation?

In cultures of mammalian fibroblast cell lines the frequency of spontaneous TG- and AG-resistant mutants is usually lower than has been found in human lymphocyte cultures. However, according to the model of the random occurrence of mutation, the frequency of mutants at the time of sampling should have no typical value. Results from fluctuation tests show that in a model system like the peripheral blood cells the frequency of mutants can be expected to vary between persons. Depending on when a mutational event occurs, on the stem cell level, in a cell generation preceding the peripheral lymphocyte stage, or in a peripheral lymphocyte, the mutant frequency can vary a lot when the test is made at the end. It follows that in older persons one can expect a higher frequency of mutants due to accumulation of mutants with time in the stem cell pool. As shown in Fig.

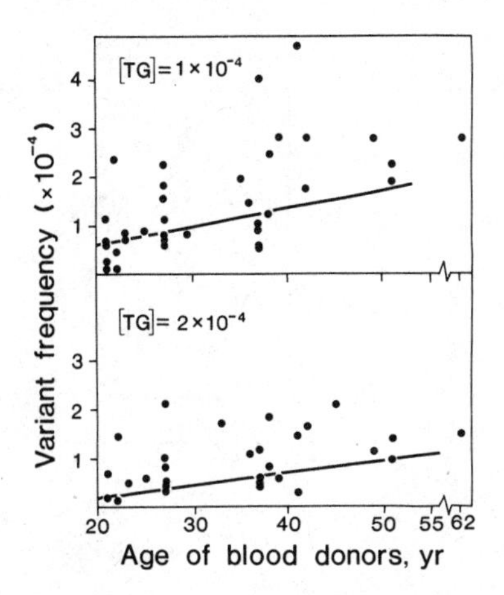

Fig. 7. The variant frequency at 1×10^{-4} M and 2×10^{-4} M TG as a function of the age of the blood donors.

7 we registered higher Vf values in older persons than in younger ones, but the differences were not great. A similar age dependence for Vf was shown by Evans and Vijayalaxmi [9], but no evidence for higher Vfs at higher age was found by Strauss and Albertini [8].

A positive correlation between Vf and age is counteracted by selection against the HGPRT deficient cells. That such a selection occurs is indicated by the fact that in women heterozygous at the HGPRT locus the proportion of TG^r variants is only a few percent while the expectation is 50%. It is probable that the selection against HGPRT deficient cells is strong on the stem cell level where DNA synthesis is required for cell divisions. In nondividing peripheral lymphocytes such a selection should not be so pronounced.

As seen in Fig. 4 the variant frequency in control persons was dependent on the concentration of TG in the medium. Figure 8 also shows that there was a considerable variation between different persons in the Vf registered at two TG concentrations. This might reflect differences in the HGPRT activity. The inhibiting effect of TG on the lymphocytes should be dependent on the amount of product formed by HGPRT, the phosphorylated purine analog. Furthermore, the amount of toxic product formed should be determined by the concentrations of HGPRT and TG. It follows that cells with less HGPRT activity should be less sensitive for a given concentration of TG. For persons with inherent low HGPRT activity this will give a higher Vf at a certain TG concentration, compared to that of persons with higher HGPRT activity. However, if the concentration of TG is high enough to inhibit also cells with low HGPRT activity, the selection curve would reach a plateau. In the experiments by Strauss and Albertini [8] and by Evans and Vijayalaxmi [9] such a plateau was

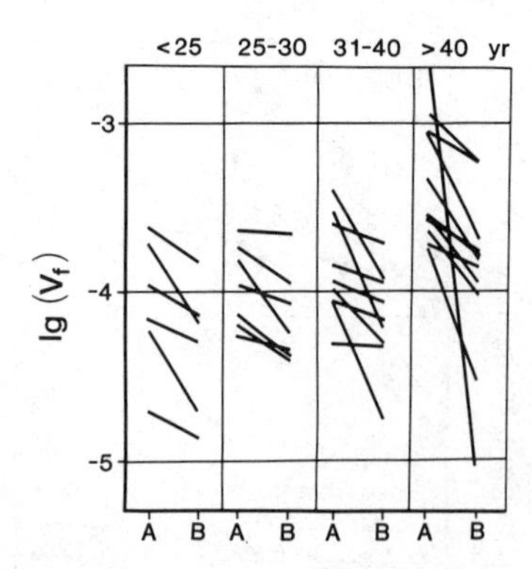

Fig. 8. The variant frequency determined at two TG concentrations, 1×10^{-4} M (A) and 2×10^{-4} M (B), for control persons of different age.

indicated at about 1×10^{-4} M TG and AG, respectively. Also we obtained selection curves of a similar shape in cultures from many persons but for some persons the selection curve never reached a plateau with increasing TG concentration.

4.3. Is the Heterogenous Population of Lymphocytes Used in the Test an Acceptable Model System?

It is known [23] that the fraction of mononuclear blood cells isolated by gradient centrifugation is indeed a mixture of subpopulations of cells. For example, it has been shown with immunological methods that T lymphocytes can be grouped into different subpopulations. Different mitogens are known to stimulate different populations of lymphocytes, PHA being supposed to initiate division of mainly T lymphocytes. It may be that the spectrum of T lymphocyte populations varies from person to person and that this complicates attempts to standardize factors like cell cycle length, HGPRT activity and the overall sensitivity to TG or AG. Different treatments, e.g., with cytostatics, might change the spectrum of subpopulations amenable to study with the methods described.

Albertini et al. [24], found that a small proportion of the lymphocytes were in S-phase already at the start of the culture, indicating that these cells were dividing in vivo. Furthermore, it was shown that many of these cycling cells were resistant to TG. Freezing the blood samples before culturing caused a synchronization of the cycling cells so that upon thawing and culturing they rapidly passed their next S-phase. When the labeling with ^{3}H-TdR was made after these cells had gone through their S-phase the registered Vf was found to be an order of magnitude lower than in cultures from unfrozen samples. The authors consider these lower Vf values, found in control persons as well as in cancer patients treated with cytostatics, to be more realistic supposing that the TG resistance of variant cells is due to somatic mutation.

The very high Vf values found in unfrozen samples from patients treated with cytostatics or PUVA [8, 10] indicates that, if the interpretation of the freezing effect is valid [24], the treatments cause an increase of the number of in vivo cycling lymphocytes resistant to TG. However, we have preliminary data from PUVA-treated psoriasis patients with elevated Vf showing no increase of cycling cells in unstimulated lymphocyte cultures. More data are needed before the origin and nature of TG resistant cells induced by cytostatics or PUVA can be deduced.

4.4. Are Induced Phenocopies of any Importance in Risk Assessment?

Phenocopies like the TG resistant cells present in such a high frequency (10^{-2}-10^{-1}) after treatments with cytostatics or PUVA might represent an important physiological condition which could become critical for the cell, the tissue, and for the organism in which it is present. There is a complicated interaction between genes and gene products on the regulatory level. Models have been proposed [25] for how metabolic pathways may be irreversibly shut off, or alternatively become constitutively opened by changing metabolite patterns. It is not known whether such non-mutational permanent changes are of importance for neoplastic transformation or cancer promotion. Figure 6 shows that the elevated frequencies of TG resistant cells after treatment with PUVA were only partly reversible.

5. FUTURE DEVELOPMENT

Albertini and Borcherding [26], Albertini [27], and Strauss [28] have reported that TG resistant lymphocytes can be cloned with the help of T-cell growth factors (TCGF = interleukin-2) and X-irradiated allogenic peripheral blood mononuclear feeder cells. In the cloning assay it was found that Vf in control persons was of the order of 10^{-6}-10^{-5}, which corresponds well with the values found in autoradiographic assessment of Vf in cultures started from frozen blood samples. The cloning assay proved that HGPRT defective variant cells are of a stable phenotype and most likely the result of somatic mutation. The nature of the mutational alterations can be investigated when a DNA probe of the HGPRT gene is available.

If the methods described above can be correlated with the cloning assay it seems likely that these methods would prove to be more useful in epidemiological investigations where the number of tests required will be high and simple and inexpensive methods might be preferred.

Other genetic markers in lymphocytes may be used in developing tests for the study of the effects of genotoxic agents. It seems likely that immunological methods, using monoclonal antibodies in combination with flow cytometry, will be further developed and applicable to tests with human lymphocytes.

ACKNOWLEDGEMENT

Work supported by Grants 79/1069, 80/1375, and 81/2122 from the Swedish Council for Planning and Coordination of Research and by Grant B-2468-101 from the Swedish Natural Science Research Council. I thank Dr. Irene M. Jones, Biomedical Sciences Division, Lawrence Livermore National Laboratory, University of California, for critical reading of this manuscript.

REFERENCES

1. K. C. Atwood, The presence of A_2 erythrocytes T in A_1 blood, Proc. Natl. Acad. Sci. USA, 44:1054-1057 (1958).
2. K. C. Atwood and S. L. Scheinberg, Somatic variation in human erythrocyte antigens, J. Cell Comp. Phys., 52:97-123 (1958).
3. H. E. Sutton, Monitoring of somatic mutations in human populations, in: Mutagenic Effects on Environmental Contaminants (E. H. Sutton and M. I. Harris, eds.), Academic Press, New York, pp. 121-128 (1972).
4. H. E. Sutton, Somatic cell mutations, in: Birth Defects; Proceedings 4th International Conference (A. G. Motulsky and W. Lenz, eds.), Exerpta Medica, Amsterdam, pp. 212-214 (1974).
5. T. Papayannopoulou, T. C. McGuire, G. Lim, E. Garzel, P. E. Nute, and G. Stamatoyannopoulos, Identification of haemoglobin S in red cells and normoblasts, using fluorescent anti-Hb S antibodies, Br. J. Haematol., 34:25-31 (1976).
6. W. L. Bigbee, E. W. Branscomb, H. B. Weintraub, Th. Papayannopoulou, and G. Stamatoyannopoulos, Cell sorter immunofluorescence detection of human erythrocytes labeled in suspension with antibodies specific for hemoglobin S and C, J. Immunol. Methods, 45:117-127 (1981).
7. G. H. Strauss and R. J. Albertini, 6-Thioguanine resistant lymphocytes in human peripheral blood, in: Progress in Genetic Toxicology (D. Scott, B. A. Bridges, and F. H. Sobels, eds.), Elsevier/North Holland, Amsterdam, pp. 327-334 (1977).
8. G. H. Strauss and R. J. Albertini, Enumeration of 6-thioguanine-resistant peripheral blood lymphocytes in man as a potential test for somatic cell mutations arising *in vivo*, Mutat. Res., 61:353-379 (1979).
9. H. J. Evans and Vijayalaxmi, Induction of 8-azaguanine resistance and sister chromatid exchange in human lymphocytes exposed to mitomycin C and X rays *in vitro*, Nature, 292:601-605 (1981).

10. H. Amnéus, P. Matsson, and G. Zetterberg, Human lymphocytes resistant to 6-thioguanine: Restrictions in the use of a test for somatic mutations arising in vivo studied by flow-cytometric enrichment of resistant cell nuclei, Mutat. Res., 106: 163-178 (1982).
11. C. T. Caskey and G. D. Kruh, The HPRT locus, Cell, 16:1-9 (1979).
12. R. DeMars, Genetic studies of HGPRT-deficiency and the Lesch-Nyhan syndrome with cultured human cells, Fed. Proc., 30:944-955 (1971).
13. R. DeMars, Resistance of cultured human fibroblasts and other cells to purine and pyrimidine analogoues in relation to mutagenesis detection, Mutat. Res., 24:335-364 (1974).
14. H. N. Kelley, Enzymology and biochemistry, A. HG-PRT deficiency in the Lesch-Nyhan syndrome and gout, Fed. Proc., 27:1047-1052 (1968).
15. P. Stutts and R. W. Brockman, A biochemical basis for resistance of L 1210 mouse leukeran to 6-thioguanin, Biochem. Pharmacol., 12:97-104 (1963).
16. G. B. Elion and G. H. Hitchings, Metabolic basis for the actions of analogs of purines and pyrimidines, in: Advances in Chemotheraphy (A. Goldin, F. Hawkin, and R. J. Schnitzer, eds.), Vol. 2, pp. 91-97, Academic Press, New York (1965).
17. G. B. Elion, Biochemistry and pharmacology of purine analogs, Fed. Proc., 26:898-904 (1967).
18. R. P. Miech, R. E. Parks, Jr., J. H. Anderson, Jr., and A. C. Sartorelli, A hypothesis on the mechanism of action of 6-thioguanine, Biochem. Pharmacol., 16:2222-2227 (1967).
19. J. Arly-Nelson, J. W. Carpenter, L. M. Rose, and D. J. Adamson, Mechanisms of action of 6-thioguanine, 6-mercaptopurine, and 8-azaguanine, Cancer Res., 35:2872-2878 (1975).
20. M. Lesch and W. L. Nyhan, A familial disorder or uric acid metabolism and control nervous system function, Am. J. Med., 36:561-570 (1964).
21. J. E. Seegmiller, F. M. Rosenbloom, and W. N. Kelley, Enzyme defect associated with a sex-linked human neurological disorder and excessive purine synthesis, Science, 155:1682-1684 (1967).
22. C. H. M. M. de Bryun, Hypoxanthine-guanine phosphoribosyltransferase deficiency, Human Genet., 31:127-150 (1976).
23. L. B. Vogler, C. E. Grossi, and M. D. Cooper, Human lymphocyte subpopulations, Progress in Hematology, 11:1-45 (1979).
24. R. J. Albertini, E. F. Allen, A. S. Quinn, M. R. Albertini, Human somatic cell mutation: In vivo variant lymphocyte frequencies as determined by 6-thioguanine resistance, in: Population and Biological Aspects of Human Mutation (E. B. Hook and J. H. Porter, eds.), pp. 235-263, Academic Press, New York (1981).
25. R. P. Wagner and H. K. Mitchell, in: Genetics and Metabolism, 2nd ed., John Wiley and Sons, Inc., New York (1964).

26. R. J. Albertini and W. R. Borcherding, Cloning in vitro of human 6-thioguanine resistant (TG^r) peripheral blood lymphocytes (PBL's) arising in vivo, Abstract 13th Ann. Mtg. Environ. Mutagen Society, Environ. Mutag. (in press) (1982).
27. R. J. Albertini, STudies with T-lymphocytes: An approach to human mutagenicity monitoring, in: Indicators of Genotoxic Exposure (B. A. Bridges, B. E. Butterworth, and I. B. Weinstein, eds.), Banbury Report, 13:393-412, Cold Spring Harbor Laboratory (1982).
28. G. H. S. Strauss, Direct mutagenicity testing: The development of a clonal assay to detect and quantitate mutant lymphocytes arising in vivo, in: Indicators of Genotoxic Exposure (B. A. Bridges, B. E. Butterworth, and I. B. Weinstein, eds.), Banbury Report, 13:423-441, Cold Spring Harbor Laboratory (1982).

DISCRIMINATION BETWEEN SPONTANEOUS AND INDUCED MUTATIONS IN HUMAN CELL POPULATIONS BY USE OF MUTATIONAL SPECTRA

Thomas R. Skopek*, Phaik-Mooi Leong,
and William G. Thilly

Toxicology Group
Department of Nutrition and Food Science
Massachusetts Institute of Technology
Cambridge, Massachusetts 02139

*Department of Molecular Biophysics and Biochemistry
Yale University
P. O. Box 1937, Yale Station
New Haven, Connecticut 06520

INTRODUCTION

The central premise of genetic toxicology is that environmental agents are responsible for genetic changes in man. This premise has not been tested and we thus have not confirmed if mutations in the population are predominantly spontaneous or environmentally induced. We propose in this paper a method that would permit us to discriminate between spontaneous and induced mutations in cells isolated from a population and thus diagnose the major mutagenic stimuli in humans.

DETERMINATION OF MUTATIONS IN HUMANS: POSSIBLE APPROACHES

One approach to assaying mutation in human cells would be to obtain cell samples from individuals and to determine the frequency of recognizable mutations in each. The frequency of mutations could be used as a parameter for determining if mutations were spontaneous or induced; we would expect higher frequencies of mutations in those persons exposed to mutagens and lower levels in those spared exposure if the central premise were correct. However, who has been spared exposure to environmental mutagens and what is the level of spontaneous mutations? What degree of variability of spontaneous mutations from one individual to another would we expect? One would predict that the greater the variability, the greater the induced level of mutations would need to be to permit us to unambiguously differentiate between spontaneous or non-spontaneous etiology.

In this paper, we will discuss another approach which should permit us to discriminate between mutations of spontaneous and non-spontaneous origin. We propose that we could diagnose the major mutagenic stimuli in a population by the spectrum of mutations present. This spectrum could be generated by analysis of the kinds and frequencies of gene locus mutation in human blood cell populations. Spontaneous and induced mutants are predicted to exhibit unique mutational spectrum, as mutations are not randomly distributed [2].

GENERATION OF MUTATIONAL SPECTRA

1. From Analysis of Frequencies of Mutations at Nucleotide Level

Review of the evidence in literature have demonstrated the presence of unique mutational spectrum for spontaneous and induced mutations in bacteriophage T4, lac I system in *E. coli* and cI system of prophage lambda in *E. coli* (P.-M. Leong et al., submitted to Am. J. Human Genetics; 1) This spectrum was obtained by examination and illustration of number of occurrences of mutations at each nucelotide. Analysis of Miller's data in lac I gene in *E. coli* have shown that spontaneous mutants yield a unique and different mutational spectrum from chemically induced ones. Using this analysis, it is even possible to discriminate between chemically induced mutations, such as mutations induced by EMS and MNNG (P.-M. Leong et al., submitted to Am. J. Human Genetics).

The presence of unique mutational spectrum is the result of base pairs which are more mutable than others termed "hotspots" [2]. Since these more mutable base pairs are relatively infrequent (less than 10% of the whole gene), one would then expect to still be able to detect the "hot spot" nature of the base pair when analyzing blocks of nucleotides. Blocks of nucleotides containing 1 or more "hotspots" would have higher frequency of mutations; those not containing a "hotspot" would have low frequency of mutation. Analysis of data from lac I system have confirmed this prediction. Indeed, unique mutational spectra were generated from spontaneous and EMS induced mutants when mutations were analyzed in blocks of 30 nucleotides. Nine of these 32 blocks show greater than 10-fold difference between spontaneous and EMS induced mutants [1].

TABLE 1. Summary of Mutational Responses Using Specific Selection Systems in CHO Cells

Selection system	Mutant Fraction		
	Spontaneous	EMS induced	Conc. x Time (hr)
Inhibitors of protein synthesis			
Cryptopleurine	1.5×10^{-7} (3)	55×10^{-7} (3)	350 ug/ml x 17
Emetine	2×10^{-7} (5)	70×10^{-7} (5)	350 ug/ml x 17
Tylocrebine	2.5×10^{-7}(3)	60×10^{-7} (5)	300 ug/ml x 20
Trichodermin	6×10^{-7} (8)	1.1×10^{-6} (8)	350 ug/ml x 20
Inhibitors of DNA/RNA metabolism			
DRB	4.4×10^{-7}(5)	15.6×10^{-6} (5)	300 ug/ml x 20
Cycloleucine	5.3×10^{-6} (4)	49×10^{-6} (4)	400 ug/ml x 16
Inhibitor of oxidative phosphorylation			
Venturicidin	8.9×10^{-6} (9)	1.5×10^{-5} (9)	300 ug/ml x 18
Inhibitor of Na^+/K^+ ATPase pump			
Ouabain	0.9×10^{-6} (7)	1820×10^{-6} (4)	300 ug/ml x 20

2. From Analysis of Mutation Frequencies at Genetic Markers by Selection of Mutants Resistant to Drugs Inactivating Essential Proteins

If a distinct spectrum could be generated by analysis of blocks of nucleotides contiguously within a gene, there is no reason why we should restrict ourselves to one gene. We would expect to be able to generate unique spectra by analysis of mutations at blocks of nucleotides located on different genes. Hence, we need a system whereby we could probe for mutations occurring at limited number of base pairs.

Such systems exist in cell culture. These include selection for drug resistant mutants to selective agents which bind to and inactivate essential proteins. One would expect that only mutations which permit a protein to be refractile to toxin binding yet retains its enzymatic properties to be detected. Due to nature of mode of resistance, the genetic target is putatively restricted to a limited number of codons which can mutate to prevent toxin binding but which have little effect on the enzyme's catalytic activity. These systems, then, should effectively be monitoring for mutations at a limited number of base pairs and be sensitive to the specificity of mutagens.

Several examples of this specific type of forward mutation assay have been developed and characterized in mammalian cells, most frequently for Chinese hamster ovary (CHO) cells. Table 1 is a summary of the results from several laboratories. The selection systems all involve selective agents which inactivate an essential protein vital to survival of the cell. They include protein synthesis (ribosome) inhibitors such as cryptopleurine, trichodermin and tylocrebine, and inhibitor of mRNA production, 5,6-dichloro-(b)D-ribofuranosylbenzimidazole an oxidative phosphorylation inhibitor, venturicidin and an inhibitor of Na^+/K^+ ATPase, ouabain.

These selection assays are distinctly different from selection systems using drugs which inactivate non-essential enzymes. One can imagine that any mutations which results in inactivation or loss of gene product would render the cell resistant to the drug. Hence, the genetic target would encompass the whole gene (target size of approximately 1000 base pairs) and specificity of mutagens would not be reflected.

Since frequencies in spontaneous populations are predicted to differ from that in induced ones and mutagens differ in potency to induce mutations, the results obtained from the different populations should be normalized for comparison purposes. To generate the mutational spectrum, we have chosen to express the results as a ratio of mutation frequency at small target (sensitive to specificity of mutagens) to frequency at larger target (not sensitive to specificity

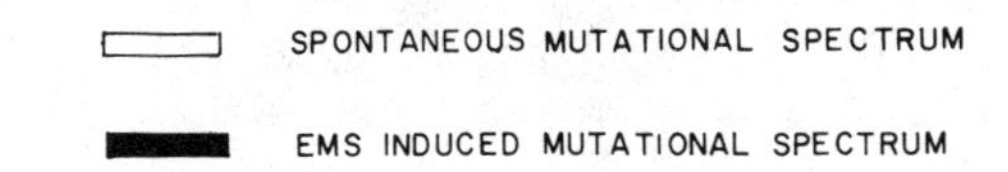

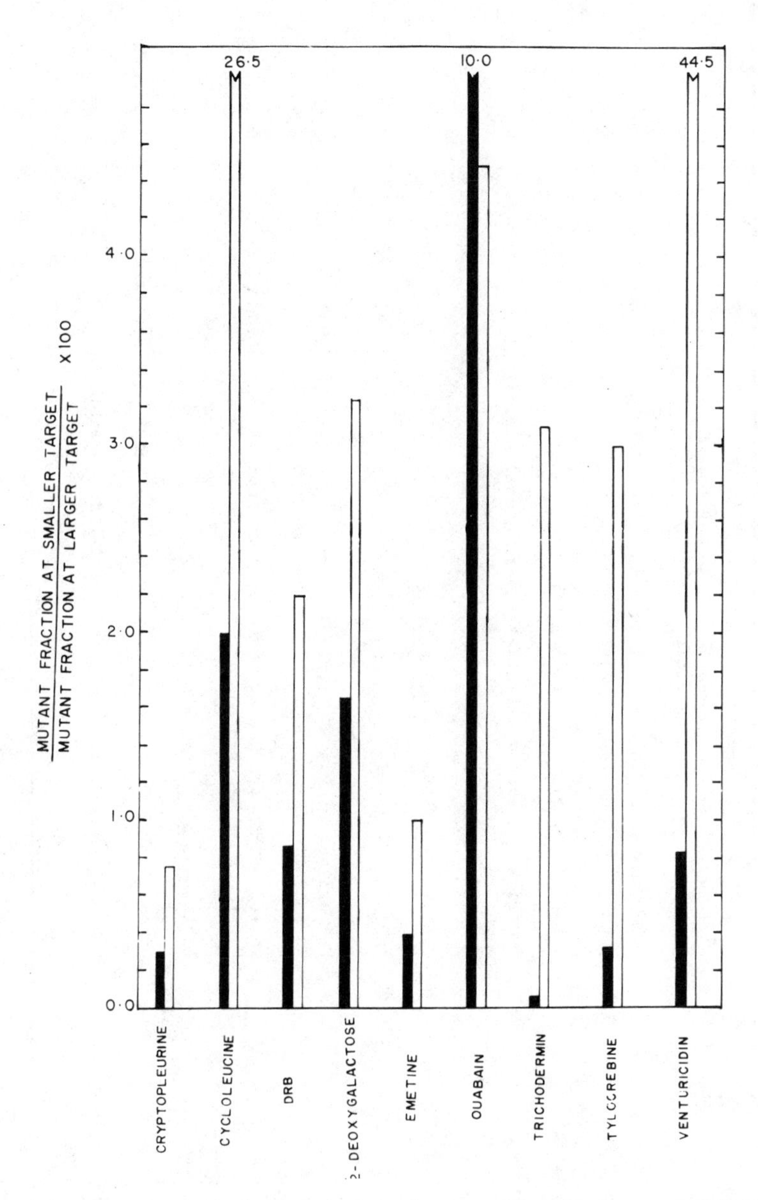

Fig. 1. Spontaneous and EMS induced mutational spectra of CHO cells. The mutational spectra of CHO cells were generated from data in the literature by expression of mutant frequencies as a ratio of mutant frequencies obtained at small target to one at large target. DRB: 5,6-dichloro(b)D-ribofuranosylbenzimidazole.

TABLE 2. List of Compounds Being Screened for Potential Selective Agents

COMPOUND	MODE Of INHIBITION
	Protein synthesis inhibitors
Anguidine	Primarily inhibits initiation; also causes partial polysome stabilization at high drug concentration (12)
Cycloheximide	Inhibits translocation at 60s ribosomal subunit (13)
Emetine	Inhibits ribosome movement at 40s ribosomal subunit (13)
Gougerotin	Inhibits transfer of amino acids from aminoacyl s-RNA to protein (14)
	Inhibitors of DNA/RNA metabolism
Cycloleucine	Inhibits S-adenosyl methionine (SAM) biosynthesis and RNA maturation (15)
Daunomycin	Primarily inhibits RNA synthesis; also interfere DNA synthesis and is postulated to inhibit DNA-dependent DNA polymerase (16)
DRB	Inhibits mRNA synthesis; postulated to inhibit initiation of chromosomal heterogenous RNA synthesis (17)
	Inhibitors affecting microtubules
Colchicine	Binds tightly and specifically to tubulin resulting in disruption of microtubules (18)
Podophyllotoxin	Binds to tubulin and prevents microtubule assembly (18)
	Inactivators of essential enzymes
Ouabain	Binds ot Na^{+}/K^{+} ATPase pump resulting in its inactivation (19)

of mutagen). The spontaneous mutational spectrum generated from ratios of mutation frequencies in the above mentioned sites was found to be significantly different from that of EMS induced ones (Fig. 1). Three of the selection systems, cycloleucine resistance, trichodermin resistance, and venturicidin resistance, displayed a greater than 10-fold difference in the ratios.

These data using CHO cells from the literature support the feasibility of generating mutational spectra by determination of mutation frequencies at different loci. In order to apply this system to a human population, we would need to develop a set of such selection systems using cells of human origin. Unfortunately, these systems are presently limiting in these cells. Preliminary results from just two loci, by selection for ouabain resistance or 6-thioguanine resistance, are promising. Just from measurements at these two loci, we could discriminate spontaneous populations from induced ones and one chemically induced population from another. Mutation frequencies at these 2 loci have been measured in our laboratory using human lymphoblasts. Spontaneous responses have been found to be significantly different from chemically induced ones; a ratio of 0.16 ± 0.02 was seen in spontaneous population compared to 6 ± 3 in EMS induced or 0.08 in ICR-191 induced ones [11].

Clearly what is needed now is a set of markers which would be sensitive to the specificity of mutagens. Selection systems using compounds which inactivates essential proteins and thus are toxic to human lymphoblasts are being explored in our laboratory (Table 2). Once these selections have been developed, we can then determine if mutational spectra unique to different mutagenic stimuli could be generated in human lymphoblasts.

APPLICATION OF MUTATIONAL SPECTRA TECHNIQUE TO CELLS ISOLATED DIRECTLY FROM A POPULATION

If mutational spectra could be generated by analyzing mutation frequencies at a set of specific drug resistance markers in cell culture, one should then be able to apply these markers to cells isolated directly from a subject. In applying this technology to the human population, one would expect the availability of cells to be a limiting factor. We should determine frequency of mutations in a population of cells which are easily accessible. Only markers whose mutation frequencies are at levels which are detectable in the system used would be useful.

Human peripheral T-lymphocytes are a promising source of cells for determination of mutations in a person. A blood sample, which can easily be isolated directly from an individual, yields approximately 1×10^6 T-lymphocytes per ml of blood [20]. Hence, we could expect to be able to screen up to 1×10^8 T-lymphocytes for mutations by drug resistance at one site per sampling (100 ml) from any individual. Since several markers are required to generate a useful spectrum, these cells could be propagated using the new T-lymphocyte cloning technique of Albertini [21] to obtain sufficient cells for determination at all the markers. Alternately, one could screen for a smaller number of cells per marker at the expense of lower sensitivity.

We would expect mutation frequencies of at least as great as cell cultures to be present in humans if there is little selective pressure operating either for or against these traits. One would expect little selective pressure if the detectable mutations are non-essential characteristics. Thus, we could expect mutant frequencies at range of 10^{-5} to 10^{-6} for larger target and 10^{-7} to 10^{-8} for the smaller specific ones.

Using mutation frequency of 3×10^{-6} as an example at larger target one can expect the presence of 300 mutants in a population of 10^{8} lymphocytes, with upper 95% confidence value of 335. If we assumed two small specific targets, A and B, to have spontaneous mutant fractions of 5×10^{-8} and 10^{-7}, we could predict the presence of 5 and 10 mutants respectively. Since the 95% confidence values of 5 are 1.6 and 11.7; and that of 10 are 4.7 and 18.4, one could expect the spontaneous mutational spectrum, generated from the ratios of mutant frequencies at small target to large one (95% confidence), to have the following characteristics:

Ratio at A: 1.7×10^{-2} $(0.48 \times 10^{-2}, 4.4 \times 10^{-2})$
Ratio at B: 3.3×10^{-2} $(1.4 \times 10^{-2}, 6.9 \times 10^{-2})$

Let us imagine that this population was subjected to a test regimen which resulted in an induced mutant fraction of 21×10^{-6} in larger target and 10^{8} lymphocytes were screened for presence of mutations. This would imply that a spectrum yielding a ratio of 6×10^{-2} or 8×10^{-2} at A or B respectively, would be significantly different (95% confidence) from the spontaneous population and hence permit us to decide if the population has been exposed to mutagens. These ratios translates to induced mutant frequencies in the range of twenty fold that of spontaneous at the smaller target.

However, the ratios of mutant frequencies need not be significantly different at all the markers. Since we would generate the mutational spectrum by using a series of markers, one would expect that a significant difference in ratio in any one marker should be sufficient to produce a unique spectrum. The mutational spectrum obtained should reflect the identity of mutagenic stimulus(i) of the population and thus, would permit us to diagnose the major causes of genetic changes in the population.

SUMMARY

We have proposed that generation of mutational spectrum by analysis of kinds and frequencies of gene locus mutations in human peripheral T-lymphocytes should permit us to discriminate between spontaneous and induced mutations. This approach is based on the following assumptions:

1. That mutations are non-randomly distributed when determined at the nucleotide level.

2. The specific pattern of mutations could be determined when mutations at blocks of nucleotides are probed for by specific drug resistant selection systems.

3. The selection system developed in cell culture could be applied to human peripheral T-lymphocytes isolated directly from the population.

ACKNOWLEDGEMENT

This work was supported by the U.S. Department of Energy Contract DE-AC02-77-EV04267 and was prepared as an adjunct to oral presentation at this meeting. It is not intended as an original presentation of this discussion topic.

REFERENCES

1. W. G. Thilly, P.-M. Leong, and T. R. Skopek, Potential of mutational spectra for diagnosing the cause of genetic change in human cell populations, in: Banbury Report 14 Indicators of Genotoxic Exposure in Man and Animals, Proceedings: Banbury Conference, April 1982 (in press).
2. S. Benzer, On the topography of the genetic fine structure, Proc. Natl. Acad. Sci. (U.S.A.), 47:403-415 (1961).
3. R. S. Gupta and L. Siminovitch, Mutants of CHO cells resistant to protein synthesis inhibitors, cryptopleurine, and tylocrebine, Genetic and biochemical evidence for common site of action of emetine, cryptopleurine, tylocrebine, and tubulosine, Biochemistry, 16:3209-3214 (1977).
4. M. Caboche and P. Mulsant, Selection and preliminary characterization of cycloleucine-resistant CHO cells affected in methionine metabolism, Som. Cell Genetics, 4:407-421 (1978).
5. R. S. Gupta and L. Siminovitch, Genetic markers for quantitative mutagenesis studies in CHO cells, Characteristics of some recently developed selection systems, Mut. Res., 69:113-129 (1980).

6. C. D. Whitfield, B. Buchsbaum, R. Bostedor, E. H. Y. Chu, Inverse relationship between glactokinase activity and 2-deoxygalactose resistance in Chinese hamster ovary cells, Som. Cell Genetics, 4:699-713 (1978).
7. C. L. Crespi, The temporal induction of 6-thioguanine and ouabain resistance in synchronous Chinese hamster ovary cells by 5-bromo-2-deoxyuridine and ethyl methanesulfonate, M.S. Thesis, Department of Nutrition and Food Science, Massachusetts Institute of Technology (1979).
8. R. S. Gupta and L. Siminovitch, Genetic and biochemical characterization of mutants of CHO cells resistant to the protein synthesis inhibitor Trichodermin, Som. Cell Genetics, 4:355-374 (1978).
9. H. E. Lagarde and L. Siminovitch, Studies on CHO mutants showing multiple cross resistance to oxidative phosphorylation inhibitors, Som. Cell Genetics, 5:847-871 (1979).
10. C. E. Campbell and R. G. Warton, Evidence obtained by induced mutation frequency analysis for functional hemizygosity at emt locus in CHO cells, Som. Cell Genetics, 5:51-75 (1979).
11. W. G. Thilly, J. G. DeLuca, E. E. Furth, H. Hoppe, D. A. Kaden, J. J. Krolewski, H. L. Liber, T. R. Skopek, S. A. Slapikoff, R. J. Tizard, and B. W. Penman, Gene locus mutation assays in diploid human lymphoblast lines, in: Chemical Mutagens, Vol. 6 (F. J. de Serres and A. Hollaender, eds.), pp. 331-361 (1980).
12. T. W. Doyle and W. T. Bradner, Trichothecanes, Anticancer agents based on natural product models, p. 43 (1980).
13. S. Pestka, Inhibitors of ribosome functions, Ann. Review of Microbiology, 25:487 (1971).
14. J. M. Clark, Jr., Gougerotin, Antibiotics, Vol. 1 (D. Gottlieb and P. D. Shar, eds.), Springer-Verlag, New York, p. 278 (1967).
15. M. Caloche and J. P. Bachellerie, RNA methylation and control of eukaryotic RNA biosynthesis, Effects of cycloleucine, a specific inhibitor of methylation, on ribosomal RNA maturation, Eur. J. Biochem., 74:19 (1977).
16. A. DiMarco, Daunomycin and related antibiotics, Antibiotics, Vol. 1 (D. Gottlieb and P. D. Shar, eds.), Springer-Verlag, New York, p. 190 (1967).
17. E. Egyhezi, A tentative initiation inhibitor of chromosomal heterogeneous RNA synthesis, J. Mol. Biol., 84:173 (1974).
18. L. Wilson, J. R. Bamburg, S. B. Mizel, L. M. Grishan, and K. M. Greswell, Interaction of drugs with microtubule proteins, Fed. Proc., 33(2)-158 (1974).
19. T. Reichstein and H. Reich, The chemistry of steroids, Ann. Review of Biochemistry, 15:155, 1946:403 (1977).
20. A. C. Guyton, Textbook of Medical Physiology, Fourth Edition, Part II, Blood Cells, Immunity, and Blood Clotting, W. B. Saunders Co., Philadelphia, pp. 98-146 (1971).

21. R. J. Albertini and R. B. Wayne, Cloning *in vitro* of human 6-thioguanine resistance ($6TG^R$) peripheral blood lymphocytes (PBL's) arising *in vivo*, Program and Abstracts, 13th Annual Meeting, Environmental Mutagen Society, Environmental Mutagen Society, Washington, D.C., p. 134 (1982).

SHORT TERM TESTS ON BODY FLUIDS

Claes Ramel

Division of Toxicological Genetics
Wallenberg Laboratory
University of Stockholm
S-106 91 Stockholm, Sweden

ABSTRACT

Investigation of the mutagenicity of body fluids constitutes a valuable source of information to elucidate the fate of genotoxic chemicals in the body and to reveal the exposure to such chemicals. The biotransformation and transport of chemicals in body fluids can be studied experimentally by perfusion systems. *In vitro* perfusion of rat liver, combined with *Salmonella* or mammalian cells as indicator systems for mutagenicity, has been a useful device to study the activation of promutagens by an intact liver and the excretion of metabolites via the bile. The *in vivo* exposure of experimental animals and humans to genotoxic chemicals can be investigated by means of mutagenicity studies of body fluid samples. The predominant and most useful material for such studies has been urine samples, combined with bacterial tests for mutagenicity. Concentration of urine by means of non polar resins, the addition of β-glucuronidase for splitting conjugates and the use of fluctuation test with bacteria have increased the sensitivity and the application of these urine assays.

Model studies particularly of urine samples from smokers have indicated the usefulness of these tests for monitoring human populations and for the identification of mutagenic components. For occupational exposures mutagenicity tests of urine samples can be used to establish the exposure to genotoxic chemicals among workers. When the exposure situation involves complex mixtures of chemicals, it is of importance to combine urine tests with other genetic and chemical analyses in order to identify hazardous chemicals and processes in the work environment. This is exemplified by a Finnish-

Swedish project on the identification of carcinogenic and mutagenic hazards in the rubber industry.

INTRODUCTION

Regulatory actions against mutagens and carcinogens usually are based on more or less insufficient scientific information. Individual epidemiological and experimental data are rarely adequate for a reasonably accurate risk evaluation. Decision must therefore be based on many different pieces of evidence - both experimental data and observations on exposed people. All the methods employed have their limitations, but the combination of the data hopefully can form a reliable picture of the risk situation. The notorious shortage of sufficient data on mutagenic and carcinogenic effects of chemicals makes it imperative to use all available possibilities to collect data particularly on exposed human populations. In the first place epidemiological data is desirable, which in reality means cancer data. The possibilities of collecting any epidemiological data on induced hereditary diseases are very remote. Although cancer data on exposed human populations is particularly useful in order to make quantitative risk evaluation, there are many obvious obstacles in obtaining such data - such as the long latency period for most cancer forms and the difficulties of obtaining cohorts of sufficient size for a discrimination of induced incidence against the spontaneous background level. Furthermore the damage is of course already a fact when epidemiological data is available. In order to prevent carcinogenic and mutagenic effects of chemicals their genotoxic actions have to be caught far ahead of the point when the chemicals have started to cause epidemiological disorders.

However, exposed persons can also be used for monitoring of mutagenic effects of chemicals. This can principally be done in two ways. The most obvious way is to use the cells of the exposed person himself as indicators of genetic effects, such as chromosome aberrations, sister chromatid exchange, and point mutations. However, the presence of mutagenic chemicals or metabolites in the body can also be studied by exposing sensitive genetic indicators, usually bacteria, to body fluids from exposed people. The advantage of such an approach is the fact that one can use well established mutagenicity systems for point mutations such as Ames histidine revertant system of *Salmonella*. Point mutation systems with human cells *in vivo* are still in the process of development and will anyway hardly reach the sophistication of microbial systems. Furthermore, bacterial test systems are particularly employed for testing excretion products in urine and feces and they therefore supplement tests on cells *in vivo*.

I will give a short review of some literature data on the test of urine samples and some other experimental data of some relevance in connection with screening of body fluids. The use of feces for mutagenicitiy screening is a different and important field, which I, however, will not be able to touch upon. The large variation in mutagenicity of feces, presumably due to the diet and biotransformation by the gut and microflora is of great interest in itself. However, at the present stage feces mutagenicity does not easily seem to lend itself for monitoring of exposures to xenobiotics.

BIOTRANSFORMATION

Before discussing testing procedures with body fluids it may be appropriate to make a short survey of the fate of chemicals in the body with special reference to body fluids. The intake of chemicals via the gut or the lungs is followed by a transportion in body fluids between different compartments, where the chemicals may be metabolized in different ways. The primary compartments for detoxication and activation of compounds are the liver and the lungs. Particularly the liver functions as the center for those processes and for studies in chemical carcinogenesis much attention has consequently focused on the performance and the effect of the liver. The transport of metabolites from the liver by the blood and the occurrence of mutagenic compounds in the circulatory system in the body, is therefore of course of primary interest for the evaluation of mutagenic and carcinogenic risks. In the present context with mutagenicity testing of body fluids, blood samples taken _in vivo_, are, however, of limited value, because of the low volume and small amount of mutagens available in samples. Blood samples, however, constitute an important source of information for chemical investigations and for dose determination by measuring alkylation of hemoglobin [1, 2]. For _in vivo_ samples excretion products, both urine and feces, have instead been used for mutagenicity studies. The urine contains conjugated compounds, which can be analyzed for mutagenicity after cleavage of the conjugation. The feces are of importance as it also may reflect the metabolic conversion performed by the intestinal flora.

EXPERIMENTAL BODY FLUID SYSTEMS

Although mutagenicity analysis _in vitro_ of samples from the circulatory fluid is difficult, _in vivo_ studies by exposing test organisms inside mammals can be done. The host mediated assay of Garbridge and Legator [3] was the first attempt to investigate the formation and transport of reactive metabolites in the mammalian body by using bacteria as a genetic indicator system. Although the host mediated assay did respond both to activation by the mixed function oxygenase system and by glycosylases in the gut flora, that test system was insensitive and not suitable for detailed analyses of metabolic conversions. The metabolism and transport of reactive products can, however, be followed experimentally in perfusion systems. Such perfusion systems have primarily been employed to study the liver and the lungs, which are the organs of particular capacity and relevance for metabolic detoxication as well as activation of promutagens and procarcinogens. Although perfusion can be performed both _in vivo_ and _in vitro_, the _in vitro_ system allows a higher degree of control and manipulation of the perfused organ.

For the purpose of studying the function and transport of mutagenic metabolites we have at our laboratory employed an _in vitro_ liver perfusion system, developed by Dr. Brita Beije, in combination with _Salmonella_ or V79 hamster cells [4, 5]. This experimental system has given useful information concerning the activation, deactivation and transport of mutagens via the mammalian liver. A brief account of this test system may be of some interest in the present connection.

Bovine blood is used as perfusate and samples can be taken of the perfusate at different distances from the liver and also from the bile, which is collected during the perfusion. The bile has rarely been used in mutagenicity testing, but the importance of testing the excretion of metabolic products in the bile in some cases may be illustrated by 1,2-dichloroethane and 1,2-dibromoethane. Rannug and Beije [6] tested the mutagenicity of these compounds in the liver perfusion system with _Salmonella_ and found that the perfusion fluid gave no or only a marginal mutagenic effect. But the bile gave a dramatic increase of mutagenicity. The background of this effect of dichloroethane and dibromoethane has been analyzed in detail by Rannug et al. [7]. These two compounds are activated by conjugation to glutathione, which normally is an efficient detoxication pathway. In this case the conjugation, however, gives rise to a highly potent mutagen, which is excreted with the bile. To what estent the mutagenic glutathione conjugates of dichloroethane and dibromoethane are reabsorbed through the intestine and can recirculate is not clear.

The liver perfusion system has also been combined with mutagenicity studies on hamster cells [5, 6]. The cells can be exposed to the test compounds at different distances from the liver in the perfusion system. This arrangement has made it possible to estimate the life length of reactive metabolites. Thus the active metabolites of dimethylnitrosamine had a life length of around 2-3 seconds. The metabolite of styrene on the other hand was quite stable and gave a mutagenic response also in the periferal chamber of the perfusion system. The data also indicated that the major reactive metabolite of styrene cannot be styrene oxide as usually suggested. The mutagenicity of a corresponding concentration of styrene oxide was not sufficient to explain the mutagenicity of styrene by conversion to styrene oxide [8].

MUTAGENICITY OF URINE

The analysis of body fluids - in practice mostly urine - from animals and humans exposed in vivo, is of particular relevance in order to trace mutagens and carcinogens in the human environment. The use of urine for mutagenicity analyses has been explored by several investigators. Durston and Ames [9] treated rats with 2-acetylaminofluorene and analyzed the excretion of mutagenic metabolites in the urine by means of the Salmonella/microsomal system and the strain TA 1538. After addition of β-glucuronidase the administration of as little as 1.6 mg/kg of the test compound to rats results in a significant mutagenic response.

Another model substance, hycanthone, was investigated in a similar way in mice by Legator et al. [10]. Revertants of Salmonella TA 1538 showed a dose related increase, but only when the urine samples had been treated with β-glucuronidase indicating that the metabolite of hycanthone was excreted in conjugated form. A fractionation of the urine into acid, basic and neutral fractions showed a mutagenicity in the acid fraction. Legator et al., also performed a test of the urine of hycanthone treated mice with hamster ovary cells (CHO). Anaphase bridges, chromosome lagging, multiple spindles and stickiness were recorded. A dose related increase of abnormal anaphases was observed. This effect did not require the addition of β-glucuronidase. The cells themselves evidently have sufficient amount of enzymes to split the conjugates.

The mutagenicity test results of urine in animal experiments clearly showed the potential use of the same technique for screening humans exposed to mutagenic and carcinogenic compounds. Several investigations have also demonstrated the usefulness of such an approach. Mutagenicity tests of urine samples have been performed both for individual chemicals and for complex mixtures. Many examples of excretion of mutagenic metabolites from exposed people have been revealed. Minnich et al. [11] on the other hand could show that the urine of unexposed persons does not seem to exhibit any mutagenic effect. Of over 1000 human urine samples only 8 showed any mutagenicity - in 4 cases dependent on metronidazole and in 2 cases on chemo-therapeutic agents.

Concerning single chemicals Siebert [12] and Siebert and Simon [13, 14] demonstrated a mutagenic effect on yeast by the urine from patients treated with cyclophosphamide. Legator et al. [17] found that urine from patients treated with the drugs metronidazole and niridazole, used against Trichomonas vaginalis, was mutagenic to Salmonella.

ANALYSES OF URINE FROM SMOKERS

The use of urine tests is also a highly useful tool to identify genotoxic components in complex chemical mixtures. Particularly detailed and interesting analyses have been made on tobacco smokers by Yamasaki and Ames [15] and by Putzrath et al. [16]. Yamasaki and Ames introduced the use of nonpolar resins such as XAD-2 column for the adsorption of metabolites from the urine. Eluation of the column was performed with acetone. The mutagenicity content was concentrated about 200 folds in this way. That technique was applied for mutagenicity screening of urine from smokers and non-smokers. A mutagenic response in smokers was obtained with the Salmonella strain TA 100, TA 98, and TA 1538, but not TA 1535, indicating the presence of frame shift mutagens. An increase of the mutagenic effect occurred with S9. A large scale investigation of the urine of smokers and non-smokers revealed a considerably higher mutagenicity with smoker's urine than with non-smokers. Smokers, who did not inhale the smoke had no increased mutagenicity over the control. The mutagenicity of smokers was about twice as high in the evening as in the morning, but even in the morning there was a significantly elevated mutagenicity in some cases, which points to the occurrence of long lived mutagenic metabolites and components. The elevated incidence of bladder cancer among smokers is probably a reflection of this excretion of mutagenic compounds in the urine.

Putzrath et al. [16] have recently carried this analysis of smoker's urine to further sophistication involving more detailed characterization of the mutagenic components. The concentration procedure with XAD-2 column was essentially the same as by Yamaski and Ames and this was combined with high performance liquid chromatography analysis of the urine. The HPLC fractions were tested separately for mutagenicity. Mutagenic components were relatively non polar and appeared in about 15 consecutive fractions. Although the pattern of peaks varied between smokers the mutagenic activity occurred in the same fractions. Urine from non-smokers did not show any mutagenicity with the exception of the non-smoking spouse of a smoker, who evidently exhibited some mutagenicity from passive smoking exposure.

After fractionation of the urine by extraction with an acid and a base, most mutagenicity appeared in the acid extract. As it was shown by Yamasaki and Ames that the mutagenic components in the smoker's urine was activated by the S9 microsomal fraction Putzrath et al., studied the effect of different inducers of P 450 and the S9 used in the mutagenicity tests. They found that both 3-methylcholanthrene and Arochlor induced livers gave a higher mutagenic response than phenobarbital induced livers. This information in combination with the fact that the mutagenicity is of the frame shift type suggests that polyaromatic compounds are involved.

URINE ANALYSES FOR OCCUPATIONAL EXPOSURES

The technique of using urine for analyses of mutagenicity has a particularly promising field of application for occupational exposure to chemicals. There are principally two types of situations to be investigated in that connection. In one situation people are handling a known carcinogen and the question arises whether they are exposed to that carcinogen or not. In the other type of situation people are exposed to complex mixtures of chemicals of more or less unknown composition and the problem is whether this mixture of chemicals contains genotoxic components. Both these situations can be investigated by means of tests of urine.

As an example of the exposure to known carcinogens and mutagens a study on epichlorhydrin may be mentioned. Killian et al. [17] collected urine from workers exposed to epichlorhydrin. Six workers exposed to 0.8-4 ppm of epichlorhydrin did not show any increase of mutagenicity to _Salmonella_ from their urine. However, two other workers were accidentally exposed to over 25 ppm and their urine collected 9 hours after the exposure revealed a significantly increased mutagenicity. That case illustrates the value of taking advantage of exceptional exposures - accidental or otherwise. The sensitivity of the ordinary test procedure with plate test on bacteria is often not sufficient to reveal any mutagenicity, but the sensitivity may be increased by using the fluctuation test system according to the procedure described by Green et al. [21]. In the fluctuation test the bacteria are treated in suspension and the suspension is divided up in many small vials. The dilution of bacteria is such that less than one revertant bacteria per vial occurs in the control. In those vials, which contain a revertant bacteria there will be a bacterial growth and after some days such vials can be recognized by a special staining. An increase of vials exhibiting the staining reaction will thus indicate an increased mutagenicity. This system has been reported to be more sensitive than the plate technique in general and it has been successfully used by Sorsa, Vanio and their collaborators at the Institute of Occupational Health in Helsinki to study the mutagenicity of urine samples after occupational exposures. With that fluctuation technique they found for instance that nurses handling cytostatics in hospitals showed a significantly increased mutagenicity of their urine, indicating an exposure to the cytostatics [19].

In the situation when one wants to know whether the exposure to complex mixtures of chemicals may imply a risk for carcinogenic and mutagenic compounds, urine mutagenicity tests have been used to reveal such risks in industries. However, in order to actually localize and identify riskful industrial processes and chemicals compounds in such complex exposures it is of importance to have a wider analysis of the exposure, genetical as well as chemical. The urine analysis only constitutes one part of such a battery of investigations. As an example of such an approach I may briefly describe a project concerning genotoxic hazards in the rubber industry. This is a Finnish-Swedish collaborative project with the Institute of Occupational Health in Helsinki and Wallenberg Laboratory at the University of Stockholm. The choice of the rubber industry for this investigation was based on the fact that several epidemiological investigations have shown that workers in rubber industries have an increased risk of cancer [20]. Furthermore, a survey by Holmberg et al. [21] of chemicals used in the Swedish rubber industry included many known and suspected mutagenic compounds. On the Swedish side we have been responsible for short term mutagenicity tests with different assays of individual chemicals and samples of various complex mixtures, both model samples and samples collected at industrial sites. The Finnish group under Dr. Sorsa and Dr. Vainio have dealth with epidemiological studies and monitoring of rubber workers in Finland, including chromosomal aberrations and SCE in lymphocytes [22] and mutagenicity of the urine [23, 24].

The urine assay was performed with fluctuation tests both with *Escherichia coli* WP2, uvrA for base substitutions and *Salmonella typhimurium* TA 98 for frame shift. The urine was tested after adsorbtion to XAD-2 resin. In order to account for variations in the urine concentration the mutagenicity was expressed in relation to the creatinine content.

Both the bacterial assays showed an increased mutagenicity in the rubber workers as compared to controls. For the frame shift mutagenicity with *Salmonella* there was synergistic effects with smoking, which was n ot noticeable with the *coli* base substitution strain, as expected from the frame shift mutagenicity of urine from smokers.

If different categories of workers in the rubber plant are compared the highest mutagenicity is found among the weighers and mixers but also vulcanizers show an increased mutagenicity. The high exposure of the mixers is also supported by an increased SCE [22]. In one set of data two cleaners of the mixing department exhibited particularly high mutagenicity. That data also showed that workers with high mutagenicity during their working time had a control level of mutagenicity after returning from vacations. It must be stressed that only glucuronic and sulfate conjugates are accessible for mutagenicity tests in the urine in this way. Apparently the addition of a liver microsomal fraction (S9) was sufficient for the relase of mutagenic species from such conjugates. However, an important detoxication pathway for reactive metabolites is also conjugation with glutathione and they will not be split and detected in the mutagenicity tests. A way to measure glutathione conjugates in the urine is to analyze the content of thioethers [25]. Although the experience from such measurements in connection with occupational exposures is limited, this method seems to constitute a valuable additional tool to study the exposure to reactive chemicals and metabolites, as was shown by Vainio and coworkers [26]. They found a significant increase of thioether excretion in the urine of rubber workers in Finland.

Although we do not know which chemicals and mixtures are responsible for this mutagenicity among rubber workers, the short term tests by Agneta Hedenstedt-Rannug at our lab have revealed several candidates. Among the accelerators there is a series of dithiocarbamates, which exhibit mutagenic effects in bacterial assays [27]. Particularly for workers in manufacturing processes gases from vulcanization can be suspected to be of special importance. By laboratory vulcanization the most hazardous processes could be revealed [28, 29]. It turns out that vulcanization of natural rubber does not give rise to any noticeable increase of mutagenicity. Various synthetic rubber mixtures, however, bring about different spectrums of mutagenic effects, evidently depending on the composition used. So far only a rough separation by different solvents has been done in order to identify the chemicals or group of chemicals of particular importance for the mutagenicity of the vulcanization. Finally collection of vulcanization gases at the actual industrial sites has revealed a high mutagenicity from a couple of vulcanization sites, and this has been helpful in order to localize areas of potential risk in the plants [29].

In conclusion it may be said that mutagenicity tests of urine has important practical applications to detect occupational exposures to mutagens and carcinogens. It should, however, also be stressed that the method has obvious limitations, particularly when dealing with exposures to complex mixtures of chemicals. The conjugates excreted in the urine constitute detoxified chemicals and metabolites. The relation between the conjugation pathway and covalent binding to macromolecules in the body is not known and can be influenced by many factors. Furthermore, the method does not cover all relevant excretion products. It is from that point of view important that tests of the mutagenicity of body fluids are combined with other experimental searches for genotoxic chemicals and emissions in the work environment.

REFERENCES

1. L. Ehrenberg and S. Osterman-Golkar, Alkylation of macromolecules for detecting mutagenic agents, Teratogenesis, Carcinogenesis, and Mutagenesis, 1:105-127 (1980).
2. S. Jensen, Alkylation of hemoglobin for monitoring dose of genotoxic exposure, This symposium (1982).
3. M. G. Garbridge and M. S. Legator, A host-mediated microbial assay for the detection of mutagenic compounds, Proc. Soc. Biol. Med., 130:831-834 (1969).

4. D. Jenssen, B. Beije, and C. Ramel, Mutagenicity testing on Chinese hamster V79 cells treated in the in vitro liver perfusion system, Comparative investigation of different in vitro metabolizing systems with dimethylnitrosamine and benzo(a)-pyrene, Chem.-Biol. Interactions, 27:27-39 (1979).
5. B. Beije, D. Jenssen, E. Arrhenius, and M.-A. Zetterqvist, Isolated liver perfusion - a tool in mutagenicity testing for the evaluation of carcinogens, Chem.-Biol. Interactions, 27:41-57 (1979).
6. U. Rannug and B. Beije, The mutagenic effect of 1,2-dichloroethane on Salmonella typhimurium, II. Activation by the isolated perfused rat liver, Chem-Biol. Interactions, 24:265-285 (1979).
7. U. Rannug, A. Sundvall, and C. Ramel, The mutagenic effect of 1,2-dichloroethane on Salmonella typhimurium, I. Activation through conjugation with glutathione in vitro, Chem.-Biol. Interactions, 20:1-16 (1978).
8. B. Beije and D. Jenssen, Investigation of styrene in the liver perfusion/cell culture system, No indication of styrene-7,8-oxide as the principal mutagenic metabolite produced by the intact rat liver, Chem.-Biol. Interactions, 39:57-76 (1982).
9. W. E. Durston and B. N. Ames, A single method for the detection of mutagens in urine: Studies with the carcinogen 2-acetyl-aminofluorene, Proc. Natl. Acad. Sci., U.S.A., 71:737-741 (1974).
10. M. S. Legator, L. Truong, and T. H. Connor, Analysis of body fluids including alkylation of macromolecules for detection of mutagenic agents, in: Chemical Mutagens, Principles, and Methods for Their Detection (A. Hollaender, ed.), Vol. 5, 1-23 (1980).
11. V. Minnich, M. E. Smith, D. Thompson, and S. Kornfield, Detection of mutagenic activity in human urine using mutant strains of Salmonella typhimurium, Cancer, 38:1253-1258 (1976).
12. D. Siebert, A new method for testing genetically active metabolites: Urinary assay with cyclophosphamide (Endoxan, Cytoxan) and Saccharomyces cerevisiae, Mutation Res., 17:307-314 (1973).
13. D. Siebert and U. Simon, Cyclophosphamide: Pilot study of genetically active metabolites in the urine of a treated human patient. Induction of mitotic gene conversions in yeast, Mutation Res., 19:65-72 (1973).
14. D. Siebert and U. Simon, Genetic activity of metabolites in the ascitic fluid and in the urine of a human patient treated with cyclophosphamide: Induction of mitotic gene conversion in Saccharomyces cerevisiae, Mutation Res., 21:257-264 (1973).
15. E. Yamasaki and B. N. Ames, Concentration of mutagens from urine by adsorbtion with the nonpolar resin XAD-2: Cigarette smokers have mutagenic urine, Proc. Natl. Acad. Sci., U.S.A., 74:3555-3559 (1977).
16. R. M. Putzrath, D. Langley, and E. Eisenstadt, Analysis of mutagenic activity in cigarette smokers urine by high performance liquid chromatography, Mutation Res., 85:97-108 (1981).

17. D. J. Killian, T. O. Pullin, T. H. Connor, and M. S. Legator, Mutagenicity of epichlorohydrin in the bacterial assay system: Evaluation by direct in vitro activity and in vivo activity of urine from exposed humans and mice, Abstr. 8th Ann. Meeting of the Environmental Mutagen Soc., Aa-12 (1977).
18. M. H. L. Green, B. A. Bridges, A. M. Rogers, O. Horspool, W. J. Muriel, J. W. Bridges, and J. R. Fry, Mutagen screening by a simplified bacterial fluctuation test: Use of microsomal preparations and whole liver cells for metabolic activation, Mutation Res., 48:287-294 (1977).
19. K. Falck, P. Gröhn, M. Sorsa, H. Vainio, E. Heinonen, and L. Holsti, Mutagenicity in urine of nurses handling cytostatic drugs, Lancet, 1250-1251 (1979).
20. R. R. Monson and L. J. Fine, Cancer mortality and morbidity among rubber workers, J. Natl. Cancer Inst., 61:1047-1053 (1978).
21. B. Holberg, B. Sjöström, and S. Olsson, A toxicological survey of chemicals used in the Swedish rubber industry, Investigation Report, 1977:19. National Board of Occupational Safety and Health, Stockholm, Sweden (1977).
22. M. Sorsa, J. Mäki-Paakkanen, and H. Vainio, Identification of mutagen exposures in the rubber industry by the sisterm chromatid exchange method, Cytogenet, and Cell Genetics, 33:68-73 (1982).
23. K. Falck, M. Sorsa, and H. Vainio, Mutagenicity in urine of workers in rubber industry, Mutation Res., 79:45-52 (1980).
24. M. Sorsa, K. Falck, and H. Vainio, Detection of worker exposure to mutagens in the rubber industry by the use of the urinary mutagenicity assay, in: Environmental Mutagens and Carcinogens, Proc. 3rd Int. Conf. on Environmental Mutagens, Tokyo (T. Sugimura, S. Kondo, H. Takebe, eds.), 323-329 (1982).
25. R. van Doorn, C. M. Leidekkers, R. P. Bos, R. M. E. Brouns, and P. T. Henderson, Detection of human exposure to electrophilic compounds by assay of thioether detoxication products in urine, Ann. Occup. Hyg., 24:77-92 (1981).
26. H. Vainio, H. Savolainen, and I. Kilpikari, Urinary thioether of employees of a chemical plant, Brit. J. Industrial Med., 35: 232-234 (1978).
27. A. Hedenstedt, U. Rannug, C. Ramel, and C. A. Wachtmeister, Mutagenicity and metabolism studies on 12 thiuram and dithiocarbamate compounds used as accelerators in the Swedish rubber industry, Mutation Res., 68:313-325 (1979).
28. A. Hedenstedt, C. Ramel, and C. A. Wachtmeister, Mutagenicity of rubber vulcanization gases in Salmonella typhimurium, J. Toxicol. and Environm. Health, 8:805-814 (1981).
29. A. Hedenstedt-Rannug and C. Ostman, Application of mutagenicity test for detection and some assessment of genotoxic agents in the rubber work atmosphere, in: Short-Term Bioassays in the Analysis of Complex Environmental Mixtures (M. D. Waters, S. S. Sandhu, Y. Lewtas, L. Claxton, N. Chernoff, eds.), Plenum Press, 541-553 (1983).

CHEMICAL ANALYSIS OF HUMAN SAMPLES
IDENTIFICATION AND QUANTIFICATION
OF POLYCHLORINATED DIOXINS AND DIBENZOFURANS

Christoffer Rappe and Martin Nygren

Department of Organic Chemistry
University of Umeå
S-901 87 Umeå, Sweden

1. INTRODUCTION

The polychlorinated dioxins (PCDDs) and dibenzofurans (PCDFs) are two series of tricyclic aromatic compounds with similar chemical, physical, biological, and toxicological properties. The basic structures and numbering are given below:

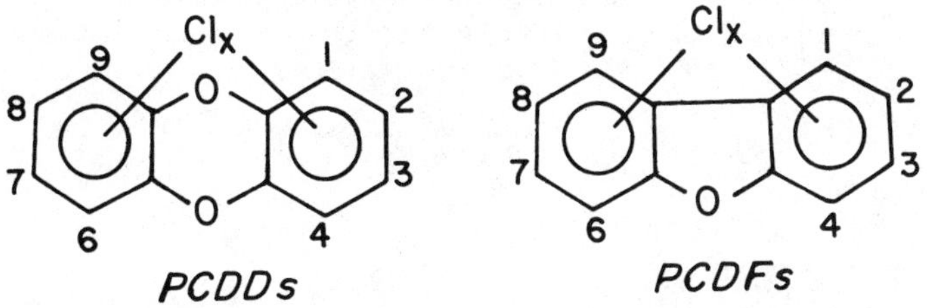

PCDDs *PCDFs*

The number of chlorine atoms can vary between one and eight, and the number of positional isomers is quite large; in all there are 75 PCDDs and 135 PCDFs.

Some of these compounds have extraordinarily toxic properties and have been the subject of much concern. They are known as highly stable contaminants in chlorinated phenols, phenoxy acids and in PCBs. They have also been identified in fly ash and other products from municipal incinerators and accidental fires.

There is pronounced difference in biological effects between different PCDD and PCDF isomers. A factor of 10^3-10^4 can be found for so closely related isomers as 2,3,7,8- and 1,2,3,8-tetra-CDD. The isomers with the highest acute toxicity have 4-6 chlorine atoms and they have their four lateral positions substituted for chlorine. They all have LD_{50}-values in the range 1-100 µg/kg for the most sensitive animal species.

Human exposure to PCDDs or PCDFs may be due to either specific exposure, mainly of occupational origin, accidental exposure or due to general exposure of the public. A general exposure to the public by 2,3,7,8-tetra-CDD has previously been discussed in relation to the herbicide spraying program in Vietnam in the 1960s and the explosion in a chemical plant near Seveso in N. Italy, July 1976. The Yusho oil accidents in Japan in 1968 and in Taiwan in 1979 resulted in an exposure to the public to various PCDFs.

Due to the extreme toxicity of some of the PCDD and PCDF isomers, very sensitive and highly specific analytical techniques are required. The different isomers of PCDDs and PCDFs is known to vary greatly in their biological and toxicological properties, consequently the separation, identification, and quantification of the most toxic isomers becomes important.

In recent years a number of analytical methods have been developed for trace analyses of PCDDs and PCDFs in a variety of matrices. The most specific, sensitive and selective of these methods are utilizing high resolution gas chromatography - mass spectrometry.

In the present study we are discussing levels of PCDDs and PCDFs in human blood plasma and tissue samples. Special attention is paid to variation in individual levels and the various reasons for these variations.

2. EXPERIMENTAL

2.1. Cleanup Procedure

2.1.1. Blood Plasma

The cleanup of the blood plasma samples is based on partitioniong between n-hexane and acetonitrile.

n-Hexane (10 ml) is added to 10 ml of blood plasma. The sample is spiked with 0.1 ng of $^{13}C_{12}$-2,3,7,8-tetra-CDD. Acetonitrile saturated with n-hexane (20 ml) was added (three times) the mixture shaken and centrifugated. The combined acetonitrile phases were treated with 10 ml of aqueous sodium sulfate (1%) and extracted (three times) with 20 ml of n-hexane. The combined n-hexane layers were dried (sodium sulfate) and concentrated until 2 ml and added to a column (1 g of alumina). The column was eluted first with 10 ml of 2% methylene chloride in n-hexane and thereafter with 10 ml of

50% methylene chloride in n-hexane. The first eluate was discarded, the second concentrated and analyzed by GC/MS.

2.1.2. Tissue Samples

The cleanup of the tissue samples have been performed using methods described elsewhere [1, 2].

2.2. GC/MS Analyses

The purified extracts were used directly for the final analyses using a Finnigan 4021 gas chromatograph mass spectrometer (GC/MS) system equipped with a PROMIM multiple ion monitoring unit and negative chemical ionization options. Aliquots (1-2 μg) corresponding to up to 1 g of blood plasma or tissue sample were injected splitlessly onto a glass capillary column (OV-17, Silar 10 c) or a fused silica column (SP 2100, SE 54) leading directly into the ion source. Mass specific detection (mass fragmentography) was used by selective monitoring M, M + 2 and /or M + 4 ions. The recovery of the clean-up procedure was studied from the internal standard.

The quantification was based on peak area measurements using external standards making the assumption that all isomers of a particular PCDD or PCDF (e.g., the tetrachloro isomers) have the same response. The isomer identification is based on retention time studies using polar and unpolar capillary columns and the qualitative standards available, about 40 PCDDs and 70 PCDFs. The 2,3,7,8-tetra-CDD isomer was separated from the other tetra-CDD isomers using a 55 m narrow bored Silar 10 c glass capillary column [3].

3. RESULTS AND DISCUSSION

3.1. Variation in Individual Levels Due to Nature of Exposure

In earlier studies we have investigated the levels of PCDDs and PCDFs in samples of blood plasma taken from workers occupationally exposed to chlorinated phenols in the saw-mill, leather, and textile industry [4, 5]. The levels were normally below 100 ppt (100 pg/g blood plasma). In Table 1 we have collected values from this investigation illustrating the variation of individual levels due to different type of exposure.

We have found that the highest values for both chlorophenol in the urine and the PCDDs and PCDFs in the blood were found for the workers directly exposed to liquid chlorophenol formulations (textile and tannery) or aqueous solutions of chlorophenates (saw-mill). In the textile industry we observed a 100-fold difference between workers in different job categories. For epidemiological studies

TABLE 1. Levels of Chlorophenols in Urine and PCDDs and PCDFs in Blood Plasma from Workers Exposed to Chlorinated Phenols

	Cl-Phenols in urine µg/ml	PCDDs (pg/g in blood plasma)		PCDFs (pg/g in blood plasma)	
		octa-	hepta-	octa-	hepta-
Saw-mill, loader [a)]	0.04	5	2	40	< 3
-"- -"- [a)]	0.03	18	10	18	< 3
-"- packager [a)]	< 0.05	< 3	< 2	7	< 3
-"- cleaner [a)]	< 0.02	5	2	30	< 3
Textile, mixer	3.12	304	59	10	33
-"-	0.16	30	6	< 1	< 1
-"-	< 0.01	3	< 1	< 1	< 1
-"-	0.42	105	15	< 1	< 1
Tannery [a)]	0.04	80	30	7	18
-"- [a)]	0.03	12	4	< 3	< 3

a) Sampling 6-8 months after last exposure to chlorophenols

TABLE 2. Levels of PCDDs and PCDFs in Blood Plasma for Workers Exposed During the Manufacturing of Chlorophenol

Years	PCDD (pg/g in blood plasma)		PCDF (pg/g in blood plasma)	
	octa-	hepta-	octa-	hepta-
3	ND	ND	ND	47
3	ND	ND	ND	46
13	ND	ND	ND	132
6+3	ND	ND	ND	184
< 0.5	ND	ND	ND	3
18	ND	ND	ND	197
5	ND	ND	ND	20
1	ND	ND	ND	191
2	ND	ND	ND	19
4	ND	ND	ND	42
0.5	ND	ND	ND	5

ND = Not Detected

TABLE 3. Levels of PCDDs and PCDFs (pg/g) in Blood Plasma from Workers in the Textile Industry

Worker	Octa-CDD	Hepta-CDDs	Octa-CDF	Hepta-CDFs
1 A	14	4	1	5
B	6	1	ND	1
2 A	6	3	1	4
B	2	ND	ND	ND
3 A	304	59	10	33
B	35	7	ND	5
4 A	3	ND	ND	ND
B	3	1	ND	ND
5 A	10	1	ND	ND
B	1	ND	ND	ND
6 A	105	15	ND	ND
B	25	3	ND	1
7 A	30	6	ND	4
B	5	1	ND	1

A = 5 months after exposure to heavily contaminated product

B = 16 -"- -"- -"- -"- -"- -"-

ND= Not Detected

it must be of great importance to separate between workers highly exposed and those only slightly exposed.

3.2. Variation in Individual Levels Due to Different Time of Exposure

We have also performed a study of a group of workers manufacturing a chlorophenol formulation, and the values are given in Table 2. In the table we have also included how many years the workers have been occupied by this production the highest levels were found for those workers having the longest exposure.

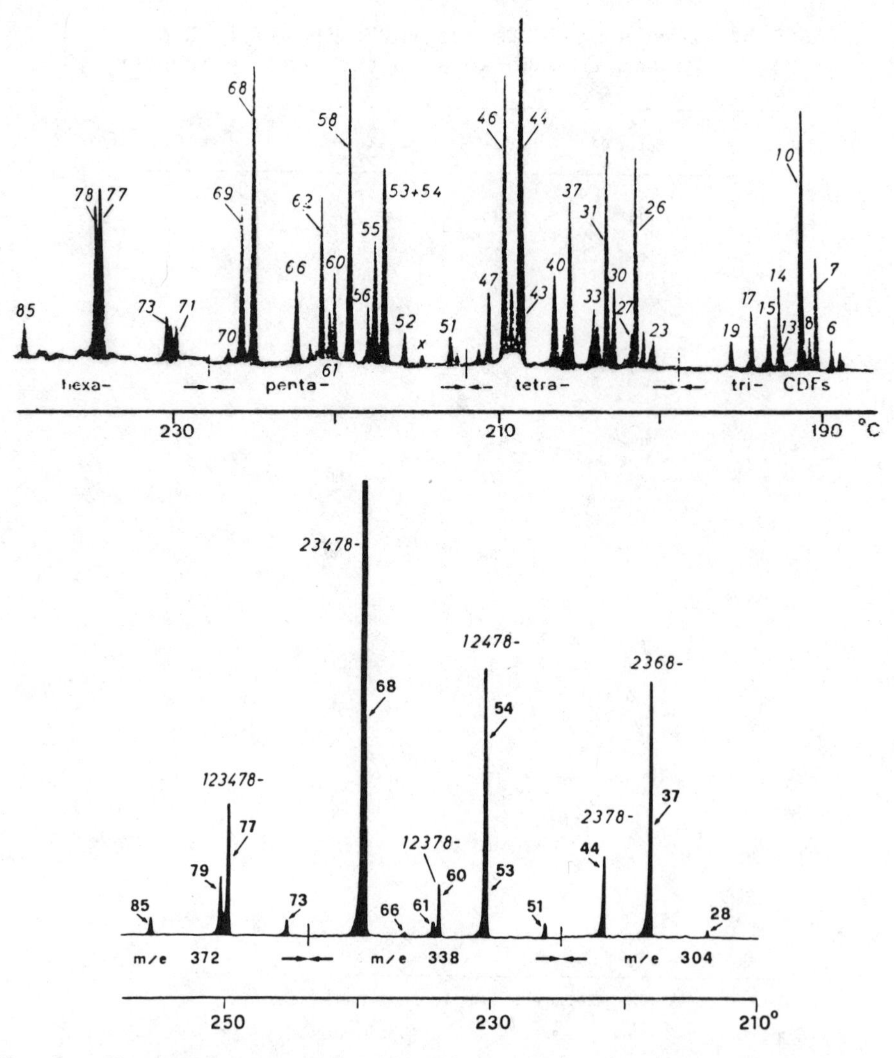

Fig. 1. Mass fragmentation of Yusho oil (top) and a liver sample (bottom).

3.3. Variation in Individual Levels Due to Different Chemical Exposure

Examining Tables 1 and 2 a difference can be found between on one side the saw-mill and chemical workers and on the other side the textile and leather workers. The saw-mill workers and chemical workers were all exposed to a 2,3,4,6-tetrachlorophenate formulation where the PCDFs were the major impurities. These workers had in general higher levels of PCDFs than of PCDDs.

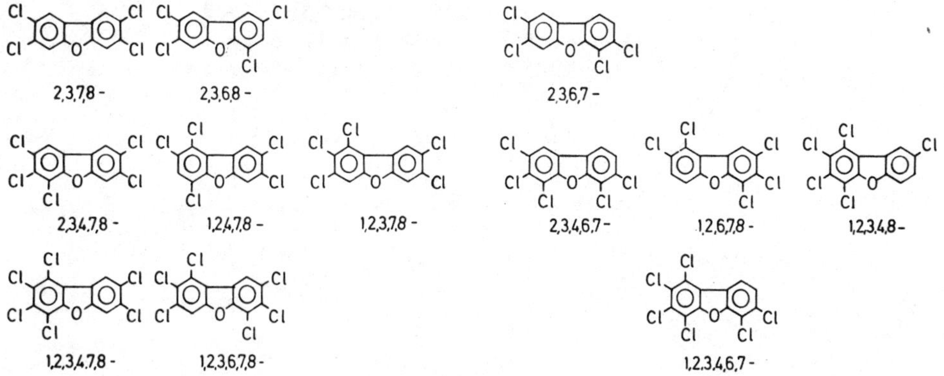

Fig. 2. PCDF isomers retained (left) and excreted (right) in Yusho oil.

In the textile and leather industry pentachlorophenols were used, and the PCDDs were the major impurities in these formulations. The blood plasma of these workers were mainly contaminated by PCDDs.

3.4. Variation in Individual Levels Due to Metabolism or Elimination

3.4.1. Textile Industry

We have followed the workers in the textile industry for a longer period. Samples of blood plasma were taken at two occasions, 5 months and 16 months after the exposure to the contaminated pentachlorophenol formulation, see Table 3. It was found that the reduction of the blood levels during the 9 months' period between the two sampling occasions were 60-90%, the values for the most exposed workers were 88% and 76% respectively.

3.4.2. Yusho Episodes Japan 1968 and Taiwan 1979

In 1968 more than 1500 persons in south-west Japan were intoxicated by consuming a commercial rice oil accidentally contaminated by PCBs, PCDFs, and PCQs. In 1979 a similar episode was reported from Taiwan, the number of persons involved approaching 2000.

Earlier analyses have proven that the Japanese rice oil contained more than 40 PCDF isomers, tri- to hexa-CDFs [6]. Analysis of liver samples taken from the Japanese patients about 18 months after the exposure showed a dramatic decrease in the number of PCDF isomers, see Fig. 1 [7]. Apparently most of the PCDF isomers have been metabolized or excreted during the period between exposure and sampling.

TABLE 4. Levels of PCDFs (pg/g) in Blood Plasma from Yusho Patients (values in parenthesis are the detection limits)

Isomer	Japan		Taiwan	
	A	B	C	D
Tetra-CDFs	< 3	< 3	< 30[a)]	< 30[a)]
1,2,4,7,8-Penta-CDF	ND	ND	60	40
1,2,3,7,8- -"-	ND	ND	30	20
2,3,4,7,8- -"-	3(3)	3(3)	120	80
1,2,3,4,7,8-Hexa-CDF	< 6	< 6	150	60

a) = high detection limits due to large amounts of overlapping PCBs and other chlorinated compounds

ND = Not Detected

TABLE 5. Levels of PCDDs and PCDFs (pg/g) in Tissue Samples (values in parenthesis are the detection limits)

Isomer	Liver	Kidney
2,3,7,8-Tetra-CDD	2(1)	5(1)
1,3,6,8- -"-	1(1)	1(1)
Penta-CDDs	< 2	< 2
1,2,3,6,7,8-Hexa-CDDs	4(4)	2(2)
1,2,3,4,6,7,8-Hepta-CDD	72(5)	8(4)
Octa-CDD	350(8)	15(4)
Tetra-CDFs	< 1	< 1
2,3,4,7,8-Penta-CDF	10(2)	7(2)
Hexa-CDFs	55(2)	8(2)
1,2,3,4,6,7,8-Hepta-CDF	100(4)	6(4)
Octa-CDF	< 3	< 3

TABLE 6. Levels of PCDFs (pg/g) and PCBs (ng/g) in Tissue Samples of Yusho Baby from Taiwan

Isomer	Adipose	Liver	Muscle	Omentum	Diaphragm
2,3,7,8-Tetra-CDF	17	60	ND	ND	ND
1,2,4,7,8-Penta-CDF	14	42	ND	ND	ND
1,2,3,7,8- -"-	44	194	ND	ND	ND
2,3,4,7,8- -"-	68	91	ND	ND	ND
1,2,3,4,7,8-Hexa-CDF	88	193	ND	ND	ND
PCBs	316	27	38	64	46

ND = Not Detected

A comparison between the PCDF isomers found in the Yusho oil and in the liver samples revealed an interesting relationship between the isomers retained in the liver and those isomers excreted, see Fig. 2. All the PCDF isomers excreted had two vicinal unchlorinated C-atoms in at least one of the two aromatic rings of the PCDF system. On the contrary, none of the isomers retained had two vicinal unchlorinated C-atoms in any of the two aromatic rings. Most of the latter isomers had all lateral (2-, 3-, 7-, and 8-) positions substituted for chlorine.

We have also analyzed a set of four blood samples taken from Yusho patients, two collected in Japan in 1979 which means 11 years after the exposure, and the two others were collected in Taiwan in 1980, one year after the exposure. The results are collected in Table 4. Low levels of 2,3,4,7,8-penta-CDF could be identified in all samples, the levels in the Japanese samples were just above the detection limit.

3.5. Variation in Levels in Different Organs

We have analyzed samples of the liver and the kidney from a man occupied in the production of phenoxy acid herbicides like 2,4-D and MCPA and 2,4,6-trichlorophenol. The exposure ceased 3-4 years before the death of the man (in a pancreatic tumor). The results of our analyses are given in Table 5. In general the levels were much higher in the liver than in the kidney, however for the highly toxic 2,3,7,8-tetra-CDD the ratio was reversed.

We have also analyzed various organs of a "Yusho Baby" from Taiwan. In Table 6 we have collected the results of these PCDF analyses together with the PCB levels.

It is interesting to notice that the PCB levels seem to parallel the fat levels in the tissue. However, dibenzofurans could only be detected in the adipose tissue and in the liver. The highest PCDF values were found in the liver, but amazingly here we found the lowest PCB values.

Discussing individual isomers it is interesting to point out that in the liver sample the dominating isomer is the 1,2,3,7,8-penta-CDF, while in adipose tissue the highest value was found for the 2,3,4,7,8-penta-CDF.

REFERENCES

1. D. L. Stalling, J. D. Petty, L. M. Smith, C. Rappe, and H. R. Buser, Isolation and Analysis of Polychlorinated Dibenzofurans in Aquatic Samples, in: Chlorinated Dioxins and Related Compounds (O. Hutzinger, R. W. Frei, E. Merian, and F. Pocchiari, eds.), Pergamon, Oxford, pp. 77-85 (1982).
2. C. Rappe, H. R. Buser, D. Stalling, L. M. Smith, and R. C. Dougherty, Identification of Polychlorinated Dibenzofurans in Environmental Samples, Nature, 292:524-526 (1981).
3. H. R. Buser and C. Rappe, High-Resolution Gas-Chromatography of the 22 Tetrachlorodibenzo-p-dioxin Isomers, Anal. Chem., 52:2257-2262 (1980).
4. C. Rappe and H. R. Buser, Occupational Exposure to Polychlorinated Dioxins and Dibenzofurans, in: Chemical Hazards in the Workplace, Measurements and Control (G. Choudhary, ed.), ACS Symposium Series, Vol. 149, ACS, Washington, pp. 319-342 (1981).
5. C. Rappe, M. Nygren, H. R. Buser, and T. Kauppinen, Occupational Exposure to Polychlorinated Dioxins and Dibenzofurans, in: Chlorinated Dioxins and Related Compounds (O. Hutzinger, R. W. Frei, E. Merian, and F. Pocchiari, eds.), Pergamon, Oxford, pp. 495-513 (1982).
6. H. R. Buser, C. Rappe, and A. Garå, Polychlorinated Dibenzofurans (PCDFs) in Yusho Oil and Used Japanese PCB, Chemosphere, 7:439-449 (1978).
7. C. Rappe, H. R. Buser, H. Kuroki, and Y. Masuda, Identification of Polychlorinated Dibenzofurans (PCDFs) Retained in Patients with Yusho, Chemosphere, 8:259-266 (1979).

HEMOGLOBIN AS A DOSE MONITOR OF ALKYLATING AGENTS

DETERMINATION OF ALKYLATION PRODUCTS OF N-TERMINAL VALINE

S. Jensen, Margareta Törnqvist,
and L. Ehrenberg

Stockholms Universitet
Strålningsbiologiska Institutionen
S-106 91 Stockholm, Sweden

INTRODUCTION

Most mutagens and cancer initiators ("genotoxic agents" [1]) are electrophilic, mainly alkylating, reactants, or are metabolized to such [2]. Large numbers of pollutants in air, water, foods (also natural constituents [3]) are potentially genotoxic. The induction of genotoxic effects is a stochastic process with a raised risk at very low levels or doses [1]. Due to a low resolving power [4] of epidemiological studies and laboratory experiments, the identification and estimation of genotoxic risks to man becomes a difficult problem [5]. To this adds our imperfect knowledge of the metabolism of foreign compounds, leading to an uncertainty with regard to the relationship between level of exposure and *in vivo* dose of ultimate genotoxic reactants.

Evidence is accumulating to show that initial events leading to heritable damage and cancer consist in alkylation (or other type of chemical change) of DNA of the respective target cells [6]. The degree of alkylation (etc.) of DNA in a tissue after exposure to an electrophilic compound is directly proportional to the dose, i.e., the time integral of the concentration of the compound in the compartment of DNA. A quantitation of adducts to DNA by chemical or immuno-chemical methods would therefore in principle give a sensitive measure of genotoxic activity. However, DNA is less suitable for monitoring purposes because of chemical instability and biochemical repair of lesions, and because only small amounts are available in conveniently obtained samples. The possibilities of using adducts to blood proteins, especially hemoglobin, for monitoring of tissue doses of electrophilic compounds have therefore been investigated [7, 8].

The relationship between hemoglobin alkylation and DNA alkylation has been studied in model experiments with rodents, using radiolabelled compounds (for a compilation see Ref. 9). It has been shown that for low-molecular weight compounds with a sufficiently high stability and solubility in both lipids and water, such as ethylene oxide and methyl methanesulfonate, the dose determined by hemoglobin alkylation gives a good estimate of the dose to DNA (determined as guanine-N-7 alkylation). For more complicated molecules effects of compartmentalization influences the correspondence between hemoglobin alkylation and DNA alkylation. Short-lived compounds such as chloroethylene oxide (reactive metabolite of vinyl chloride) or carbonium ions (e.g., CH_3^+ or $CH_3N_2^+$; reactive intermediates of dimethylnitrosamine) give a high dose close to the site of formation, although hemoglobin can still be used for dose monitoring [8, 10, 11]. It has been demonstrated experimentally that low degrees of alkylation do not affect the life span of the erythrocytes (126 days in man, 40 days in the mouse) and that, therefore, stable reaction products of hemoglobin accumulate in a predictable way [8, 12].

Analytical procedures for the chemical determination of the degree of alkylation of cysteine and histidine in hemoglobin have been worked out for a few alkylating agents [13, 14, 15]. Briefly, these methods involve the isolation of globin from the red cells, addition of radiolabelled internal standard, hydrolysis of the protein and enrichment of the alkylated product by ion-exchange chromatography. The product is then determined directly on the amino acid analyzer or is derivatized for analysis by gas chromatography-mass fragmentography. These methods have been used to monitor tissue doses in rats after exposure to ethylene oxide [16], propylene oxide [15], dimethylnitrosamine [14] and tissue doses in man after occupational exposure to ethylene oxide [13] and propylene oxide (unpublished data).

The desired and achievable resolving power, respectively, of tests for genotoxic activity could be illustrated by expressing data in the reference scale of the genotoxic potential of ionizing radiation, notably γ-radiation. The detection limit (i.e., the lowest dose the effect of which could be detected with acceptably small α- and β-errors) [4] of most biological test systems is of the order of 100 rad [9]. For comparison, the permissible annual dose limits for radiological workers and for members of the public are set at 5 rad and 0.5 rad, respectively. The resolving power of the analysis of the hitherto available techniques for determination of hemoglobin alkylation can at present be judged from analysis of 3-(2-hydroxyalkyl)histidines by GC-MS (Ref. 13; Osterman-Golkar and Farmer/current work). The method permits theoretically the detection of the order of 0.1-0.01 pmol adduct. However, "noise" factors, partly formed during hydrolysis, reduce the sensitivity of the method to a detection limit of about 0.1 nmol per g hemoglobin, corresponding in

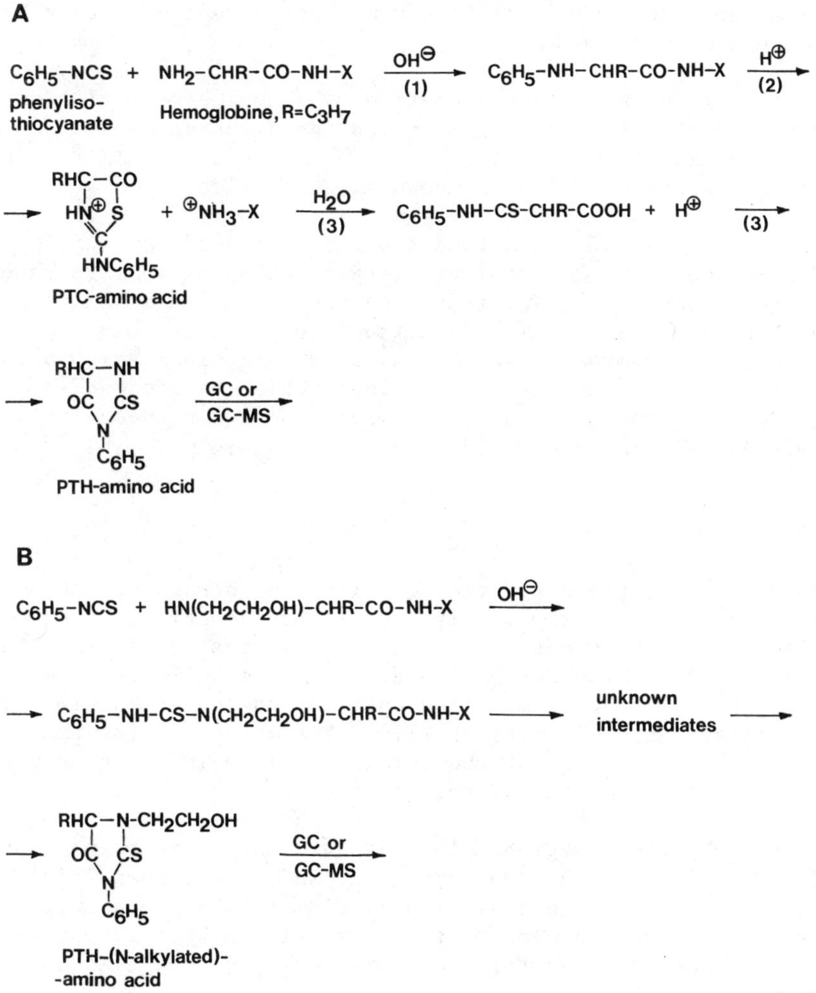

A

C_6H_5-NCS + NH_2-CHR-CO-NH-X $\xrightarrow[(1)]{OH^{\ominus}}$ C_6H_5-NH-CHR-CO-NH-X $\xrightarrow[(2)]{H^{\oplus}}$

phenyliso-thiocyanate; Hemoglobine, R=C_3H_7

→ RHC—CO / $HN^{\oplus}$ S / C / HNC_6H_5 (PTC-amino acid) + $^{\oplus}NH_3$-X $\xrightarrow[(3)]{H_2O}$ C_6H_5-NH-CS-CHR-COOH + $H^{\oplus}$ $\xrightarrow[(3)]{}$

→ RHC—NH / OC CS / N / C_6H_5 (PTH-amino acid) $\xrightarrow[GC\text{-}MS]{GC\ or}$

B

C_6H_5-NCS + HN(CH_2CH_2OH)-CHR-CO-NH-X $\xrightarrow{OH^{\ominus}}$

→ C_6H_5-NH-CS-N(CH_2CH_2OH)-CHR-CO-NH-X ⟶ unknown intermediates ⟶

→ RHC—N-CH_2CH_2OH / OC CS / N / C_6H_5 (PTH-(N-alkylated)--amino acid) $\xrightarrow[GC\text{-}MS]{GC\ or}$

Fig. 1. A) Normal Edman degradation of native protein including the coupling (1), cleavage (2), and conversion reactions (3). B) Edman degradation of a protein containing a terminally N-alkylated amino acid. All steps proceed here in the initial alkaline reaction medium. This contrast to the normal Edman degradation is used for the isolation of the alkylated PTH derivative.

the case of ethylene oxide to the risk from about 20 mrad per week or 1 rad per year. It would therefore be desirable to increase the sensitivity of the method by at least one order of magnitude to permit the monitoring of factors contributing with only a fraction of this over-risk. However, with radioisotope-labelled compounds, in animal experiments, the sensitivity may be increased further by several orders of magnitude.

The procedures for determination of hemoglobin alkylation based on total hydrolysis of the protein and ion-exchange separation are tedious and expensive. Furthermore, they are not suitable for the detection and determination of unknown electrophiles.

A development of the method towards increased sensitivity, reduced time and costs and applicability to unknown products should therefore be based on a specific cleavage of alkylated amino acids or adducts from the protein. Efforts were therefore directed towards the specific determination of N-alkylated N-terminal amino acids (in α- and β-chains of hemoglobin: valine), which besides N^{Im}-alkylhistidines is a major product following reaction of hemoglobin *in vivo* with alkylating agents [9, 16].

METHOD

A method for the estimation of alkylated hemoglobin should be mild, rapid, and sensitive. With the method described below it is possible for one person to carry out 10 analyses per day and the detection limit is below 0.1 pmol adduct. It is built on a modification of the well-known Edman procedure for the determination of amino-acid sequence of proteins [17]. The method estimates specifically the alkylation of the terminal valine in hemoglobin. Thus no acid hydrolysis of the protein is needed.

The hemolyzed hemoglobin is dissolved in a mixture of 2-propanol and 0.5 M potassium hydrogen carbonate, and excess phenyl isothiocyanate is added. In this alkaline medium (pH 9) the phenyl isothiocyanate is then coupled to the terminal $-NH_2$ (see formula). In the Edman procedure, the terminal amino acid is then cleaved off in acidic medium.

The present procedure is based on the observation that, when the NH_2 of terminal valine is alkylated, the amino-acid isothiocyanate complex is split off and ring-closed already in the alkaline medium. This reaction was found to be completed within 1 hour at 45°C for different alkyl adducts. The phenylthiohydantoin (PTH) formed (see formula, Fig. 1) can then be extracted with toluene.

The fact that alkylated residues are extracted specifically without interference of non-alkylated valine reduces the ensuing cleaning-up to a minimum.

After the addition of water to the reaction mixture the alkylated valine-PTH is extracted in toluene together with the excess of the phenyl isothiocyanate.

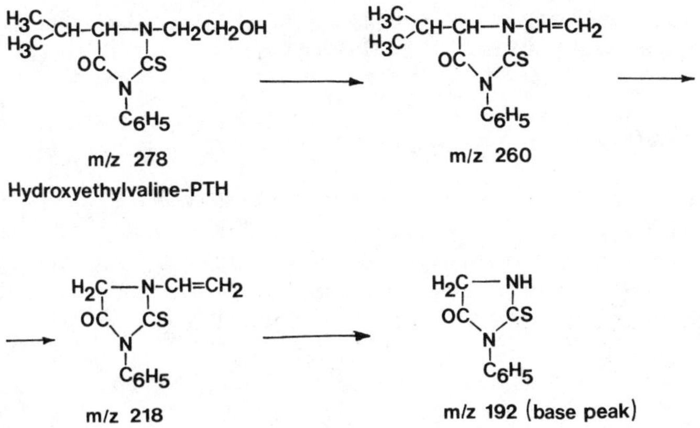

m/z 278
Hydroxyethylvaline-PTH

m/z 260

m/z 218

m/z 192 (base peak)

Fig. 2. Mass fragments from hydroxyethylvaline - PTH used for multiple ion detection in GS-MS.

The extract is then purified on one g of silica gel dry packed in a disposable pipet. The PTH is eluted with 10% acetone in benzene and the fraction is eventually concentrated and injected on the gas-chromatograph fitted with an electron capture detector (ECD). A 4 m 4% SE 54 silicone column will elute hydroxyethylvaline-PTH in 15 min in a program starting at 100°C and with a final temperature of 280°C (15°C/min). The detection limit for a peak four times the noise level is at present 30 pg corresponding to 0.1 pmol hydroxyethylvaline-PTH.

Alternatively the extract is injected on a gas chromatograph-mass spectrometer fitted with a 25 m SE 54 capillary column. For hydroxyethylvaline-PTH four characteristic mass numbers will be found (see Fig. 2), namely 278 (M), 260, 218, and 192 (base peak). With multiple ion detection (MID) of these four ions a high degree of sensitivity and certification of structure will be reached.

With this last method the presence of hydroxyethylvaline in hemoglobin from persons without any known exposure to ethylene oxide or other hydroxyethylating agents has been established.

FUTURE APPROACHES

With Fig. 2 in mind it can be seen that the method is well adaptable for the identification of hitherto unknown alkylating agents. Furthermore, it can give us a tool by which we can estimate the total alkylation of valine, independently of the structures of the alkylating agents. The unit with mass number 192 is common to all adducts. This means that after direct injection on the MS of an extract the height of the 192 peak will be a measure of the total degree of alkylation of valine.

ACKNOWLEDGEMENT

We are grateful to Siv Osterman-Golkar for valuable discussions and skillful help and to Jacques Mowrer for running the mass-spectrometer. The work was supported financially by the National Swedish Environment Protection Board, the Swedish Work Environment Fund, and the Swedish Natural Science Research Council.

REFERENCES

1. C. Ramel, ed., Ambio Special Report No. 3 (1973).
2. E. E. Miller and J. A. Miller, Pharmacol. Rev., 18:805-838 (1966).
3. J. A. Miller, Naturally occurring substances that can induce tumors, in: Toxicants Occurring Naturally in Foods, Natl. Acad. Sci., 2nd ed., Washington, D.C. (1973).
4. L. Ehrenberg, Aspects of Statistical Inference for Genetic Toxicity, in: Handbook of Mutagenicity Test Procedures (B. Kilbey, et al., eds.), Elsevier/North Holland, 420-458 (1977).
5. L. Ehrenberg, Acta Biol. Iugosl., Series F, Genetika 6:367-398 (1974).
6. P. Brookes, Role of covalent binding in carcinogenicity, in: Biological Reactive Intermediates (D. J. Jollow, et al., eds.), Plenum Press, New York, 470-480 (1977).
7. S. Osterman-Golkar, Studies on the reaction kinetics of biologically active electrophilic reagents as a basis for risk estimates, Thesis, Stockholm University, Sweden (1975).
8. S. Osterman-Golkar, et al., Mutat. Res., 34:1-10 (1976).
9. L. Ehrenberg and S. Osterman-Golkar, Teratogenesis, Carcinogenesis and Mutagenesis, 1:105-127 (1980).
10. S. Osterman-Golkar, et al., Biochem. Biophys. Res., Commun., 76:259-266 (1977).
11. L. Ehrenberg, et al., Report for ICPEMC, Mutat. Res., 123:121-182 (1983).
12. D. Segerbäck, et al., Mutat. Res., 49:71-82 (1978).
13. C. J. Calleman, et al., J. Environ. Pathol. Toxicol., 2:427-422 (1978).
14. E. Bailey, et al., Cancer Res., 41:2514-2519 (1981).
15. P. B. Farmer, et al., Biomed. Mass Spectrom., 9:69-71 (1982).
16. D. Segerbäck, Chem.-Biol. Interactions, 45:139-151 (1983).
17. P. Edman and A. Henschen, Sequence Determination, in: Protein Sequence Determination (S. Needleman, ed.), p. 232 (1975).

VARIATIONS IN SENSITIVITY AND DNA REPAIR IN HUMAN CELLS EXPOSED TO GENOTOXIC AGENTS*

James E. Cleaver

Laboratory of Radiobiology and Environmental Health
University of California
San Francisco, California 94143

SUMMARY

The human population contains a few rare genetic disorders involving major increases in cellular sensitivity to radiation or carinogens, and a large number of disorders with small increases in cellular sensitivities. Major hypersensitivities are caused by DNA repair deficiencies in xeroderma pigmentosum cells, or by abnormalities in DNA replication in xeroderma pigmentosum variant, ataxia telangiectasia, and Cockayne syndrome cells. Small hypersensitivities are probably caused by a variety of biochemical abnormalities that are not directly related to DNA repair. Measurements of DNA repair currently available for use in large populations all have technical problems. The use of scintillation counting of repair synthesis in the presence of hydroxyurea needs to be carefully evaluated in relation to the inhibitory effects of hydroxyurea on the polymerization and ligation steps of repair.

Within human populations there is a considerable range of clinical and cellular sensitivities to radiations and carcinogens due to differences in many biochemical processes involved in responses to environmental agents. If damage to DNA is considered paramount, then differences in cellular responses can involve differences in activating or detoxifying enzymes that are involved in causing damage, and differences in DNA repair, DNA replication, cellular proliferation kinetics, etc. Recent work has tended to overemphasize the importance of DNA repair as the underlying cause of most differences in radiation and carcinogen sensitivity (Bech-Hansen et al.,

*This work supported by the U.S. Department of Energy, Contract No. DE-AM03-76-SF01012.

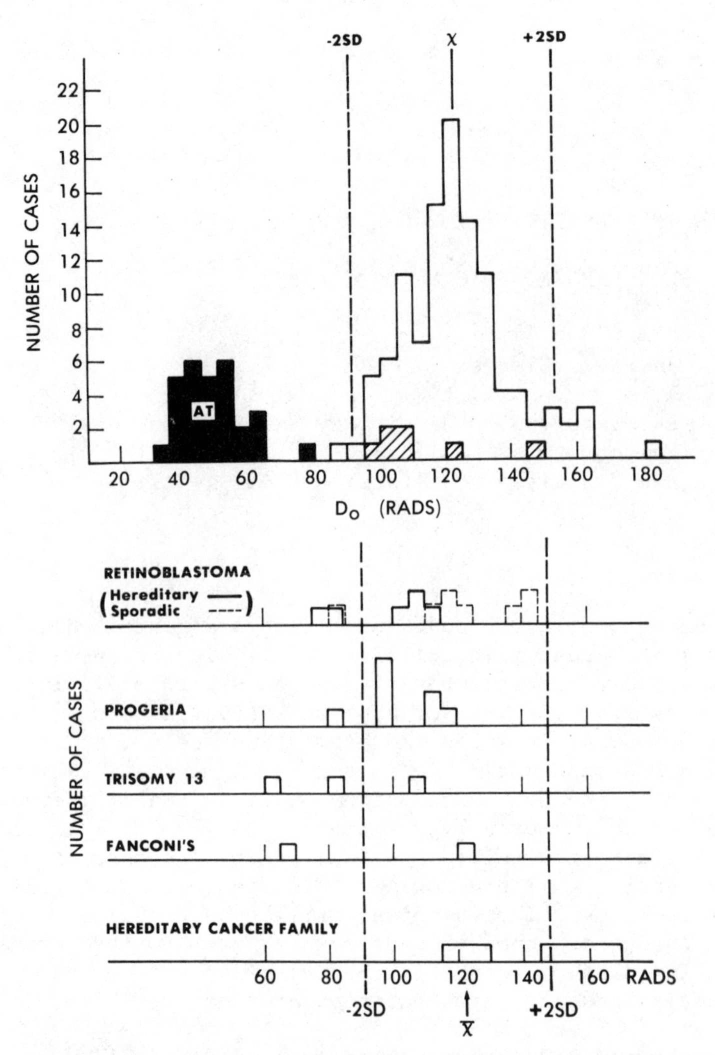

Fig. 1. D_0 (rads) of skin fibroblasts from normal individuals and those with various disorders. Data from Arlett and Harcourt (1978), Cox and Masson (1980), Little et al. (1980), Paterson et al. (1976, 1979), and Weichselbaum et al. (1980). AT (Solid histogram), ataxia telangiectasia; hatched histogram, ataxia telangiectasia heterozygotes. For some data the D_0 values were corrected so that the average D_0 for normal cells corresponded to the value of Cox and Masson of 122 ± 17 rads. Dashed vertical lines indicate the range encompassing ±2 standard deviations.

1981; Little et al., 1980; Kinsella et al., 1982; Scuderio et al., 1981). In fact, when cells from patients with a clinical disorder exhibit a D_0 different from normal, it is the exception for this to be attributable to a defect in DNA repair.

When fibroblasts from various human disorders and from clinically normal individuals are assayed for radiation sensitivity, even the normals show a considerable variation (Fig. 1). Normal fibroblasts show a distribution of D_0's around a mean of approximately 120 rads, with a standard deviation of ±17 rads. Significant differences from normal should therefore not be considered truly biologically or clinically important until differences of about two standard deviations from the population average can be demonstrated. Fibroblasts from homozygous individuals with ataxia telangiectasia are therefore clearly X-ray sensitive, with D_0's of 40-60 rads. Fibroblasts from many other putative radiosensitive diseases, however, are within the range of variation exhibited by clinically normal individuals. Retinoblastoma is therefore not a clearly radiosensitive disorder (Weichselbaum et al., 1977; Cox and Masson, 1980) and attempts to identify repair defects in this disorder have been unsuccessful (Cleaver et al., 1982).

The disorders that have been investigated with respect to radiation or carcinogen sensitivities can be put into two classes consisting of those that exhibit either major or minor hypersensitivities (Table 1). Those diseases exhibiting major hypersensitivities include the excision repair deficient disease xeroderma pigmentosum (XP), and the XP variant, ataxia telangiectasia (AT), Cockayne syndrome and Fanconi's anemia (FA). Strictly speaking XP is the only DNA repair defective disease; the other diseases have more varied and complex biochemical abnormalities. In addition, XP and at least the XP variant, AT and Cockayne syndrome all share abnormalities of various kinds in replication of DNA after exposure to X-rays or UV light (Lehmann et al., 1977, 1979; Cleaver et al., 1980, 1983a; Painter and Young, 1980; Kaufmann and Cleaver, 1981). Major increases in sensitivity to DNA damaging agents therefore predomimantly involve biochemical abnormalities in repair or replication of DNA. The variety of clinical symptoms encompassed by these disorders highlight the importance of the integrity of DNA in maintaining normal differentiation, immune function, neurological capacity and resistance to carcinogenesis.

The disorders exhibiting minor hypersensitivities are characterized by small variable, and inconsistent increases in sensitivities that are within a factor or two of normal (Table 1). In none of these have the sensitivity increases been shown to be correlated with any biochemical abnormality in DNA repair, and many other causes can be envisaged. For example, since radiation sensitivity changes around the cell cycle, if cell cultures from a particular genetic disorder have an altered cell cycle distribution the radiation sensi-

TABLE 1. Human Diseases Showing Hypersensitivity to Radiations or Carcinogens

Disease	Agent	D_o ratio	Reference
Major Hypersensitivities[1]			
Xeroderma pigmentosum (A,C,D)	UV (Chem. Car.)	5-10	Andrews et al. (1978)
XP variant	UV	1-6	Andrews et al (1978)
Ataxia telangiectasia	X rays (gamma)	2.9-3.5	Peterson et al. (1976,1979)
Cockayne syndrome	UV	4.6	Wade and Chu (1979)
Fanconi's anemia	mitomycin C	2-15	Fujiwara et al. (1977)
Minor Sensitivities			
AT heterozygotes	X rays (gamma)	0.9-1.2	Peterson et al. (1979)
Cockayne Heterozygote	UV	1.8	Wade and Chu (1979)
Retinoblastoma	X rays	1.2-1.5	Weichselbaum et al. (1980)
Huntington's chorea	X rays	1.25-2.0	Arlett (1980)
Huntington's chorea	MNNG	2.1	Scudiero et al. (1981)
Partial trisomy 13	X rays	1.6-2.0	Weichselbaum et al. (1980)
Progeria	X rays	1.1-1.6	Weichselbaum et al. (1980)
Werner's syndrome	X rays	1.1-1.6	Weichselbaum et al. (1980)
Gardner's syndrome	UV light	2.3	Little et al. (1980)
Gardner's syndrome	X rays	1.4	Little et al. (1980)
Chediak-Higashi syndrome	UV light	2.2	Kanaka and Orii (1980)

[1]The XP variant is included in this category because it has a major biochemical abnormality in DNA repliation after UV. Although this does not cause large increases in sensitivity does cause a large increase in mutability.

tivity of an asynchronous population could be different from normal for this trivial reason. Also, changes in nucleotide pool sizes can exert a large influence on the sensitivity of cells to alkylating agents (Meuth 1981). Most likely, the small differences in sensitivity seen in these disorders are the consequence of alterations in one or more biochemical or regulatory pathways that only indirectly affect the survival of cells under stress. This is not to say, however, that these alterations are unimportant because they are likely to be far more important for the general public and affect more people than frank DNA repair defects.

The mere observation of a difference in D_0 is obviously insufficient to claim that a difference in DNA repair capacity exists. It is important therefore to have methods by which to assay large numbers of individuals for biochemical parameters that have a bearing on their sensitivities to DNA damaging agents.

TECHNIQUES FOR ASSESSING DNA REPAIR

Repair of DNA damage involves a sequence of biochemical steps that can be studied with a variety of techniques designed to detect (a) the formation and removal of damage bases, (b) the insertion of new bases during repair synthesis, and (c) the formation and sealing of DNA breaks during excision. Few of these techniques are amenable to rapid, simple, and inexpensive use in screening large numbers of individuals. The main cell type available in population studies, peripheral lymphocytes, also presents difficulties because its repair capacity is depressed in comparison to many somatic cell types, probably because it is low in DNA polymerase alpha (Scudiero et al., 1976).

Available techniques include alkaline sucrose gradients for single strand breaks, isopycnic gradients and BrdUrd photolysis for repair replication, direct measurement of loss of adducts with labeled carcinogens and HPLC. These are all time consuming, need specialized skills and can be expensive (Cleaver 1974). Alkaline elution has potential and multiple parallel setups are conceivable (Petzold and Svenberg 1978) but is subject to complex artifacts and interpretations are rarely simple.

Probably the only method applicable for screening purposes is measurement of the patching step of repair - unscheduled synthesis. Various methods exist for determining this, most notably autoradiography of cells in G_1 and G_2 phases of the cell cycle, and scintillation counting of ^{3}H-thymidine incorporated into DNA in cell populations in which there is negligible semiconservative replication (Cleaver 1974). Extensive studies have been made using various kinds of DNA damage (Stich et al., 1975) and it is capable of good and accurate measurements. Autoradiography is both qualitative, because it reveals that repair is occurring outside of the normal S phase, and quantitative, although grain counting is a tedious and unpleasant assay to do full-time (Cleaver and Thomas, 1981). There is, however, the possibility that it can be automated.

One of the most practical methods for measuring repair synthesis is to label a cell population with ^{3}H-thymidine under conditions in which semiconservative DNA synthesis makes only a small contribution to the overall radioactivity incorporated. Scintillation counting of cells labeled with ^{3}HdThd in the presence of inhibitors of DNA replication such as hydroxyurea is the most practical, but this is not without serious technical problems. This method merely registers a net change in counts without any qualitative parameter to guarantee that the increase is due to repair or its absence is due to the lack of repair. Use of confluent cultures or non-proliferating tissues such as lymphocytes or liver cultures that can activate precursors of ultimate carcinogens (Williams 1976) raises the possible complication that these cell populations may contain organ-specific repair systems that differ in some respects from proliferating cells in culture. In tissue culture, plateau phase and proliferating cells appear similar in repair of pyrimidine dimers (Kantor et al., 1980) but this may not always be true for all cell populations. The repair capacity of peripheral lymphocytes changes considerably after stimulation with phytohemagglutinin, because of repressed nucleotide synthesis and low DNA polymerase levels in the unstimulated lymphocytes (Scudiero et al. 1976). Also differences in ^3H uptake between individuals may be due not only to DNA repair differences but to differences in thymidine kinase, phosphorylase and pyrimidine nucleotide pools (Cleaver 1969).

A common protocol is to label exposed cell populations with ^3H-thymidine in the presence of 1-10 mM hydroxyurea to suppress semiconservative DNA replication preferentially. Although low levels (i.e., 1 mM) of hydroxyurea do discriminate repair from semiconservative synthesis (Cleaver 1968), higher concentrations can interfere with repair. Above 2 mM hydroxyurea in human fibroblasts, an apparent absence of effect on repair synthesis is actually due to a combination of increased patch size, delayed ligation and a reduced number of sites involved in repair (Francis et al. 1979). In fact, it was recognized over 10 years ago (Ben-Hur and Ben-Ishai 1971) that hydroxyurea delays ligation of excision-repair patches. This effect is readily detectable in alkaline elution studies of human fibroblasts (Fig. 2). In this method the elution rate of DNA strands through filters is a function of the single strand break frequency. Soon after irradiation with UV light, rapid excision of pyrimidine dimers results in a fast elution rate because of the relatively large number of breaks. Later, the rate of excision declines and few breaks are open, unless cells are grown in hydroxyurea. Even 2 mM hydroxyurea results in a large increase in the number of breaks over those normally seen. Studies involving repair measurements in the presence of hydroxyurea must therefore pay careful attention to the concentration used. Since the effect of hydroxyurea can be counteracted by high levels of nucleotides (Ben-Hur and Ben-Ishai 1971), the pool sizes in different cell types will alter their responsiveness to hydroxyurea.

A striking example of the kind of misleading information generated using hydroxyurea is illustrated in recent studies of the role of poly-ADP-ribose polymerase in DNA repair. This polymerase is stimulated by DNA breaks, but inhibited by benzamide derivatives.

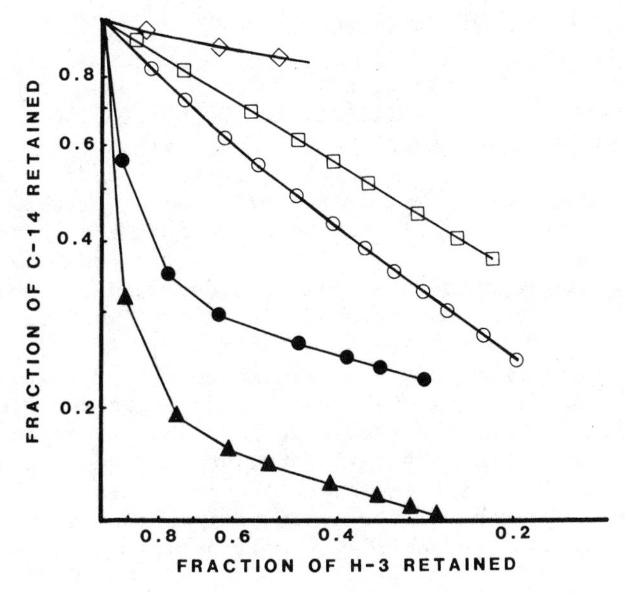

Fig. 2. Inhibition of repair of ultraviolet damage by hydroxyurea in human fibroblasts. ◇) Control; ○) harvested 5 min after 13 J/m^2; □) harvested 2 hr after 13 J/m^2; ●) grown in 2 mM hydroxyurea for 2 hr after 13 J/m^2; ▲) grown in 10 mM hydroxyurea for 2 hr after 13 J/m^2 (J. E. Cleaver and W. F. Morgan, unpublished experiments).

In a recent study, repair synthesis of UV damage was apparently stimulated by the inhibition of poly-ADP-ribose polymerase (Sims et al., 1982), but this is most likely artifactual. The study was done in the presence of 10 mM hydroxyurea which inhibits repair and increases the number of breaks present during UV repair thereby stimulating excess poly-ADP-ribose synthesis; a role for poly-ADP-ribose was therefore created that does not usually occur during UV repair (Cleaver et al., 1983b; Charles and Cleaver, 1982; Morgan and Cleaver, 1982).

There are no simple answers to these technical problems, and although repair synthesis is a rapid and convenient assay to use, it must be used with care, caution and with due appreciation of the number of factors, in addition to repair capacity, that can alter the incorporated radioactivity. When individuals or population groups are identified that apparently have markedly altered repair synthesis, these should be investigated with independent methods, such as cellular sensitivity, strand breakage, or excision of DNA damage, to determine whether repair is genuinely involved.

USE OF DNA REPAIR AS A SCREENING PRINCIPLE

The general principle of the use of DNA repair to detect differences in carcinogen sensitivity in the population should be viewed with reservations. DNA repair will always present a problem as a screening principle for one important reason: certain kinds of DNA alterations that lead to mutation and/or cancer either do not stimulate repair processes at all, or do so to such a small extent that detection is not practicable. Mutagenesis from chemicals such as the intercalating agents, ethidium bromide, acriflavine, actinomycin D, and adriamycin does not involve DNA repair synthesis even at high doses (Cleaver 1968; Painter 1978; Painter and Howard 1978). Similarly, X-rays are mutagenic and carcinogenic, but induce very little repair synthesis (Painter and Young, 1972; Regan and Setlow, 1974). Also, some metals that are suspected carcinogens do not elicit detectable repair (Painter 1981). Thus, while detection of repair synthesis is useful after exposure to some mutagens or carcinogens, resolution of important differences within a population that are relevant to carcinogen sensitivity will not always be possible by this method.

DNA repair is also a poor indicator of mutagenic risk because of its generally error-free character, although reductions in repair could indicate a tendency for accumulation of unrepaired damage. Of major interest is the effect of DNA damage that persists and causes disturbances in DNA replication, generates sister chromatid exchange, chromosome aberrations, mutations, and carcinogenic transformation. The association of abnormalities in semiconservative DNA replication with carcinogen sensitivities in many disorders (Table 1), implies that methods for assessing replication capacity are also required to complement those already in use for DNA repair (Painter and Howard, 1978).

REFERENCES

A. D. Andrews, S. F. Barrett, and J. H. Robbins, Xeroderma pigmentosum neurological abnormalities correlate with colony-forming ability after ultraviolet irradiation, Proc. Natl. Acad. Sci., U.S.A., 75:1984-1988 (1978).

C. F. Arlett, Survival and mutation in γ-irradiated human cell strains from normal or cancer-prone individuals, in: Radiation Research (S. Okada, M. Imamura, T. Terasima, and H. Yamaguchi, eds.), Japanese Association of Radiation Research, 596-602 (1980).

N. T. Bech-Hansen, W. A. Blattner, B. M. Sell, E. A. McKeen, B. C. Lampkin, J. F. Fraumeni, Jr., and M. C. Paterson, Transmission of *in vitro* radioresistance in a cancer-prone family, Lancet, 1:8234, 1335-1337 (1981).

E. Ben-Hur and R. Ben-Ishai, DNA repair in ultraviolet-light irradiated HeLa cells and its reversible inhibition by hydroxyurea, Photochem. Photobiol., 13:337-345 (1971).

W. C. Charles and J. E. Cleaver, Comparison of nucleoid and alkaline sucrose gradients in the analysis of inhibition of DNA repair in human fibroblasts, Biochem. Biophys. Res. Common, 107:250-257 (1982).

J. E. Cleaver, Repair replication of mammalian DNA: effects of compounds that inhibit DNA synthesis or dark repair, Rad. Res., 37:334-348 (1968).

J. E. Cleaver, Thymidine metabolism and cell kinetics, North Holland Publ. Co., Amsterdam, The Netherlands (1969).

J. E. Cleaver, Repair processes for photochemical damage in mammalian cells, in: Advances in Radiation Biology, Vol. 4 (J. T. Lett, H. Adler, and M. Zelle, eds.), Academic Press, New York and London, 1-75 (1974).

J. E. Cleaver, R. M. Arutyunyan, T. Sarkisian, W. K. Kaufmann, A. E. Greene, and L. Coriell, Similar defects in DNA repair and replication in the pigmented xerodermoid and the xeroderma pigmentosum variants, Carcinogenesis, 1:647-655 (1980).

J. E. Cleaver and G. H. Thomas, The measurement of unscheduled synthesis by autoradiography in Handbook of DNA repair techniques, Vol. 1B (E. C. Friedberg and P. C. Hanawalt, eds.), M. Dekkar, Inc, New York, 277-288 (1981).

J. E. Cleaver, D. Char, W. C. Charles, and N. Rand, Repair and replication of DNA in hereditary (bilateral) retinoblastoma cells after X-irradiation, Canc. Res., 42:1343-1347 (1982).

J. E. Cleaver, W. K. Kaufmann, L. N. Kapp, and S. D. Park, Replicon size and excision repair as factors in the inhibition and recovery of DNA synthesis from ultraviolet damage, Biochim. Biophys. Acta, 739:207-215 (1983a).

J. E. Cleaver, B. Zelle, and W. S. Bodell, Effects of 3-aminobenzamide in damaged cells: evidence for a greater role for poly-ADP-ribose synthesis in cells damaged by an alkylating agent than by ultraviolet light, J. Biol. Chem., 258:9059:9068 (1983b).

R. Cox and W. K. Masson, Radiosensitivity in cultured human fibroblasts, Int. J. Rad. Biol., 38:575-576 (1980).

A. A. Francis, R. D. Blevins, W. L. Carrier, D. P. Smith, and J. D. Regan, Inhibition of DNA repair in ultraviolet-irradiated human cells by hydroxyurea, Biochim. Biophys. Acta, 563:385-392 (1979).

Y. Fujiwara, M. Tatsumi, and M. S. Sasaki, Crosslink repair in human cells and its possible defects in Fanconi's anemia cells, J. Mol. Biol., 113:635-650 (1977).

H. Kanaka and T. Orii, High sensitivity but normal DNA-repair activity after UV irradiation in Epstein-Barr virus-transformed lymphoblastoid cell lines from Chediak-Higashi syndrome, Mut. Res., 72:143-150 (1980).

G. J. Kantor, R. S. Petty, C. Warner, D. J. H. Phillips, and D. R. Hull, Repair of radiation-induced DNA damage in nondividing populations of human diploid fibroblasts, Biophys. J., 30: 399-414 (1980).

W. K. Kaufmann and J. E. Cleaver, Inhibition of replicon initiation and DNA chain elongation by ultraviolet light in normal human and xeroderma pigmentosum fibroblasts, J. Mol. Biol., 149:171-187 (1981).

T. J. Kinsella, J. B. Little, J. Nove, R. R. Weichselbaum, F. P. Li, R. J. Meyer, D. J. Marchetton, and W. P. Patterson, Heterogeneous response to X-ray and ultraviolet light irradiation of cultured skin fibroblasts in two families with Gardner's syndrome, J. Natl. Canc. Inst., 68:697-701 (1982).

A. R. Lehmann, S. Kirk-Bell, C. F. Arlett, M. C. Paterson, P. H. M. Lohman, E. A. DeWeerd-Kastelein, and D. Bootsma, Xeroderma pigmentosum cells with normal levels of excision repair have a defect in DNA synthesis after UV-irradiation, Proc. Natl. Acad. Sci., U.S.A., 72:219-223 (1975).

A. R. Lehmann, S. Kirk-Bell, and L. Mayne, Abnormal kinetics of DNA synthesis in ultraviolet light irradiated cells from patients with Cockayne's syndrome, Cancer Res., 39:4237-4241 (1979).

J. B. Little, J. Nove, and R. R. Weichselbaum, Abnormal sensitivity of diploid skin fibroblasts from a family with Gardner's syndrome to the lethal effects of X-irradiation, ultraviolet light and mitomycin C, Mutat. Res., 70:241-250 (1980).

M. Meuth, Role of deoxynucleoside triphosphate pools in the cytotoxic and mutagenic effects of DNA alkylating agents, Somat. Cell Genet., 7:89-102 (1981).

W. F. Morgan and J. E. Cleaver, 3-Aminobenzamide synergistically increases sister-chromatid exchanges in cells exposed to methylmethane-sulfonate but not to ultraviolet light, Mutat. Res., 104:361-366 (1982).

R. B. Painter, DNA synthesis inhibition in mammalian cells as a test for mutagenic chemicals, in: Short term tests for chemical carcinogens (H. F. Stick and R. H. San, eds.), Springer-Verlag, New York, pp. 59-64 (1981).

R. B. Painter, DNA synthesis inhibition in mammalian cells as a test for mutagenic chemicals, Proc. Workshop on short term tests for chemical carcinogens, Vancouver, B.C. (in press) (1979).

R. B. Painter and R. Howard, A comparison of the HeLa DNA synthesis inhibition test and the Ames test for screening of mutagenic carcinogens, Mut. Res., 54:113-115 (1978).

R. B. Painter and B. Young, Repair replication in mammalian cells after X-irradiation, Mut. Res., 14:225-235 (1972).

R. B. Painter and B. Young, Radiosensitivity in ataxia telangiectasia: a new explanation, Proc. Natl. Acad. Sci., U.S.A., 77:7315-7317 (1980).

M. C. Paterson, B. P. Smith, P. H. M. Lohman, A. K. Anderson, and L. Fishman, Defective excision repair of γ-ray-damaged DNA in human (ataxia telangiectasia) fibroblasts, Nature, 260:444-447 (1976).

M. C. Paterson, A. K. Anderson, B. P. Smith, and P. J. Smith, Radiosensitivity of cultured fibroblasts from ataxia telangiectasia heterozygotes: defective colony forming ability and reduced repair replication after hypoxic irradiation, Cancer Res., 39: 3725-3734 (1979).

G. L. Petzold and J. A. Swenberg, Detection of DNA damage induced in vivo following exposure of rats to carcinogens, Cancer Res., 38:1589-1594 (1978).

J. D. Regan and R. B. Setlow, Two forms of repair in the DNA of human cells damaged by chemical carcinogens and mutagens, Cancer Res., 34:3318-3325 (1974).

D. Scudiero, A. Novin, P. Karran, and B. Strauss, DNA excision-repair deficiency of human peripheral blood lymphocytes treated with chemical carcinogens, Cancer Res., 36:1397-1403 (1976).

D. A. Scudiero, S. A. Meyer, B. E. Clatterbuck, R. E. Tarone, and J. H. Robbins, Hypersensitivity to N-methyl-N'-nitro-N-nitrosoguanidine in fibroblasts from patients with Huntington's disease, familial disautonomia, and other primary neuronal degenerations, Proc. Natl. Acad. Sci., U.S.A., 78:6451-6455 (1981).

J. L. Sims, G. W. Sikorski, D. M. Catino, S. J. Berger, and N. A. Berger, Poly(adenosine-diphosphoribose) polymerase inhibitors stimulate unscheduled deoxyribonucleic acid synthesis in normal human lymphocytes, Biochem., 21:1813-1820 (1982).

H. F. Stich, P. Lam, L. W. Lo, D. J. Koropatnick, and H. C. San, The search for relevant short term bioassays for chemical carcinogens: the tribulation of a modern sisyphus, Genet. Cyt., 17:471-492 (1975).

M. H. Wade and E. H. Y. Chu, Effects of DNA damaging agents on cultured fibroblasts derived from patients with Cockayne's syndrome, Mut. Res., 59:49-60 (1979).

R. R. Weichselbaum, J. Nove, and J. B. Little, X-ray sensitivity of diploid fibroblasts from patients with hereditary or sporadic retinoblastoma, Proc. Natl. Acad. Sci., U.S.A., 75:3962-3964 (1978).

R. R. Weichselbaum, J. Nove, and J. B. Little, X-ray sensitivity of 53 human diploid fibroblast cell strains from patients with characterized genetic disorders, Cancer Res., 40:920-925 (1980).

G. M. Williams, Carcinogen-induced DNA repair in primary rat liver cell cultures: a possible screen for chemical carcinogens, Cancer Lett., 1:231-236 (1976).

UNSCHEDULED DNA SYNTHESIS INDUCED BY N-ACETOXY-2-ACETYLAMINOFLUORENE AS AN INDICATOR OF RISK FROM GENOTOXIC EXPOSURES

Ronald W. Pero

Department of Biochemical and Genetic Ecotoxicology
University of Lund
The Wallenberg Laboratory
Box 7031, S-220 Lund 7, Sweden

1. INTRODUCTION

The evidence is strong now that metabolic parameters and DNA repair competence are probably the most important factors determining cancer incidence. It has been estimated that most human cancers are caused from exposures to genotoxic agents in our environment [1]. Furthermore, most environmental mutagens are not active as DNA damaging agents until they have been metabolized by the mixed function oxygenases (MFO). The MFO system are inducible enzymes [2] and this form of altered metabolism may relate to high affinity absorption of xenobiotics [3]. Clearly, any effective human assay system for assessing risk from hazardous environmental exposures must take into account the cellular mechanisms governing uptake, transport, activation and degradation of potentially genotoxic agents.

The capacity of human cells to repair lesions in their DNA is also a well-established factor affecting individual risk to cancer [4]. The best documented example concerns individuals with Xeroderma pigmentosum, who develop cutaneous tumors upon exposure to sunlight, because their cells are deficient in removing UV-induced lesions in their DNA [5]. Therefore, it is equally important to consider interindividual variation in DNA repair mechanisms when assessing the effectiveness of any assay procedure for estimating risk in humans from mutagen/carcinogen exposures.

2. THE NA-AAF METHOD

In order to evaluate simultaneously the contributions of both DNA repair and metabolism to the interindividual variation from genotoxic exposures, we have decided to quantify the response of human mononuclear blood cells to a standardized dose of N-acetoxy-2-acetylaminofluorene (NA-AAF). The major advantages of this approach are:

2.1. The reactive site for binding of NA-AAF to DNA is well characterized being primarily in the number 8 position of guanine [6].

2.2. NA-AAF can bind to DNA directly without any metabolic activation, and thus, any alterations in the cellular uptake and transport mechanisms which would in turn effect the level of NA-AAF binding to DNA, could be estimated.

2.3. When NA-AAF binds to DNA it induces a "large patch type" DNA excision repair [7]. This fact sensitizes the quantification of DNA repair when unscheduled DNA synthesis (UDS) is being measured, because there are more opportunities for the unscheduled incorporation of radioactive deoxyribonucleoside precursors into DNA.

2.4. If NA-AAF induced UDS is quantified in cells that are exposed in culture to a standardized dose of NA-AAF which does not give a saturated UDS response, then the level of NA-AAF induced UDS is sensitive to regulation by both cellular metabolic events that influence the level of initial DNA damage inflicted (i.e., a type of regulation of DNA damage in cells we have referred to as mutagen sensitivity, Ref. 8) as well as DNA repair proficiency.

2.5. There is a very low level background of scheduled DNA synthesis in mononuclear blood cells which allows the calculation of UDS in primary cell cultures without the necessity of synchronizing cells out of S-phase.

The most serious disadvantages of the NA-AAF method are concerned with the selection of mononuclear blood cells as the target cells in this assay procedure. Of course, blood cells have the important advantage of being one of the few cell types ethically available for sampling in humans. On the other hand, mononuclear blood cells are a very heterogeneous group of cells composed of monocytes and various subclasses of lymphocytes such as T-cell subsets and B-cells. In addition, resting lymphocytes are known to be deficient in DNA repair and MFO metabolism when compared to other cell types [9-11].

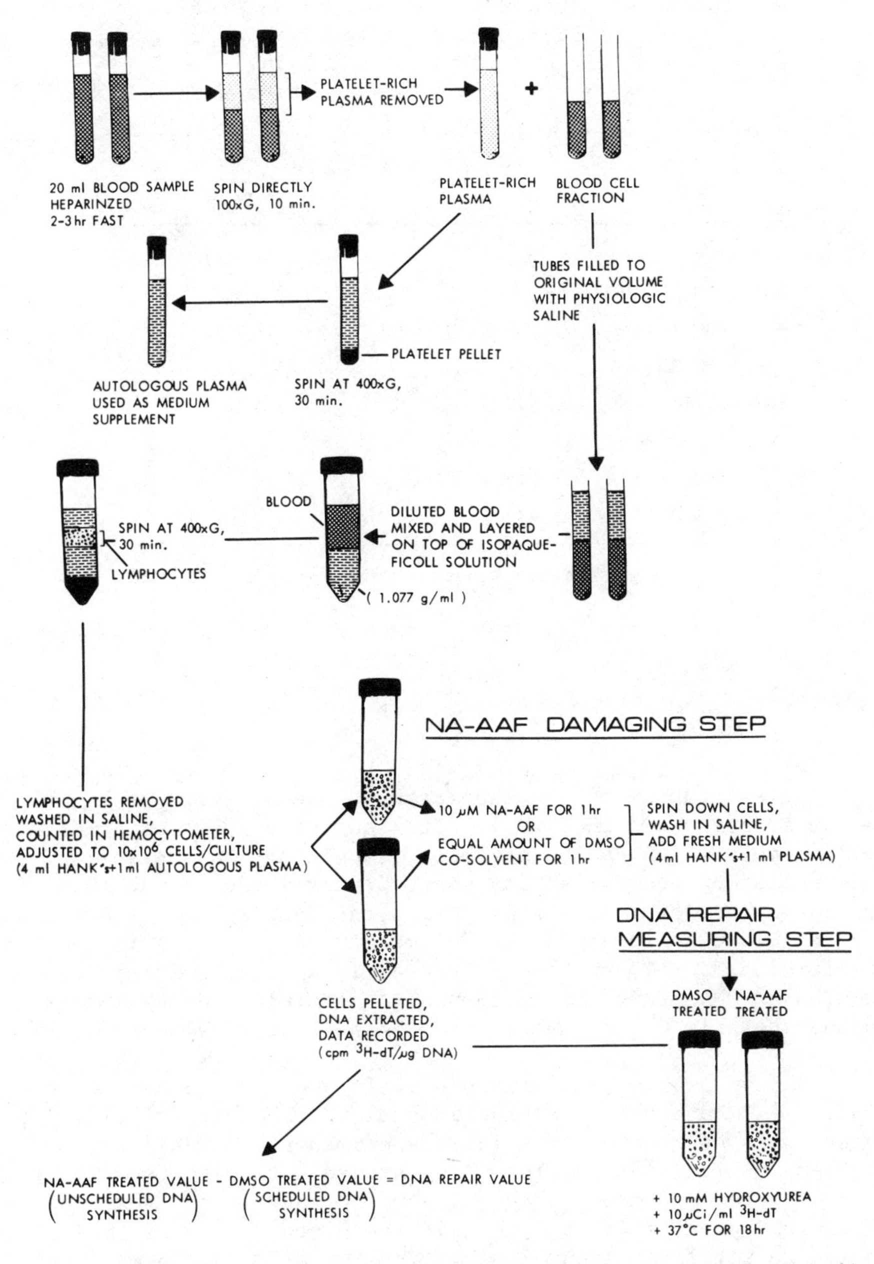

Fig. 1. The steps involved in the assay for determining an individual's level of NA-AAF induced UDS.

TABLE 1. Reproducibility of NA-AAF Induced UDS Estimations in Lymphocytes from a 23 yr Old Male Individual Sampled for 8 Consecutive Weeks

Sampling date [1]	Value	Mean ± S.D.	Coefficient variation
820217	365		
820225	334		
820304	306		
820310	302		
820316	273		
820324	199		
820331	310		
820406	401	311 ± 60	19.3 %

[1] The samples were collected and assayed according to the protocol outlined in Fig. 1.

A flow diagram of the laboratory procedure for the NA-AAF method is presented in Fig. 1. It requires a 20 ml heparinized blood sample which can be processed from start to finish in 3 days. One laboratory technician can conveniently manage about 10 samples per day. The level of UDS in peripheral lymphocytes (i.e., mononuclear blood cells are about 20% monocytes, Ref. 12) induced by a standard dose of 10 μM NA-AAF is quantified biochemically as cpm ^{3}H-dThd/μg DNA minus the influence of scheduled DNA synthesis. Further details of the method have been reported elsewhere [13].

We have examined the reproducibility of the NA-AAF method previously [13] and the interexperimental variability (±15.7%) was about twice as great as the intraexperimental variability (±9.0%). However, our earlier studies were carried out using fetal calf serum as a culture medium supplement instead of autologous human plasma. Here, we report in Table 1 that if the protocol in Fig. 1 is followed the interexperimental variation for a 23 yr old male subject over 8 consecutive weeks of repeated sampling was ±19.3% when expressed as a coefficient of variation. These data indicate the ability of the NA-AAF method to distinguish sub-groups in a reproducible manner provided an adequate sample size is examined.

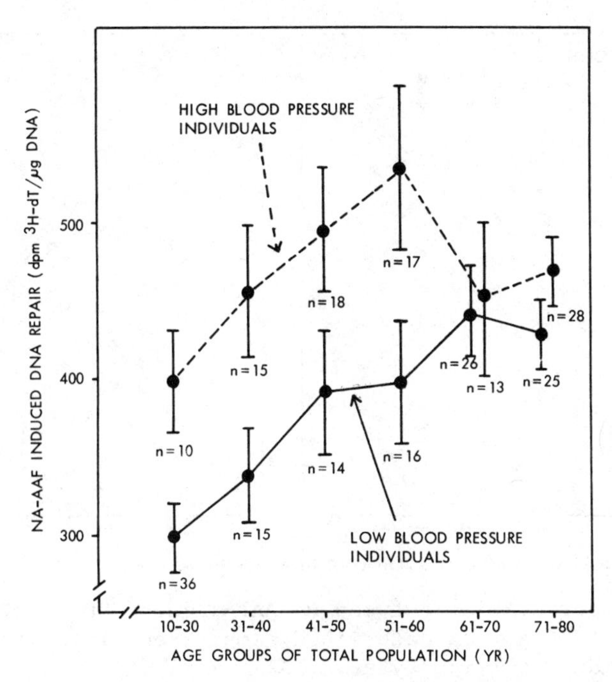

Fig. 2. The effect of blood pressure, after age adjustment, on mutagen sensitivity as estimated by the NA-AAF-induced UDS synthesis (repair) method. Lymphocytes from individuals with elevated diastolic blood pressure for 10-30 years (>70 mm Hg), 31-40 years (>85 mm Hg), 41-50 (>90 mm Hg), 51-60 years (>95 mm Hg), 61-70 years (>95 mm Hg), and 71-80 (>95 mm Hg) were compared to lymphocytes from individuals with normal blood pressures for their age group in an effort to ascertain the effect of high blood pressure on NA-AAF-induced UDS. The NA-AAF-induced UDS values for high- and low-blood pressure individuals in each group were compared directly by Student's t-test. The high- and low-blood-pressure individuals for the age groups 10-30, 31-40, 41-50, and 51-60 years were significantly different from each other ($P < 0.05$). Means ±SE are shown. (From Environmental Res., 24:409-424 (1981).)

3. INTERINDIVIDUAL VARIATION IN MUTAGEN SENSITIVITY

We have used the NA-AAF method to examine the interindividual variation in NA-AAF-induced UDS that occurs naturally in the general population of Dalby, Sweden (population = 20,856). A total of 266 individuals were selected from the reference population to represent the normal distribution of age, sex, and blood pressure [8]. The results reported in Fig. 2 indicate there is at least a 3-fold inter-

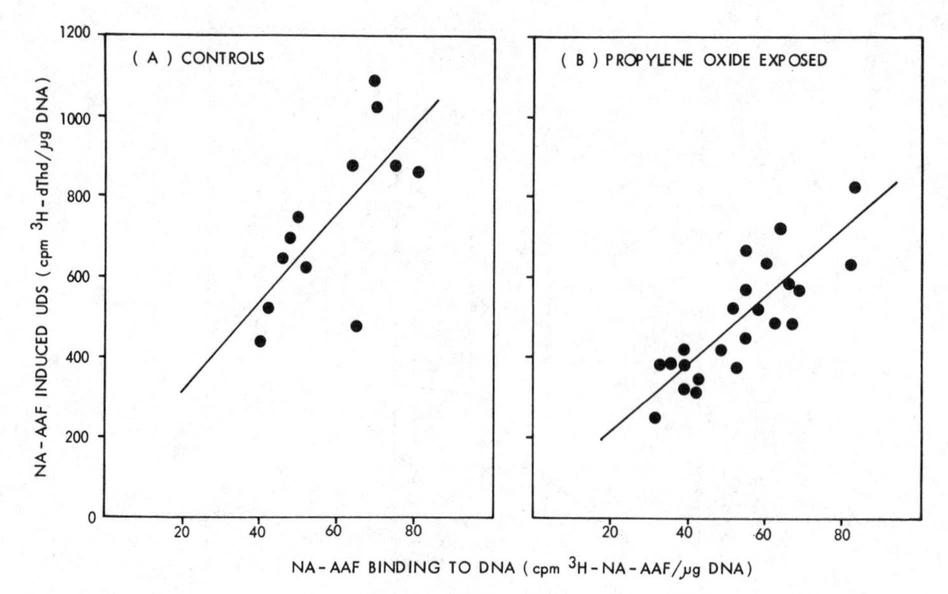

Fig. 3. The correlation between NA-AAF-induced UDS and NA-AAF binding to DNA when determined simultaneously on the same individual's lymphocytes. (A) Y = 11 (x) + 88, r = 0.70, $p < 0.02$; (B) Y = 8.5 (X) + 42, r = 0.82, $p < 0.0001$. (From Mutation Res., 104:193-200 (1982).)

individual variation in NA-AAF induced UDS and the individual values increased significantly to corresponding increases in either age or blood pressure for the individuals sampled.

Previously, we have shown that individuals with high NA-AAF induced UDS values correlated to high _in vivo_ levels of chromosome aberrations [14]. These data were taken as evidence that lymphocytes coming from individuals with old age or high blood pressure had high NA-AAF induced UDS values because their genomes became proportionally damaged more by the same standardized _in vitro_ NA-AAF treatment (i.e., increased mutagen sensitivity rather than having increased DNA repair proficiency).

Recently, we have been able to acquire NA-AAF in radioactive form (^{3}H-NA-AAF, Midwest Research Institute, St. Louis, Mo.). This has allowed us to make more direct comparisons between the level of UDS induced by NA-AAF and the degree to which ^{3}H-NA-AAF binds to DNA when the determinations are carried out on the same individual's lymphocytes. Figure 3 demonstrates that the level of UDS induced by NA-AAF is strongly dependent on how much NA-AAF binds to DNA initially (i.e., 30 min exposure). Clearly, these data confirm that an individual's lymphocytes sensitivity to suffer DNA damage can vary, and this variation is monitored by NA-AAF-induced UDS.

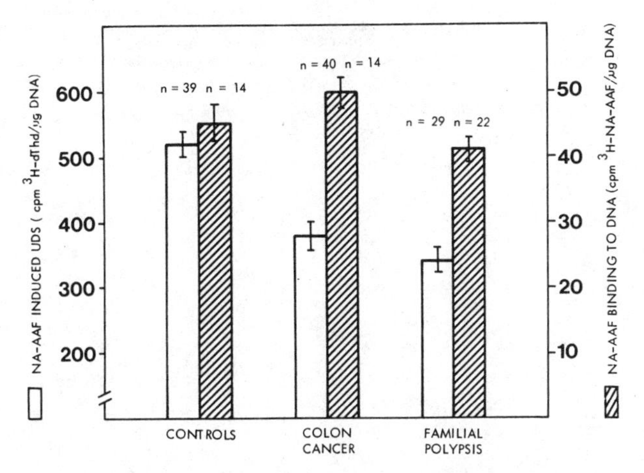

Fig. 4. A reduced capacity for NA-AAF-induced UDS in lymphocytes from male patients with either a history of or the hereditary predisposition for colorectal cancer. The difference in NA-AAF binding to DNA was not significant between the 3 groups. A t-test comparison of NA-AAF-induced UDS values in controls to the levels in the colon cancer and the familial polyposis groups gave $p < 0.001$ and $p < 0.001$, respectively. Individuals with familial colon cancer were also included in the familial polyposis group. The colon cancer group included only individuals with an earlier diagnosis of adenocarcinoma of the large bowel, and who had undergone a curative surgical resection of the primary tumor, so that there was no evidence of disease at the time of sampling.

4. INTERINDIVIDUAL VARIATION IN DNA REPAIR PROFICIENCY

Although it is evident from the above discussion that mutagen sensitivity is the main source of variation for NA-AAF-induced UDS when considering the general population, these data do not exclude the possibility that the NA-AAF method may also be useful in distinguishing sub-groups in the population which have a reduced capacity to carry out DNA repair synthesis. In order to assess this possibility we have decided to study individuals who have had colon cancer. The reason for our choice was because there is strong epidemiologic evidence that diet plays an important role in the incidence of colon cancer [15], and thus, if we select matched controls who have had no colon cancer, then we are assured they get the necessary environmental exposure by having similar diets but yet not the disease. In such a way, we have reasoned that perhaps we could define a sub-group of individuals that may be genetically predis-

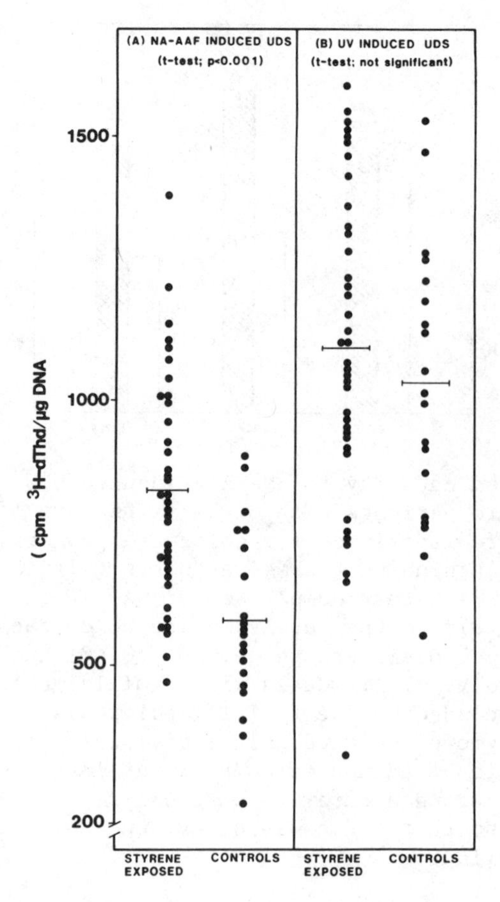

Fig. 5. Interindividual variation in the values for NA-AAF-induced UDS and UV induced UDS collected on lymphocytes from 38 subjects exposed to 1-40 ppm styrene and from 20 unexposed controls. (From Carcinogenesis, 3(6):681-685 (1982).

posed to colon cancer because they have reduced DNA repair proficiency.

We have measured both NA-AAF-induced UDS and NA-AAF binding to DNA in resting lymphocytes of 69 patients with either a history of or the hereditary predisposition for colon cancer, and then, compared their values with those obtained for matched controls. This study was carried out as a collaborative effort with clinical material organized by the Strang Clinic and the Memorial Sloan Kettering Cancer Center in New York City. The data presented in Fig. 4 show that NA-AAF-induced UDS is considerably reduced in the lymphocytes of patients associated to colon cancer when compared to control

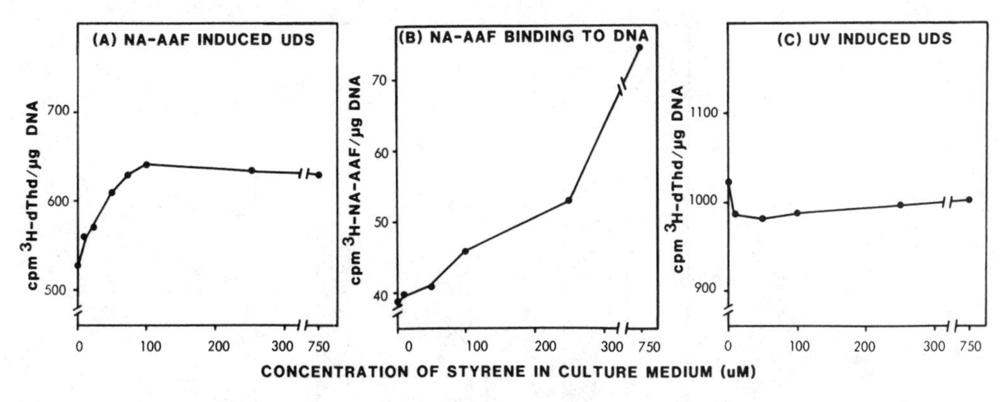

Fig. 6. The effect of _in vitro_ exposure to styrene on the levels of NA-AAF-induced UDS, NA-AAF binding to DNA and UV induced UDS calculated using human resting lymphocytes. Lymphocytes cultured in Hank's Eagle medium supplemented with 20% autologous plasma were exposed to styrene concentrations ranging from 0 µM to 750 µM for 15 min at 37°C, and then damaged directly with either 10 µM NA-AAF or 10 J/m^2 UV for the quantitative estimation of UDS or covalent binding to DNA as previously described [13, 16]. (A) NA-AAF-induced UDS, 0-100 µM range of styrene exposure, $Y = 0.89 (x) + 553$, $r = 0.62$, $p < 0.001$; (B) NA-AAF binding to DNA, 0-750 µM range of styrene exposure, $Y = 0.05 (x) + 39$, $r = 0.96$, $p < 0.001$; (C) UV induced UDS, no significant relationship to styrene exposure. All data points are the mean values of 3 determinations. (From Carcinogenesis, 3(6):781-685 (1982).

levels. However, this reduction in UDS was not accompanied by a corresponding reduction in NA-AAF binding to DNA suggesting a true reduction in DNA repair synthesis and not an altered mutagen sensitivity. As a result, we have concluded that there are sub-groups within the general population which indeed may have a reduced DNA repair proficiency that can be detected by the NA-AAF method.

5. ANALYSIS OF POPULATIONS OCCUPATIONALLY EXPOSED TO MUTAGENS

As we have shown that the NA-AAF method is sensitive to regulation by factors that influence both mutagen sensitivity and DNA repair proficiency, the next question to be answered was whether environmental exposures to potential mutagens were detectable by this method. Mutagens are known to chemically interact with DNA and stimulate UDS as does NA-AAF, and thus, _in vivo_ exposures of lymphocytes

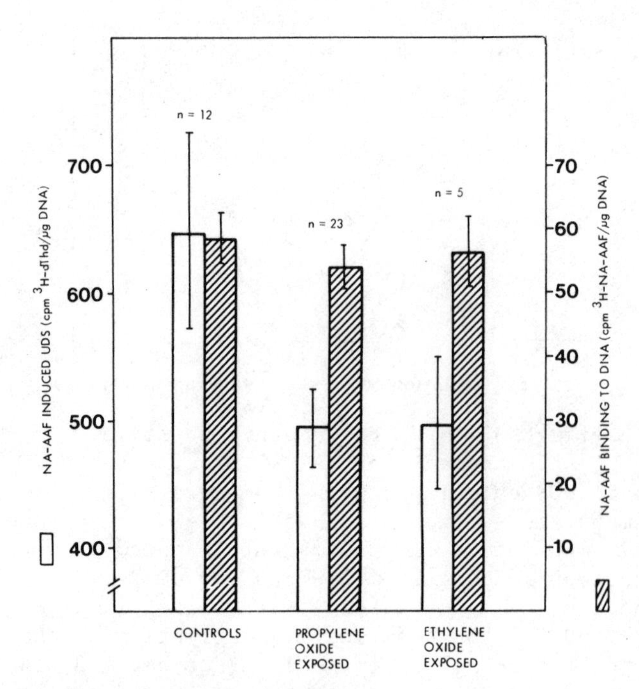

Fig. 7. A reduced capacity for NA-AAF-induced UDS in lymphocytes from individuals exposed to <12 ppm propylene oxide (PO) or ethylene oxide (EO). NA-AAF binding to DNA was not significantly different between the 3 groups. A t-test comparison of NA-AAF induced UDS values in controls to the levels observed in the combined PO- and EO-exposed groups gave $p < 0.001$.

to environmental mutagens may predispose the cells to an induced alteration in the normal pattern of NA-AAF-induced UDS estimations.

We have examined the lymphocytes of 38 workers occupationally exposed to styrene concentrations in the factory air of 1-40 ppm by the NA-AAF method [16]. The results reported in Fig. 5 show that NA-AAF-induced UDS is significantly increased for the styrene exposed group when compared to matched controls, but no effect from styrene exposure was observed when UV induced UDS was measured simultaneously. In order to resolve this apparent discrepancy we have exposed lymphocytes _in vitro_ to styrene, and then, estimated in parallel samples NA-AAF induced UDS, UV-induced UDS and NA-AAF binding to DNA. It is apparent from Fig. 6 that styrene exposure increases the mutagen sensitivity in lymphocytes rendering them susceptible to accumulate NA-AAF-induced DNA damage. In this case, DNA repair synthesis _per se_ estimated by UV induced UDS was unaffected by styrene exposure.

On the other hand, we have also obtained evidence from occupational exposures to ethylene oxide (EO) and propylene oxide (PO) that the DNA repair proficiency in lymphocytes can be reduced by exposure to <12 ppm of these alkylating agents [17]. It can easily be seen in Fig. 7 that while the level of NA-AAF-induced UDS is significantly reduced in the lymphocytes from individuals occupationally exposed to EO or PO, the level of NA-AAF binding to DNA in duplicate lymphocyte samples was unaffected. These data support the conclusion that PO or EO do not affect mutagen sensitivity, but rather inhibit the capacity of cells to carry out DNA repair synthesis.

6. REGULATION OF DNA DAMAGE INDUCED BY NA-AAF IN BLOOD CELLS

Now that we have demonstrated the potential usefulness of the NA-AAF method in assessing risk from genotoxic exposures, the next step is to better understand what regulates the interindividual variation observed in both mutagen sensitivity and DNA repair proficiency. Two recent developments in my laboratory have suggested some potential mechanisms of regulatory control.

Firstly, it has occurred to us that perhaps we could assess if there exists a high affinity uptake and transport of NA-AAF by lymphocytes. This could be done because NA-AAF is a direct acting carcinogen and dependent only on cellular uptake and transport mechanisms in order to bind covalently to DNA. Thus, we can determine affinity transport of NA-AAF by taking advantage of the law of mass action and assessing if ^{3}H-NA-AAF binding to DNA in cultured leukocytes could be inhibited by addition of competing concentrations of unlabeled NA-AAF or other potential antagonists. Using this approach, we have determined that a 100-fold excess concentration of unlabeled NA-AAF, 2-acetyl-aminofluorene, sex steroids or melatonin reduced the level of ^{3}H-NA-AAF binding to DNA in leukocyte cultures to 70% or more of the unchallenged, original value. Furthermore, a double-reciprocal plot of the inhibition of NA-AAF binding to DNA in the presence or absence of the antagonists indicated that both 2-acetylaminofluorene and melatonin were competitive-type inhibitors for the "receptors" that is responsible for the high affinity transport of NA-AAF into the nucleus where it can bind covalently to DNA (Fig. 8).

The affinity of the NA-AAF "receptor" was evaluated by competing various structural analogues of NA-AAF for the "receptor" by exposing leukocyte cultures to a dose range of competitor up to 200-fold molar excess, and then, immediately adding ^{3}H-NA-AAF and quantifying if the level of ^{3}H-NA-AAF bound to DNA was reduced. The close structural analogues of NA-AAF, fluorene and aminofluorene, were totally ineffective in competing for the NA-AAF "receptor". The

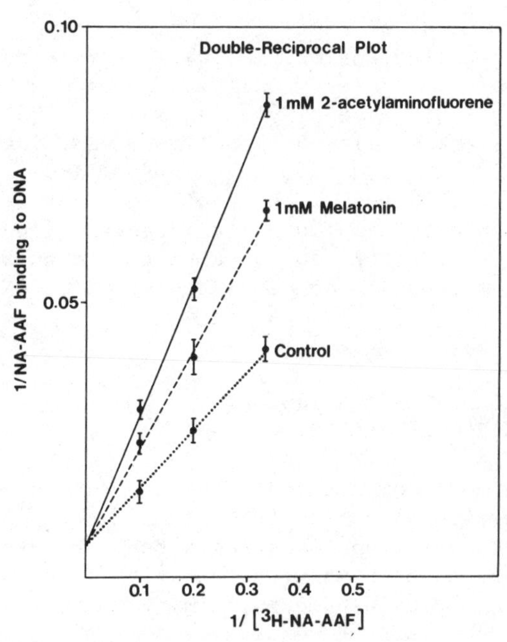

Fig. 8. Determination that 2-acetylaminofluorene and melatonin reduce the covalent binding of NA-AAF to DNA because they compete for a "receptor" that is responsible for transporting NA-AAF into the nucleus. Viable resting lymphocyte cultures were incubated in the presence of either 1 mM 2-acetylaminofluorene or 1 mM melatonin for <5 min, and then immediately ^{3}H-NA-AAF was added for 30 min at 37°C at doses of 3 μM, 5 μM, or 10 μM. The cells were harvested by centrifugation at 400 XG for 10 min, and the DNA was extracted, purified and quantified as cpm ^{3}H-NA-AAF/μg DNA as described elsewhere [13, 16, 17]. Neither 2-acetylaminofluorene nor melatonin are known to bind covalently to DNA directly, but melatonin is known to have a cytosolic receptor [19]. All data points represent duplicate determinations.

melatonin analogues, serotonin, and 5-methoxy-tryptamine, as well as the glucocorticoids and the mineralcorticoids were poor competitors for NA-AAF DNA binding sites in cultured leukocytes. This high degree of specificity strongly suggest an important regulatory role for affinity uptake and transport of NA-AAF in leukocytes. The implications are that individual mutagen sensitivity may be determined by the presence or absence of high affinity absorption sites ("receptors") in leukocytes.

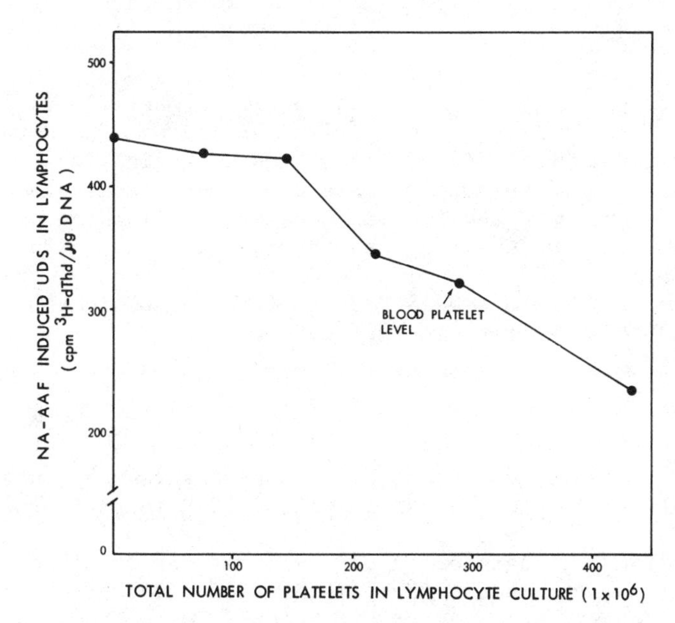

Fig. 9. The inhibitory effect of platelet concentration in lymphocyte cultures on UDS. About 10×10^6 platelet-free lymphocytes were treated for 1 hr in 20% platelet-poor plasma (PPP) supplemented culture medium containing 10 μM NA-AAF. The cells were harvested by centrifugation and fresh medium supplemented with 20% PPP, was added. Platelets were concentrated above whole blood levels by centrifugation of platelet-rich plasma in an Isopaque-Ficoll gradient (1.077 gm/ml) for 25 min at 400 XG and then using the platelet-rich interphase. UDS was estimated as in Fig. 1. (From Carcinogenesis, 2(11):1103-1110 (1981).)

On the other hand, we have also recently reported that human platelets release a substance that inhibits UDS in mononuclear blood cells but it does not affect mutagen sensitivity [18]. Platelet to mononuclear blood cell ratios that exist in whole blood were found to be sufficient to inhibit UDS *in vitro* (Fig. 9). Therefore, the platelet derived inhibitor of UDS at least has the potential to be an independent regulator of NA-AAF-induced UDS via its regulatory influence on DNA repair synthesis only and not because of any effects on mutagen sensitivity. Together the above mentioned data suggest that an individual's risk to genotoxic exposures is under a complex control system involving at least 2 independent points of regulation at the cellular level; i.e., DNA repair capacity and mutagen sensitivity.

ACKNOWLEDGEMENTS

I am deeply grateful to my co-workers Lisbeth Boström, Carl Bryngelsson, Tomas Bryngelsson, Kristin Holmgren, Åke Nordén, Anders Olsson, Margareta Pero, Cathy Vopat, and Catharina Östlund, who carried out many of the experiments reported in this study. This work was supported by the Swedish Council for Planning and Coordination of Research in 'Chemical Health Risks in our Environment,' by the Swedish Workers' Protection Fund, and by a special grant to the Dalby Community Care Sciences Program from the National Board of Health and Social Welfare in Sweden.

REFERENCES

1. R. Doll and I. Vodapija (eds.), Host environment interactions in the etiology of cancer in man, IARC Scientific Publication, No. 7, pp. 464 (1973).
2. A. H. Conney, Pharmacological implications of microsomal enzyme induction, Pharmacol. Rev., 19:317-366 (1967).
3. D. W. Nebert, H. J. Eisen, M. Negishi, M. A. Lang, and L. M. Hjelmeland, Genetic mechanisms controlling the induction of polysubstrate monooxygenase (P-450) activities, Ann. Rev. of Pharm. Toxicol., 21:431-462 (1981).
4. R. B. Setlow, Repair deficient human disorders and cancer, Nature, London, 271:713-717 (1978).
5. J. E. Cleaver, Xeroderma pigmentosum: A human disease in which an initial stage of DNA repair is defective, Proc. Natl. Acad. Sci., Washington, D.C., 63:428-435 (1969).
6. E. Kriek, Carcinogenesis by aromatic amines, Biochem. Biophys. Acta, 355:177-203 (1974).
7. J. D. Regan and R. B. Setlow, Two forms of repair in the DNA of human cells damaged by chemical carcinogens and mutagens, Cancer Res., 34:3318-3325 (1974).
8. R. W. Pero and Å. Nordén, Mutagen sensitivity in peripheral lymphocytes as a risk indicator, Environmental Res., 24:409-424 (1981).
9. J. P. Whitlock, H. L. Cooper, and H. U. Gelboin, Aryl hydrocarbon (benzopyrene) hydroxylase is stimulated in human lymphocytes by mitogens and benz(a)anthracene, Science, Washington, D.C., 177:618-619 (1972).
10. Z. Darzynkiewicz, Radiation-induced DNA synthesis in normal and stimulated human lymphocytes, Exp. Cell Res., 69:356-360 (1971).
11. R. Lewensohn, D. Killander, U. Ringborg, and B. Lambert, Increase of UV-induced DNA repair synthesis during blast transformation of human lymphocytes, Exp. Cell Res., 123:107-110 (1979).
12. S. Soman and L. S. Kaplow, Monocyte contamination in Ficoll-Hypaque mononuclear cell concentrates, J. Immunol. Methods, 32:215-221 (1980).

13. R. W. Pero, C. Bryngelsson, F. Mitelman, T. Thulin, and Å. Nordén, High blood pressure related to carcinogen-induced unscheduled DNA synthesis, DNA carcinogen binding, and chromosomal aberrations in human lymphocytes, Proc. Natl. Acad. Sci., Washington, D.C., 73(7):2496-2500 (1976).
14. R. W. Pero and F. Mitelman, Another approach to the in vivo estimation of genetic damage in humans, Proc. Natl. Acad. Sci., Washington, D.C., 76:462-462 (1979).
15. M. Lipkin, P. Sherlock, and J. J. DeCosse, Risk factors and preventive measures in the control of cancer of the large intestine, in: Current problems in Cancer (R. C. Hickey, ed.), Vol. 17, No. 10, Chicago-London, Yearbook Medical Publishers, Inc., pp. 1-57 (1980).
16. R. W. Pero, T. Bryngelsson, B. Högstedt, and B. Åkesson, Occupational and in vitro exposure to styrene assessed by unscheduled DNA synthesis in resting human lymphocytes, Carcinogenesis, 3(6):681-685 (1982).
17. R. W. Pero, I. Bryngelsson, B. Widegren, B. Högstedt, and H. Welinder, A reduced capacity for unscheduled DNA synthesis in lymphocytes from individuals exposed to propylene oxide and ethylene oxide, Mutation Res., 104:193-200 (1982).
18. R. W. Pero and C. Vopat, A human platelet-derived inhibitor of unscheduled DNA synthesis in resting lymphocytes, Carcinogenesis, 2(11):1103-1110 (1981).
19. M. Cohen, D. Roselle, B. Chabner, T. J. Schmidt, and M. Lippman, Evidence for a cytoplasmic melatonin receptor, Nature, London, 274:894-895 (1978).

INDIVIDUAL VARIATION IN DNA REPAIR IN HUMAN PERIPHERAL BLOOD MONOCYTES

Robert S. Lake

Pathology-Toxicology
Pharmaceutical Research Division
Schering-Plough Corporation
Lafayette, New Jersey 07848

1. INTRODUCTION

The growing list of genetic and biological markers presently applied for predicting risk of human beings to manifest irreversible chemically induced disease is now legion. It includes at the cellular level such diverse endpoints as cytogenetics [1], mutagen/carcinogen metabolism, DNA damage and repair [2, 3], mutability, transformability [4], promotability, radiosensitivity, immunokinetics, and cell physiology. Yet, the list of suitable and accessible human tissues available is small, limited in practice to heterogeneous cultured skin cells of fibroblastic or epithelial morphology, peripheral blood lymphocytes [5], bone marrow, and organ autopsy or biopsy specimens. Each cell system has a common limitation in that baseline parameters which may not be stable or intrinsic are measured after some intervening culture manipulation. To circumvent this limitation, the studies to be discussed in this presentation have focused on the peripheral blood monocyte as a potential fresh _ex vivo_ cell type in which to study selected endpoints [6, 7, 8]. Specifically, the extent of benzo[a]pyrene metabolism and unscheduled DNA synthesis after several DNA damaging treatments have been examined. These studies have lead to the realization that _in vitro_ individuality in response to genetic damage is a complex function of both extra-genetic response to the agent(s) inflicting damage and possible genetically based peculiarities in processing of resultant lesions.

2. METHODS

2.1. Benzo[a]pyrene (BP) Metabolism in Monocytes

Mononuclear cells, monocytes, and phytohemagglutin stimulated lymphocytes were obtained from heparinized whole blood as previously described [6]. Aqueous acetone-soluble BP metabolites were separated from whole cultures using the extraction procedure outlined [6].

2.2. DNA Repair Synthesis in Monocytes

Unscheduled [^{3}H]dThd incorporation in monocytes (adherent fraction of mononuclear cells) was determined by autoradiography and direct liquid-scintillation counting as previously described [7].

3. XENOBIOTIC BIOTRANSFORMATION IN MONOCYTES

Toward development of a practical assay for determining inter- and intraindividual variations in metabolic response to environmental carcinogens and drugs, the methodological aspects of measuring the whole-cell metabolism of benzo[a]pyrene to water-soluble metabolites in peripheral blood monocytes was systematically examined [6]. It was found that, unlike benzo[a]pyrene metabolism in phytohemagglutinin-stimulated lymphocytes, the specific metabolic activity (SMA) of benzo[a]pyrene in human monocyte cultures is independent of phytohemagglutinin stimulation (Table 1), culture age up to three days, fetal bovine serum levels from 5 to 20% in the culture medium, and the initial input monocyte concentration in each culture.

Under standard conditions chosen on the basis of the above features, monocyte SMA assay with five ml of starting whole blood can be reproducibly performed in three days. One human volunteer assayed at one time by ten independent tests, at four times on one day, and eight times over a three-month period yielded a monocyte SMA coefficient of variation of 7.3, 9.03, and 11.8%, respectively. When three individuals were assayed at different times over a two-month period, the coefficient of variation with each individual was 13 to 19% in monocytes and 28 to 31% in lymphoblasts (Table 2). Hence the monocyte SMA assay seemed very sensitive, convenient, and less variable than the similar lymphoblast assay but still subject to fluctuation caused by donor status at the time of blood sampling. And more importantly, intraindividual variation was comparable to variation in the standardized method. Only marginal interindividual differences, no greater than 2-fold, were observed in monocyte metabolism.

TABLE 1. Relative Rate of Extent of BP Conversion to Water-Soluble Metabolites in Various Peripheral Blood Mononuclear Cell Cultures[a]

Culture type[a]	Phytohemagglutinin-P added at 3 hr[b]	Viable cell count at: 3 hr	Viable cell count at: 144 hr	BP converted in each culture in 3 days (pmoles)[c]	SMA (pmoles BP/10^6 cells/3 days)
Total MNC	+		9.8×10^5	1921	2189
	-		0	516	
Three-hr adherent cells	+	1.2×10^5		123	1024
(monocytes)	-	1.1×10^5		196	1902
Nonadherent cells[d] (lymphocytes)	+		8.7×10^5	2059	2347
	-		0	40	

[a]From Lake et al.,(6) with permission from Waverly Press, Inc.

[b]Three hr, length of time for monocyte adherence.

[c][^{3}H]BP was added to monocyte cultures at 3 hr and incubated for 3 days; [^{3}H]BP was added to cultures of MNC and nonadherent cells at 72 hr and incubated for 3 days.

[d]Nonadherent cells were derived from the 1st shake-off of dishes of adherent cells.

4. DNA-REPAIR SYNTHESIS IN HUMAN PERIPHERAL BLOOD MONOCYTES

Fresh *ex vivo* cultures of normal human peripheral blood monocytes, which are non-replicative and known to possess cytochrome P-450 associated mixed-function oxidase activity, were used to assay unscheduled DNA synthesis (UDS) manifested as augmented [^{3}H]thymidine (dThd) incorporation following treatment in culture with diverse mutagenic carcinogens [7]. Untreated monocyte cultures established from pools of three to six normal donors incorporated a low level of cytoplasmic [^{3}H]dThd throughout a majority of the cells during an 18 hr incubation. This background incorporation into whole cells due to mitochrondrial DNA synthesis [7] was 80-90% inhibited by hydroxyurea (HU) at concentrations greater than 5 mM. Dose-related increases in the cumulative 18 hr [^{3}H]dThd incorporation in monocytes were studied following treatments with UV, N-methyl-N'-nitro-N-nitrosoguanidine (MNNG), methyl methanesulfonate (MMS), mitomycin C(MMC), N-acetoxy-acetylaminofluorene (NA-AAF), aflatoxin B_1 (AFB_1) benzo-

TABLE 2. Monocyte and Lymphoblast SMA of 3 Human Donors[a]

Donor	Date	% adherence	Yield of monocytes/ml of blood	SMA[a] (pmoles/10^6 cells/3 days) Monocytes	Lymphoblasts
A	7/6				7558
	7/16	7.8	7.0 X 10^4	3580	5693
	7/27	6.9		3690	3129
	8/6	9.0	1.0 X 10^5	2551	7934
	8/17	2.8	5.5 X 10^4	4101	5989
				3480 ± 658[b]	6060 ± 1903[b]
				18.9%[c]	31.4%[c]
B	7/6	6.0	6.7 X 10^4	5711	3433
	7/16	9.4	9.3 X 10^4	4193	2978
	7/27	7.8		5620	1851
	8/6	5.0	5.25 X 10^4	4831	2106
	8/17	6.6	2.0 X 10^5	4545	
				4980 ± 666[b]	2592 ± 739[b]
				13.3%[c]	28.5%[c]
C	7/6	12.0	1.1 X 10^5	4121	2635
	7/16	6.1	5.5 X 10^4	4416	5350
	7/27	11.9	9.9 X 10^4	4675	5212
	8/6	12.0	1.0 X 10^5	5551	5689
	8/17	10.6	8.0 X 10^4	3689	
				4490 ± 696[b]	4724 ± 1406[b]
				15.5%[c]	29.7%[c]

[a]From Lake et al.,[6] with permission from Waverly Press, Inc.

[b]Mean ± S.D.

[c]Coefficient of variation (C.V.) = S.D./ $\overline{X}$ X 100.

[a]pyrene (BaP), and dimethylnitrosamine (DMN). The presence of HU during chemical treatment and throughout this 18 hr period of incubation with [^{3}H]dThd did not influence the dose-response curves obtained with UV, MMS, NA-AAF, and BaP but it increased the input dose of MNNG, MMC, DMN, and AFB_1 required to give peak repair in-

TABLE 3. UDS Response of Peripheral Blood Monocytes to MNNG on Separate Occasions

Donor	Occasion	dpm $[^3H]$dThd per 10^5 monocytes
A	1	1,441 ± 780
	2	8,720 ± 351
	3	738 ± 58
B	1	1,202 ± 151
	2	1,131 ± 344
	3	2,938 ± 147

[a]Cultures were treated with 2 μg/ml MNNG and harvested at 28-24 hr. post addition of $[^3H]$dThd.

corporation. When HU was added to cultures following MNNG damage no interference with repair response was observed. HU apparently influenced the extent of DNA damage by direct reactivity with these chemicals or their endogenously generated metabolites rather than inhibiting DNA-repair processes. These results provided evidence that monocytes are enzymatically proficient in base and nucleotide excision pathways and have endogenous capacity to metabolize BaP, AFB_1, and DMN to DNA-damaging metabolites. As such, the monocyte is a potentially useful human cell type for detecting genotoxic chemicals and studying individuality in chemical-biological interactions.

5. INTERINDIVIDUAL REPAIR VARIATION IN RESPONSE TO MNNG AND NA-AAF

In order for UDS response in fresh *ex vivo* monocytes to be effectively used to monitor for intrinsic genetically based differences, intra- and interindividual variability under standard conditions must be statistically manageable. To circumscribe this variability, studies with normal individuals were conducted using MNNG (small patch repair) and/or NA-AAF (long patch repair).

Two individuals were assayed on three occasions. As shown in Table 3, responses to a standard dose of MNNG varied as much as 10-fold in one individual and 2.6-fold in another. Variable status of the individual is presumed to be the source of variation.

TABLE 4. Background and Carcinogen Stimulated DNA Synthesis in Peripheral Blood Monocytes of 10 Normal Human Donors

Donor	MNC Per ml blood	% Adherence	DNA Synthesis per 10^5 Monocytes ± S.D.: 0.01% Acetone (Bkg)	MNNG 2 μg/ml	NA-AAF 2 μg/ml	MNNG/NA-AAF
1 TG	$.86 \times 10^6$	26	1,132 ± 186	2,548 ± 413	2,720 ± 296	.93
2 AG	1.20×10^6	24	1,015 ± 93	4,649 ± 745	3,648 ± 653	1.27
3 JH	1.10×10^6	25	1,583 ± 320	3,562 ± 389	2,529 ± 412	1.40
4 JB	$.73 \times 10^6$	21	1,263 ± 255	2,724 ± 444	2,039 ± 417	1.33
5 KO	1.58×10^6	43	813 ± 128	1,974 ± 227	2,079 ± 94	.94
6 MH	$.96 \times 10^6$	45	770 ± 95	2,123 ± 276	1,280 ± 255	1.65
7 SH	1.20×10^6	20	981 ± 216	2,887 ± 482	3,585 ± 452	.80
8 JA	1.18×10^6	15	4,590 ± 1201	9,733 ± 2,563	7,925 ± 2,251	1.22
9 SM	1.08×10^6	49	437 ± 201	1,800 ± 284	1,437 ± 171	1.25
10 MP	$.72 \times 10^6$	24	1,288 ± 277	3,866 ± 517	2,676 ± 529	1.44
POOL		27	1,010 ± 437	3,703 ± 467	2,478 ± 181	1.49
AVERAGE OF ALL DONORS		29	1,259	3,411	2,924	1.16

To examine the range of responses across individuals, ten donors and a pool of these donors' monocytes were assayed with MNNG and NA-AAF as the DNA damaging agents. As depicted in Table 4, responses to MNNG ranged from 1800 to 9733 dpm per 10^5 monocytes. Likewise, response to NA-AAF ranged from 1280 to 7925 dpm per 10^5 monocytes; 5 to 6-fold differences. Hence, variation within individuals was comparable to variation between individuals under a single set of standard conditions. The magnitude of responses was positively correlated ($r = 0.72$, and $r = 0.69$) with the amount of background DNA synthesis. Interestingly, the response of pooled monocytes was nearly equal to the arithmetic mean of individual responses, indicating no interaction between donor monocytes. One attractive possibility explaining this variability in background and induced synthesis is the participation of platelets or derived factors in modulating DNA synthesis [9]. It is not known whether the level of platelet contamination in the adherent mononuclear cells is sufficient to cause the differences in background synthesis observed in Table 4. An equally likely explanation for wide differences in absolute synthesis is cell population heterogeneity. Monocyte precursors from bone marrow still engaged in DNA synthesis exist to various degrees in each person [10]. Three to five percent of MNNG treated monocytes have been found by autoradiography to be hyperresponsive, yielding 15 to 40 grains per cell [7]. These are speculated to be fresh peripheralized bone marrow precursors which have a higher constitutive ability to undergo excision repair synthesis.

Given this capricious variability in absolute response within and between individuals with no real understanding of its source, the ratio of MNNG and NA-AAF response was calculated. Individuals differed as much as 2-fold in relative (normalized) response, Table 4. Although these individuals were not re-sampled it is likely that such a ratio could provide meaningful comparisons in terms of individual differences in response to different types of genetic damage. For instance, a panel of chemical agents causing crosslinks, large monoadducts, base intercalation, dimers, small monoadducts, base damage, or strand decomposition could be used to detect unique relative sensitivity of monocytes within individuals.

6. DISCUSSION

Interindividual differences in monocyte metabolism of benzo[a]-pyrene was reproducible from occassion to occassion. Unscheduled DNA synthesis ellicited by treatment with MNNG and NA-AAF was highly variable (2-6 fold variation interindividual and 3-10 fold intra-individual). In the absence of objective knowledge of the source of this intraindividual variation more extensive definition or newer methodology for isolation, culture, and measurement of DNA damage [11] should be applied before screening is attempted in selected human populations.

It is concluded on the basis of the variability of the monocyte present UDS system that this human cell would be most advantageously used to examine drugs for their ability to inflict repairable genetic damage rather than detecting individuality in response. Progenotoxic drugs showing induction of UDS in the rat hepatocyte UDS [12] system for instance could be subjected to monocyte assay to confirm or refute the human counterpart activity. If pathways of human monocyte metabolism are unique and reflect the potential for adverse metabolism in various human organs as appears to be the case [8], this approach may aid in extrapolating subhuman data to predict human hazard.

One limitation to the use of monocytes versus lymphocytes for studying genetic damage is that fresh <u>ex vivo</u> peripheral monocytes are largely non-replicative. This precludes effective application of cytogenetic analysis. In view of the wide array of other genetic damage endpoints; such as poly(ADP-ribose) polymerase levels [13]; carcinogen DNA damage and analysis for DNA strand breaks [14, 15] which are available, this limitation appears to be minor.

In the context of this workshop, if the relative response to a panel of agents inflicting fundamentally different types of DNA damage is reproducible across individuals giving variable absolute responses, the much sought [16] utility for finding genetically based variability could be achieved.

REFERENCES

1. C. F. Arlett and A. R. Lehmann, Human disorders showing increased sensitivity to the induction of genetic damage, Ann. Rev. Genet., 12:95-115 (1978).
2. R. T. Barfknecht and J. B. Little, Abnormal sensitivity of skin fibroblasts from familial polyposis patients to DNA alkylating agents, Cancer Res., 42:1249-1254 (1982).
3. N. G. J. Jaspers, J. de Wit, M. R. Regulski, and D. Bootsma, Abnormal regulation of DNA replication and increased lethality in Ataxia Telangiectasia cells exposed to carcinogenic agents, Cancer Res., 42:335-341 (1982).
4. T. Kakunaga and J. D. Crow, Cell variants showing differential susceptibility to ultraviolet light-induced transformation, Science, 209:505-507 (1980).
5. R. W. Pero and C. Ostlund, Direct comparison, in human resting lymphocytes, of the interindividual variations in unscheduled DNA synthesis induced by N-acetoxy-2-acetylamino fluorene and ultraviolet irradiation, Mut. Res., 73:349-361 (1980).
6. R. S. Lake, M. R. Pezzutti, M. L. Kropko, A. E. Freeman, and H. J. Igel, Measurement of benzo[a]pyrene metabolism in human monocytes, Cancer Res., 37:2530-2537 (1977).

7. R. S. Lake, M. L. Kropko, S. McLachlan, M. R. Pezzutti, R. H. Shoemaker, and H. J. Igel, Chemical carcinogen induction of DNA-repair synthesis in human peripheral blood monocytes, Mut. Res., 74:357-377 (1980).
8. M. W. Kahng, M. W. Smith, and B. F. Trump, Aryl hydrocarbon hydroxylase in human bronchial epithelium and blood monocyte, J. Natl. Cancer Inst., 66:227-232 (1981).
9. R. W. Pero and C. Vopat, A human platelet-derived inhibitor of unscheduled DNA synthesis in resting lymphocytes, Carcinogenesis, 2:1103-1110 (1981).
10. A. Volkman, in: Immunobiology of the Macrophage (D. S. Nelson, ed.), pp. 291-322, Academic Press, New York (1976).
11. R. W. Pero and H. O. Sjogren, The influence of monocyte (adherent cell) content in human mononuclear blood populations on the estimation of individual levels of N-acetoxy-2-acetylaminofluorene induced unscheduled DNA synthesis, Carcinogenesis, 3:39-43 (1982).
12. G. M. Williams, Detection of chemical carcinogens by unscheduled DNA synthesis in rat liver primary cell cultures, Cancer Res., 37:1845-1851 (1977).
13. C. J. Skidmore, M. I. Davies, P. M. Goodwin, O. Omidiji, A. Zia'ee, and S. Shall, in: Novel ADP-Ribosylations of Regulatory Enzymes and Proteins (Smulson and Sugimura, eds.), pp. 197-205, Elsevier North Holland, Inc., Amsterdam (1980).
14. K. W. Kohn and R. A. Grimek-Ewig, Alkaline elution analysis, a new approach to the study of DNA single-strand interruptions in cells, Cancer Res., 33:1849-1853 (1973).
15. H. C. Birnboim and J. J. Jevcak, Fluorometric method for rapid detection of DNA strand breaks in human blood cells produced by low doses of radiation, Cancer Res., 41:1889-1892 (1981).
16. E. S. Vesell, Intraspecies differences in frequency of genes directly affecting drug disposition: the individual factor in drug response, Pharm. Revs., 30:555-563 (1978).

INFLUENCE OF DEMOGRAPHIC FACTORS ON UNSCHEDULED DNA SYNTHESIS AS MEASURED IN HUMAN PERIPHERAL LEUKOCYTES AND FIBROBLASTS

John J. Madden[1,2], Arthur Falek[1],
David A. Shafer[1], Robert M. Donahoe[1],
Deborah C. Eltzroth[1], Felicia Hollingsworth[1],
and Peter J. Bokos[3]

[1]Dept. of Psychiatry
Emory University
Atlanta, Georgia 30322
and Human and Behavioral Genetics Research Laboratory
Georgia Mental Health Insitute
Atlanta, Georgia 30306

[2]Dept. Biochemistry
Emory University
Atlanta, Georgia 30322

[3]Interventions
1313 S. Michigan Avenue, Suite 602
Chicago, Illinois 60605

ABSTRACT

In a population study of leukocytes from 140 human subjects, far UV-induced unscheduled DNA synthesis (UDS) variation correlated with at least three demographic factors - smoking tobacco, season of sample obtainment and "street opiate" addiction. The ability to correlate UDS variability with these factors demonstrates the importance of environmental influences in control of DNA repair processes and the need to include the effect of demographics in genetic risk assessment models. In addition, mutagen-induced UDS patterns for mononuclear leukocytes from a single subject did not match the UDS pattern for fibroblasts grown from a forearm skin biopsy from the same individual which suggests that UDS may be tissue specific.

INTRODUCTION

Along with providing the initial indications of genetically-based DNA repair defects, human peripheral leukocytes have also been used in assays to determine which demographic components affect DNA repair capacity. To date several such factors, including tobacco use [1], age [2], drug abuse [1], and Down's syndrome [3], have been identified as lowering mutagen-induced DNA repair. This diminished repair capacity, whether from genetic or environmental factors, corresponds to a potential increase in genetic risk within these groups, and as such represents a significant health concern for populations with these defineable factors. In trying to extrapolate from the population level to individual risk assessment, however, many other factors which affect DNA repair become significant including the homogeneity of the cell population [4-7], the ratio of cells in G_0 to cells in S phase of the growth cycle [8], the concentration of platelets [9], nucleotide pool sizes [10], and the variable effects of DNA synthesis inhibitors used in many of the repair assays [10, 11]. While inter-group variation of many of these factors is insignificant between most populations, variability introduced by these factors makes quantitative comparison of repair capacity between individuals exceedingly difficult based on a single repair screening assay such as the unscheduled DNA synthesis assay (UDS).

In this report, we present information about some of the demographic factors which correlate with altered repair of far UV damage of DNA in a population of 140 individuals. We also demonstrate that the pattern of repair capacity found in the leukocytes need not parallel that found in a fibroblast culture from the same individual.

METHODS AND RESULTS

Subjects

The methods and materials are as described in Madden et al. [1]. The subjects consisted of 140 volunteers between the ages of 18 and 62 who signed consent forms and provided demographic information by questionnaire.

Assay

The UDS assay consists of diluting the whole blood in a mixed salt, minimal carbon source buffer (ABG) formulated as follows: 0.54 mM KCl, 0.0145 M tris, 0.145 M NaCl, 50 μM $CaCl_2$, 0.1 mM $MgCl_2$, 0.01% D-glucose, pH 7.6. We feel that the use of minimal medium and autologous serum is critical to maintaining significant nucleotide imbalances which might be the result of such demographic factors as smoking or age, and which might be abolished by the use of enriched media,

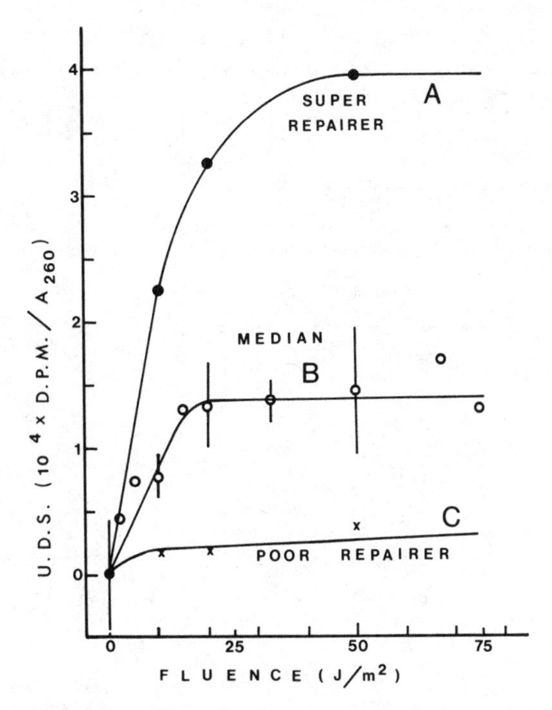

Fig. 1. Typical fluence/UDS response curves for 3 individuals: A, a super repairer (-●-); B, a median repairer (-o-); and C, a poor repairer (-×-). The median individual was tested for UV-induced UDS at 7 different times over a period of 1.5 years, and the mean ± standard deviation for the various time points plotted. B was used as the principal control subject for the other experiments.

e.g., RPMI-1640 with fetal calf serum. Duplicate leukocyte cultures are irradiated at a rate of 1 $J/m^2/sec$ to fluences of 0, 5, 10, 20, and 50 J/m^2. Immediately (within 2 min) after irradiation [12], hydroxyurea (HU) and tritated thymidine (^{3}H-dT) are added to final concentrations of 4 mM and 0.05 μM respectively, with the thymidine held at a specific activity of 2 μCi/ml. The cells are incubated at 37°C for 2 hr with constant slow rotation (5 rev per min), pelleted, and frozen at -80°C. DNA is extracted by the procedure of Prashad and Cutler [13] and quantitated by diphenylamine [14]. DNA synthesis is reported as thymidine incorporation (measured by liquid scintillation counting) per unit of DNA absorbance at 260 nm (D.P.M./ A_{260}nm). The DNA synthesis value obtained at 0 J/m^2 (residual semiconservative replication) is subtracted from the values for the UV-irradiated cells, and the difference reported as unscheduled DNA synthesis (UDS).

TABLE 1. UDS Values Grouped by Age, Use of Tobacco, Heroin Addiction and Season When the Blood was Drawn. Winter-Spring Corresponds to the Period Between December and May Inclusive While Summer-Autumn Corresponds to the Period June through November

	SEASON		
	Winter-Spring	Summer-Autumn	All Seasons
	Mean (S.D.) n	Mean (S.D.) n	Mean (S.D) n
All Controls	9542 (7106) 59	15488 (11094) 47	12178 (8874) 106
Controls:			
Non Smokers:			
All	13604 (8468) 25	17733 (13850) 26	15709 (11605) 51
≤ 35 Years Old	14959 (10239) 11	16879 (13039) 15	16067 (11751) 26
> 35 Years Old	12539 (6998) 14	18898 (15458) 11	15337 (11681) 25
Smokers:			
All	7497 (3255) 18	12537 (6134) 15	9788 (5347) 33
≤ 35 Years Old	7016 (3348) 14	11840 (4616) 5	8286 (4195) 19
> 35 Years Old	9179 (2572) 4	12886 (6976) 10	11827 (6183) 14
Heroin Addicts:			
Tobacco Smokers	4028 (4056) 21	5657 (3965) 10	4554 (4035) 31

The UDS value for each individual was the maximum incorporation value obtained regardless of the UV fluence used to obtain this value. This method eliminates the need for very accurate dosimetry of the absorbed UV fluence which is required when a single UV fluence is used as a criterion of repair capacity. As can be seen in Fig. 1, the typical response is represented by the median repairer where the 5, 10, and 20 J/m^2 fluences produced increases in incorporation over the preceding dose while the 50 J/m^2 fluence was identical to the 20 J/m^2 fluence. However, response curves for some individuals plateaued at 5 J/m^2 while others produced increased incorporation up to 50 J/m^2 fluence (Fig. 1).

Demographic Factors

Evaluation of the results did not show a correlation between UDS and any of the following variables: gender, the use of alcohol,

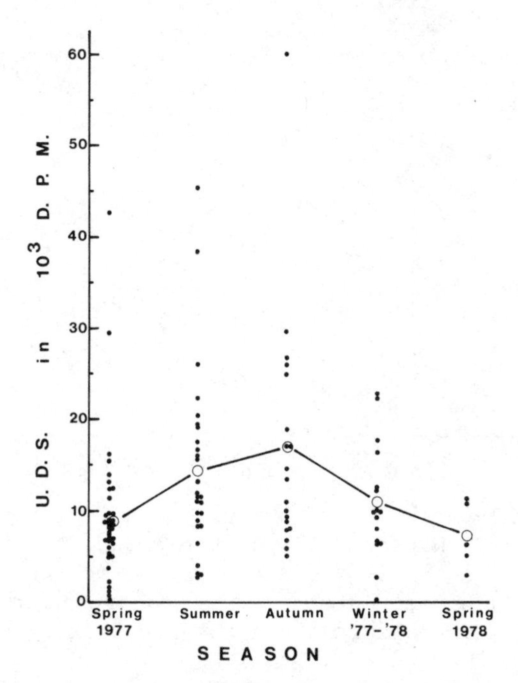

Fig. 2. Seasonal variation in UV-induced UDS. The maximum UDS values for 106 individuals (●) determined for leukocytes obtained between Spring, 1977 and Spring, 1978 and the mean for each group (○).

caffeine, or marihuana. While Lambert et al. [2] showed a decrease in UV-induced UDS in senescent individuals, we did not find a correlation of UDS with age in the age range represented (18-62 years). We did find a significant inverse correlation ($F = 3.15$, $p \leqslant 0.01$) between tobacco smoking and UDS (Table 1). The overlap in UDS values between smoking and non-smoking groups was large and the inverse correlation became apparent only on the basis of a statistical analysis of 84 subjects. The depression in UDS caused by heroin addiction was far more pronounced than that found for tobacco smoking [1]. A one way analysis of variance between heroin addicts (who smoke tobacco) and control tobacco smokers was significant at the 0.01 level (Table 1).

As first reported by Pero at this meeting (this volume), lymphocyte samples obtained in the fall showed a higher level of NAAF-induced UDS than matched samples obtained in the spring. Subsequent reanalysis of our data on repair of UV damage showed an identical pattern of seasonal variation (Fig. 2). Samples collected between March, 1977 and May, 1978 were divided into 4 seasonal groups: spring (March-May), summer (June-August), autumn (September-November) and winter (December-February). The summer and autumn mean UDS values

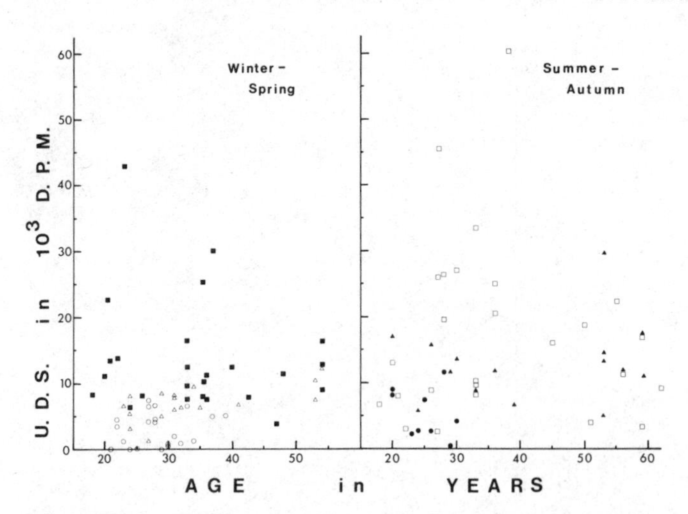

Fig. 3. Seasonal variation of UV-induced UDS with age. Groups include: 1) Winter-Spring: non-smoker controls (■), tobacco smoking controls (△), and tobacco smoking heroin addicts (○). 2) Summer-Autumn: non-smoker controls (□), tobacco smoking controls (▲), and tobacco smoking heroin addicts (●).

(14486 ± 9806 and 16964 ± 12993) were significantly higher in a one way analysis of variance than the combined 1977 and 1978 spring value (8869 ± 7254) at the 0.01 level of significance. Winter (11171 ± 6446) was similar to the spring value, but not significantly lower than autumn ($p \leqslant 0.06$).

Multivariate analysis to determine the relative contributions of age, smoking, and season of sample donation reaffirmed the significant contributions of the latter two to UDS variance while age was not significant (Table 1). For this multivariate analysis, we combined the winter and spring samples into one category and summer and fall into another in order to maintain an adequate number of subjects in each category. We feel that this was justified on the basis of the closeness of the sample means for the groups involved (winter: spring:: 11171: 8869; and summer: autumn:: 14486: 16964, respectively) and because of a gap in samples between mid-November and early January which clearly separated the autumn and winter samples. Using these groupings, further multivariate analysis to include stree opiate addiction demonstrated that lymphocytes from addicts had a significant decrease in UV-induced UDS when matched to control subjects for tobacco smoking and season of sample collection (Table 1). Even when common factors, like heroin addiction, smoking, or season, were correlated with repair capacity the most striking feature of the data was the great interindividual variance in the sample set (Fig. 3) which suggests the existence of repair control factors not yet identified.

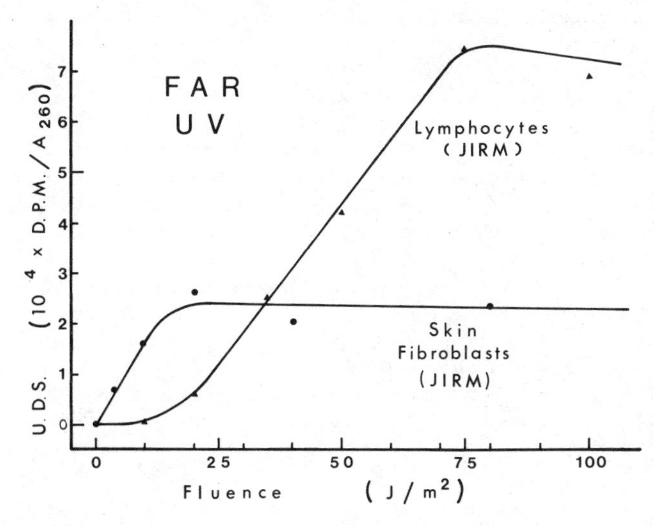

Fig. 4. Far UV-induced UDS in leukocytes (▲) and confluent, early passage fibroblasts (●) from the same subject. UV-induced UDS in the fibroblasts was measured by irradiating confluent cells after replacing the growth medium (MEM - 10%) with ABG. The UDS assay was as described for the leukocytes. The leukocytes used in this experiment were isolated on Ficoll-paque and washed extensively to remove platelets [6].

Comparison of Fibroblasts to Lymphocytes

Mononuclear leukocytes (principally lymphocytes) obtained by Ficoll-paque separation (Pharmacia) from a subject from the UDS screening program exhibited an increase in UDS at fluences up to 100 J/m^2. Conversely, mononuclear leukocytes from the same individual did not exhibit UDS after treatment with either 8-methoxypsoralen-near UV light (PUVA) or ethylmethane sulfonate (EMS) at doses which produced significant UDS in mononuclear leukocytes from a number of control subjects. Trypan blue dye exclusion as a measure of cell viability and sister chromatid exchange frequency also confirmed an increased sensitivity of these mononuclear leukocytes to chemical mutagens (unpublished data). Because of these unusual responses, a fibroblast explant cell culture (JIRM) was established from a 3-4 mm forearm-skin biopsy to determine whether similar UDS response patterns could be obtained from an alternate cell source from the same individual.

The fibroblasts were subcultured 4-5 times at near confluence in Eagle's Minimal Essential Medium, 10% fetal calf serum and antibiotics (penicillin and streptomycin). For the mutagen treatments, the fibroblasts were grown to 100% confluence, the media decanted, and the cells

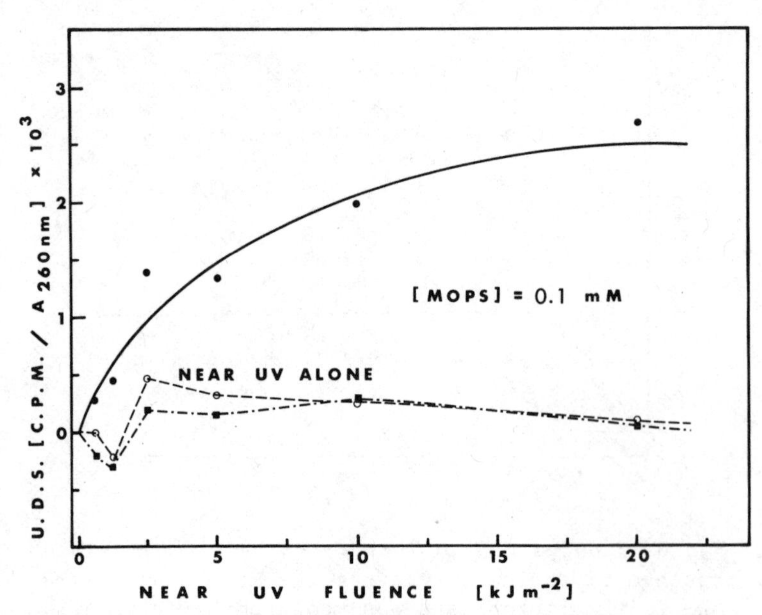

Fig. 5. PUVA-induced UDS in control leukocytes and JIRM fibroblasts. PUVA treatment consisted of incubating either Ficoll-separated leukocytes or confluent fibroblasts in 0.1 mM 8-methoxyl psoralen (MOPS) for an hour in the dark at 37°C. The cells were then irradiated with near UV light (Sylvania BLB/F15T8) at 20 $J/m^2/sec$. UDS curves shown include: leukocytes from control subject B (Fig. 1) (-●-); leukocytes from B irradiated with near UV, but with no MOPS present (--o-); JIRM fibroblasts treated with PUVA (·■·-). Not shown are control leukocytes and JIRM fibroblasts treated with MOPS alone, and JIRM leukocytes treated with PUVA, all of which demonstrated no measurable UDS.

gently washed with 5 ml ABG. A fresh 5 ml ABG was then added and the cells UV irradiated from above through the UV-transparent media or chemical mutagens added for one (1) hr in this media. At the end of the treatment period, the ABG was removed and the cells washed for a second time with 5 ml ABG. Five ml McCoys 5A Medium, containing HU and ^{3}H-dT adjusted to the same concentrations as for the lymphocyte experiments was added, and the cells incubated at 37°C for 2 hr. The media was removed and the cells washed with 5 ml ABG. The cells were harvested with a rubber policeman in a minimal amount of ABG (<3 ml), frozen at -80°C and then extracted as in the lymphocyte experiments.

Treatment of the confluent JIRM-fibroblasts with UV, PUVA, and EMS and subsequent measurement of UDS produced a repair pattern dis-

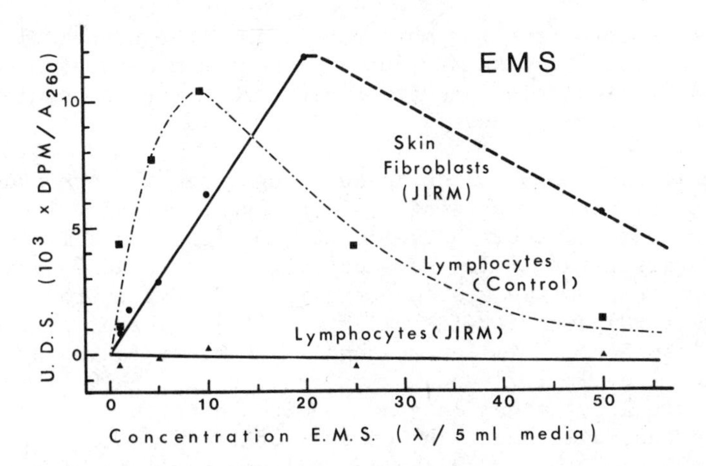

Fig. 6. EMS-induced UDS in JIRM leukocytes (-▲-) and fibroblasts (-●-) and control leukocytes (·-·■·). Cells were incubated for 1 hr in EMS, washed twice with ABG to remove unreacted EMS, and then incubated in ABG with hydroxyurea and tritiated thymidine for 2 hr at 37°C. UDS was measured as described (Methods and Results).

tinctly different from that found in leukocytes from the same individual (JIRM-leukocytes) tested at the same time and drawn within a month of the skin biopsy. The JIRM-fibroblasts produced a linear increase in UDS at far UV fluences of up to only 20 J/m^2 while the JIRM-leukocytes were able to repair fluences up to 100 J/m^2 (Fig. 4). PUVA elicited no UDS in either JIRM-leukocytes or fibroblasts at doses which produced UDS in the mononuclear leukocytes of other individuals including the control subject shown (Fig. 5). Treatment of either JIRM-leukocytes or JIRM-fibroblasts or control lymphocytes with either near UV alone or MOPS alone also produced no measurable UDS. EMS, however, elicited high levels of UDS in the JIRM-fibroblasts and the control leukocytes but elicited no response in the JIRM-leukocytes (Fig. 6). Thus the ability to repair DNA damage was mutagen specific when fibroblasts were compared to leukocytes from the same individual.

DISCUSSION

The ability to correlate demographic components and UDS variability provides evidence that environmental factors play an important role in control of DNA repair capacity. In our study tobacco smoking, season, and heroin addiction caused significant alterations in UDS. While others have reported that age correlated with UDS changes [10, 12], we could not confirm these findings. These demographic factors affect far more people than the genetically-based DNA repair defects, like Xeroderma pigmentosum, and as such must represent critical elements for inclusion in future population-based risk assessment models.

The mechanisms by which these demographic factors modulate repair capacity can, at present, only be hypothesized. The seasonal variation, for example, in UV-induced UDS found in all comparison groups parallels a similar seasonal variation found in another mutagen-related enzyme system - the induced aryl hydrocarbon hydroxylase (AHH) activity in leukocytes [16]. These results are also reminiscent of similar seasonal variations found for circulating hormone levels [17] and hormone receptors [18-19]. To the extent that leukocytes can respond to circulating hormones, UDS and AHH are potentially under the control of these factors [20]. Since smoking [21] and opiate addiction [22] can modulate both UDS capacity and circulating hormone concentrations, this possible interrelationship between hormone level and repair capacity is an important one for future study of UDS control.

The comparison between mononuclear leukocytes and the fibroblasts suggests the need for caution in the use of leukocytes as a whole body indicator of genetic risk. As reported by other workers, the repair capacity of leukocytes and lymphoblast lines is apparently controlled by the synthesis of poly(ADP)-ribose which is in turn controlled by the concentration of NAD^+ [23-25]. Thus far, however, poly(ADP)-ribose has not been conclusively linked to repair in other types, e.g., fibroblasts [26]. It may be that control of repair in leukocytes is a tissue specific phenomenon and that additional cell types (e.g., fibroblasts or sperm) need to be tested in conjunction with leukocyte analysis to establish genetic risk for the human organism as a whole. The idea of tissue-specific repair abilities is supported by the recent review by Van Buul [27] in which he discussed the implications of data showing that mammalian peripheral blood lymphocytes and stem-cell spermatogonia differ in their chromosomal radiosensitivity. A final caution must be noted in comparing repair results between tissues and between cell lines. Mortlemaus et al. [28] have shown that culture media conditions can vary the activity of the human fibroblast DNA repair enzyme, DNA photolyase, in ways not fully understood. No one has yet fully defined the optimal culture media conditions for unscheduled DNA repair and shown that these conditions are similar for a variety of fibroblast lines.

In summary, the repair capacity of human peripheral leukocytes is correlated with a number of genetic and environmental (demographic) factors which leads to much interindividual variability. The repair capacity of mononuclear leukocytes appears to be tissue specific, possibly because of the unique control of leukocyte repair by poly(ADP)-ribose synthesis.

The authors acknowledge Dr. Elmer C. Hall and Dr. Mike Lynn (Department of Statistics, Emory University) for their statistical assistance and Gloria Sproles and Joe Fain (GMHI Audio-Visual Dept.) for their aid in preparing the figures. This study was funded in part by National Institute of Drug Abuse, DA 01451.

REFERENCES

1. J. J. Madden, A. Falek, D. A. Shafer, and J. H. Glick, Effects of opiates and demographic factors on DNA repair synthesis in human leukocytes, Proc. Natl. Acad. Sci., U.S.A., 76:5769-73 (1979).
2. B. Lambert, U. Ringborg, and L. Skoog, Age-related decrease in UV light-induced DNA repair synthesis in human peripheral leukocytes, Cancer Res., 39:2792-95 (1979).
3. B. Lambert, K. Hansson, T. H. Bui, F. Funes-Cravioto, J. Lindsten, M. Holmberg, and R. Strausmanis, DNA repair and frequency of x-ray and UV-induced chromosome aberrations in leukocytes from patients with Down's Syndrome, Ann. Hum. Genet., London, 39:293-303 (1976).
4. M. Frey-Wettstein, R. Longmire, and C. G. Craddock, Deoxyribonucleic acid (DNA) repair replication of ultraviolet (UV) irradiated normal and leukemic leukocytes, J. Lab. Clin. Med., 74:109-18 (1972).
5. F. H. Yew and R. T. Johnson, Human B and T lymphocytes differ in UV-induced repair capacity, Exp. Cell Res., 113:227-31 (1978).
6. F.-H. Yew and R. T. Johnson, Ultraviolet-induced DNA excision repair in human B and T lymphocytes. III. Repair in lymphocytes from chronic lymphocytic leukaemia, J. Cell Sci., 39: 329-37 (1979).
7. R. W. Pero and H. O. Sjogren, The influence of monocyte (adherent cell) content in human mononuclear blood cell populations on the estimation of individual levels of N-acetoxy-2-acetylaminofluorene induced unscheduled DNA synthesis, Carcinogenesis, 3:39-43 (1982).
8. T. J. Lampidis and J. B. Little, Enhancement of UV-induced unscheduled DNA synthesis by hydroxyurea, Exp. Cell Res., 110: 41-46 (1977).
9. R. W. Pero and C. Vopat, A human platelet-derived inhibitor of unscheduled DNA synthesis in resting lymphocytes, Carcinogenesis, 2:1103-10 (1981).
10. A. R. S. Collins, S. L. Schor, and R. T. Johnson, The inhibition of repair in UV irradiated human cells, Mutat. Res., 42: 413-32 (1977).
11. K. Erixon and G. Ahnstrom, Single strand breaks in DNA during repair of UV-induced damage in normal human and xeroderma pigmentosum cells as determined by alkaline DNA unwinding and hydroxylapatite chromatography. Effects of hydroxyurea, 5-fluorodeoxyuridine and 1-β-D-arabinofuranosylcytosine on the kinetics of repair, Mutat. Res., 59:257-71 (1979).

12. R. W. Pero and C. Ostlund, Direct comparison, in human resting lymphocytes, of the interindividual variations in U.D.S. induced by N-acetoxy-2-acetylaminofluorene and UV irradiation, Mutat. Res., 73:349-61 (1980).
13. N. Prashad and R. Cutler, Percent satellite DNA as a function of tissue and age of mice, Biochem. Biophys. Act., 418:1-23 (1976).
14. K. W. Giles and A. Myers, An improved diphenylamine method for the estimation of DNA, Nature, 4979:93 (1965).
15. R. W. Pero and F. Mitelman, Another approach to in vivo estimation of genetic damage in humans, Proc. Natl. Acad. Sci., U.S.A., 76:462-3 (1979).
16. B. Paigen, E. Ward, A. Reilly, L. Horton, H. L. Gurtoo, J. Minowada, K. Steenland, M. B. Havens, and P. Sartori, Seasonal variation of aryl hydrocarbon hydroxylase activity in human lymphocytes, Cancer Res., 41:2757-61 (1981).
17. E. W. Van Carter, E. Virasoro, R. Le Clercq, and G. Copinschi, Seasonal, circadian, and episodic variations of human immunoreactive β-MSH, ACTH, and cortisol, Int. J. Pept. Protein Res., 17:3-13 (1981).
18. P. B. Curtis-Prior, Seasonal variation in the responses of adipose tissue to a serum fat-mobilizing factor may be explained in terms of reduced availability of fat cell receptors, Horm. Metab. Res., 13:686-88 (1981).
19. E. E. Codd and W. L. Byrne, Seasonal variation in the apparent number of binding sites for ^{3}H-opioid agonists and antagonists, Life Sci., 28(2577-83 (1981).
20. R. J. McDonough, J. J. Madden, A. Falek, D. A. Shafer, M. Pline, D. Gordon, P. Bokos, J. C. Kuehnle, and J. Mendelson, Alteration of T and null lymphocyte frequencies in the peripheral blood of human opiate addicts: in vivo evidence for opiate receptor sites on T lymphocytes, J. Immunol., 125:2539-43 (1980).
21. A. Melander, E. Nordenskjold, B. Lundh, J. Thorell, Influence of smoking on thyroid activity, Acata Med. Scan., 209(1-2): 41-3 (1981).
22. A. Beaumont and J. Hughes, Biology of opioid peptides, Ann. Ref. Pharmacol. Toxicol., 19:245-67 (1979).
23. N. A. Berger, G. W. Sikorski, S. J. Petzold, and K. K. Kurohara, Defective poly (adenosine diphosphoribose) synthesis in xeroderma pigmentosum, Biochem., 19:289-93 (1980).
24. H. Juarez-Salinas, J. L. Sims, and M. K. Jacobson, Poly(ADP-ribose) levels in carcinogen-treated cells, Nature, 282:740-1 (1979).
25. B. W. Durkacz, O. Omidiji, D. A. Gray, and S. Shall, (ADP-ribose)$_n$ participates in DNA excision repair, Nature, 283:593-96 (1980).
26. J. D. Roberts and M. W. Lieberman, Deoxyribonucleic acid repair synthesis in permeable human fibroblasts exposed to ultraviolet radiation and N-acetoxy-2-(acetylamino) florene, Biochem., 18: 4499-4505 (1979).

27. P. P. W. Van Buul, Absence of correlation between the chromosomal radiosensitivity of peripheral blood lymphocytes and stem-cell spermatogonia in mammals, Mutat. Res., 95:69-77 (1982).

INDIVIDUAL VARIATION IN BENZO(A)PYRENE METABOLISM AND ITS ROLE IN HUMAN CANCER*

Hira L. Gurtoo[1], Beverly Paigen[2], and Jun Minowada[3]

[1]Department of Experimental Therapeutics and Grace Cancer Drug Center
[2]Department of Molecular Biology
[3]Department of Immunology
Roswell Park Memorial Institute
New York State Department of Health
Buffalo, New York 14263

ABBREVIATIONS

AHH = Aryl hydrocarbon hydroxylase; BP = Benzo(a)pyrene; MC = 3-Methylcholanthrene; DBA = Dibenz(a,h)anthracene; BA = Benzo(a)-anthracene; PAHs = Polycyclic Aromatic Hydrocarbons; BP-7,8-diol = 7,8-Dihydroxy-7,8-dihydrobenzo(a)pyrene; BP-9,10-diol = 9,10-Dihydroxy-9,10-dihydrobenzo(a)pyrene; BP-4,5-diol = 4,5-Dihydroxy-4,5-dihydrobenzo(a)pyrene; TCDD = 2,3,7,8-Tetrachlorodibenzo-p-dioxin; HPLC = High Pressure Liquid Chromatography; PWM = Pokeweed mitogen; PHA = Phytohemagglutinin; Con A = Concanavalin A; LPS = Bacterial Lipopolysaccharide; FCS = Fetal Calf Serum; MZ = Monozygotic; DZ = Dizygotic.

INTRODUCTION

A strong association between the smoking of cigarettes and the development of human lung cancer has been suggested by various reports [1, 2]. Experimental evidence in animals has incriminated

*Supported by USPHS Grants CA-13038, CA-17538, CA-18542, CA-14413, CA-17609, CA-25362; USPHS Contract CP-55629; Grants from Council for Tobacco Research U.S.A., CTR-1253 and 1080; and a Grant from American Cancer Society, BC-303.

polycyclic aromatic hydrocarbons (PAHs) present in cigarette smoke as the causative agents in the development of lung cancer [3-5]. In addition to cigarette smoke, PAHs are found in various other environmental situations including emissions from automobiles, from burning refuse, coal, and wood furnaces and from the making of coke or the refining of oil [6]. Among the various PAHs present in cigarette smoke, the metabolism of benzo(a)pyrene (BP) by both animal and human tissues and cells has been extensively investigated, mainly because of the availability of a sensitive fluorometric assay, i.e., aryl hydrocarbon hydroxylase (AHH) assay, which essentially measures the phenolic metabolites. PAHs _per se_ are biologically inactive, and are activated _in vivo_ by substrate inducible cytochrome P-450-dependent microsomal mixed function oxidases found in most mammalian tissues. Cytochrome P-450 exists in multiple forms [7, 8], their proportion being determined by the type of inducer and by the species and strain of animal studied. Benzo(a)pyrene appears to be preferentially metabolized by the form of cytochrome P-450 that is inducible by PAHs. Various reports prior to 1974 suggested that the activated metabolites of PAHs are K-region epoxides. Largely as a consequence of the improved resolution of various BP metabolites by HPLC, during the recent years incontrovertible evidence has evolved to suggest that the activation of PAHs, especially BP, proceeds to a large extent via the formation of bay region diol-epoxides [9]. Initially, benzo(a)pyrene is converted to epoxides that are formed at various sites in the molecule. These epoxides undergo one or more of the following reactions: enzymatic and nonenzymatic conjugation with glutathione; nonenzymatic rearrangement to phenols; NADPH and cytochrome P-450-mediated reduction to the parent molecule; and conversion, in the presence of epoxide hydrase, to the corresponding dihydrodiols. Both phenols and dihydrodiols can undergo detoxification by conjugation with glucuronic acid or with sulfate, and dihydrodiols, especially 7,8-dihydroxy-7,8-dihydrobenzo(a)pyrene (BP-7,8-diol), can alternatively be recycled by the mixed function oxidase system to form bay region diol epoxides (e.g., BP-7,8-diol-9,10-oxides) which are believed to be the ultimate metabolites that bind DNA and are also mutagenic and carcinogenic.

It is a widely held view that majority of human cancers derive from the chemical contamination of the environment [10, 11]. However, this does not mean that genetic factors are irrelevant in chemical carcinogenesis. The fact that a majority of individuals in a given population, who are exposed to the same chemical environment, e.g., cigarette smokers, fail to develop cancer indicates that the differential susceptibility to chemical carcinogens may be partially due to individual differences in genetic trait(s) that contribute to the development of such cancer. Since metabolic activation of carcinogens is the first initial step in a multistep process leading to the emergence of cancer, and since both _in vivo_ and _in vitro_ studies have demonstrated variable levels of mixed function oxidases in humans that participate in carcinogen metabolism [12,

13], an evaluation of the levels and the regulation of these enzymes is a logical first step in understanding the factors that affect an individual's susceptibility to cancer.

Epidemiological evidence suggests that familial and genetic factors play a role in the development of human lung cancer, since both smoking and nonsmoking relatives of lung cancer patients are more likely to acquire lung cancer than is the general population [14]. In the same vein, genetic variability in the ability of various inbred strains of mice to induce AHH is paralleled by the appearance of PAH-induced cancers; inducible strains being more susceptible than are the non-inducible strains [15]. For instance, AHH inducibility in the off-spring of C57BL/6 × DBA/2 mice is expressed almost exclusively as an autosomal dominant trait and inducible parents and off-springs have been shown to be more susceptible to PAH-induced cancers than the non-inducible mice [5, 16].

In 1972, AHH activity was reported in mitogen-activated lymphocytes [17, 18]; this observation provided an opportunity for evaluating the role of AHH in the development of human lung cancer associated with cigarette smoke. Kellermann et al. [19] reported a genetic polymorphims for lymphocyte AHH inducibility which suggested the existence of three distinct categories of low, intermediate, and high AHH incubility phenotypes. Subsequently, these investigators reported [20] the presence of a disproportionate number of intermediate and high lymphocyte AHH inducibility phenotypes among lung cancer patients but not among patients with other neoplasms. These studies indicated that lymphocyte AHH inducibility is a single gene autosomal trait with additive inheritance between the two alleles and suggested that the allele for high inducibility conferes an increased risk for lung cancer. Considerable interest among cancer researchers in different laboratories was generated by these reports. A collaborative research effort was initiated at Roswell Park Memorial Institute; this collaboration involved three groups with expertise in lymphocyte cell culture, genetics, and biochemistry of mixed function oxidase system.

Initially, in an effort to confirm the reports of Kellermann et al. [19, 20] we attempted to repeat their studies; however, our attempts were foiled by the relative insensitivity of the lymphocyte AHH assay used by Kellerman et al. [18-20]. Because of a need to develop a sensitive and a reproducible lymphocyte AHH assay, the possibility that AHH assay may not be a good indicator of the formation of DNA-binding metabolites from benzo(a)pyrene, and the possibility that lymphocytes may not be a suitable representative tissue for measuring the effects of PAHs on the lung tissue, we decided to carry out detailed investigations of various aspects of human BP metabolism. These studies included investigations of lymphocyte culture conditions and AHH assay for the purpose of developing a sensitive and a reproducible lymphocyte AHH assay, studies of the metabolism of BP by lymphocytes and other human tissues, and studies on the heritability of lymphocyte AHH and its role in the development of the cancer of lung, larynx, and bladder. This report summarizes the results of these studies that have extended over a span of eight years.

METHODS

Details of various methods employed for measuring lymphocyte AHH, for HPLC analysis of BP metabolism by lymphocytes and placentas, for DNA-binding of BP catalyzed by placenta, and for the metabolism of BP by the lymphocyte cell line RPMI-1788 and by placenta have been published previously [21-24].

Briefly, lymphocyte AHH activity was analyzed as follows: 20-30 ml of heparinized blood was drawn from each person and used immediately to isolate lymphocytes on Hypaque-ficoll gradients. The washed lymphocytes at a density of 1×10^6 cells/ml were suspended in RPMI-1640 medium containing 10% heat - inactivated fetal calf serum, 1% pokeweed mitogen, 1% phytohemagglutinin, and penicillin and streptomycin. Eight ml aliquots were distributed into four or six Falcon T-flasks. After 48 hrs of culture, methylcholanthrene in acetone (or another AHH inducer) was added to half the cultures, the remaining cultures received solvent only. After an additional 24 hr of incubation, cells were harvested and assayed for AHH and DNA content. Basal and induced AHH activity is expressed as pmole equivalents of 3-hydroxybenzo(a)pyrene/min/10^6 cells, unless otherwise noted. AHH inducibility is given as the ratio of induced/basal activity. AHH activity was determined by measuring fluorescence of the NaOH extract and quantitated with a standard curve developed with 3-hydroxybenzo(a)pyrene. The fluorescence of BP-phenols equivalent to that of one p-mole of 3-hydroxybenzo(a)pyrene produced by 10^6 cells/min is defined as one unit of AHH activity.

Lymphocyte AHH Activity and Factors Influencing It

Because lymphocyte AHH assay described by Busbee et al. [18-20] presented a number of difficulties we carried out detailed studies of lymphocyte culture conditions and the enzymology of AHH activity. These investigations were carried out to determine optimal conditions for the measurement of lymphocyte AHH activity and the development of a reproducibility assay [21, 25]. Various parameters investigated in studies on lymphocyte culture conditions included: type of culture container, cell density at the time of seeding cultures, storage of blood versus lymphocytes, reliability of cell counting versus DNA determinations in calculating AHH specific activity, length of exposure to mitogens, culture medium ingredients including types and quantities of different mitogens, relationship between blastogenesis as measured by ^{3}H-thymidine and ^{14}C-amino acid incorporation and AHH activity, and the kinetics of the appearance of AHH during culture.

Comparison between Falcon T-30 culture flasks and 15 ml glass culture tubes revealed that a larger surface area provided in the Falcon flasks supported a vigorous growth of lymphocytes as well as allowed attainment of optimal AHH activity, both basal and induced, at 72 hrs of culture rather than at 96 or 120 hr attained in the tube cultures and also reported by others [26]. Furthermore, the kinetics of AHH activity in tube cultures was more variable. Studies in which cultures were started with different initial cell numbers ranging from 0.5×10^6 to 1.5×10^6 cells/ml culture demonstrated dependence of AHH activity and inducibility on initial cell density [25]. Storage of blood led to lower lymphocyte AHH activity; this was independent of the type of anticoagulant used during blood collection. In contrast, lymphocytes isolated fresh and stored for 24 hr at room temperature in RPMI-1640 medium retained their full potential for the expression of AHH activity and inducibility during culture. This observation proved useful in epidemiologic studies where transport of blood over long distances was required. Duplicate cultures prepared at the same time from the same sample of blood were found to have nearly identical (within 10%) AHH activities, cell numbers and DNA content, and while affecting AHH activity to different extents in different people, the presence of inducer (1.0 µm 3-methylcholanthrene) affected the other parameters less than 10%. Consequently, calculation of AHH-inducibility based either on AHH activity related to whole culture or to DNA, was not affected; however, if dissimilar cultures were compared, because of the cell clumping induced by mitogens specific activity related to DNA was found to be more reliable. Kouri et al. [26] have considered the reliability of AHH activities when related to DNA content and to NADH-cytochrome C reductase activity, and they have reported a high degree of correlation ($r = 0.96$; $p < 0.01$) between AHH/DNA and AHH/cytochrome C reductase. Different lots of pokeweed mitogen (PWM) and phytohemagglutinin (PHA) did not differ significantly in their effect on lymphocyte AHH activity; only an occasional lot of mitogens and RPMI-1640 medium failed to support AHH activity. In contrast, different lots of fetal calf serum (FCS) had diifferent effects on both the basal and the induced activity, but the basal activity was generally more sensitive. These studies led to the conclusion that some endogenous factors in FCS were affecting the lymphocyte AHH activity. Gielen and Nebert [27] have reported the presence of unknown factors in FCS that induced AHH in mammalian liver cell culture system and Atlas et al. [28] and Kouri et al. [26] have made similar observations.

Different mitogens, namely PHA, PWM, concanavalin A (con A) and bacterial lipopolysaccharide (LPS), were used alone or in combination to (a) select the dose and type of mitogen(s) that produced optimal AHH activity, (b) evaluate mitogens of known specificity for

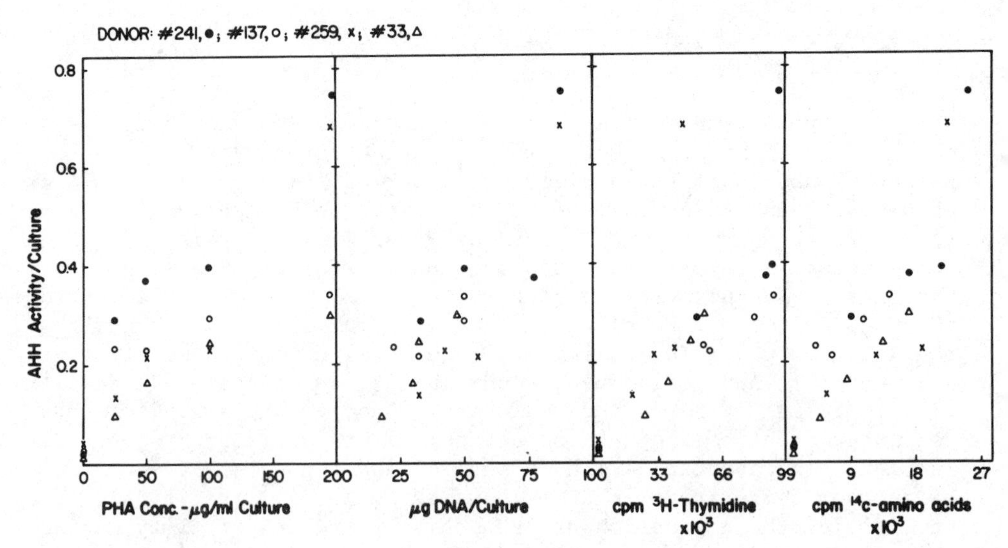

Fig. 1. Dose-response relationships between AHH activity, cellular DNA concentration, and blastogenesis of lymphocytes cultured in the presence of different doses of PHA. Lymphocytes were isolated from fresh blood and distributed into several identical cultures in a medium devoid of mitogens. Various concentrations of PHA were added to duplicate culture, which also received 2.5 μCi [^{3}H] thymidine and 0.25 μCi [^{14}C] amino acid mixture. Cultures were incubated for 72 hr (in the absence of inducer), and then the cells were harvested and assayed for AHH activity, cellular DNA concentration, and [^{3}H] thymidine and [^{14}C] amino acid incorporation. Each value is the mean of duplicated determinations. Points correspond to values obtained at varying doses of the mitogen. Reproduced from the J. Natl. Cancer Inst. (Ref. 25).

various lymphocyte subpopulations, especially T and B cells, in a mixed lymphocyte culture, (c) determine the relationship between AHH activity and blastogenic response measured by the incorporation of ^{3}H-thymidine and ^{14}C-amino acids into the acid insoluble cell fraction, and (d) determine the length of exposure of lymphocytes to mitogens required for optimal AHH activity and least toxicity observed as cell clumping [25]. Exposures of lymphocytes for 18-24 hr to a combination of PHA + PWM produced optimal AHH activity and least toxicity. However, inconsistent nature of this effect, lack of the effect of the length of mitogen exposure on AHH inducibility ratio, and the cell loss that results from washing the cells to remove mitogens did not confer any added advantages. On the contrary, these additional steps further complicated the lymphocyte culture system. Therefore, cells were cultured routinely for 72 hrs in the presence of mitogens.

Both LPS and PWM produced weak blastogenic response and the AHH activity was also very low; furthermore, no meaningful correlation was observed between AHH versus mitogen dose, AHH versus DNA, AHH versus ^{3}H-thymidine or ^{14}C-amino acid incorporation when PWM was the only mitogen used. On the other hand, excellent correlations between these paired parameters were observed when PHA or Con. A were used as the mitogen. These mitogens also produced significant AHH activity and blastogenesis in lymphocytes (Fig. 1). Generally, a combination of PHA + PWM (1:100 dilution of each) was better than either PWM alone or PHA alone. Therefore, this mitogen combination was routinely used in all our studies. Because of the order of potency of the mitogens tested (PHA + PWM $\geqslant$ PHA $>$ Con. A $\geqslant$ PWM = LPS), and the results of AHH analysis of lymphoid cells from patients with T and B-type leukemias reported earlier by us [29], it was concluded that the primary responding cells in mixed lymphocyte cultures are the T-cells, though interaction between T and B cells or between PHA and PWM during blastogenesis may be responsible for the expression of AHH in mixed lymphocyte cultures.

Studies on the enzymology of lymphocyte AHH activity revealed that the activity, both basal and induced, was linear with the time up to 30 min, and with cell number up to 17 × 10^{6} cells, that it was enhanced by 30% when NADH (1.3 mM) was added to the NADPH-fortified incubation mixture, and that the pH optimum of the reaction was 8.5 rather than 7.4 noted for optimal AHH activity in other mammalian cells [21]. The activity remained fairly stable at -70°, and was induced optimally in cells in culture exposed to 1 μM DBA or MC for a period not exceeding 24-30 hrs. These modifications when combined provided a lymphocyte AHH assay that on the average was 17-fold (range 10-60 fold) more sensitive (Table 1) than the previously described assay [18-20].

Since validation of reproducibility of the assay intended to be employed in epidemiologic studies for the measurement of a biochemical phenotype is an essential prerequisite, initially employing a small group of donors (8 individuals) we measured the reproducibility of lymphocyte AHH activity, both basal and induced as well as the inducibility ratio, in repeat determinations on the same individual during a period extending from 3 to 4 months in summer and fall [25] (as will be discussed later, season of the year for such studies is important for obtaining reliable results). While the mean intraindividual coefficients of variation for basal and induced activities were 36 ± 5% and 33 ± 6%, respectively, the AHH inducibility ratio was much less variable (16 ± 2%) [25]. In subsequent studies on a population of 39 donors AHH inductibility measurements were carried twice on each donor. The difference from the mean averaged 11%.

Thirty two of the 39 donors gave repeat measurements with an average difference from the mean of less than 15%. The statistical

TABLE 1. Comparison between the RPMI Method and the Method Used by Kellermann, et al. [18-20] for the Measurement of AHH Activity in Cultured Lymphocytes

		3-OH-BP formed (pmoles)					
		Method used by Kellermann et al.			RPMI Method		
Subject	Culture	Total activity/ culture/30 min	Specific activity/ 10^6 cells/min	Induced: control	Total activity/ culture/30 min	Specific activity/ 10^6 cells/min	Induced: control
1	Control	2.90	0.018	1.51	74.7	1.51	1.41[a]
	MC	4.38	0.027		104.7	2.11	
2	Control	1.32	0.006	1.67	76.2	1.64	1.86[a]
	MC	2.20	0.010		141.5	3.04	
3	Control	3.11	0.093	1.40	65.6	1.43	2.21[a]
	MC	4.35	0.132		145.3	1.67	
4	Control	3.85	0.042	1.10	94.1	1.35	1.78[a]
	MC	4.26	0.047		167.6	2.41	
5	Control	3.00	0.060	2.67	25.9	0.90	1.43[a]
	MC	8.00	0.167		37.1	1.00	
6	Control	12.5	0.119	1.81	112.1	1.35	1.56[a]
	MC	22.6	0.215		174.7	2.10	
7	Control	4.52	0.049	1.19	70.6	0.36	3.34[b]
	MC	5.38	0.057		236.1	1.21	
8	Control	19.9	0.247	1.05	183.3	0.74	2.16[b]
	MC	21.0	0.260		395.8	1.59	
9	Control	6.27	0.192	1.00	61.1	0.92	1.07[b]
	MC	6.16	0.188		65.6	0.98	
10	Control	4.33	0.071	0.87	79.2	0.83	1.73[b]
	MC	3.75	0.062		137.2	1.43	
11	Control				130.0	1.21	1.90[b]
	MC				247.8	2.31	
12	Control				107.2	1.15	1.52[b]
	MC				163.7	1.76	

[a]Blood sedimented once on Hypaque-Ficoll to purify lymphocytes for this method.
[b]Blood sedimented twice on Hypaque-Ficoll to purify the lymphocytes for this method.

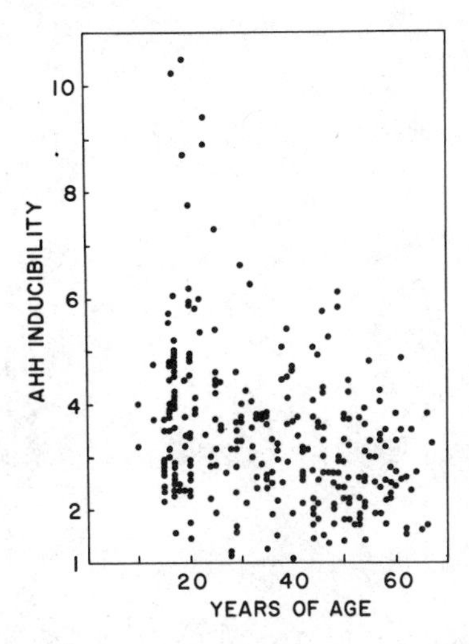

Fig. 2. Effect of age on AHH inducibility. A total of 283 individuals were tested and each closed circle represents AHH inducibility of one individual. Other details are described in the text. Reproduced with permission from Raven Press, New York (Ref. 53).

treatment of these data indicated that two tests would suffice to determine AHH inducbility for an individual within a 95% confidence limit if the assay is restricted to the season of the year when AHH is high and if individuals with more divergent repeat measurements are tested at least 4 times [30]. In practice, it was regarded prudent to consider all subjects with repeat measurements varying greater than 15% from the mean as requiring additional testing; therefore, for various population genetics studies some subjects were tested repeatedly.

Since subjects employed in population genetics studies would by necessity be of different sexes, ages, and social backgrounds, it was considered necessary to evaluate the effects of smoking, age, and sex on lymphocyte AHH inducibility.

AHH activity and inducibility in 60 smokers and 106 nonsmokers were examined using a nonparametric one-way analysis of variance that tested whether the distribution of values was the same without assuming normality of distribution (Kruskal-Wallis H statistic). For this test AHH values from the lowest to the highest were arranged and assigned a rank order of 1 through 166, and subsequent statistics were done on the rank orders. The mean rank of induced AHH activity was 78.4 for smokers and 86.4 for nonsmokers, indicating no significant difference between the groups ($p = 0.3$). Likewise, the mean ranks of AHH inducibility were 81.1 and 84.9 for smokers and non-

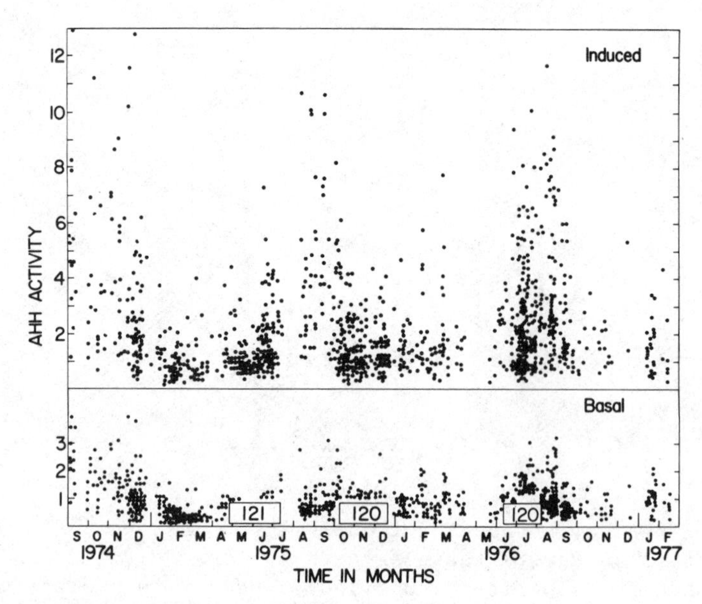

Fig. 3. Annual rhythm of basal and induced AHH activity over 30 months. Each point represents either the basal or induced activity of one person on a particular day (based on duplicate or triplicate determinations). Approximately equal numbers of males and females are represented. A total of 977 samples were tested; some persons were tested more than once, and a few were tested frequently over the entire time span. The age of donors ranged from 10 to 81 years, but the majority were between 20 and 50 years. The enclosed areas in basal AHH activity represent areas with so many values (number in box) that individual points were obscured. Letters on abscissa, month designations beginning with September. Reproduced with permission from Cancer Research [Paigen et al., Cancer Res., 41:2757-2761 (1981)].

smokers, respectively, indicating no significant difference ($p = 0.6$). We also compared 84 males and 82 females and found no significant difference between them in either induced AHH activity or AHH inducibility (mean ranks of induced AHH activity were 83.5 for both males and females; mean ranks of AHH inducibility were 86.0 and 81.0 for males and females, giving a p value of 0.5).

The question of whether age affected AHH inducibility was tested in a larger population of 282 individuals ranging in age from 10 to 69 years. AHH inducibility values above 6 were rarely obtained after age 30, but they occurred with some frequency in the population of individuals less than 30 years old (Fig. 2). To evaluate the effect of age in persons over 30 years of age, the adult population was analyzed at 10 year age intervals. Chi square analysis showed that age had no effect on AHH inducibility from ages 30 to 69.

During the course of our investigation it became apparent that seasonal changes were accompanied by significant changes in lymphocyte AHH activity in the same individual. Detailed studies established the existence of a definite pattern of seasonal effects on lymphocyte AHH (Fig. 3).

The induced AHH activity, averaged over all subjects measured in a given week, ranged from a low of 0.77 units in 1975 and 0.80 units in 1976 to a high of 6.4 (1974), 7.9 (1975), and 4.0 (1976) units. Thus, the variation in induced AHH activity can be as high as 10-fold. Several possible explanations for the seasonal effect on the activity were considered. The seasonal changes do not appear to be due to differences in the mitogen response of lymphocytes. The average mitogen response ±S.E., measured by [^{3}H]thymidine uptake into the lymphocyte DNA, of 15 individuals tested in August was $61.4 \times 10^3 \pm 6.5$ cpm/culture flask; the mitogen response on the same 15 subjects tested in February was $59.8 \times 10^3 \pm 7.7$ cpm which was not significantly different from the value obtained in August. We do not think the seasonal changes can be accounted for by changes in cell growth since the number of cells or DNA content did not change from winter to summer. The seasonal changes were not accompanied by changes in laboratory reagents or air or water supply. In one study we measured 4 subjects with high AHH activity levels during the low winter months. Some of these people had just returned from a midwinter vacation in Miami, Florida. In 4 of 8 individuals on whom we had measurements before and after their vacation, induced AHH activity immediately after their return from Florida was characteristic of late summer. A few weeks later their induced AHH activity had decreased to values more typical of winter. The changes in AHH activity in these individuals could have been due to changes in environmental factors known to affect biological rhythms. These might include changes in light intensity or in daily rhythms that may occur in people on vacation. While not observed in all regions of this country, the seasonal effects initially reported by us [30] were later confirmed by some other laboratories [31]. It was reasoned that changes in AHH activity in lymphocytes might be due to altered plasma levels of hormones known to have seasonal variations. Accordingly, a series of concentrations of each of the following hormones were tested for their effect on AHH activity: corticosterone (0.001 to 0.01 μM), follicle-stimulating hormone (0.03 to 0.13 unit), adrenocorticotropic hormone (0.0001 to 0.01 unit), melatonin (0.001 to 0.1 μM), serotonin (0.001 to 0.1 μM), thyroxine (0.001 to 0.1 μM), and d-aldosterone (0.001 to 0.01 μM). These exogeneous hormones were in addition to those endogenous hormones contained in fetal calf serum used to culture lymphocytes. None of these added hormones affected basal or induced AHH activity. These experiments were done in May, which is the beginning of the transition time from winter to summer activity.

Humans have biological rhythms that vary with time on daily, monthly, and perhaps even weekly cycles [32]. One must take into account the possible influence of these other rhythms when trying to model the data for a seasonal variation. It is possible that humans might have a circadian rhythm in AHH activity since circadian rhythms in microsomal mixed function oxidases have been reported in experimental animals [33-35].

AHH in a Stable Lymphocyte Cell Line (RPMI-1788) and Its Comparison with AHH in Mitogen-Activated Fresh Lymphocytes

Since PAH-induced cytochrome P-450 (cytochrome P-448/P_1-450) in rodent liver and other tissues is spectrally, immunologically and enzymatically different from the constitutive enzyme [7, 8, 36-38], the need for distinguishing the PAH-induced AHH from the AHH in non-induced lymphocytes was recognized. In view of the complexity of the lymphocytes cell culture system required for AHH analysis, it was deemed important to identify an intrinsic qualitative trait(s) of the induced enzyme that might distinguish it from the uninduced enzyme. We reasoned that should such a difference exist, then it might be possible to identify the inducible phenotype from the non-inducible phenotype on the basis of the qualitative differences in such characteristics as inhibitor sensitivity, Km, heat denaturation, enzyme half-life, etc. This objective dictated a need for a stable source of large batches of human lymphocytes. Since fresh lymphocytes were not available in the required quantities, we turned out attention to established lymphocyte cell lines with the hope of finding one or more of these lines for detailed studies on the spectral, thermodynamic, and enzymological characteristics of AHH, on the induction of AHH by various inducers, and on the inhibition by various inhibitors known to differentiate PAH-induced rodent liver AHH from the non-induced enzyme [7, 8, 36-39]. From a screen of over 100 lymphocyte cell lines derived from normal donors and leukemia patients, three cell lines were found to possess measurable AHH activity quantitatively comparable to that present in mitogen-activated fresh lymphocyte cultures. One of these cell lines (RPMI-1788) was selected for detailed studies [24, 40, 41].

RPMI-1788 is a stable, immunologically defined human B-lymphocyte line from a normal donor. Without mitogenic prestimulation, optimization of various parameters influencing AHH enzyme activity and inducibility led to the attainment of enzyme levels and inducibility ratios comparable to those found in mitogen-activated fresh lympocytes. Under standardized and strictly controlled culture conditions, the average AHH-inducibility ratio of seven stock cultures was 3.2 ± 0.7 (S.D.). The average basal and induced (24 hr with 0.3 μM dibenz(a,h)anthracene) AHH activities were 0.11 ± 0.02 and 0.33 ± 0.11 pmoles/min/10^6 cells, respectively. Maximal basal and induced enzyme activities and cell viability occurred at

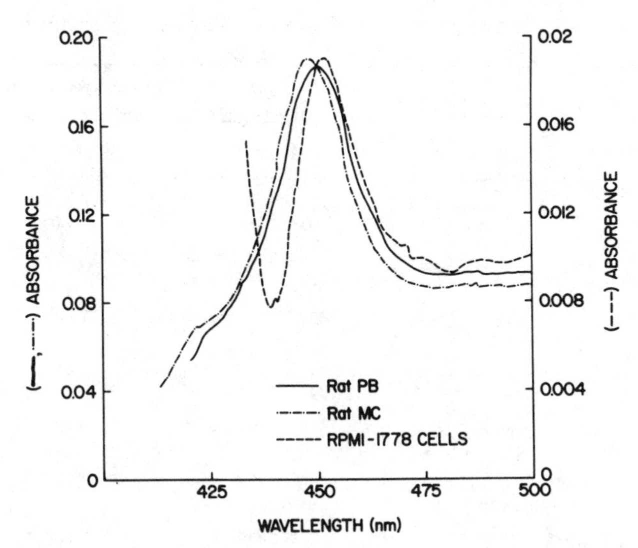

Fig. 4. Difference spectra of cytochrome P450 in RPMI-1788 cells. Cells (4×10^9) were suspended in 60 ml of 0.25 M sucrose-1 mM EDTA and homogenized, using a glass homogenizer, fitted with a Teflon pestle. The homogenate was centrifuged twice at 15,000 × g for 15 and 5 minutes, respectively. The 15,000 × g supernatant was centrifuged in a Beckman L5-50 Model ultracentrifuge at 105,000 × g for 60 minutes. The resulting microsomal pellet was suspended in 15 ml of 0.1 M potassium phosphate buffer (pH 7.4) and was centrifuged again at 105,000 × g for 60 minutes. The resulting pellet was suspended in 7 ml of the above buffer (8 mg protein/ml) containing 1 mM sodium cyanide. Three ml were transferred to two cuvettes. The baseline of equal light absorbance was recorded on an Aminco DW/2 spectrophotometer previously calibrated with homium oxide filter. A few mg of sodium hydrosulfite were added to both cuvettes; carbon monoxide was bubbled for 30 sec only through the contents of the sample cuvette and the spectrum was recorded. Using an extinction coefficient of 91 mM^{-1} cm^{-1} for the cytochrome P-450, the concentration was calculated to be 0.2 pmole cytochrome P-451/10^6 cells. For rat liver microsomes from phenobarbital or 3-methylcholanthrene-treated rats, spectrum was recorded under identical conditions except that cyanide was omitted. The concentration of rat liver microsomal protein used was as follows: phenobarbitol (PB) 0.6 mg/ml; 3-methylcholanthrene (MC) 0.8 mg/ml. Reproduced with permission from Biochemical Pharmacology [Gurtoo, et al., Biochem. Pharmacol., 27:2659-2662 (1978)].

TABLE 2. Inhibition of AHH Activity in RPMI-1788 Lymphocyte Cell Line by Various Compounds

Inhibitor	Cell Type	AHH activity as % of the activity of the uninduced or induced cells		
		1 μM	10^2 μM	10^3 μM
α-NF	Uninduced Cells	72	15	36
	Induced Cells	65	13	14
β-NF	Uninduced Cells	73	68	61
	Induced Cells	68	62	39
Metyrapone	Uninduced Cells	103	103	89
	Induced Cells	100	92	75
SKF-525A	Uninduced Cells	118	123	111
	Induced Cells	111	116	82
TCPO	Uninduced Cells	109	107	96
	Induced Cells	90	95	91
CHO	Uninduced Cells	110	80	99
	Induced Cells	95	95	96

[a]NF, naphthoflavone; TCPO, 1,1,1-trichloropropene 2,3-oxide; CHO, cyclohexene oxide; SKF-525A, 2-diethylaminoethyl-2,2-diphenylvalerate.

48 hr of culture, while the cell viability frequently declined by 72 hr. Optimal time of exposure for induction with DBA was 24 hrs. While the requirement for mitogens was not obligatory, PHA and Con. A increased AHH specific activity while at the same time decreasing cell viability. On the other hand, both PWM and LPS were without any effect on AHH activity but were toxic to the cells. AHH activity was influenced by the batch of fetal calf serum used and dose-response studies demonstrated that basal activity was affected more by FCS than was the induced activity. Difference spectral studies indicated the presence of cytochrome P-450-form with maximal absorbance at 451 nm in the reduced-CO complexed form (Fig. 4).

More than 22 inducers were tested for the induction of AHH and over the dose ranges tested and on a molar basis the order of potency of the most active inducers was TCDD > DBA > MC ⩾ BA. Several characteristics of the basal AHH were compared with those of enzyme optimally induced by DBA. Several of these were found to be virtually identical. Both had similar pH curves (optima at 8.25) and inhibitor specificity (α- and β-naphthoflavones, metyrapone, and 2-diethylaminoethyl-2,2-diphenylvalerate in decreasing potency) (Table 2). The induced and basal enzymes exhibited similar half-lives (41, 46 hr), apparent activation energies (16.7, 16.6 kcal/mol), temperature optima (37-38, 38-39°), temperature-dependence of denaturation (range, 42-50°), and apparent Km's with benzo(a)pyrene (40, 41). HPLC analysis of BP metabolites produced by the DBA-induced enzyme indicated several similarities with those produced

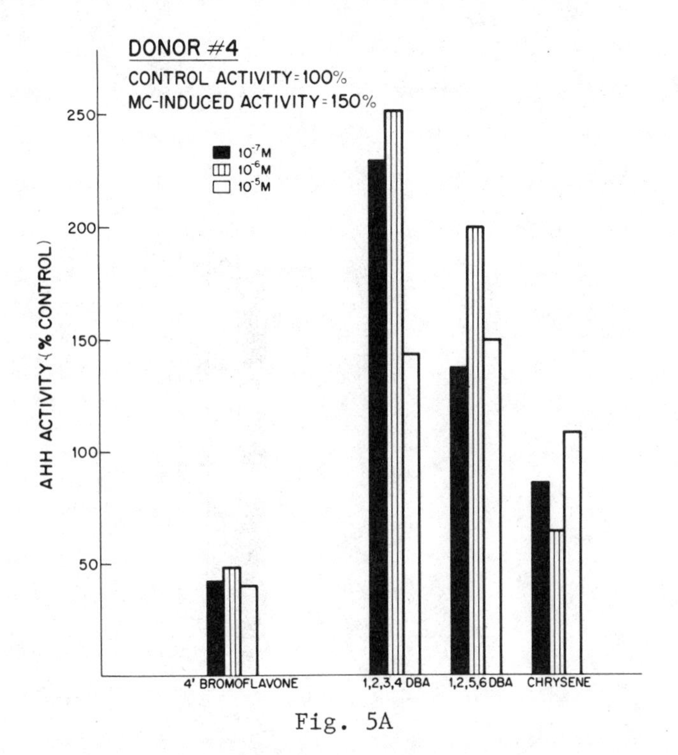

Fig. 5A

Fig. 5. Effects of inducers on lymphocyte AHH activity. Lymphocytes of different donors were harvested in large batches and suspended at 10^6/ml in FCS-fortified RPMI-1640 Medium containing 1 in 100 dilution each of pokeweed mitogen and phytohemagglutinin. Eight ml aliquots of the suspension were distributed to separate flasks and incubated for 72 hr as described in the text. At 24 hr before harvest, a pair of flasks ineach set received a given concentration of the inducer. At the same time, a pair of control flasks received only the solvent and another pair received 3-methylcholanthrene (MC). The cells were harvested 24 hr later and assayed for AHH. Each bar is a mean of 2 determinations. The data obtained with various inducers are illustrated in A through F. Each chart represents the data obtained on the cells derived from the same donor, unless otherwise indicated. A) 4'-bromoflavone, 1,2,3,4-DBA, 1,2,5,6-DBA, and chrysene; B) α- and β-naphthoflavone; C) benzo(a)anthracene; D, p-p-1, 1,1-trichloro-2,2-bis(p-chlorophenyl)ethane, phenanthrene, 7,12-dimethylbenz(a)-anthracene, pyrene, DL-norepinephrine, B(e)P, and lindane; E) cholecalciferol, n-octylamine, DL-isoproternol, testosterone, and PPO; F) 17β-estradiol and metyrapone. Reproduced with permission from Cancer Research (Ref. 41).

(continued)

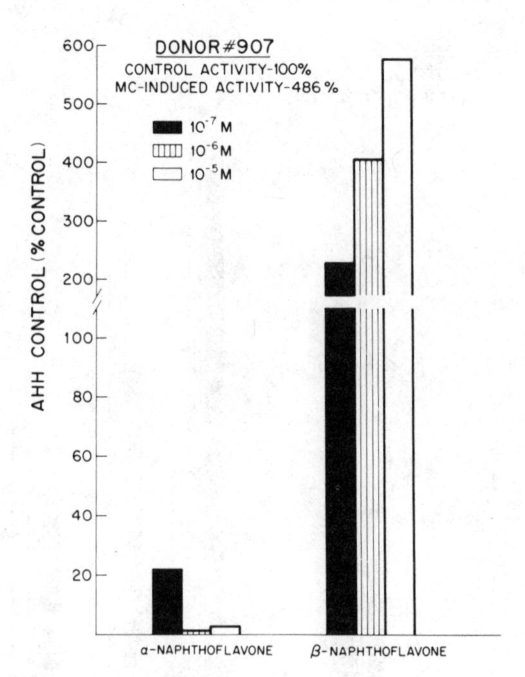
DONOR#907
CONTROL ACTIVITY-100%
MC-INDUCED ACTIVITY-486%
10⁻⁷ M
10⁻⁶ M
10⁻⁵ M
AHH CONTROL (% CONTROL)
600
500
400
300
200
100
80
60
40
20
α-NAPHTHOFLAVONE
β-NAPHTHOFLAVONE

Fig. 5B

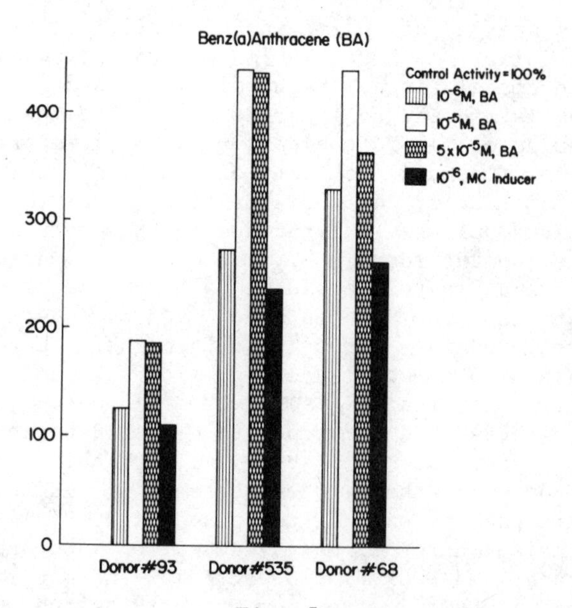
Benz(a)Anthracene (BA)
Control Activity = 100%
10⁻⁶M, BA
10⁻⁵M, BA
5 x 10⁻⁵M, BA
10⁻⁶, MC Inducer
400
300
200
100
0
Donor #93
Donor #535
Donor #68

Fig. 5C

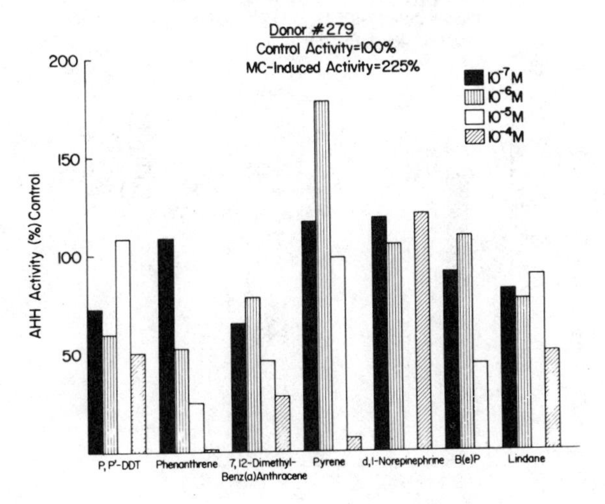
Donor #279
Control Activity=100%
MC-Induced Activity=225%
10^{-7}M
10^{-6}M
10^{-5}M
10^{-4}M
AHH Activity (%) Control
200
150
100
50
P, P'-DDT
Phenanthrene
7, 12-Dimethyl-Benz(a)Anthracene
Pyrene
d,l-Norepinephrine
B(e)P
Lindane

Fig. 5D

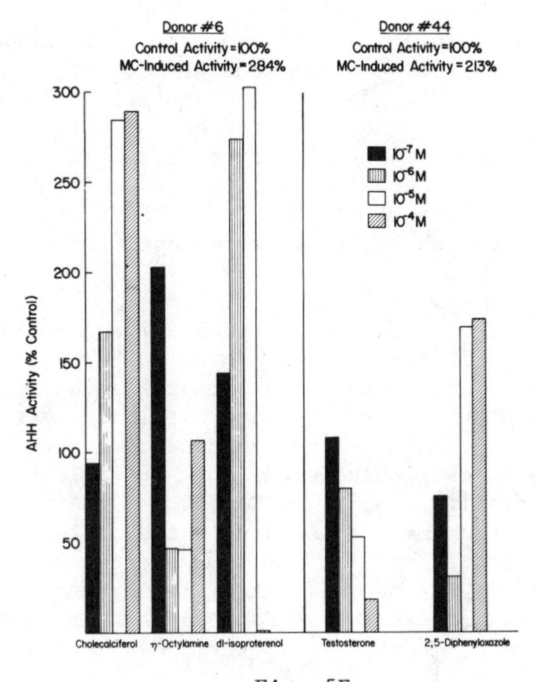
Donor #6
Control Activity=100%
MC-Induced Activity=284%
Donor #44
Control Activity=100%
MC-Induced Activity=213%
10^{-7}M
10^{-6}M
10^{-5}M
10^{-4}M
AHH Activity (% Control)
300
250
200
150
100
50
Cholecalciferol
η-Octylamine
dl-isoproterenol
Testosterone
2,5-Diphenyloxazole

Fig. 5E

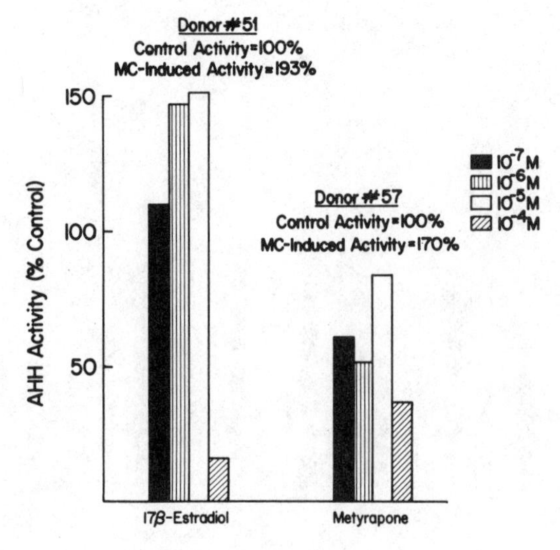

Fig. 5F

by mitogen-stimulated short-term cultured human lymphocytes isolated from fresh blood and with those produced by rodent liver microsomes. The major metabolite produced was 3-hydroxy BP; 9-hydroxy BP, quinones and BP-9,10-diol were also formed, but BP-7,8-diol and BP-4,5-diol were produced in only small amounts. Mitogen-activated fresh lymphocyte cultures produce the same BP metabolites except that BP-7,8-diol is produced by these cells in somewhat larger amounts [22]. The results furnished by our studies on the effects of FCS and the comparison of various characteristics of the basal and the induced enzymes led to the conclusion that the "so called" basal enzyme is an artifact of culture conditions and appears to be a consequence of induction of the enzyme by some unidentified endogenous factors in FCS.

Studies on the induction and inhibition specificities of AHH in mitogen-activated short-term lymphocyte cultures and comparison of basal and induced enzyme in these cultures revealed several characteristics that were similar to those obtained for the enzyme in RPMI-1788 cells. Of the 24 chemicals tested, TCDD, DBA, BA, MC, β-naphthoflavone, cholecalciferol, and DL-isoproterenol were good inducers (inducibility ratio, >2.0). Other chemicals produced effects ranging from mild induction to inhibition. Some of the results are illustrated in Fig. 5. TCDD was the most potent inducer, followed by DBA $\geqslant$ BA $>$ MC [41]. Additional data also suggested that the latter four inducers can be used interchangeably as they appear to activate common, genetically determined factors involved in the induction of AHH [41].

The order of potency among the inhibitors was α-naphthoflavone > β-naphthoflavone, followed by 2-diethylaminoethyl-2,2-diphenylvalerate, metyrapone, 1,1,1-trichloropropene oxide, and cyclohexene oxide (Fig. 6). Depending upon the concentration used, the latter four inhibitors produced moderate stimulation to moderate inhibition; however, the inhibition pattern for the basal and the PAH-induced AHH was indistinguishable. The half-life of the enzyme during cell culture and the Km values of the AHH in uninduced and PAH-induced cells were also similar. From these and other studies [40, 41] it was concluded that: (a) TCDD, DBA, BA, and MC have a common mechanism of action, differing only in the degree of induction of AHH produced; and (b) basal and induced AHH are qualitatively similar and differ only quantitatively in comparable uninduced and PAH-induced cells.

Because of several similarities between the AHH enzyme in RPMI-1788 cells and that in fresh mitogen-activated lymphocyte, and the availability of RPMI-1788 cells in large quantities, these cells could be used for the purification of cytochrome P-450 and in studies on the biochemistry and molecular biology of human lymphocyte AHH.

Metabolism of BP by Lymphocytes and Monocytes

Because the AHH assay measures essentially phenolic detoxification, AHH measurements as indicators of susceptability to cigarette smoke-associated cancers appeared questionable. To evaluate any significance AHH measurements may have, we decided to determine the relationship between AHH activity and the formation of BP phenols and procarcinogenic BP-7,8-diol, and relate the results to in vivo hepatic drug metabolism measured as plasma antipyrine half-life or urinary 4-hydroxyantipyrine half-life. To be able to measure various lymphocyte BP metabolites, culture and assay conditions were improved to measure various BP metabolites (resolved by HPLC) produced by as little as 10-15 × 10^6 lymphocytes and monocytes [22]. Both of these cell types produced BP-7,8-dihydrodiol, BP-quinones, 3-hydroxy, and 9-hydroxy phenols. Neither cell type produced BP-4,5-diol and only lymphocytes produced an early eluting polar metabolite peak believed to contain BP-9,10-diol and conjugation, reduction and hydrolysis products of BP metabolites [22]. Selkirk et al. [42] also failed to detect BP-4,5-diol formation by lymphocytes. However, these investigators also failed to detect formation of BP-9,10-diol and BP-7,8-diol after 30 min of incubation, although these metabolites were detected by these investigators after 24 hr of incubation of lymphocytes with BP. Selkirk et al. [42] have also reported the formation of other unidentified metabolites of BP eluting in the diol region; we failed to detect these metabolites either in short-term or in long-term incubations. As expected, inclusion of 1,1,1-trichloropropene oxide in the incubation resulted in the loss of BP-7,8-diol peak in both cell types. Trichloropropene oxide is an inhibitor of epoxide hydrase [43]. α-Naphthoflavone, a relatively spe-

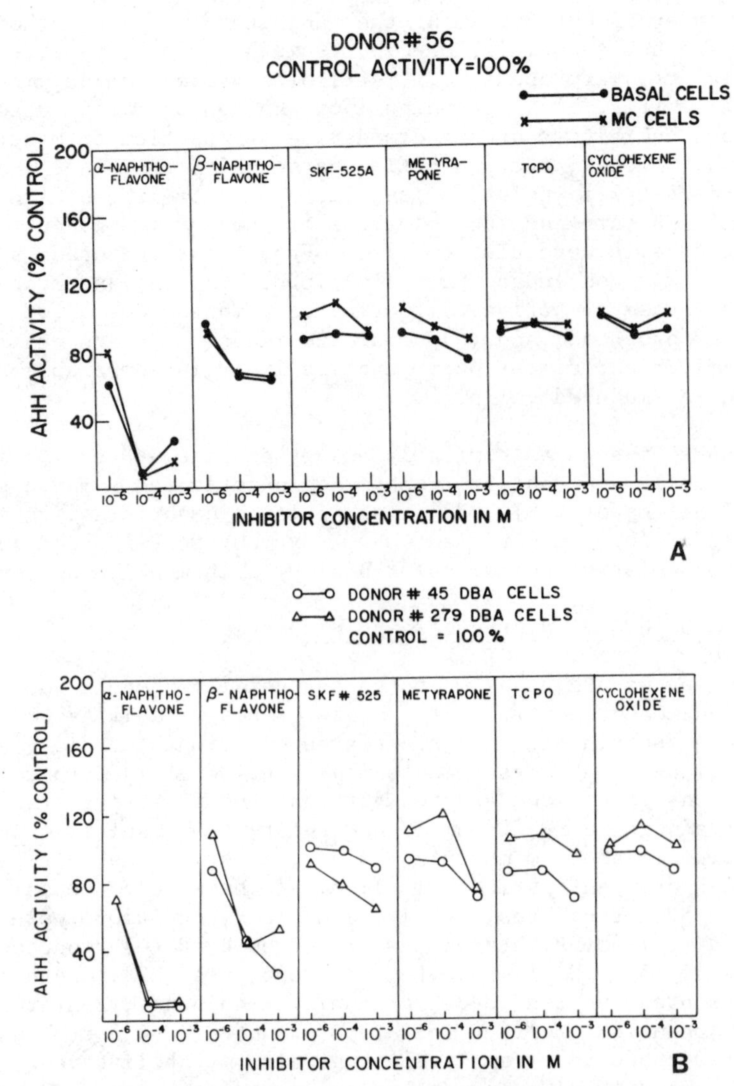

Fig. 6. Effect of inhibitors on lymphocyte-AHH activity. Mitogen-activated lymphocytes cultured in the presence or absence of the indicated inducer during the last 24 hr were used. The cells were harvested, resuspended in 0.25 M sucrose and equal aliquots containing 7×10^6 cells were distributed to various AHH assay tubes. Inhibitor at the indicated concentration was added just before the start of the reaction with BP. In the control tube the inhibitor was replaced by the solvent. Each point is a mean of 2 determinations. A and B represent data obtained on the cells derived from the indicated donor(s). A) Lymphocytes from Donor 56; o) basal cells; ×) 3-methylcholanthrene-induced cells; B) lymphocytes

cific inhibitor of PAH-inducible cytochrome P-450, produced an overall inhibition of BP metabolism [22].

In other studies, lymphocytes from six healthy donors, whose AHH inducibility ratios ranged from 2.4 to 4.6 and whose plasma antipyrine half-lives ranged from 8 to 17 hrs, were evaluated for the metabolism of BP to phenols, quinones and BP-7,8-diol [44]. BP metabolites elaborated by uninduced (control) and BA-induced lymphocytes were qualitatively similar among the six donors. A good correlation (r = 0.79) was found between the AHH inducibility ratios for the donors, as determined by the conventional fluorometric AHH assay, and the induction of BP-phenol production quantitated from HPLC data. HPLC results also indicated that the induction of BP-7,8-diol did not parallel BP phenol induction. Furthermore, the data also indicated a good negative correlation between AHH inducibility and plasma antipyrine or urinary 4-hydroxyantipyrine half-lives (r = -0.88 or -0.91), respectively. Kellerman et al. [12, 45] have previously reported a high negative correlation between lymphocyte AHH inducibility and plasma antipyrine half-lives. The excellent correlation obtained between the half-lives of plasma antipyrine and urinary 4-hydroxyantipyrine (r = 0.98) strongly support the contention that plasma antipyrine half-lives are reflections of the *in vivo* mixed function oxidase-mediated metabolism of antipyrine.

Distribution of Lymphocyte AHH Inducibility in Normal Humans

Kellerman et al. [19] have reported a trimodal distribution of lymphocyte AHH inducibility in a human population allowing the establishment of low, intermediate and high AHH inducibility groups. This pattern of population distribution is consistent with regulation by a single autosomal gene locus with codominance between the two alleles. Using our improved method of AHH analysis, we studied the distribution of lymphocyte AHH inducibility ratios in 53 healthy adults [30] and found that AHH inducibility ranged from 0.93 to 4.99 with a mean of 2.48. Males did not differ significantly from females in mean AHH inducibility or in the cumulative distribution. The 36 males had a mean of 2.53 ± 1.14, and the 17 females had a mean of 2.38 ± 0.78.

No distinct separation into low, intermediate, and high AHH-inducibility groups could be recognized. A clean separation into

---

Fig. 6 (cont.)
from Donors, 45 and 279; o) DBA-induced cells of Donor 45;) DBA-induced cells of donor 279. Control activity in basal cells is set at 100%. 1,1,1-Trichloropropene oxide (TCPO). Reproduced with permission from Cancer Research (Ref. 41).

different classes would not be expected even if the three groups were to exist, given the small range of the lymphocyte AHH inducibility ratios. In different studies on "cured" lung and laryngeal cancer patients [46], "cured" bladder cancer patients [47], progeny of lung [48], and of bladder cancer patients [47], we also studied the distribution of lymphocyte AHH inducibility in matched controls of each population and also investigated the distribution in monozygotic and dizygotic twins [49]. No evidence of trimodal distribution was obtained in these studies which involved several hundred normal donors. Similar negative results have also been reported by Atlas et al. [28].

Heritability of Lymphocyte AHH Activity and Inducibility

Heritability determinations of a genetic trait are useful as they allow estimation of the contributions environmental and genetic factors make to the phenotypic expression of the trait. Also, heritability estimates provide some indication of the susceptibility of offspring to a condition influenced or determined by a genetic trait. Conversely, if the heritability of the trait is significant, then progeny (and sibling) studies to evaluate the risk-relationship confered on the parental phenotype are meaningful. Heritability estimates involve comparison of concordance and discordance between monozygotic (MZ) and dizygotic (DZ) twins for the same phenotypic characteristic. The implicit assumption being that dizygotic twin variance is a consequence of the combined contribution of the environmental and genetic factors, whereas monozygotic twin variance represents essentially the contribution from environmental factors alone.

To evaluate the heritability of AHH activity (both basal and induced) and the inducibility ratio, control and induced lymphocyte AHH activities were measured for 18 pairs of MZ and 30 pairs of DZ twins [49]. MZ twins had similar AHH activities and inducibility; the average difference being the same as the average difference of simultaneous assays of a single, divided blood sample. Whereas most DZ twins were as similar as were MZ twins, a few DZ twins showed more variability. Two pairs of DZ twins contributed 61% of the variance for the basal AHH activity; four pairs contributed 70% of the variance for the induced AHH activity, and two pairs contributed 62% of the variance for the AHH inducibility. Heritability (H = Variance DZ-variance MZ/variance DZ) estimates yielded heritability of 0.8 to 0.9 for basal AHH activity; 0.66 to 0.95 for the induced AHH activity, and about 0.73 for the AHH inducibility ratio [49]. Two other studies have reported on the heritability of human AHH. Atlas et al. [28] reported that AHH inducibility in human lymphocytes is heritable (H = 0.8) but the basal and induced activities are not. This is in partial disagreement with our findings and the discrepancy between the two reports may be related either to the age of the subjects or the sensitivity of the AHH assay used in the two studies. From studies on AHH activity in monocytes of twins, Okuda et al. [50] concluded that while induced AHH activity is highly heritable, basal activity and inducibility ratios are not.

An interesting observation in our studies was a high degree of concordance of the results in the majority of DZ twins. In classical twin studies showing significant heritability, MZ twins are more alike than DZ twins because quantitative traits, such as weight, and height, are polygenic characteristics. However, an enzyme activity is most likely the product of a single gene locus. When single gene differences exist in the human population, the probability is high that two sibs, whether DZ twins or not, carry the same genotype and thus resemble MZ twin pairs. Therefore, in such studies it is not surprising to find that only a few DZ twins are discordant [49]. The same argument can be used to support the suggestion that because of a high degree of concordance between the results of a majority of DZ twins, human lymphocyte AHH activity (basal and induced) and/or inducibility are determined by one or at the most a few gene loci.

Role of Lymphocyte AHH in Human Cancer

It has been reported [20, 51] that the distribution of lymphocyte AHH-inducibility was dramatically shifted toward the high end of the range in a group of lung and laryngeal cancer patients. The interpretation was given that intermediate and especially high AHH inducibility may confer an increased susceptibility to lung cancer. Because of the significance of these observations to cancer control programs geared to identify sensitive individuals, we carried out our own investigations of the lymphocyte AHH inducibility of patients with bronchogenic carcinoma. We found inadequate expression (very low) of AHH activity in at least half of the 62 patients tested. In addition, even though the same number of lymphocytes were isolated from the blood of lung cancer patients as were isolated from the blood of normal donors, the lymphocytes from lung cancer patients failed to grow in culture, with the result that cell count at harvest was also low. The ability to obtain successful cultures was not related to the stage of the disease, to cancer-cell type or to patient survival time after testing. Studies on ^{14}C-amino acid incorporation and DNA analysis revealed impaired label incorporation and decreased cell growth (up to 50%) in half of the 12 patients tested in these studies. These observations of decreased cell growth of lymphocytes from lung cancer patients are paralleled by reports of other investigators who have found impaired blastogenesis in lymphocytes from lung cancer patients [52]. Because of the difficulties encountered in culturing cancer patient's lymphocytes, we decided on two other approaches involving analysis of AHH activity in the lymphocytes of (a) the progeny of lung cancer patients, and (b) "apparently cured" lung and laryngeal cancer patients.

Studies on the Progeny of Lung Cancer Patients

Since both we [49] and Atlas et al. [28] reported a high heritability for the lymphocyte AHH inducibility in humans, it was considered valid to estimate the AHH genotypes of lung cancer patients by testing their progeny. Theoretically, the progeny should have a distribution of AHH phenotypes which falls halfway between that of the patients and the normal population. The progeny test does not depend on the genetic mode of additive expression of two alleles at a single locus. Almost any type of inheritance should produce a distribution halfway between patients and normal population, provided that the parental (patient) population is different from the normal population.

Because of the possibility that the narrow range for the human lymphocyte AHH inducibility may obscure small differences in AHH inducibility between the cancer patient's progeny and the control population, it was considered important to match controls for age, sex, social background and food habits, even though we had previously reported that a number of these factors, e.g., sex, smoking, and age over 30 yrs, do not affect lymphocyte AHH inducibility [53]. In order to meet these criteria, all analyses were carried out during the high activity season of the year and the spouse of each offspring was used as the control. The 57 lung cancer patients indirectly tested by use of their progeny had a distribution of 40% squamous-cell carcinoma, 18% adenocarcinoma, 11% small-cell, 5% large cell, 5% undiffereniated, 11% unknown and 10% other types. The use of the spouse provided, in our opinion, well matched control because of the similarity in age, other background and environment such as diet and neighborhood of residence. Although the population under study consisted of both sexes, we have previously shown that males and females do not differ in their AHH levels or inducibility ratios [48, 53], and the progeny population was not therefore biased. The average age was 44.1 years for the progeny and 43.8 years for controls; 17 of the progeny and 18 of the controls were smokers, and 26 of the progeny and 31 of the controls were male.

Figure 7 shows the distribution of AHH inducibility in the controls and in the progeny and persons with lung cancer. No obvious difference can be observed between the two populations. The mean ratio was 2.95 for both groups. The 50% and 90% divisions for the control population, shown by the dotted lines in Fig. 7, were used to separate the progeny population into three groups. These arbitrary divisions were chosen because they approximated the distribution of lows, intermediates, and highs reported by Kellermann et al. [19, 20] for AHH inducibility. A Chi square analysis of the groups thus formed indicated no difference between the progeny and the control populations ($p = 0.9$).

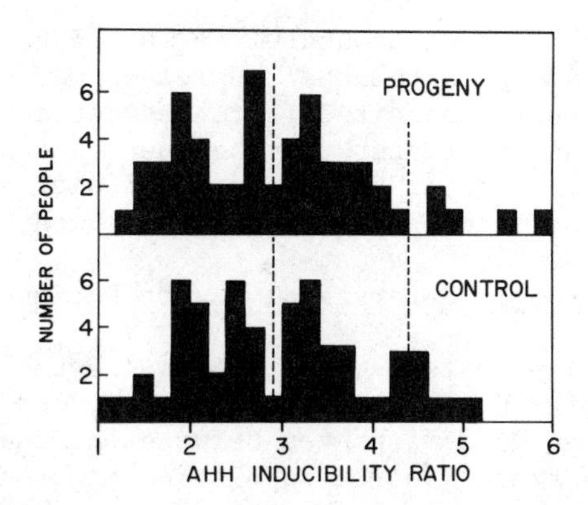

Fig. 7. AHH inducibility in the progeny of patients with lung cancer and matched controls. The dotted lines are drawn at the 50% and 90% points in the distribution of AHH inducibility of the control population and continued through the progeny population. Reproduced with permission from the New England Journal of Medicine (Ref. 48).

Since Guirgis et al. [54] have reported that lung cancer patients have higher induced AHH activity than controls, we questioned that induced AHH activity levels may be more important than the ratio in determining cancer risk. Therefore, the induced AHH levels of the progeny of lung cancer patient were compared with those of the matched control [48]. These data also showed no significant differences between the two groups.

Several members of the control population had some history of cancer in their families. As a further test, those controls with no family history of cancer were separated out and compared to those progeny with a strong family history, that is, they had more than one first-degree relative with cancer. These controls were also compared to those progeny who themselves had a history of cancer. No significant differences between the controls and the progeny were found. We also found no difference between AHH inducibility in progency of parents with squamous cell carcinoma or with adenocarcinoma.

To further investigate the effect of environmental factors, such as living in the same dwelling and eating the same diet, that might have obscured any difference in the lymphocyte AHH inducibility between the progeny and the control, the data obtained for each male were matched with the data for an unrelated female tested on the same day. If environmental factors affected these measurements, then the data for 57 husbands and wives should have looked more alike than that for the unrelated couples we matched on a random basis. However, the intrapair differences between unrelated pairs were not significantly greater than between husband and wife, indicating that environmental factors such as diet and neighborhood of residence do not affect AHH in cultured lymphocytes.

Studies on Cured Lung and Laryngeal Cancer Patients

Since several reports [20, 51, 54] suggested that AHH was important in determining cancer susceptibility, and because of the possibility that small differences between the lung cancer patient's progeny and the matched control population might have been masked by some unknown and uncontrollable variable, an additional study involving previous lung and laryngeal cancer patients who were apparently free of the disease was carried out. In this study, the spouse of each cancer patient was used as the control and where no spouse was available (17 cases) a control matched for age was selected. The "cured" lung and laryngeal cancer patients were grouped together because of the necessity to get meaningful numbers and because Trell et al. [51] had reported a distribution of AHH inducibility in laryngeal cancer patients similar to that reported [20] earlier for lung cancer patients. The present study, therefore, involved 32 "cured" lung and 30 "cured" laryngeal cancer patients and 57 matched controls.

The histologic diagnosis for the patients with lung cancer was squamous-cell in 11 patients, adenocarcinoma in 9, bronchioalveolar in 5, large-cell in 3, anaplastic in 2 and clear-cell in 1, while 1 patient was described as having carcinoma *in situ*, cell type unknown. All patients with laryngeal cancer had squamous-cell type. All histological types of lung cancer and squamous-cell carcinoma of the larynx are highly associated with cigarette smoking. Patients and their spouses did not differ in the number of first-degree relatives who had cancer of all types. However, patients reported 11 first-degree relatives with lung or head and neck cancer while controls reported only 6 such relatives. This increase would be expected because first-degree relatives of patients with lung cancer have an increased risk of lung cancer [14].

Patients had had their surgery from 1 to 10 years prior to testing, with 26 subjects operated on within 2 years, 12 from 2-5 years, and 21 more than 5 years before sampling. All were thought to be free of disease at the time of sampling. Long-term disease-free status was not critical, however, for this study, since we were primarily interested in selecting patients whose clinical status at the time of sampling permitted lymphocyte response in culture similar to that of normal individuals.

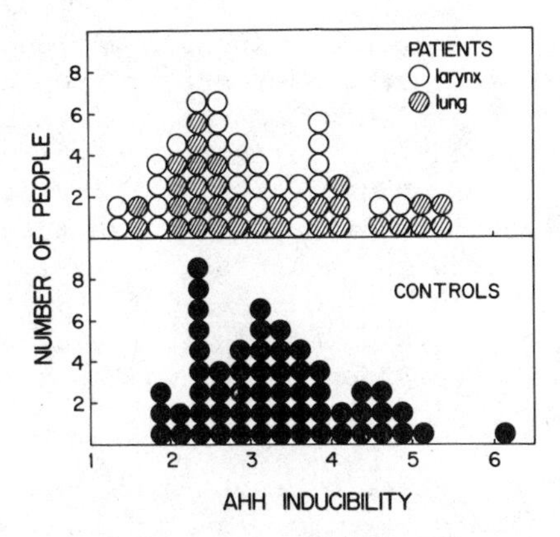

Fig. 8. Distribution of AHH inducibility in patients who had lung and laryngeal cancer and matched controls. Reproduced with permission from International J. of Cancer, Geneva, Switzerland (Ref. 46).

Earlier we reported [30, 48, 53] that lymphocytes from a higher proportion of lung cancer patients (50%) express low levels of AHH during culture and this on the average has led us to reject 5% of the normal subjects and about 50% of the subjects with lung cancer. In this study 3 patients with lung cancer (9%), 2 patients with laryngeal cancer (7%) and 3 controls (5%) were rejected on the basis of this criterion.

In Fig. 8 are compared the AHH inducibility ratios of controls and of patients who had lung or laryngeal cancer. No shift towards higher AHH inducibility was observed in the patient population. The mean inducibility ratio ±S.E. was 3.30 ± 0.20 for patients who had lung cancer, 2.96 ± 0.18 for patients who had laryngeal cancer, and 3.29 ± 0.04 for all controls.

The 50% and 90% divisions in the control population were used to separate the patient population into three groups. A Chi square analysis of the groups also indicated no difference between the patient and control populations.

Induced enzyme activity was similar for patients and controls. Mean induced AHH activity was 0.25 pmoles 3-hydroxy-BP/min/10 μg DNA for patients who had laryngeal and lung cancer and 0.27 for pooled controls. A histogram of induced activity showed no difference in the distribution of induced activity between patients and controls.

We compared patients of younger age at onset of disease with those of advanced age at onset because the group which developed cancer early in life would tend to include patients with genetic predisposition to cancer. No difference was found when mean AHH inducibility (±S.E.) of patients less than 57 years old at diagnosis (3.15 ± 0.23) was compared with patients over 57 years old at diagnosis (2.95 ± 0.19), or when the mean AHH inducibility of patients who developed lung cancer when they were less than 50 years old (3.14 ± 0.39) was compared to patients who developed lung cancer after age 60 (3.27 ± 0.42). Although genetic susceptibility may influence age at onset, this effect is likely to be complicated by the known dose-response relationship between cigarette consumption and respiratory tumors.

Patients who have a genetic trait which increases susceptibility to cancer would likely have an elevated incidence of cancer in their families. We identified patients with a strong family history of cancer in two ways; those patients who reported one or more relatives with lung or head and neck cancer, and those patients with a history of cancer of any type in two or more first-degree relatives. The mean inducibility ratios (±S.E.) of these two groups (3.52 ± 0.41 and 3.45 ± 0.35, respectively) was not different from the mean inducibility ratios of control patients with no history of cancer in their family (3.34 ± 0.18). Thus, positive family history of cancer does not identify a subgroup of patients with elevated AHH inducibility.

Since some earlier reports [20, 30, 48, 51], several conflicting reports [55-57] measuring either induced lymphocyte AHH activity or inducibility ratio have appeared concerning the role of AHH in the development of cigarette smoke-associated human lung and laryngeal cancer.

Studies on Bladder Cancer Patients and Their Progeny

While association between cigarette smoking and bladder cancer has been suggested [1], most identified bladder carcinogens are believed to be aromatic amines and nitrosamines [11, 58]. These chemicals are initially hydroxylated by the microsomal mixed function oxidases and subsequently conjugated with sulfate or glucuronic acid for excretion in the urine. In the bladder these conjugates can decompose or be split by hydrolytic enzymes, such as urinary glucuronidase or sulfatase, to release the active carcinogen [11, 59, 60]. Interindividual variability in any of these enzymatic pathways might explain individual susceptibility to bladder cancer. Such susceptibility may have a genetic component since clustering of bladder cancer patients have been reported in some families [61, 62]. It has also been proposed that the relative potency of bladder carcinogens depends upon the extent to which they are metabolized by the hepatic mixed function oxidases [63]. Since lymphocyte AHH activity and inducibility have been reported to vary in the human population [19, 20], it was of interest to evaluate their role in human bladder cancer development.

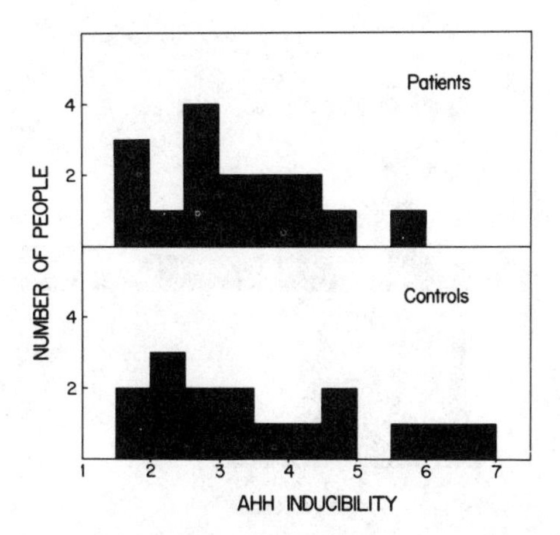

Fig. 9. Distribution of AHH inducibility in patients with bladder cancer and matched controls. AHH inducibility in 16 patients and 16 matched controls was determined as described in the text. Reproduced with permission from International J. of Cancer, Geneva, Switzerland (Ref. 47).

Sixteen ostensibly cured bladder cancer patients, 53 progeny of the bladder cancer patients, and matched controls for each group were analyzed for lymphocyte AHH activity and inducibility. The 53 bladder cancer patients tested indirectly had the following distribution of bladder cancer by cell type: 1 undifferentiated carcinoma and 52 transitional-cell carcinomas; 42 of the patients were males and 11 females; 36 were smokers; 16 non-smokers and 1 with unknown smoking history. A single offspring of each of the 53 cancer patients was tested and spouses were used as controls. If a spouse was not available (two cases), the subject selected a person of similar age and neighborhood to serve as a control. The average age was 50 years for both progeny (range 28-76) and control (range 26-69); 19 progeny and 14 controls were smokers; 27 progeny and 24 controls were male.

Among the 16 "cured" bladder cancer patients 14 were males and 2 were females with an average age of 69 years for all (range 54-80 years). The spouse of each patient was used as a matched control. The average age of the spouse was 64 years (range 41-77 years). The patients had been first diagnosed as having bladder cancer an average of 6.7 years (range 1 to 16 years) prior to this analysis. Of these patients 7 were current smokers, 7 were previous smokers and 2 were nonsmokers. Half of the patients were past the 5-year survival time.

In Fig. 9 is compared the distribution of lymphocyte AHH inducibility ratios of "cured" bladder cancer patients and matched

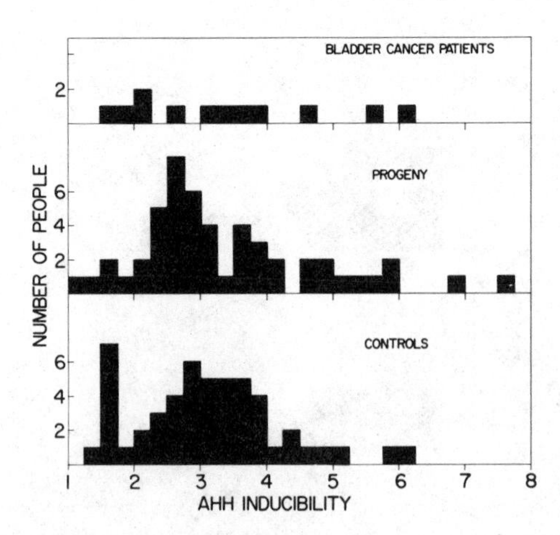

Fig. 10. Distribution of AHH inducibility in the bladder cancer patients and in the progeny of bladder cancer patients and their matched controls. Details described in the text.

controls. No significant difference was observed between the two groups; the distribution of induced AHH activity was also comparable for patients with bladder cancer and their matched controls [47].

In Fig. 10 is illustrated the AHH inducibility distribution of the progeny of 53 bladder cancer patients and their matched control. The two populations (progeny and matched control) did not differ from each other. The existence of a few more individuals at the high end of the range in the progeny population was not statistically significant. The mean AHH inducibility ratio of the progeny was 3.32 ± 0.19 (S.E.) and the mean ratio of the control population was 3.08 ± 0.15 (S.E.). The difference between the two means was not significant at the 0.05 level (paired t-test). A non-parametric Kouskal-Wallis test of rank orders revealed a mean of 54.8 for the progeny and 52.2 for the control. This difference was not significant at the 0.05 level. Statistical analysis also revealed no significant difference between the progeny and the control populations in the induced AHH activity.

If the shift in AHH inducibility were real, it might be more pronounced for those patients with an early age at onset of bladder cancer since genetic factors are generally stronger for early age of onset. We separated the progeny into two groups based on whether the age at onset of bladder cancer in the parent was above or below 60 years of age. The AHH inducibility of the two groups was not significantly different at the 0.05 level (AHH inducibility for early age at onset, 3.11 ± 0.33; AHH inducibility for later age at onset, 3.38 ± 0.22).

As a further test, controls with no familial history of cancer were separated out (23 cases, AHH inducibility 2.98 ± 0.21) and compared to progeny with a strong family history of cancer - that is more than 1 parent, sib, or grandparent with cancer (24 cases, AHH inducibility 3.31 ± 0.33). These means were not significantly different at the 0.05 level (paired t-test).

Benzo(a)pyrene Metabolism by Human Placenta

Since placenta is capable of carrying out many drug (chemical) metabolism reactions (including oxidations, hydrolytic reactions, reduction and conjugation reactions [64]), it can, most likely, activate a number of chemicals to the detriment of the fetus. A number of PAHs have been reported to act as transplacental carcinogens in rodents [65-67] and smoking in humans during pregnancy results in an increased incidence of respiratory disease as well as in smaller newborns [1, 68]. Since PAHs require activation to become carcinogenic, smoking by inducing placental mixed function oxidases may cause an enhancement in the activation of PAHs present in the cigarette smoke and in other sources. Indeed, AHH activity and the associated metabolism of BP to various metabolites is several fold higher in the placenta of smokers than of nonsmokers [23, 69, 70].

In comparison to lymphocytes, and despite obvious limitations, placenta affords several advantages for studies on PAH metabolism. Placental tissue can be obtained in large quantities without undesirable surgical or physical interventions, and the analysis of placental tissues allows evaluation of at least some of the *in vivo* effects of smoking on the metabolism of PAH by humans. Because of these advantages and the complexity of lymphocyte cell culture systems, placental tissue was selected for investigating the effects of smoking on BP metabolism, the nature of the BP metabolites produced and the qualitative similarity to those produced by lymphocytes, and the quantitative distributon of placental BP metabolism in a given population.

In the first series of experiments term placenta from 8 women who did not smoke and from 13 who smoked from 1 to 40 cigarettes/day were collected and analyzed for epoxide hydrase activity using styrene-oxide as the substrate, for AHH activity and for the transformation of BP to various metabolites (diols, quinones, and phenols) as resolved by HPLC [23]. In addition, three placental samples from nonsmokers and three with high AHH activity from smokers were analyzed for the formation of DNA-binding metabolites and BP metabolite-nucleoside adducts separated by Sephadex LH-20 chromatography.

For the epoxide hydrase activity, which presumably participates in the formation of the procarcinogenic BP-7,8-diol metabolite, no difference between the smokers and nonsmokers was found. Use of

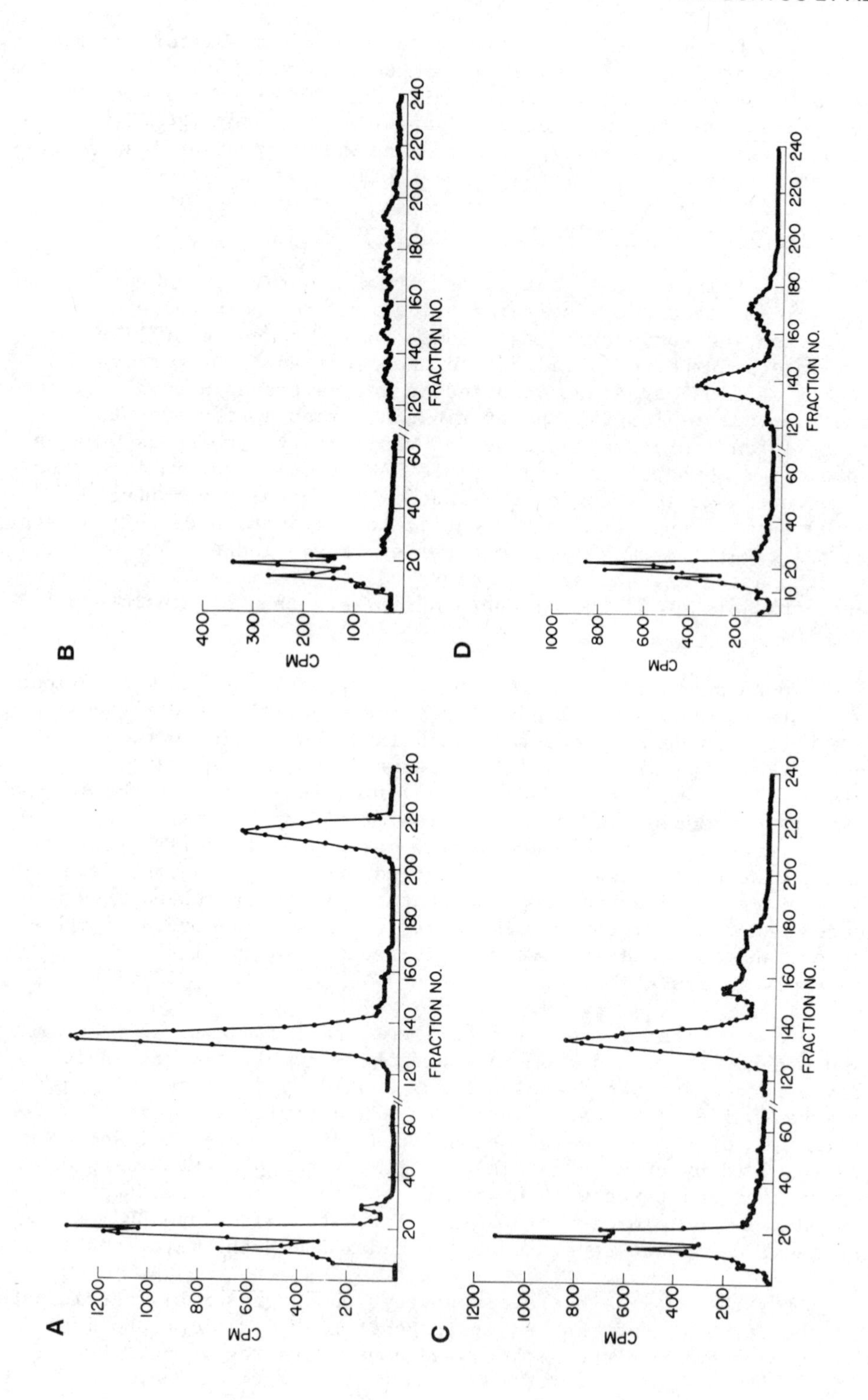
A
B
C
D
CPM
FRACTION NO.
1200
1000
800
600
400
300
200
100
10
20
40
60
120
140
160
180
200
220
240

Fig. 11. Sephadex LH-20-profile of BP-DNA adducts formed by placental microsomes of smokers and non-smokers. The DNA samples isolated from the incubations containing placental microsomes, NADPH, ^{3}H-BP and calf thymus DNA were sequentially digested with DNAse, phosphodiesterase, and alkaline-phosphatase. The digest was then subjected to LH-20 column chromatography. Details have been described elsewhere [23]. A) Liver microsomes from methylcholanthreene-treated rat; B) placental microsomes of a nonsmoker; C and D) placental microsomes of smokers. Reproduced with permission from Cancer Research (Ref. 23).

styrene oxide as the substrate may be adequate since Kapitulnik et al. [71] have shown a good correlation between human liver epoxide hydrase activities involved in the hydration of 11 different epoxide substrates. In contrast to the epoxide hydrase activity, placental AHH activity was several-fold higher in smokers. The average activity among the nonsmokers was 0.06 units with a range of 0.04 to 0.09; smokers had an average AHH activity of 4.55 units (75-fold higher) with a range of 0.28 to 11.02 units. Resolution of BP metabolites by HPLC revealed a profile qualitatively similar to that produced by lymphocytes [22, 23]. While phenols, quinones and both BP-9,10- and BP-7,8-diols were produced, BP-4,5-diol was produced in negligible quantities. BP-7,8-diol, a major metabolite, was produced in amounts equivalent to phenols.

In considering the average enzyme activities involved in the formation of BP metabolites by nonsmokers versus smokers, the dihydrodiols were induced to a much greater extent than were quinones and phenols. Quinones and phenols were on the average 4.6- and 6.2-fold higher, respectively, in smokers, whereas BP-9,10-diol and BP-7,8-diol were 13.8- and 26.5-fold, respectively, higher in smokers compared to nonsmokers. Regression analysis of the data obtained for smokers revealed an excellent correlation between placental AHH activity and BP-phenol formation ($r = 0.82$), and between BP-phenol and BP-7,8-diol formation ($r = 0.90$), suggesting that either (a) numerous chemicals in cigarette smoke cause a parallel induction of various mixed function oxidase enzymes (cytochrome P-450s) involved in the formation of different BP metabolites, or (b) cigarette smoke induces a form of cytochrome P-450 that catalyzes conversion of BP via different pathways. These results differ from those for lymphocytes where a single inducer (benzanthracene) in culture resulted in unparallel induction of the metabolism of BP to phenols and BP-7,8-diol [44]. Placentas from smokers also demonstrated higher activities for the metabolism of BP to DNA binding metabolites. Enzymatic digestion followed by Sephadex LH-20 chromatography of the modified DNAs yielded BP metabolite-nucleoside adducts which, based on comparisons with the rodent model system, appeared to be nucleoside adducts of BP-7,8-diol-epoxide(s); no such adducts were found in assays utilizing microsomes from the placentas of nonsmokers (Fig. 11).

Because a very good correlation was found between BP-phenols (or AHH activity) and BP-7,8-diol formation, and between AHH activity and the formation of DNA-binding metabolites when placentas from smokers were analyzed [23, 72], it was considered adequate to use AHH activity as an indicator of the overall metabolism, including the activation, of BP by placenta. Therefore, subsequently term placentas from ten nonsmokers and 67 smokers were collected and analyzed for AHH activity [72]. The data were assigned to different groups depending upon the similarities in the smoking histories of various individuals, and the mean of each group with the range were plotted to construct a dose-response curve (Fig. 12).

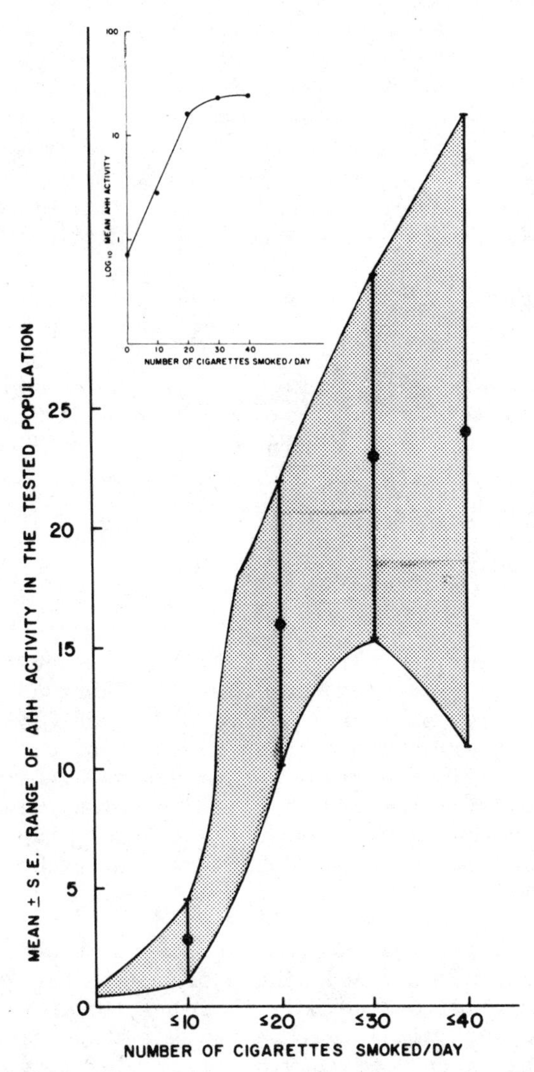

Fig. 12. Dose-response curve demonstrating relationship between AHH activity of placental microsomes and the number of cigarettes smoked/day. Details described elsewhere [72]. The estimates concerning the number of cigarettes smoke/day should be considered best approximations based on the smoking histories of the individuals. These individuals are believed not to have interupted their normal pattern of smoking. Inset shows semi-log plot of mean AHH activities. This work is described in detail in a manuscript submitted recently to Cancer Research.

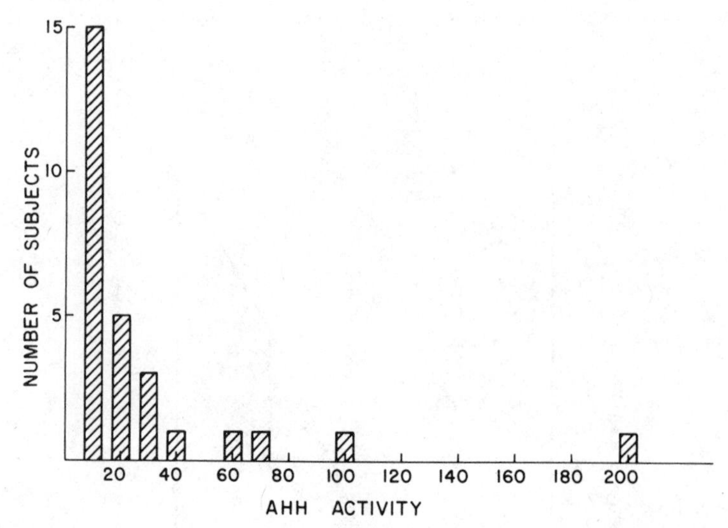

Fig. 13. Frequency distribution of placental AHH activity of smokers smoking 30 cigarettes per day. Details described elsewhere [72].

As reported previously [23], activity in placentas from non-smokers was very low, whereas considerably higher activity was found in the placentas from smokers. While considerable interindividual variability was observed among individuals with comparable smoking histories, AHH activity showed a definite dose-response relationship. Mean activity increased with the number of cigarettes smoked and the dose-response curve assumed a sigmoidal shape which became linear (inset Fig. 12) when the data were plotted on a semi-log scale.

As the number of cigarettes smoked/day increases, mean AHH activity appears to increase reaching a plateau after 20-25 cigarettes/day (Fig. 12). While considerable interindividual variability in AHH activity was observed within each group, greater variability was observed within the groups of individuals smoking more than 20 cigarettes/day and the least variable was the group of individuals smoking the lowest number of cigarettes (⩽10/day). Further analysis revealed considerable clustering of the individuals near the low AHH activity values whether the population was examined as one unit or separately by groups defined by the number of cigarettes smoked. In the whole population, 65% of the tested individuals (39 out of 67) had less than 10 units of activity; 76% had less than 20 units of activity; and 87% (58 out of 67) had less than 30 units of activity. Only 13% of the subjects had greater than 30 units of activity. Further analysis of the whole population as well as of each group separately revealed an assymetrical distribution of AHH activity with a right tail that did not appear to depart from a unimodal distribution (Fig. 13).

Analysis of placentas obtained from each of the six twin fetuses with dichorionic dizygous placentation and analysis of maternal blood thiocyanate levels in six other smokers suggested that genetic variability constitutes a major component of the variability among individuals with comparable smoking histories [72].

CONCLUSIONS

Optimization of lymphocyte cell culture and enzyme assay conditions led to the development of a very sensitive and a reproducible assay for the measurement of AHH activity and inducibility in lymphocytes and monocytes. The same assay was adapted to investigate the metabolism of BP to various metabolites formed by lymphocytes, monocytes and placenta. The HPLC profiles of BP metabolism produced by lymphocytes, monocytes and placenta were qualitatively similar; 3-hydroxy BP was the major metabolite and BP-7,8-diol was produced in significant quantities by all the three tissues. Placental studies also revealed the formation of DNA nucleoside-BP metabolite adduct(s) believed to be derived from BP-7,8-diol-9,10-oxides.

A very interesting observation was that lymphocyte AHH activity and inducibility in the human population varied with the season of the year, being high during summer and early fall seasons and low during other times. Studies on mono-, and dizygotic twins demonstrated that lymphocyte AHH activity and inducibility are heritable traits and that AHH induction is regulated by a few gene loci, possibly by one or at the most two. However, population distribution profile of lymphocyte AHH activity or inducibility was not trimodal as reported by Kellerman and coworkers [19, 20]. Studies on cured lung and laryngeal cancer patients, progeny of lung cancer patients, cured bladder cancer patients and their progeny did not show a population distribution profile of either lymphocyte AHH activity or its inducibility, that was different from the normal populations; these data suggest that lymphocyte AHH activity and inducibility, though significantly heritable, do not determine the sensitivity of an individual to cigarette smoke associated cancers and that parents do not pass on to their offsprings the sensitivity purported to be associated with high AHH inducibility.

Investigations of placental BP metabolism demonstrated a sigmoidal (log-linear) dose-response relationship between AHH activity and the number of cigarettes smoked, although considerable interindividual variability in AHH activity was observed among maternal smokers with similar smoking histories. These investigations suggested that genetic factors are responsible for a major fraction of interindividual variability that is independent of the dose-response effects.

Since experimental evidence in mice has clearly demonstrated a positive relationship between Ah responsiveness (i.e., AHH inducibility) and sensitivity to PAH-induced cancers, inspite of various conflicting reports it is necessary to continue investigations in humans to define the role of AHH induction in the development of cigarette smoke-associated cancers, especially those of the respiratory tract. It is possible that several factors have contributed to conflicting observations reported on the relationship between lymphocyte AHH inducibility and lung cancer risk. Some of these factors may be related to (a) intricacies of lymphocyte culture conditions; (b) the possibility that lymphocytes may not be representative of other human tissues, especially lung (and other respiratory tract components) which is the target for cigarette smoke associated cancers; and (c) that disease process itself may alter the enzyme activity.

To resolve some of these questions it may be necessary to purify PAH-inducible cytochrome P450 from humans, employ newer molecular biology techniques to dissect various problems at the genome level, and/or select other tissues for epidemiologic studies that may be suitable for analysis of in vivo effects of smoking, such as skin biopsies, macrophages, and placenta.

ACKNOWLEDGMENTS

The author wishes to thank Miss Karen Marie Schrader for her help in the preparation of this Chapter.

REFERENCES

1. United States Public Health Service, The Consequences of Smoking, A Report of the Surgeon General, Publication (HSU) 71-7513, p. 458.
2. J. M. Hopkins and H. J. Evans, Cigarette Smoke-Induced DNA Damage and Lung Cancer Risk, Nature, London, 283:388-390 (1980).
3. M. B. Shimkin and G. D. Stoner, Lung Tumors in Mice: Application to Carcinogenesis Bioassay, Adv. Cancer Res., 21:1-58 (1975).
4. D. Hoffmann and E. L. Wynder, A Study of Tobacco Carcinogenesis, XI. Tumor Initiators, Tumor Accelerators and Tumor Promoting Activity of Condensate Fractions, Cancer, 27:848-864 (1971).
5. R. E. Kouri, Relationship between Levels of Aryl Hydrocarbon Hydroxylase Activity and Susceptibility to 3-Methylcholanthrene and Benzo(a)pyrene-Induced Cancers in Inbred Strains of Mice, in: Polynuclear Aromatic Hydrocarbons: Chemistry, Metabolism,

and Carcinogenesis, Vol. 1, pp. 139-151 (R. Freudenthal and P. W. Jones, eds.), Raven Press, New York (1976).
6. Particulate Polycyclic Organic Matter, Washington, D.C., National Academy of Sciences (1972).
7. A. Y. H. Lu and S. B. West, Multiplicity of Mammalian Microsomal Cytochrome P450, Pharmacol. Rev., 31:277-295 (1980).
8. F. P. Guengerich, Isolation and Purification of Cytochrome P450 and the Existence of Multiple Forms, Pharmacol. Ther., 6:99-121 (1979).
9. H. V. Gelboin, Benzo(a)pyrene Metabolism, Activation, and Carcinogenesis: Role and Regulation of Mixed Function Oxidases and Related Enzymes, Pharmacol. Reviews, 60:1107-1166 (1980).
10. S. S. Epstein, Environmental Determinants of Human Cancer, Cancer Res., 34:2425-2435 (1974).
11. E. C. Miller, Some Current Perspectives on Chemical Carcinogenesis in Humans and Experimental Animals: Presidential Address, Cancer Res., 38:1479-1496 (1978).
12. G. Kellerman, M. Luyten-Kellerman, M. B. Horning, and M. Stafford, Correlation of Aryl Hydrocarbon Hydroxylase Activity of Human Lymphocyte Cultures and Plasma Elimination Rates for Antipyrine and Phenylbutazone, Drug Metab. Dispos., 3:47-50 (1975).
13. J. Kapitulnik, P. J. Poppers, and A. H. Conney, Comparative Metabolism of Benzo(a)pyrene and Drugs in Human Liver, Clin. Pharmacol., 21:166-176 (1977).
14. G. K. Tokuhata and A. M. Lilienfeld, Familial Aggregation of Lung Cancer in Humans, J. Natl. Cancer Inst., 30:289-312 (1963).
15. R. E. Kouri, H. Ratrie, and C. E. Whitmire, Evidence of a Genetic Relationship between Methylcholanthrene-Induced Subcutaneous Tumors and Inducibility of Aryl Hydrocarbon Hydroxylase, J. Natl. Cancer Inst., 51:197-200 (1973).
16. R. E. Kouri, R. A. Salerno, and C. E. Whitmire, Relationship between Aryl Hydrocarbon Hydroxylase Inducibility and Sensitivity to Chemically Induced Subcutaneous Sarcomas in Various Strains of Mice, J. Natl. Cancer Inst., 50:363-368 (1973).
17. J. P. Whitlock, H. L. Copper, and H. V. Gelboin, Aryl Hydrocarbon (Benzo(a)pyrene) Hydroxylase is Stimulated in Human Lymphocytes by Nitrogens and Benz(a)anthracene, Science, 177:618-619 (1972).
18. D. L. Busbee, C. R. Shaw, and E. T. Cantrell, Aryl Hydrocarbon Hydroxylase Induction in Human Leukocytes, Science, 178:315-316 (1972).
19. G. Kellerman, M. Luyten-Kellerman, and C. R. Shaw, Genetic Variation of Aryl Hydrocarbon Hydroxylase in Human Lymphocytes, Am. J. Hum. Genet., 25:327-331 (1973).
20. G. Kellerman, C. R. Shaw, and M. Luyten-Kellerman, Aryl Hydrocarbon Hydroxylase Inducibility and Bronchogenic Carcinoma, New Eng. J. Med., 289:934-937 (1973).
21. H. L. Gurtoo, N. Bejba, and J. Minowada, Properties, Inducibility, and an Improved Method of Analysis of Aryl Hydrocarbon

Hydroxylase in Cultured Human Lymphocytes, Cancer Res., 35: 1235-1243 (1975).

22. J. B. Vaught, H. L. Gurtoo, B. Paigen, J. Minowada, and P. Sartori, Comparison of Benzo(a)pyrene Metabolism by Human Peripheral Blood Lymphocytes and Monocytes, Cancer Lett., 5:261-268 (1978).
23. J. B. Vaught, H. L. Gurtoo, N. B. Parker, R. LeBoeuf, and G. Doctor, Effects of Smoking on Benzo(a)pyrene Metabolism by Human Placental Microsomes, Cancer Res., 39:3177-3183 (1979).
24. H. J. Freedman, H. L. Gurtoo, J. Minowada, B. Paigen, and J. B. Vaught, Aryl Hydrocarbon Hydroxylase in a Stable Human B-Lymphocyte Cell Line, RPMI - 1788, Cultured in the Absence of Mitogens, Cancer Res., 39:4605-4611 (1979).
25. H. L. Gurtoo, J. Minowada, B. Paigen, N. J. Parker, and N. T. Hayner, Factors Influencing the Measurement and Reproducibility of Aryl Hydrocarbon Hydroxylase in Cultured Human Lymphocytes, J. Natl. Cancer Inst., 59:787-798 (1977).
26. R. E. Kouri, R. L. Imblum, R. G. Sosnowski, D. I. Slomiany, and C. E. McKinney, Parameters Influencing Quantitation of 3-Methylcholanthrene Induced Aryl Hydrocarbon Hydroxylase in Cultured Human Lymphocytes, J. of Environ. Pathol. and Toxicol. 2:1079-1098 (1979).
27. J. E. Gielen and D. W. Nebert, Aryl Hydrocarbon Hydroxylase in Mammalian Liver Cell Culture, III. Effects of Various Sera, Hormones, Biogenic Amines and Other Endogenous Compounds on the Enzyme Activity, J. Biol. Chem., 247:7591-7602 (1972).
28. S. A. Atlas, E. S. Vesell, and D. W. Nebert, Genetic Control of Interindividual Variations in the Inducibility of Aryl Hydrocarbon Hydroxylase in Cultured Human Lymphocytes, Cancer Res., 36:4619-4630 (1976).
29. H. L. Gurtoo, N. T. Hayner, N. B. Parker, and J. Minowada, Aryl Hydrocarbon Hydroxylase (AHH) in Cultured Lymphoid Cells from Normal Humans and Leukemia Patients, Proc. Am. Assoc. Cancer Res., 17:93 (1976).
30. B. Paigen, J. Minowada, H. L. Gurtoo, K. Paigen, N. B. Parker, E. Ward, N. T. Hayner, I. D. J. Bross, F. Bock, and R. Vincent, Distribution of Aryl Hydrocarbon Hydroxylase Inducibility in Cultured Human Lymphocytes, Cancer Res., 37:1829-1837 (1977).
31. A. Richter, D. Kadar, E. Liszka-Hagmajer, and W. Kalow, Seasonal Variation of Aryl Hydrocarbon Hydroxylase Inducibility in Human Lymphocytes in Culture, Res. Commun. Chem. Pathol. and Pharmacol., 19:453-475 (1978).
32. A. Reinberg, Aspects of Circannual Rhythms in Man, in: Circannual Clocks (E. T. Pengelley, ed.), pp. 423-505, Academic Press, Inc., New York (1974).
33. P. K. Benthin and W. F. Bousquet, Long-term Variation in Basal and Phenobarbital-Stimulated Oxidative Drug Metabolism in the Rat, Biochem. Pharmacol., 19:620-625 (1970).

34. A. Jori, E. DiSalle, and V. Stantini, Daily Rhythmic Variation and Liver Drug Metabolism in Rats, Biochem. Pharmacol., 20: 2965-2969 (1971).
35. J. M. Tredger and R. S. Chhabra, Circadian Variation in Microsomal Drug Metabolizing Enzyme Activities in Rat and Rabbit Tissues, Xenobiotica, 7:481-489 (1977).
36. A. H. Conney, Pharmacological Implications of Microsomal Enzyme Induction, Pharmacol. Rev., 19:317-366 (1967).
37. A. P. Alvares, G. Schilling, W. Levin, and R. Kuntzman, Studies on the Induction of CO-Binding Pigments in Liver Microsomes by Phenobarbital and 3-Methylcholanthrene, Biochem. Biophys. Res. Commun., 29:521-537 (1967).
38. A. Y. H. Lu, S. B. West, D. Ryan, and W. Levin, Characterization of Partially Purified Cytochromes P450 and P448 from Rat Liver Microsomes, Drug Metabol. and Dispos., 1:29-39 (1973).
39. W. Levin, A. Y. H. Lu, D. Ryan, and S. West, R. Kuntzman, and A. H. Conney, Partial Purification and Properties of Cytochromes P450 and P448 from Rat Liver Microsomes, Arch. Biochem. Biophys., 153:543-553 (1972).
40. H. J. Freedman, N. B. Parker, A. J. Marinello, H. L. Gurtoo, and J. Minowada, Induction, Inhibition, and Biological Properties of Aryl Hydrocarbon Hydroxylase in a Stable Human B-Lymphocyte Cell Line, RPMI-1788. Cancer Res., 39:4612-4619 (1979).
41. H. L. Gurtoo, N. B. Parker, B. Paigen, M. B. Havens, J. Minowada, and H. J. Freedman, Induction, Inhibition, and Some Enzymological Properties of Aryl Hydrocarbon Hydroxylase in Fresh Mitogen-Activated Human Lymphocytes, Cancer Res., 39:4620-4629 (1979).
42. J. Selkirk, R. Croy, J. Whitlock, and H. Gelboin, In vitro Metabolism of Benzo(a)pyrene by Human Liver Microsomes and Lymphocytes, Cancer Res., 35:3651-3655 (1975).
43. T. A. Stoming and E. Bresnick, Heptatic Epoxide Hydrase in Neonatal and Partially Hepatectomized Rats, Cancer Res., 34: 2810-2813 (1974).
44. H. L. Gurtoo, J. B. Vaught, A. J. Marinello, B. Paigen, T. Gessner, and W. Bolanowska, High-Pressure Liquid Chromatographic Analysis of Benzo(a)pyrene Metabolism by Human Lymphocytes from Donors of Different Aryl Hydrocarbon Hydroxylase Inducibility and Antipyrine Half-Lives, Cancer Res., 40:1305-1310 (1980).
45. G. Kellerman, M. Luyten-Kellerman, M. G. Horning, and M. Stafford, Elimination of Antipyrine and Benzo(a)pyrene Metabolism in Cultured Human Lymphocytes, Clin Pharmacol. Ther., 20:72-80 (1976).
46. E. Ward, B. Paigen, K. Steenland, R. Vincent, J. Minowada, H. L. Gurtoo, P. Sartori, and M. B. Havens, Aryl Hydrocarbon Hydroxylase in Persons with Lung or Laryngeal Cancer, Int. J. Cancer, 22:384-389 (1978).
47. B. Paigen, E. Ward, K. Steenland, M. Havens, and P. Sartori, Aryl Hydrocarbon Hydroxylase Inducibility is Not Altered in Bladder Cancer Patients or Their Progeny, Int. J. Cancer, 23: 312-315 (1979).

48. B. Paigen, H. L. Gurtoo, J. Minowada, L. Houten, R. Vincent, K. Paigen, N. B. Parker, E. Ward, and N. T. Hayner, Questionable Relation of Aryl Hydrocarbon Hydroxylase to Lung Cancer Risk, New Engl. J. Med., 297:346-350 (1977).
49. B. Paigen, E. Ward, K. Steenland, L. Houten, H. L. Gurtoo, and J. Minowada, Aryl Hydrocarbon Hydroxylase in Cultured Lymphocytes of Twins, Am. J. Hum. Genet., 30:561-571 (1978).
50. T. Okuda, E. S. Vesell, E. Plotkin, R. Tarone, R. C. Bast, and H. V. Gelboin, Interindividual and Intraindividual Variation in Aryl Hydrocarbon Hydroxylase in Monocytes from Monozygotic and Dizygotic Twins, Cancer Res., 37:3904-3911 (1977).
51. E. Trell, R. Korsgaard, B. Hood, P. Kitzing, G. Norden, and B. C. Simonsson, Aryl Hydrocarbon Hydroxylase Inducibility and Laryngeal Carcinoma, Lancet, 2:140 (1976).
52. T. Han and H. Takita, Immunologic Impairment in Bronchogenic Carcinoma: A Study of Lymphocyte Response to Phytohemagglutinin Cancer, 30:616-620 (1972).
53. B. Paigen, H. L. Gurtoo, E. Ward, J. Minowada, L. Houten, R. Vincent, N. B. Parker, and J. Vaught, Human Aryl Hydrocarbon Hydroxylase and Cancer Risk, in: Carcrinogenesis, Vol. 3, Polynuclear Aromatic Hydrocarbons (P. W. Jones and R. I. Freudenthal, eds.), Raven Press, New York, pp. 429-438 (1978).
54. H. A. Guirgis, H. T. Lynch, T. Mate, R. E. Harris, I. Wells, L. Caha, J. Anderson, K. Maloney, and L. Rankin, Aryl Hydrocarbon Hydroxylase Activity in Lymphocytes from Lung Cancer Patients and Normal Controls, Oncology, 33:105-109 (1976).
55. C. G. Gahmberg, A. Sekki, T. U. Kosunen, L. R. Holsti, and O. Makela, Induction of Aryl Hydrocarbon Hydroxylase Activity and Pulmonary Carcinoma, Int. J. Cancer, 23:302-305 (1979).
56. J. R. Jett, H. L. Moses, E. L. Branum, W. F. Taylor, and R. S. Fontana, Benzo(a)pyrene Metabolism and Blast Transformation in Peripheral Blood Mononuclear Cells from Smoking and Nonsmoking Populations and Lung Cancer Patients, Cancer, 41:192-200 (1978).
57. R. Prasad, N. Prasad, J. E. Harrell, J. Thornby, J. H. Laem, P. T. Hudgins, and J. Tsuang, Aryl Hydrocarbon Hydroxylase Inducibility and Lymphoblast Formation in Lung Cancer Patients, Int. J. Cancer, 23:316-320 (1979).
58. D. B. Clayson and E. H. Cooper, Cancer of the Urinary Tract, Advances Cancer Res., 13:271-381 (1970).
59. C. C. Irving, Conjugates of N-Hydroxy Compounds, in: Metabolic Conjugation and Metabolic Hydrolysis (W. H. Fishman, ed.), Vol. 1, Academic Press, New York, pp. 53-119 (1973).
60. F. F. Kadlubar, J. A. Miller, and E. C. Miller, Hepatic Microsomal N-Glucuronidation and Nucleic Acid Binding N-Hydroxyarylamines in Relation to Urinary Bladder Carcinogenesis, Cancer Res., 37:805-814 (1977).
61. J. F. Fraumeni and L. B. Thomas, Malignant Bladder Tumors in a Man and His Three Sons, J. Amer. Med. Assoc., 201:507-509 (1967).

62. D. L. McCullough, D. L. Lamm, A. P. McLaughlin, and R. F. Gittes, Familial Transitional Cell Carcinoma of Bladder, Urology, 113:629-635 (1975).
63. J. L. Radomski and E. Brill, Bladder Cancer Induction by Aromatic Amines: Role of N-Hydroxy Metabolites, Science, 167: 992-993 (1970).
64. M. R. Juchau, Mechanisms of Drug Biotransformation Reactions in the Placenta, Fed. Proc., 31:48-51 (1972).
65. J. M. Rice, S. R. Joshi, R. E. Shenefelt, and M. L. Wenk, Transplacental Carcinogenic Activity of 7,12-Dimethyl Benz(a)-anthracene, in: Carcinogenesis: A Comprehensive Surgery, Vol. 3 (P. W. Jones and R. I. Freudenthal, eds.), Raven Press, New York, pp. 413-422 (1978).
66. L. Tomatis, Increased Incidence of Tumors in F_1 and F_2 Generations from Pregnant Mice Injected with a Polycyclic Hydrocarbon, Proc. Soc. Exp. Biol. Med., 119:743-747 (1965).
67. O. M. Bulay and L. W. Wattenberg, Carcinogenic Effects of Subcutaneous Administration of Benzo(a)pyrene during Pregnancy on the Progeny, Proc. Soc. Exp. Biol. Med., 135:84-86 (1970).
68. Committee on Environmental Hazards, American Academy of Pediatrics, Effects of Cigarette Smoking on the Fetus and Child, Pediatrics, 57:411-413 (1976).
69. D. W. Nebert, J. Winker, and H. V. Gelboin, Aryl Hydrocarbon Hydroxylase in Human Placenta from Cigarette Smoking and Non-Smoking Women, Cancer Res., 29:1763-1769 (1969).
70. O. Pelkonen and H. Saarni, Unusual Patterns of Benzo(a)pyrene Metabolites and DNA-Benzo(a)pyrene Adducts Produced by Human Placental Microsomes *in vitro*, Chem.-Biol. Interactions, 30: 287-296 (1980).
71. J. Kapitulnik, W. Levin, A. Y. H. Lu, R. Morecki, P. M. Dansette, D. M. Jerina, and A. H. Conney, Hydration of Arene and Alkene Oxides by Epoxide Hydrase in Human Liver Microsomes, Clin. Pharmacol. Therap., 21:158-165 (1977).
72. H. L. Gurtoo, C. J. Williams, K. Gottlieb, A. I. Mulhern, L. Caballes, J. B. Vaught, and A. J. Marinello, Population Distribution of Placental Benzo(a)pyrene Metabolism in Smokers, Int. J. Cancer, 31:29-37 (1983).

C-BAND VARIATION IN HYPERTENSIVE MEN

Nils Mandahl* and Ronald W. Pero†

*Department of Tumor Cytogenetics

and

†Department of Biochemical and Genetical Ekotoxicology
University of Lund
Wallenberg Laboratory, Fack 7031
S-220 07 Lund 7, Sweden

INTRODUCTION

Studies of chromosomes of various animal species by banding-, DNA- and conventional cytogenetical techniques, have demonstrated that there is a very strong, but not complete, relationship between positive C-bands, highly repeated DNA sequences and constitutive heterochromatin. The frequent findings of tandemly repeated simple DNA sequences in C-heterochromatin have led to the general idea that no structural genes reside in this type of chromatin. Another common feature of constitutively heterochromatic chromosome regions is the interchromosomal highly variable size, which contrasts to the typically invariable euchromatin. However, the heterochromatic variants are stable and obey the Mendelian laws of inheritance.

In human chromosomes the C-bands are primarily located to the centromeric regions with the notable exception of the Y chromosome. The most extensive studies of C-band polymorphism in humans concern chromosomes No. 1, 9, and 16, the autosomes with the largest C-bands and the most conspicuous variants. C-band variants, both large and small, have in several studies been found in apparently phenotypically normal individuals [1, 2]. However, some workers have reported a higher frequency of heteromorphisms, partial inversions or extreme variants of a particular chromosome in samples of individuals suffering from different kinds of clinical abnormalities as compared to healthy control samples. Correlations have for example been found to malignant diseases [3, 4, 5], increased risk of chromosome aberrations in the progeny [6], mental retardation in children with multiple malformations [7] and nondisjunction [8]. It should also be stressed that other workers have failed to find such discrepancies between affected and healthy persons. The lack of any association concerns for example mental retardation [9, 10] and some malignant diseases [11, 12].

The biological and clinical significance of C-band polymorphisms is still obscure. Those relationships, between particular C-band variants and disorders, that have been reported may be interpreted either as causative or as coincidental. Hypothesis have been advanced that relate the effects of polymorphism to increased susceptibility to chromosome breaking agents [5] or to general chromosome instability [3].

In the present study C-band variation of chromosomes No. 1, 9, and 16 has been studied in a human male material characterized with respect to blood pressure level. Individuals with high blood pressure are at elevated risk as cardiovascular disease and DNA damage are concerned [13]. The specific aim was to find out whether persons with hypertension have the same pattern of C-band polymorphism as normotensive persons or not.

MATERIAL

For this study a sample of 53 males has been selected from a population in Dalby, Sweden, which has been monitored for blood pressure for several years. Based on the diagnosis, which were in accordance to the WHO recommendations, the 53 males were divided into three groups: normotensives (group A, 24 males), suspected hypertensives (group B, 4 males) and essential hypertensives (group C, 25 males). Their ages ranged from 51 to 69 years, the mean age of the three different groups being 63.4, 58.8, and 59.9 years respectively.

The vast majority of the hypertensives were medicated, to lower their blood pressure, at the time of sampling blood for cytogenetic analysis. However, the sorting of the material into the three group was made with respect to the original diagnosis. Although data of many different parameters have been collected for the sampled males, the prime consideration in the present study has been the relation of the C-band size in chromosomes No. 1, 9, and 16 to the diagnosed blood pressure status. In the analysis of these data from the three different groups of patients (A-C) age, smoking habits and blood pressure at the time of blood sampling were also regarded.

METHODS

Conventional microcultures were initiated from peripheral whole blood. Air-dried chromosome preparations, made after 3 days of culture, were subjected to a CBG-technique for C-staining.

The C-bands were quantified by measurements on 5-7 photographic prints from each individual according to the following principles: the size of each individual C-band of chromosomes No. 1, 9, and 16 was expressed as percentage of the summarized length of the euchromatic regions of these three chromosome pairs ($C \cdot 100/\Sigma$ Eu). The individual C-bands were related to Σ Eu rather than to the size of the euchromatic portion of the chromosome to which the C-band belong. This was done in order to have the C-band sizes of all three chromosome pairs related to the same standard and to reduce the variation that may be caused by eventual diverging elongation of chromosomes as can sometimes be observed in the periphery of metaphase plates.

Whenever two homologous C-bands are close in size it could well be that the largest band in every cell not always belong to the same chromosome (maternal or paternal). Therefore, for each chromosome pair the mean sizes of the largest and the smallest bands were subjected to statistical analysis [14], to decided whether the C-bands of the two homologues were of different size or not. Significantly different C-bands were described by their individual (large and small) relative values, whereas those that did not differ significantly were both given the same value, i.e., the mean of all large and small values.

This investigation was done as a double blind study, so the code was not broken until the measurements were completed.

RESULTS

The distribution of individual C-bands of chromosomes No. 1, 9, and 16 to different size classes is shown in Table 1. When groups A and C are compared they have two differences in common for all three chromosomes. One is that the C-bands of group C have a wider range in size, both as smaller and larger variants are concerned. The other similarity is that in group C the smaller variants are relatively more abundant than in group A. This is particularly the case for chromosome No. 9. In group C more than 50% of these chromosomes have a C-band size of 2.9 relative units or less, whereas C-bands of that size are not found at all in group A. Group B, although a very small sample, is similar to group C in having some of the small C-band variants of chromosome No. 9.

TABLE 1. Frequencies (in %) of C-Bands Belonging to Different Size Classes. Group A = Normotensives, Group B = Suspected Hypertensives and Group C = Essential Hypertensives

Relative	Chromosome 1				Chromosome 9				Chromosome 16			
size (C/Eu)	A	B	C	Total	A	B	C	Total	A	B	C	Total
1.0 - 1.4	-	-	-	-	-	-	-	-	-	-	2.0	0.9
1.5 - 1.9	-	-	-	-	-	-	-	-	8.3	75.0	6.0	6.6
2.0 - 2.4	-	-	2.0	0.9	-	25.0	10.0	6.6	25.0	-	34.0	33.0
2.5 - 2.9	2.1	-	6.0	3.8	-	12.5	42.0	20.8	50.0	-	50.0	46.2
3.0 - 3.4	2.1	37.5	8.0	7.5	41.7	37.5	14.0	28.3	16.7	25.0	6.0	12.3
3.5 - 3.9	33.3	-	36.0	32.1	41.7	25.0	20.0	30.2	-	-	2.0	0.9
4.0 - 4.4	43.8	12.5	22.0	31.1	16.7	-	12.0	13.2	-	-	-	-
4.5 - 4.9	10.4	37.5	16.0	15.1	-	-	2.0	0.9	-	-	-	-
5.0 - 5.4	8.3	12.5	8.0	8.5	-	-	-	-	-	-	-	-
5.5 - 6.0	-	-	2.0	0.9	-	-	-	-	-	-	-	-
Total number of chromosomes	48	8	50	106	48	8	50	106	48	8	50	106

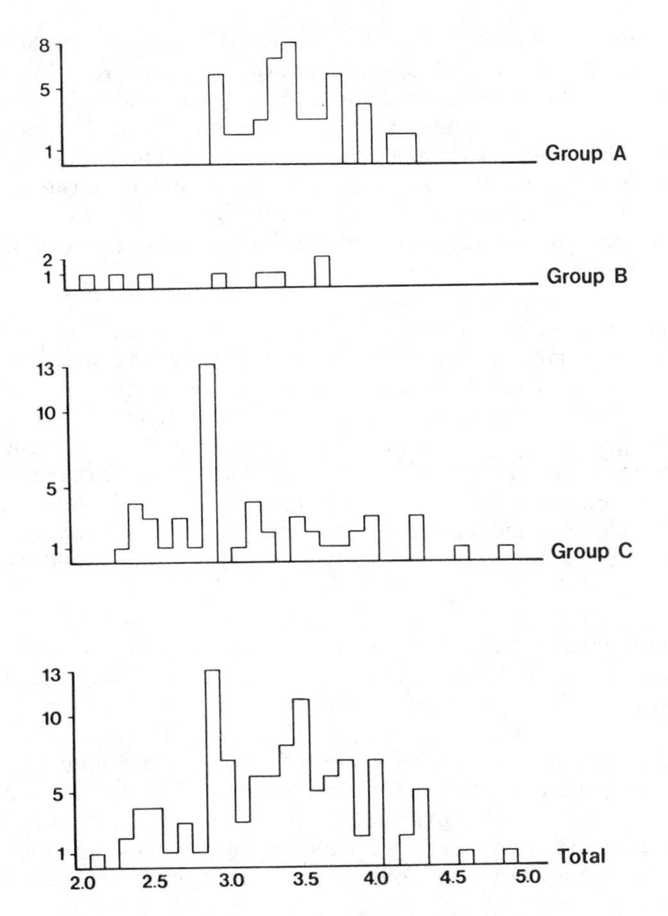

Fig. 1. Size distribution of C-bands of chromosome No. 9 in normotensives (A), suspected hypertensives (B), essential hypertensives (C) and the total material. Abscissa: relative C-band size. Ordinate: absolute numbers of chromosomes.

The limits between the different size classes were arbitrarily chosen, and as regards chromosome No. 9 the border between 2.9 and 3.0 makes the difference between groups A and C look sharper than had been the case if another class limit had been chosen. Certainly, the total absence of C-bands with the value of 3.0 in group C and of 2.9 in group A is purely by chance. The detailed distribution of the C-band size of chromosome No. 9 for groups A, B, and C is shown diagrammatically in Fig. 1. This figure shows clearly that although there may exist a greater overlap in the 2.9/3.0 region than actually is demonstrated, there are evidently much more small variants in group C than in group A. Out of the 25 hypertensive males 17 have at least one chromosome No. 9 with a C-band sized 2.9 relative units or smaller and 9 of these have such small C-bands in both homologues.

In group C the mean of all C-bands of chromosome No. 9 is significantly ($P < 0.001$) smaller than it is in group A. This is not the case for the C-bands of chromosomes No. 1 and 16. If the C-bands of chromosome No. 9 are separated into large and small bands for each individual, the mean of the small but not of the large bands differ significantly ($P < 0.001$) between groups A and C. The corresponding comparison of the small and large C-bands of chromosomes No. 1 and 16 do not show any significant differences between the same two groups.

Partly as a consequence of the high frequency of small C-band variants of chromosome No. 9 in groups C there is also a higher frequency of heteromorphic chromosome pairs (52%) than in group A (21%). The corresponding frequencies for chromosomes No. 1 and 16 are 33% and 29% for group A and 44% and 36% respectively for group C. The difference between groups A and C is weakly significant ($0.05 > P > 0.01$) for chromosome No. 9 but not for Nos. 1 and 16. Taking into account all three chromosome pairs group A has a mean number of 0.8 heteromorphisms per individual which is to be compared to 1.3 for group C.

The response of the hypertensives to medication varied from one individual to another. Some of them maintained a rather high blood pressure, whereas others were brought down to the normal level (range of diastolic blood pressure after medication: 80-110). No particular relations of the C-band size of chromosome No. 9 to the blood pressure at the time of blood sampling or to smoking habits could be detected in the present material. However, the non-smokers which had never smoked were too few, both in group A and C, compared to smokers and non-smokers that had stopped smoking, to draw any safe conclusions.

A TENTATIVE HYPOTHESIS

As mentioned earlier the significance of associations of particular C-band variants to some disorders or traits is essentially unknown. It may well be that different explanations must be considered to account for such associations. The one thing in common of such disorders like malignancy, reproductive failure and high blood pressure is that the components of the genetic determination are complex and poorly if at all understood. This fact makes it hard to verify any hypothesis by experimental evidence. Nonetheless, one possible but yet tentative hypothesis will be presented here.

In short this hypothesis postulates that there is no causative relationship between specific heteromorphisms or variants and particular disorders, but the variant serves just as a for the presence of a particular allele or set of alleles which are directly or indirectly involved in the determination of the disorder.

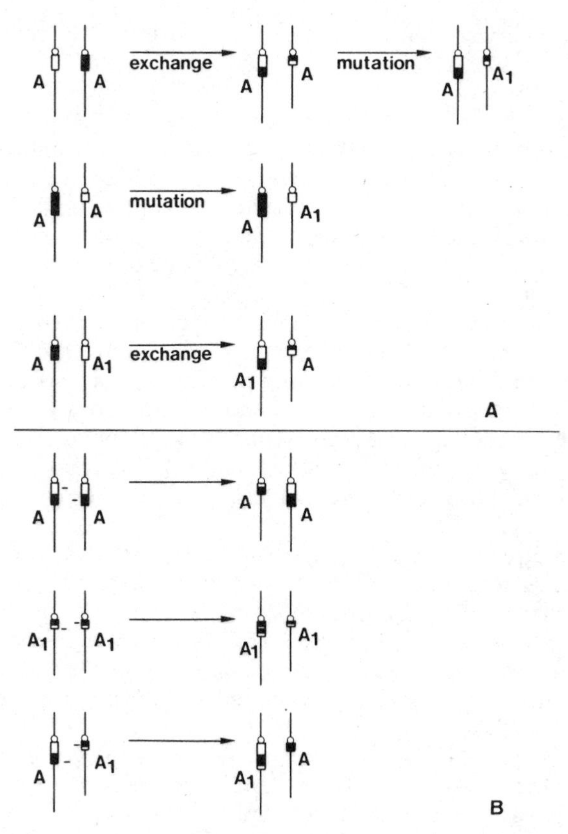

Fig. 2. Events resulting in a C-band variant being a marker for a particular allele. The drawings represent homologous chromosomes with a proximal C-band (filled and unfilled boxes) close to the centromere (small ring). The letter A denotes a gene located close to the C-band. The upper part (A) demonstrates three possibilities how a particular allele may become coupled to a small or large C-band variant by events of chromosomal exchange and/or mutation. In the lower part (B) some possible consequences of subsequent exchanges, starting with the situation depicted in the upper line of (A) are shown. Exchanges in each of the two homozygotes and in the heterozygote may result so that both alleles A and A_1 can be present together with a small or a large C-band.

The following considerations are pertinent to the hypothesis: The size of heterochromatic regions is essentially stable, but occasionally mutational events such as unequal crossing over or unequal exchange give rise to new size variants [1, 15, 16].

Meiotic crossing over does not take place in constitutive heterochromatin [17, 18].

Meiotic crossing over is probably greatly reduced in the euchromatic (gene containing) region close to a heteromorphic heterochromatin segment [19-21]. The level of reduction or prevention of crossing over may be dependent on the distance between the heterochromatin and the gene locus and also on the size difference between the heterochromatic regions of the two homologous chromosomes.

Now suppose that a particular allele is located close to the heterochromatin in a chromosome with a large or small C-band variant. This situation can result from several different alternative events (Fig. 2) including mutation and an exchange process. Once the mutant allele (A_1) has been coupled to a specific C-band variant (say a small one), this variant will serve as a marker for the presence of A_1 as a result of their strong linkage.

Subsequent premeiotic exchanges or occasional meiotic crossing over may result in large variants being linked to A_1 and small variants to the original A allele. But still, small variants linked to allele A_1 will be more frequent. This means that some individuals may be affected by the trait without having the specific C-band variant and others may not be affected despite they carry the variant. Besides, different human populations may differ from each other in displaying a particular association between a variant and a trait, due to dissimilarities in alleles and variants. Significant interracial differences in allele frequencies (e.g., blood group antigens) as well as C-band variants [22-24] have been demonstrated.

The allele has to be perpetuated in the population by having no larger selection pressure against itself than the "normal" allele has. If any selective pressure is directed against the mutant allele it is supposed to be operating after reproductive age, i.e., the fecundity of the carrier is not affected.

Thus the presence of a specific variant may have no causal relationship to the trait studied (i.e., the variant neither cause the trait nor is it caused by the trait), but may merely indicate the presence of a specific allele that actually dictates the phenotypic expression. The consequences of variable expressivity, penetrance and recessiveness/dominance on this hypothesis will not be considered here.

CONCLUDING REMARKS

The present results show that there is a significant difference in the C-band size of chromosome No. 9 between hypertensive and normotensive males in a sample from the south of Sweden. No corresponding differences could be detected in chromosomes No. 1 and 16. Although few in number the males diagnosed as suspected hypertensives closely resembled the hypertensives. Thus the work represents a new example of an association between a specific C-band variant (a small variant in chromosome No. 9) and a specific trait (essential hypertension) in man. The tentative hypothesis presented in the previous section may account for such an association, as well as for others of the same kind.

To elucidate the validity of this hypothesis it may seem less favorable to study a parameter like high blood pressure since the genetic component(s) of the determination of arterial blood tension is poorly understood. Available evidence agree well with the hypothesis that arterial pressure is polygenically inherited. In addition, it is commonly agreed that environmental factors play an important role in the determination of the blood tension. The nature of hereditary-environmental interactions are essentially unknown. Certainly, however, some humans have an inherited predispostion to a high arterial blood tension. A cytogenetic/electrophoretic study of C-variants and an allelic gene located close to heterochromatin in familial materials, might cast some informative light on the relevance of the tentative hypothesis proposed in this paper. A prospective study of C-band variants in children of normo- and hypertensive parents is also justified.

REFERENCES

1. W. H. McKenzie and H. A. Lubs, Human Q and C chromosomal variations: distribution and incidence, Cytogenet. Cell Genet., 14: 97-115 (1975).
2. H. Müller, H. P. Klinger, and M. Glasser, Chromosome polymorphism in a human newborn population, II. Potentials of polymorphic chromosome variants for characterizing the idiogram of an individual, Cytogenet. Cell Genet., 15:239-255 (1975).
3. N. B. Atkin, Chromosome 1 heteromorphism in patients with malignant disease: a constitutional marker for a high-risk group?, Brit. Med. J., I:358 (1977).
4. N. B. Atkin and V. J. Pickthall, Chromosomes 1 in 14 ovarian cancers. Heterochromatin variants and structural changes, Hum. Genet., 38:25-33 (1977).
5. F. Shabtai and I. Halbrecht, Risk of malignancy and chromosomal polymorphism: a possible mechanism of association, Clin. Genet., 15:73-77 (1979).
6. J. Nielsen, U. Friedrich, A. B. Hreidarsson, and E. Zeuthen, Frequency of 9qh+ and risk of chromosome aberrations in the progeny of individuals with 9qh+, Humangenetik, 21:211-216 (1974).
7. J. Kunze and G. Mau, A_1 and C_9 marker chromosomes in children with combined minor and major malformations, Lancet Feb. 1, 273 (1975).

8. J. H. Ford and P. Lester, Chromosomal variants and nondisjunction, Cytogenet. Cell Genet., 21:300-303 (1978).
9. J. Nielsen, U. Friedrich, and A. B. Hreidarsson, Frequency and genetic effect of 1qh+, Humangenetik, 21:193-196 (1974).
10. A. T. Tharapel and R. L. Summitt, Minor chromosome variations and selected heteromorphisms in 200 unclassifiable mentally retarded patients and 200 normal controls, Hum. Genet., 41: 121-130 (1978).
11. S. Kivi and A.-V. Mikelsaar, Q- and C-band polymorphisms in patients with ovarian or breast carcinoma, Hum. Genet., 56: 111-114 (1980).
12. L. Aguilar, R. Lisker, L. Ruz, and O. Mutchinick, Constitutive heterochromatin polymorphisms in patients with malignant diseases, Cancer, 47:2437-2439 (1981).
13. R. W. Pero and Å. Nordén, Mutagen sensitivity in peripheral lymphocytes as a risk indicator, Environ. Res., 24:409-424 (1981).
14. B. Matérn and M. Simak, Statistical problems in karyotype analysis, Hereditas, 59:280-288 (1968).
15. A. P. Craig-Holmes, F. B. Moore, and M. W. Shaw, Polymorphism of human C-band heterochromatin. II. Family studies with suggestive evidence for somatic crossing over, Am. J. Hum. Genet., 27:178-189 (1975).
16. R. B. Phillips, Inheritance of Q and C band polymorphisms, Can. J. Genet. Cytol., 19:405-413 (1977).
17. B. John, Myths and mechanisms of meiosis, Chromosoma, 54:295-328 (1976).
18. Y. Hotta and H. Stern, Absence of satellite DNA synthesis during meiotic prophase in mouse and human spermatocytes, Chromosoma, 69:323-330 (1978).
19. J. Forejt, Centromeric heterochromatin polymorphism in the house mouse. Evidence from inbred strains and natural populations, Chromosoma, 43:187-201 (1973).
20. G. L. G. Miklos and R. N. Nankivell, Telomeric satellite DNA functions in regulating recombination, Chromosoma, 56:143-167 (1976).
21. Ú. Árnason, I. F. Purdom, and K. W. Jones, Conservation and chromosomal localization of DNA satellites in balenopterid whales, Chromosoma, 66:141-159 (1978).
22. J. Matsuura, M. Mayer, and P. A. Jacobs, A cytogenetic survey of an institution for the mentally retarded, II. C-band chromosome heteromorphisms, Hum. Genet., 45:33-41 (1978).
23. M. V. Monsalve, B. Erdtmann, P. A. Otto, and O. Frota-Pessoa, The human Y chromosome: racial variation and evolution, Rev. Brasil. Genet., 3:433-446 (1980).
24. H. A. Lubs, S. A. Patil, W. J. Kimberling, J. Brown, M. Cohen, P. Gerald, F. Hecht, N. Myrianthopoulos, and R. L. Summitt, Q and C banding polymorphisms in 7 and 8 year old children: racial differences and clinical significance, in: Population Cytogenetics, Studies in Humans (E. B.Hood and I. H. Porter, eds.), Academic Press, New York (1977).

METHODS FOR EVALUATING THE EFFECTS OF ENVIRONMENTAL CHEMICALS ON HUMAN SPERMATOGENESIS*

Andrew J. Wyrobek

Biomedical Sciences Division
Lawrence Livermore National Laboratory
P. O. Box 5507, L-452
Livermore, California 94550

1. SUMMARY

Sperm tests provide a direct and effective way of identifying chemical agents that induce spermatogenic damage in man. Four human sperm tests are available: sperm count, motility, morphology (seminal cytology), and the F-body test. These sperm tests have numerous advantages over other approaches for assessing spermatogenic damage, and they have already been used to assess the effects of at least 85 different occupational, environmental, and drug-related chemical exposures. When carefully controlled, seminal cytology appears to be statistically more sensitive than the other human sperm tests and should be considered an integral part of semen analysis when assessing induced spermatogenic damage.

Human sperm studies have complex requirements and, before sampling, careful consideration should be given to exposure details, group size and makeup, as well as animal and human data that indicate spermatogenic effects. Several study designs are possible and should include questionnaires covering medical and reproductive histories as well as known confounding factors. Animal sperm tests, such as the mouse morphology test, may be used to identify the toxic

*A similar version of this paper is published in the Proceeding of Symposium on Research Needs for Evaluation of Health Effects Toxic Waste Dumps. Research Triangle Park, North Carolina, October 1981. Work performed under the auspices of the U.S. Department of Energy by the Lawrence Livermore National Laboratory under contract number W-7405-ENG-48, and by the U.S. Environmental Protection Agency.

components of a complex mixture. Animal tests may also help assess the chemical effects on fertility and reproductive outcome in cases when human data are incomplete. Further efforts are needed in these areas to (a) develop improved human sperm tests sensitive to induced spermatogenic damage, (b) develop improved animal models of induced spermatogenic damage, (c) understand the relationships among sperm changes, fertility, and reproductive outcome, and (d) develop sperm tests with express mutational endpoints.

2. INTRODUCTION

Studies with numerous chemical agents in a variety of mammalian species have shown that sperm anomalies can be used as indicators, and in certain instances, as dosimeters of chemically induced spermatogenic damage [1, 2]. Various other approaches have also been proposed to assess chemically induced spermatogenic dysfunction including testicular biopsies [3], questionnaire surveys [4], and blood levels of gonadotrophins [5]. Sperm tests are noninvasive, generally less expensive, require smaller sample sizes, and are sensitive to small changes [1, 2, 5].

A recent survey of the literature [2] showed that sperm tests have been more widely used to assess the effects of chemical exposures in man than was generally suspected; more than 100 papers involving some 89 different chemical exposures have been published. This paper briefly describes the methods and applications of the four most common human sperm tests, compares their relative sensitivities, and suggests guidelines for undertaking a new human sperm study in men exposed to toxic agents. The paper also discusses the role of animal studies, the implication of semen findings for reproductive outcome, and future research needs in these areas.

3. DESCRIPTION OF HUMAN SPERM TESTS

Human semen tests have a long history in the diagnosis of infertility [5]. Thus, it is not surprising that the early attempts to assess altered spermatogenic function in men exposed to chemicals involved measuring changes in the sperm parameters commonly used in fertility diagnosis, such as sperm concentration (counts), motility, and morphology (seminal cytology). The following is a very brief description of these methods:

Sperm count is usually reported as the number of sperm per ml of ejaculate (or as the total number of sperm ejaculated) as determined by hemocytometer [6]. The measurement is technically easy, and automated methods are also available. However, interpretation of results may be confounded by a number of factors, such as variable continence time before ejaculation, and collection of an incomplete ejaculate [7].

Sperm motility is the swimming ability of the sperm, and has been expressed in a large variety of ways [6]. Although motility may be one of the best performance evaluations of spermatogenic function in relation to fertility, it is also very sensitive to time and temperature after collection [8]. Thus semen motility is very difficult to measure in a field study, especially when samples are collected at home. Considerable emphasis has been put on automated and quantitative methods for assessing sperm motility [9].

Sperm morphology (also referred to as seminal cytology) is the visual assessment of the shapes of ejaculated sperm. Although sperm-head shape is usually emphasized, some assessments also incorporate midpiece and tail abnormalities. In general, there has been little agreement in the definition of normal shapes or in the categories of abnormal shapes. This has resulted in much interlaboratory and interscorer variability [10, 11]. However, studies of MacLeod [12], David et al. [13], Eliasson [14], and others have shown that quantitative approaches to the visual assessment of morphology can be used with considerable success. These assessments are usually made by using smears that are air-dried, fixed, and stained with a modified Papanicolaou method [15]. Sperm can be systematically assigned to shape categories. In evaluating the effects of exposure, slides of controls should be concurrently analyzed with slides of exposed men in a blind-study design. Normal ranges have been established for several unexposed populations.

We have developed a human morphology test by describing 10 classes of sperm-head shapes [16, 17] and classifying 500 sperm per individual. Through the intermittent use of coded standard slides, we have been able to assure constancy in the visual criteria for sperm morphology over a period of many years. Our experience with this test shows that visual scoring criteria can be very objective (unpublished data). We have applied this method to men occupationally exposed to carbaryl [17] and anesthetic gases [16]. The carbaryl workers showed higher proportions of sperm with shape abnormalities than controls but no differences in sperm counts. No effects of anesthetic gases on sperm were observed. In another study, men exposed to cancer-chemotherapeutic agents showed drug-related decreases in sperm counts and increases in sperm-shape abnormalities [18].

The F-body test is based on scoring the frequency of fluorescent spots in human sperm stained with quinacrine dye. Based on studies in somatic cells, it is thought that these spots represent Y chromosomes [19]. The Y-body test scores the frequency of sperm with two spots, which are thought to represent sperm with 2 Y chromosomes due to meiotic nondisjunction [20], although this remains uncertain. Unlike the other sperm tests (counts, motility, and morphology), the F-body test has no direct counterpart in the mouse or other laboratory animals. The Y-chromosomal fluorescence after quinacrine staining seems

TABLE 1. Effects of Occupational and Environmental Chemicals on Human Sperm

Agents With Adverse Effects	Agents suggestive of Adverse Effects	Agents with No Apparent Adverse Effects
Carbon disulfide	Carbaryl	Anesthetic gases
Dibromochloropropane	Kepone	Epichlorohydrin
Dibromochloropropane + ethylene dibromide		Glycerine production compounds
Lead		Polybrominated biphenyls
Toluene diamine + dinitrotoluene		

Table entries are based on studies of sperm counts, motility, morphology, and double Y-bodies. The assignment of individual agents to columns is based on the data provided in the papers reviewed by the Human Sperm Reviewing Committee of the U.S. Environmental Protection Agency (EPA) GENE-TOX Program (2). These entries are generally based on few studies and may be expected to change as more data become available.

to be unique to man and certain apes [21]. However, it should be noted that the field vole, _Microtus oeconomus_, has a unique distribution of heterochromatin, which allows visualization of the X and Y chromosome in spermatids and possibly testicular sperm [22]. Studies with several chemical agents suggest that this system may be a useful animal model for studying the induction of sex-chromosomal nondisjunction in male germ cells.

For the analyses of F-bodies in human sperm, air-dried smears can be fixed, stained, and sperm scored under a fluorescent microscope [19, 20]. The number of sperm scored depends on the statistical precision required. Each sperm is scored as OF (showing IF fluorescent body), 1F (showing fluorescent body), and 2F (showing two fluorescent bodies). We have developed the method so that we can repeatedly visualize approximately 50% of the sperm with a

TABLE 2. Effects of Experimental and Therapeutic Drugs on Human Sperm Agents or Combinations of Agents with Adverse Effects

- Acridinyl anisidide
- Adriamycin
- Aspartic acid
- Clorambucil
- Clorambucil + mechlorethamine + azathioprine
- Clomiphene citrate
- Cyclophosphamide
- Cyclophosphamide + colchicine
- Cyclophosphamide + prednisone
- Cyclophosphamide + prednisone + azathioprine
- CVP (cyclophosphamide + vincristine + prednisone)
- CVPP (cyclophosphamide + vincristine + procarbazine + prednisone)
- Cyproterone acetate
- Danazol + methyl testosterone
- Danazol + testosterone enanthate
- Enovid
- Gossypol
- Leutineizing hormone releasing factor agonist
- Medroxyprogesterone acetate
- Medroxyprogesterone acetate + testosterone enanthate
- Medroxyprogesterone acetate + testosterone propionate
- Megestrol acetate + testosterone
- Metanedienone
- Methotrexate
- MOPP (Mechlorethamine + vincristine + procarbazine + prednisone)
- MVPP (Mechlorethamine + vinblastine + prednisolone + procarbazine)
- Norethandrolone
- Norethindrone
- Norethindrone + norethandrolone + testosterone
- Norgestrel + testosterone enanthate
- Norgestrienone + testosterone
- Prednisolone
- Propafenon
- R-2323 + testosterone
- Sulphasalazine
- Testosterone
- Testosterone cyclopentyl-propionate
- Testosterone enanthate
- Testosterone propionate
- VACAM (Vincristine + adriamycin + cyclophosphamide + actinomycin D + medroxyprogesterone acetate)
- WIN 13099
- WIN 13099 + diethylstilbestrol
- WIN 17416
- WIN 18446

Table entries are based on studies of sperm counts, motility, morphology, and double Y-bodies. The assignment of individual agents to columns is based on the data provided in the papers reviewed by the Human Sperm Reviewing Committee of the U.S. Environmental Protection Agency (EPA) GENE-TOX Program (2). These entries are generally based on few studies and may be expected to change as more data become available.

TABLE 3. Effects of Experimental and Therapeutic Drugs on Human Sperm

Agents Suggestive of Adverse Effects	Agents With no Apparent Adverse Effects
Centrochroman	Bromocriptine
Cimetidine	Lysine
Colchicine	Methyltestosterone
Diethylstilbestrol	Niridazole
Methadone	Norethindrone + testosterone
Metronidazole	Orinthine
Nitrofurantoin	Tryptophan
Norethandrolone + testosterone	WIN 59,491
Trimeprimine	

Table entries are based on studies of sperm counts, motility, morphology, and double Y-bodies. The assignment of individual agents to columns is based on the data provided in the papers reviewed by the Human Sperm Reviewing Committee of the U.S. Environmental Protection Agency (EPA) GENE-TOX Program (2). These entries are generally based on few studies and may be expected to change as more data become available.

single fluorescent body (unpublished data). The F-body test is very new, only a few populations of exposed men have been analyzed [2], and its relationship to chromosomal aneuploidy remains uncertain.

4. APPLICATIONS OF HUMAN SPERM TESTS

The above methods have been applied to assess spermatogenic function in at least 89 different groups of chemically exposed men [2]. Tables 1, 2, 3, and 4 categorizes these agents into occupational and environmental chemicals (Table 1), experimental and therapeutic drugs (Tables 2 and 3), and personal drug use (Table 4). Details of the studies surveyed to generate these tables and the decision criteria used to classify each agent as one (a) with adverse

TABLE 4. Effects of Personal Drug Use on Human Sperm

Agents With Adverse Effects	Agents suggestive of Adverse Effects	Agents with No Apparent Adverse Effects
Alcoholic beverages (chronic alcoholism)	Tobacco smoke	None
Marijuana		

Table entries are based on studies of sperm counts, motility, morphology, and double Y-bodies. The assignment of individual agents to columns is based on the data provided in the papers reviewed by the Human Sperm Reviewing Committee of the U.S. Environmental Protection Agency (EPA) GENE-TOX Program (2). These entries are generally based on few studies and may be expected to change as more data become available.

effects, (b) suggestive of adverse effects, and (c) with no apparent adverse effects are published elsewhere [2]. Several agents (not listed in these tables) have been reported to improve sperm quality in some cases [2].

5. RELATIVE SENSITIVITIES OF THE HUMAN SEMEN TESTS

Experience with agents like DBCP suggests that severe spermatogenic damage may occur at doses that show no other apparent clinical signs of toxicity. Therefore, for chemical exposures it seems unlikely that analysis of somatic cells (i.e., lymphocyte) can serve as a surrogate for effects on male germ cells.

The four human sperm tests described above are technically straightforward methods. The tests for counts, morphology, and F-bodies are parameters that do not appear to be readily affected by postejaculation technical factors [2]. Motility, however, is highly sensitive to time and temperature factors. In studies where home collection is used the motility test is not practical [8].

The statistical variations of the tests for counts, morphology, and double F-bodies were recently compared for a group of control men [17]. Sperm samples from approximately 25 men were required in both the exposed and control groups to detect a 25% change in the mean proportion of abnormally shaped sperm. F-body analyses and counts required over 40 and 200 men, respectively, to detect a 25% change in means. This comparison suggests that the human sperm morphology test is statistically more sensitive to small induced changes than the other two tests. However, it is important to realize that a chemical exposure may preferentially affect any of these sperm parameters irrespective of its statistical sensitivity. At present, data are still insufficient to predict: (a) which parameter would be most sensitive to an agent, and (b) the interdependence of the parameters. Therefore, the conservative approach for assessing changes in human spermatogenic function should include tests for sperm count, morphology, F-body, and, whenever practical, motility.

6. GUIDELINES FOR NEW HUMAN SPERM STUDIES

Laboratory analyses of sperm tests represents only a small part of a human sperm study. Depending on the exposure under consideration, these studies typically require lengthy interactions with unions, local and state government, lawyers, physicians, hospital administrators, and human subject committees before any donors are contacted. The following criteria should be considered to determine if a human sperm study is warranted:

a) Are there animal data available suggesting a spermatogenic effect of the exposure under consideration? (Animal data may exist in the literature or may be obtained using the short-term animal sperm tests.), or

b) Are there human data that suggest that there may be a problem with infertility or pregnancy outcome that could be linked to the exposed male?

If the exposure under consideration meets these criteria the following additional data should be obtained to aid in study design. (Though these points are generally self-evident, they are included because they have been often overlooked in human sperm studies.)

a) What are the demographics of the exposure (size of the exposed population, geographic location, etc.)? The size of the exposed cohort is an important consideration since, as described above, the number of men sampled will be related to the statistical sensitivity for each semen parameter. The geographic dispersion of the exposed cohort is an important cost factor as well as a possible cause of sampling biases.

b) Who was exposed (how many men, what are their ages and religious backgrounds, etc.)? Such factors can be expected to affect the participation rates.

c) What are the details of the exposure (route, duration, dose, when it occurred in relation to the proposed time of semen collection, etc.)? It is well known from animal and some human studies that the occurrence of sperm anomalies is related to exposure dose. In addition, careful attention needs to be given to the time since the last exposure; since the effects of certain agents may be reversible, false negative results may appear if the time is too long.

d) Can the exposed population be divided into dose groups? Every effort should be made to group the exposed cohort by dose estimates, since a dose-related effect is extremely strong evidence for an identification of a human testicular toxin.

A questionnaire approach to assessing human problems in fertility and reproductive outcome should also be considered. This approach may be especially effective when large numbers of people of child-bearing ages have been exposed and the major exposures were many years ago. Human sperm studies are likely to be effective when smaller numbers of men are involved (see above section) and the major exposures are suspected to be recent or ongoing.

When considering sperm studies several study designs are possible. Since between-male variability in semen characteristics is high even among fertile and presumably healthy men, rather large numbers of cooperative subjects are required to establish differences between control and exposed groups in cross-sectional studies (each individual sampled only once). Longitudinal study designs may be more appropriate when fewer men are available for sampling. In this study design, repeated semen samples are collected from each man at different times in relation to the time of exposure and compared to assess chemically induced sperm defects. Since variation of sperm morphology within an individual is considerably less than variation among individuals [23], in principle, fewer people are required for induced changes to be detected. These studies, however, have some constraints: (a) repeated samplings during a period of months and perhaps years are required; (b) samples before exposure are needed (or within days of an acute exposure before any induced effects on morphology are seen); and (c) the number of men needed for an effective study is unknown.

The effects of age, smoking, illness, medication, and other possibly confounding factors especially those involving heat exposure must be considered in the analysis of all human sperm data.

7. ROLES FOR ANIMAL SPERM TESTS

The availability of both animal and human sperm tests suggests several applications of animal studies in the assessment of chemically induced spermatotoxicity, antifertility effects, and heritable genetic abnormalities in man. First, animal sperm tests (such as mouse morphology) may be used to screen large numbers of agents to establish a ranking that sets priorities for identifying exposed men. Second, animal sperm studies may also be useful in evaluating an agent or the components of a complex mixture that is suspected of affecting human sperm (such as in an occupational or environmental exposure). Third, animal breeding tests may be used to study the relationship between changes in sperm parameters, fertility changes, and heritable consequences.

Since little is known of the quantitative relationships between induced sperm abnormalities and heritable genetic damage, indirect methods may be needed to assess the genetic risk to offspring of men who show induced sperm anomalies. By combining data from short-term mutagen bioassays (e.g., *Salmonella*/microsome assay, mammalian somatic cell mutation assays), which may demonstrate mutagenic potential, with data from animal and human sperm tests, which may demonstrate activity in the testes, we may be able to evaluate whether a mutagen is active in the testes. Further studies are needed to investigate this approach.

8. GENETIC IMPLICATIONS OF CHEMICALLY INDUCED SPERM DEFECTS

8.1. Evidence from Human Studies

Although it is generally agreed that major reductions in sperm counts and motility are linked to reduced fertility, it remains unclear which sperm parameter(s), if any, is predictive of reproductive failure or heritable genetic abnormalities. Human data on this question are very limited. Infertility is seen in patients with 100% acrosomeless, round-headed spermatozoa [24] suggesting that some types of sperm shape abnormalities are associated with infertility. Human studies with DBCP showed the strong link between reduction in sperm counts and infertility [25]. Regarding reproductive outcome, Furuhjelm et al. [26] reported that a group of fathers of spontaneous abortions showed significantly higher sperm abnormalities and lower sperm counts than fathers of normal pregnancies. This finding suggests a possible link between poor semen quality and frequency of spontaneous abortions [27]. Clearly, more human studies are needed to compare exposure of the male parent, induced sperm defects, and reproductive outcome.

8.2. Evidence from Animal Studies

Most of the studies on genetic validation of induced sperm defects have been conducted with sperm morphology in mice. Several lines of evidence link induction of abnormally shaped sperm and heritable genetic abnormalities (see Ref. 1 for a review). First, it is clear that sperm shaping and the production of abnormal sperm are polygenically controlled by autosomal as well as sex-linked genes. Second, in several studies using agents that induce sperm abnormalities, sperm abnormalities were transmitted to the male offspring of the exposed mice. Third, a brief survey of the literature suggests that the mouse sperm morphology test may be an effective prescreen for the more expensive tests of heritable germ cell mutations, such as heritable specific locus, F_1 sperm morphology, heritable translocation, and dominant lethal tests in mice. False negative responses with the mouse sperm morphology test for these tests seem to be very rare or nonexistent. However, data are needed for more chemicals before this relationship can be used with confidence. Spindle poisons that may cause nondisjunction in germ cells can also be identified with the mouse morphology test. Further studies (probably best done in mice) are needed to understand the quantitative relationships among dosage regime, appearance of abnormal sperm shapes in the semen, time between exposure and conception, fertility of the exposed male, frequency of genetically abnormal offspring, and fertility of the abnormal offspring.

9. FUTURE RESEARCH NEEDS

Sperm tests have shown considerable promise in the assessment of spermatogenic damage induced by occupational and environmental exposures. But the available human tests are first-generational and carry many biases from their original applications in fertility diagnosis. More research is needed to adapt these available sperm tests and to develop the statistical criteria for the effective assessment of chemically induced spermatogenic damage. More work is also needed to: (a) develop improved animal models of human spermatogenic damage, (b) study the relationship among changes in sperm parameters, fertility, and reproductive outcome, and (c) develop new indicators of reproductive toxicity in the male, especially indicators of heritable genetic damage.

ACKNOWLEDGMENTS

The author thanks Mr. J. Cherniak and Mrs. A. Riggs for help in editing and typing this manuscript. My continuing appreciation goes to G. Watchmaker and L. Gordon without whose assistance such reviews would not be possible.

REFERENCES

1. A. J. Wyrobek, L. A. Gordon, J. G. Burkhart, M. C. Francis, R. W. Kapp, G. Letz, H. V. Malling, J. C. Topham, and M. D. Whorton, An Evaluation of the mouse sperm morphology test and sperm tests in other animals: A Report of the U.S. Environmental Protection Agency Gene-tox Program Mutation Res., 115: 1-72.
2. A. J. Wyrobek, L. A. Gordon, J. G. Burkhart, M. C. Francis, R. W. Kapp, G. Letz, H. V. Malling, J. C. Topham, and M. D. Whorton, An evaluation of human sperm as indicators of chemically induced alterations of spermatogenic function: A report of the U.S. Environmental Protection Agency Gene-tox Program Mutation Res., 115:73-148.
3. M. D. Whorton, T. H. Milby, R. M. Krauss, and H. S. Stubbs, Testicular function in DBCP exposed workers, J. Occup. Med., 21:161-166 (1979).
4. R. J. Levine, M. J. Symons, S. A. Balogh, T. H. Milby, and M. D. Whorton, A method for monitoring the fertility of workers 2, Validation of the method among workers exposed to dibromochloropropane, J. Occup. Med., 23:183-188 (1981).
5. R. Eliasson, Clincal examination of infertile men, in: Human Semen and Fertility Regulation in Men (E. S. E. Hafez, ed.), pp. 321-331, The C. V. Mosby Co., St. Louis (1976).
6. M. Freund and R. N. Peterson, Semen evaluation and fertility, in: Human Semen and Fertility Regulation in Men (E. S. E. Hafez, ed.), pp. 344-354, The C. V. Mosby Co., St. Louis (1976).
7. D. Schwartz, A. Laplanche, P. Jouannet, and G. David, Within-subject variability of human semen in regard to sperm count, volume, total number of spermatozoa, and length of abstinence, J. Reprod. Fertil., 57:391-395 (1979).
8. A. Makler, I. Zaidise, E. Paldi, and J. M. Brandes, Factors affecting sperm motility. I. In vitro change in motility with time after ejaculation, Fertil. Steril., 31:147-154 (1979).
9. J. A. Mitchell, L. Nelson, and E. S. E. Hafez, Motility of spermatozoa, in: Human Semen and Fertility Regulation in Men (E. S. E. Hafez, ed.), pp. 89-99, The C. V. Mosby Co., St. Louis (1976).
10. M. Freund, Standards for the rating of human sperm morphology, A cooperative study, Intern. J. Fertil., 11:97-118 (1966).
11. B. Fredricsson, Morphologic evaluation of spermatozoa in different laboratories, Andrologia, 11:57 (1979).
12. J. MacLeod, Effects of environmental factors and of antispermatogenic compounds on the human testis as reflected in seminal cytology, in: Male Fertility and Sterility, Proc. Serono Symposium (R. E. Mancini and L. Martini, eds.), Vol. 5, Academic Press, New York, pp. 123-148 (1974).
13. G. David, J. P. Bisson, F. Czyglik, P. Jouannet, and C. Gernigon, Anomalies morphologiques du spermatozoïde humain 1) Propositions pour un système de classification, J. Gyn. Obst. Biol. Repr. 4, Supp. 1, 17-36 (1975).

14. R. Eliasson, Standards for investigation of human semen, Andrologie, 3:49-64 (1971).
15. G. L. Humason, in: Animal Tissue Techniques, W. H. Freeman and Company, San Francisco, pp. 456-457 (1972).
16. A. J. Wyrobek, J. B. Brodsky, L. Gordon, D. H. Moore, II, G. Watchmaker, and E. N. Cohen, Sperm studies in anesthesiologists, Anesthesiology, 55:527-532 (1981).
17. A. J. Wyrobek, G. Watchmaker, L. Gordon, K. Wong, D. Moore, II, and D. Whorton, Sperm shape abnormalities in carbaryl-exposed employees, Environ. Health Perspect., 40:255-265 (1981).
18. A. J. Wyrobek, M. daCunha, L. Gordon, G. Watchmaker, B. Gledhill, B. Mayall, J. Gamble, and M. Meistrich, Sperm abnormalities in cancer patients, Proc. Amer. Assoc. Cancer Res., 21:196 (1980).
19. P. L. Pearson, M. Bobrow, and C. G. Vosa, Technique for identifying Y chromosomes in human interphase nucleus, Nature, London, 226:78-80 (1970).
20. R. W. Kapp, Detection of aneuploidy in human sperm, Environ. Health Perspect., 31:27-31 (1979).
21. H. N. Seuànez, Chromosomes and spermatozoa of the African great apes, J. Reprod. Fertil., Suppl., 28:91-104 (1980).
22. A. D. Tates, Microtus oeconomus (Rodentia), a useful mammal for studying the induction of sex-chromosome nondisjunction and diploid gametes in male germ cells, Environmental Health Perspect., 31:151-159 (1979).
23. R. J. Sherins, D. Brightwell, and P. H. Sternthal, Longitudinal analysis of semen of fertile and infertile men, in: The Testis in Normal and Infertile Men (P. Troen and H. R. Nankin, eds.), Raven Press, New York, pp. 473-488 (1977).
24. M. Nistal, A. Harruzo, and F. Sanchez-Corral, Toratozoospermia absoluta de presentacion familiar, Espermatozoides microcefalos irregulares sin acrosoma, Andrologia, 10:234 (1978).
25. D. Whorton, R. M. Krauss, S. Marshall, and T. H. Milby, Infertility in male pesticide workers, Lancet, 2:1259-1261 (1977).
26. M. Furuhjelm, B. Jonson, and C. G. Lagergren, The quality of human semen in spontaneous abortion, Int. J. Fertil., 7:17-21 (1962).
27. E. Czeizel, M. Hancsok, and M. Viczian, Examination of the semen of husbands of habitually aborting women, Orvosi. Hetilap., 108:1591-1595 (1967).

DISCLAIMER

This document was prepared as an account of work sponsored by an agency of the United States Government. Neither the United States Government nor the University of California nor any of their employees, makes any warranty, express or implied, or assumes any legal liability or responsibility for the accuracy, completeness, or usefulness of any information, apparatus, product, or process disclosed, or represents that its use would not infringe privately owned rights. Reference herein to any specific commercial products, process, or service by trade name, trademark, manufacturer, or otherwise, does not necessarily constitute or imply its endorsement, recommendation, or favoring by the United States Government or the University of California. The views and opinions of authors expressed herein do not necessarily state or reflect those of the United State Government thereof, and shall not be used for advertising or product endorsement purposes.

NEW TECHNIQUES FOR DETECTING CHROMOSOME ABNORMALITIES IN THE GERM-LINE IN MAN

M. A. Hultén*, D. A. Laurie*, R. H. Martin†,
N. Saadallah*, B. M. N. Wallace‡

*Regional Cytogenetics Laboratory
East Birmingham Hospital
Birmingham B9 5ST, England

†Alberta Children's Hospital
Medical Genetics Clinic
1820 Richmond Road
S.W. Calgary, Alberta, Canada

‡Department of Genetics
University of Birmingham
B15 2TT, England

INTRODUCTION

There are three different ways of estimating effects of genotoxic agents harmful to future generations. Investigations may be carried out on somatic or germ line cells of the exposed population, or the offspring may be studied. This paper is devoted to the second of these options and will discuss new light microscopy techniques for analyzing chromosome abnormalities in the germ line in man. These techniques are:

1. Surface spreading of first meiotic prophase.

2. Consecutive chromosome banding of air dried first metaphase spermatocytes.

3. Sperm chromosomes visualized by *in vitro* penetration of hamster oocytes.

1. Surface Spreading of Human Meiocytes

The structure formed during pairing of homologous chromosomes at first meiotic prophase, the so-called synaptonemal complex (SC) may be visualized by surface spreading. There are now a number of papers giving details of the morphology of SC:s in mammals in both the normal situation and in carriers of structural chromosome rearrangements. There is however yet much less information on the behavior of the SC:s in the human. Moses et al. [1], Fletcher [2], and Navarro et al. [3] showed that the technique is applicable to human spermatocytes, and Solari [4] karyotyped 22 selected cells at pachytene from three human males. There are no previous reports using this method in human oocytes.

Surface spreading has considerable potential for assessing meiotic chromosome abnormalities induced by genotoxic agents, but to date no studies using this technique have been performed for any exposed population and there is as yet no information on the normal interindividual variation as regards spontaneous aberrations. We here report preliminary results from five unexposed individuals, three human males and two females.

Testicular biopsies were obtained from three men, 27, 44, and 35 years of age as part of the investigations carried out in connection with obstruction of the vas deferens, infertility of unknown cause, and vasectomy, respectively. Fetal ovaries were dissected out immediately after prostaglandin induced therapeutic termination of the pregnancy in two cases; one has no history of fetal abnormality, the other was antenatally diagnosed as having the fetal karyotype 47,XX + 21. The techniques used are described in Wallace and Hultén [5]. Spreading is performed directly on a microscope slide using either 0.2 M Sucrose or a 0.03% detergent solution.

Analysis was made by light microscopy of the silver stained preparations. Suitable nuclei at late zygotene/pachytene were photographed and the SC:s traced onto a transparent acetate sheet.

In the spermatocytes and the oocytes from the normal fetus there was no hyperploidy and all cells had 23 bivalents.

In the oocytes from the fetus with Trisomy 21, there was regularly an extra chromosome the size of which corresponds to chromosome 21, but two cells contained, in addition, an extra chromosome 22. The three chromosomes 21 occurred as two separate configurations interpreted as a bivalent plus a univalent in all cells selected for counting, but a trivalent was occasionally seen among cells initially screened (Wallace and Hultén [5]).

In 6/35 oocytes and 5/44 spermatocytes supernumerary structures of unknown origin were seen (Table 1), as illustrated in Fig. 1. At

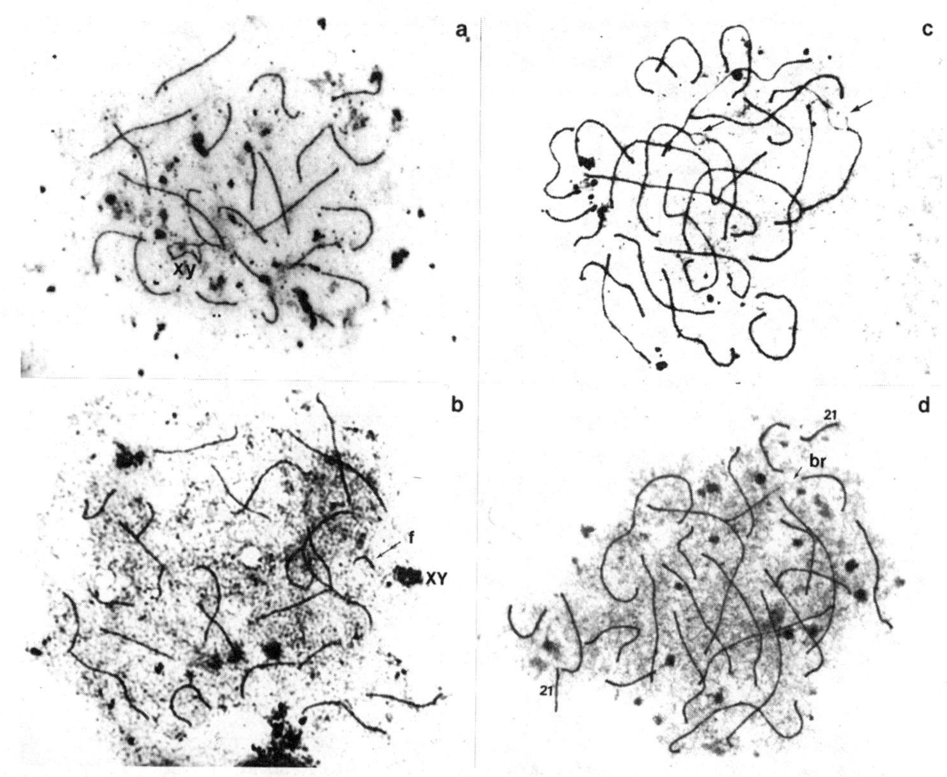

Fig. 1. Surface spreading of first meiotic prophase spermatocytes (a, b) and oocytes (c, d) with the SC's visualized by silver staining. The XY bivalent is uncondensed in spermatocyte a but condensed in b. The upper figures exemplify cells where there are no problems in counting, with both the spermatocyte a and the oocyte c having 23 bivalents. The unpaired parts of the oocyte c are interpreted to be the heterochromatic blocks of chromosomes 1 and 9 (arrows). The lower cells illustrate problems in interpretion as regards chromosome breakage. The spermatocyte b contains 23 bivalents plus a structure arrowed which is clearly too small to be a normal bivalent and which is therefore interpreted as a fragment. The oocyte d is from the fetus with the 47,XX + 21 karyotype. There are 24 bivalents and an extra chromosome 21. One of the medium sized bivalents appears to be broken (br).

present we are uncertain whether these structures are caused by breakage during spreading or whether they reflect the in vivo situation. In contrast, we did not find any indication of true chromosome rearrangements either in the spermatocytes or the oocytes. According to our preliminary experience there should be no problems in iden-

TABLE 1. Pachytene Counts

No. of Structures/ cell	23	24	25	> 25	No. of cells
Males					
HB 43	19	1	1	0	21
HB 45	7	0	0	0	7
HB 46	13	2	0	1	16
total	39	3	1	1	44
Females					
normal	15	0	0	1	16
trisomy 21	0	12	4*	3	19
total	15	12	4*	4	35

*Two cells had an extra chromosome 22 in addition to the extra 21; the other had a supernumerary structure of unknown origin.

tifying reciprocal translocations in the human male. Investigations at diakinesis/first metaphase in male translocation carriers show regular occurrence of quadrivalents (Hultén and Lindsten [6, 7]) and a preliminary analysis of pachytene spreads has clearly demonstrated that these may be easily detected also at this stage (Fig. 2). The efficiency with which other structural chromosome abnormalities can be identified remains to be investigated. It should however be pointed out that the resolving power of this technique is likely to be higher than deletion/duplication identification at mitosis as they will form loops or hairpins perpendicular to the synaptonemal complex (Moses et al. [8]; Poorman et al. [9]).

In summary we have demonstrated that the spreading technique allows sufficient separation of first meiotic prophase chromosomes for adequate counts to be performed in both the human male and female and there is rather a homogeneous picture with no true aneuploidy or chromosome rearrangements in 60 cells from four individuals with normal mitotic karyotypes. We therefore believe spreading of first meiotic prophase chromosomes may prove to be a useful technique for assessing effects of genotoxic agents at the germ line level. This technique is simple and quick, and might turn out to be particularly useful in the female as ovaries from miscarriages might be more readily available than testicular biopsies from exposed men. The resolution of this technique may be improved by sub-

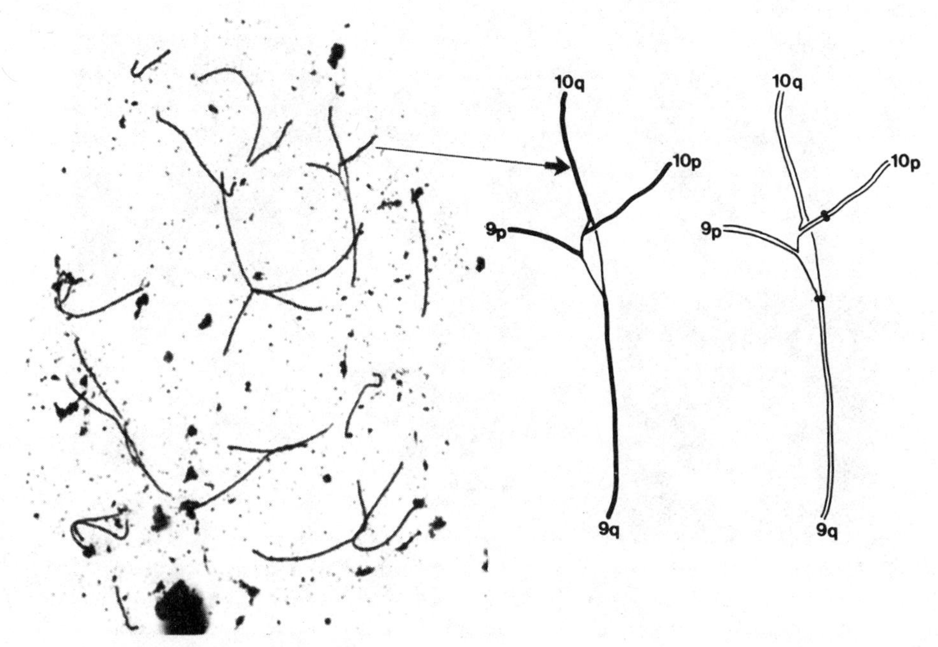

Fig. 2. Surface spreading of a first meiotic prophase spermatocyte from a human male constitutional carrier of a reciprocal translocation with the somatic karyotype 46,XY + (9;10) (p22;q24). The interpretation of the quadrivalent configuration (arrow) implies non-homologous synapsis of segments of the normal 9p with the derived 10q adjacent to the breakpoint, and some stretching of 9q.

jecting the same preparations to consecutive EM analysis (Navarro et al. [3]; Dresser and Moses [10]).

2. Sequential Staining Methods for the Analysis of Germ Line Chromosomes at First Metaphase

First metaphase chromosomes of the human male can be obtained with relative ease by testicular biopsy but in the human female there are considerable technical difficulties both in obtaining suitable material and in producing adequate chromosome preparations. As a result there is little cytogenetic information on female meiosis and this section must perforce concentrate on the male.

Meiotic preparations are made by manually releasing the cells from the testicular tubules and processing them in a way which is essentially the same as that used for harvesting cultured lymphocytes. The most common procedure is that described by Evans et al.

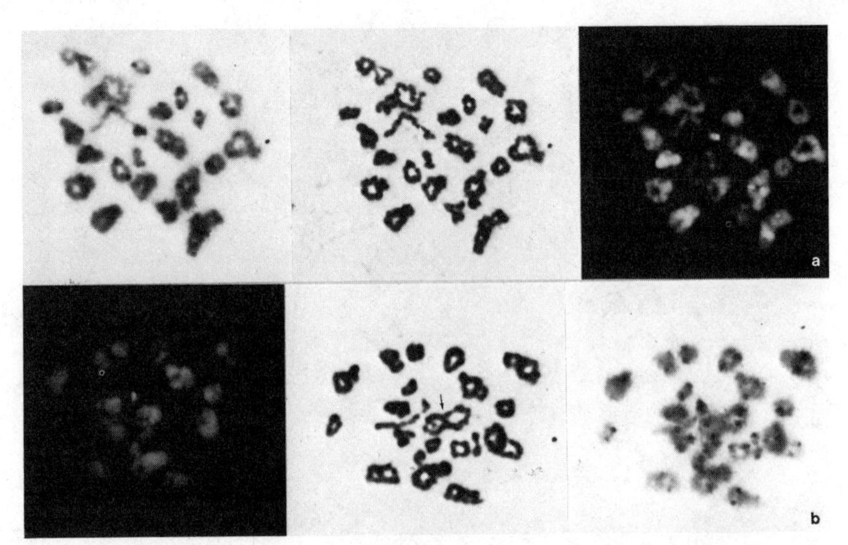

Fig. 3. Two cells sequentially stained by Q-banding, orcein staining and C-banding from a male exposed to gonadal irradiation following the removal of a testicular carcinoma. (a) Normal cell with 22 autosomal bivalents and an XY bivalent. (b) Abnormal cell with 20 bivalents, 1 quadrivalent (arrowed) and an XY bivalent.

[11] or some modification of it (Hultén et al. [12], Laurie et al. [13]) and the slides can be stained by most of the methods used for somatic chromosomes, though in our experience G-banding gives poor results. When analyzing first metaphase chromosomes it is particularly useful to combine several staining techniques as this minimizes ambiguities of interpretation and maximizes the information which can be extracted from each cell. If a sequential Q-banding, orcein staining, and C-banding regime is used then sufficiently well spread cells can usually be completely karyotyped and the frequency and distribution of chiasmata can be scored (Fig. 3a, 4a). C-banding is obligatory for this type of investigation because it is vital to know the location of the centromeres and these are not usually visible in conventionally air-dried preparations. This approach may be extended further and a quadruple staining method which incorporates Distamycin A/DAPI fluorescence after Q-banding has recently been developed in our laboratory (Saadallah and Hultén [14]).

Our understanding of the effects of genotoxic agents in the male is woefully inadequate and valuable information could be obtained by analyzing first metaphase chromosomes with these techniques The effects of hazardous agents could probably be assessed fairly accurately because meiosis in the normal human male seems to be rather invariable as far as bivalent formation, chiasma frequency

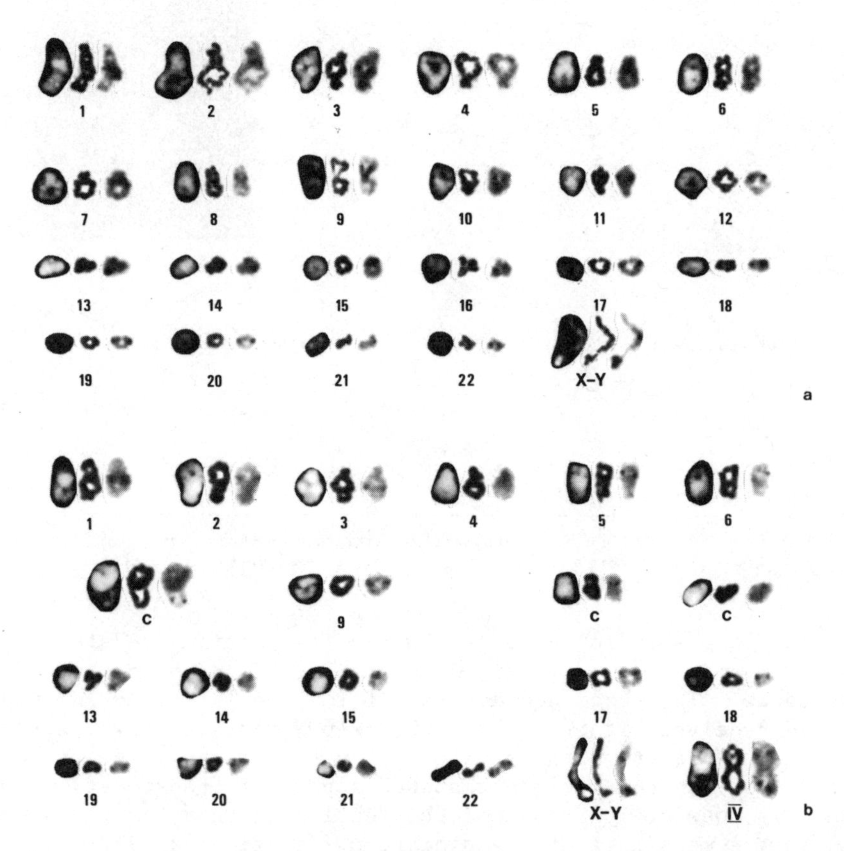

Fig. 4. Karyotypes of the two triple stained cells shown in Fig. 3. (a) Normal cell. (b) Abnormal cell with quadrivalent (IV).

and chiasma distribution are concerned (Hultén and Lindsten [7], Hultén [15], Laurie et al. [13]. Spontaneous chromosome breakage or rearrangement seems to be very rare and probably occurs at a rate of less than one per thousand cells. It is also well recognized that autosomal univalents are extremely rare and they appear to be confined to the G-group chromosomes where they occur at a probable rate of less than one per five thousand cells (Hultén [15]).

It should therefore be possible to distinguish between induced effects and naturally occurring variability. Unfortunately little work has been done to date on the effects of dangerous environmental agents on human meiosis. To our knowledge no detailed investigation has yet been made of the effects of a chemical known to cause chromosome damage in somatic cells and there is only a small amount of information on the effects of radiation.

TABLE 2. Yield of Germ-Line Quadrivalents in X-Ray Irradiated Males

Source	Dose (rad)	No. of individuals	No. of cells	No. of translocations	% of cells with translocations	Expected % of offspring with unbalanced karyotypes* 1
	0	3	200	0	0	-
Brewen et al. (1975)	78	2	371	15	4.0	0.46
	200	2	300	21	7.0	0.80
	600	2	180	11	6.1	0.70
Mean rate of translocation induction 7.7×10^{-4} translocations per cell per rad.						
PRESENT	35	1	43	1	2.3	0.26
Rate of translocation induction 6.6×10^{-4} translocation per cell per rad.						

* The mean risk for constitutional translocation carriers of having a chromosomally unbalanced offspring is estimated from antenatal diagnosis at 16 weeks gestation to be 11.4% (European Collaborative Study, August, 1981).

The long standing effects of x-ray irradiation was studied by Brewen et al. [16] who obtained testicular biopsies from irradiated volunteers. They examined a total of 1051 cells from 9 individuals who had received either 0, 78, 200, or 600 rads of 250 kV x-rays and compared these data with results from marmosets and mice. They concluded that men and marmosets showed a peak of translocation induction at a dose of 100 rads and they estimated that for these two species the mean rate of translocation induction was 7.7 × 10(-4) translocations per cell per rad at doses of 100 rads or less. The data indicated that the men, like the marmosets, were approximately twice as sensitive to radiation as the mice.

An alternative source of material for studying the effects of x-rays is males who have undergone gonadal irradiation following the removal of a testicular carcinoma. Obviously this is not an optimal source because of the possible effects of the tumor and of any accompanying chemotherapy but nevertheless we feel that this approach is worth persuing. We have examined testicular biopsies in a preliminary survey of six patients* but only one has been found to be suitable for detailed analysis with the triple staining technique. Of the remaining five one had meiotic cells of poor technical quality and the other four had severely impaired spermatogenesis (Table 2).

Meiosis in the investigated male appeared generally to be normal. The chiasma frequency was scored from 19 cells with 22 autosomal bivalents and was found to be 50.26 +/- 2.84 which is close to

*Material kindly provided by D. J. Berthelsen.

the mean for normal males (Hultén and Lindsten [6, 7], Laurie et al. [13], Hultén [15], Paris Conference [17]). Eight out of 45 cells (17.8%) showed univalent sex chromosomes and this too is within the normal range. One quadrivalent was found in the 43 karyotyped cells (Fig. 3b, 4b). No other anomalous configurations were seen and there was no evidence of non-disjunction at the spermatogonial metaphase level as there were no hyperploid diakineses. There was no chromosome breakage.

These results give an estimated rate of translocation induction of 6.64 × 10(-4) translocations per cell per rad. In view of the small sample size this figure is obviously of questionable significance but it is of the same magnitude as the 7.7 × 10(-4) translocations per cell per rad calculated by Brewen et al. [16]).

The European Collaborative Study Report [18] estimates that for male carriers of non-Robertsonian transloations the mean risk of producing chromosomally unbalanced offspring is 11.4% (the figure is calculated from the number of unbalanced karyotypes detected by antenatal diagnosis at 16 weeks gestational age). If this figure is applied to the irradiated males their risk figures are 0.26% with 35 rads, i.e., 11.4% of the 2.28% of cells containing a translocation quadrivalent (present data), 0.46% with 78 rads, 0.80% with 200 rads and 0.70% with 600 rads (Brewen et al. [16]).

Population studies indicate that 0.26% of all spontaneous abortions (1 in 400) and 0.005% of live births (1 in 20,000) have an unbalanced karyotype resulting from a de novo non-Robertsonian structural rearrantgement (Jacobs [19]). From this we can estimate that the frequency at 16 weeks gestational age would be approximately 0.0055% (1 in 18,000) as the frequency is expected to be roughly 10% higher than for live births. Thus although the risk for the irradiated males is still numerically low it has been increased by approximately 50 times at a dose of 35 rads and by approximately 150 times at a dose of 200 rads.

In view of this estimated increase in risk it is surprising that the cytogenetic study of the children of atomic bomb survivors from Hiroshima and Nagasaki, which is the only data with which these figures can be compared, did not detect more chromosome rearrangements. If we can assume that the survivors received on average a dose of 25 rads then we predict from the meiotic data that there would be approximately 12 unbalanced and 55 balanced translocation carriers in a sample of 5762 persons. If the average dose was 100 rads then the expected figures would be 50 unbalanced and 220 balanced translocation carriers. The observed figures are 0 and 14 respectively (Awa [20]).

The fact that the average age of the offspring was 19 years and that institutionalized patients were not investigated could mean that

TABLE 3

Case	Age	Seminoma Stage 1	Carcinoma in situ in other testis	X-ray Gy	Dosimetry rad	Shielding of other testis
1	40	+	+	33.12	35	+
2	43	+	+	35.2		+
3	30	+	+	33.12	122	-
4	31	+	+	35.2		+
5	35	+*	+	41.69**		+
6	32	+	-	35.2		+

* and malignant teratoma Stage II

** and chemotherapy

*** previous vasectomy

**** 1 grossly abnormal with rearranged large chromosomes and rings; 1 with Homogeneously Stained Region within chromosome 3

the frequencies of unbalanced structural rearrangements and of aneuploids were underestimated but it is curious that there was no detectable increase in the number of balanced chromosome rearrangements (Awa [21], Awa et al. [20], and [21]). The reasons for this discrepancy between the meiotic studies and the population investigations are not clear.

In summary we have found 1 quadrivalent in 43 triple stained cells from a male who had gonadal irradiation following the removal of a testicular carcinoma. This gives a rate of translocation induction of 6.64 × 10(-4) translocations per cell per rad which is in agreement with the estimate of Brewen et al. [16]. More data should be collected on the radiosensitivity of meiotic cells in order to substantiate this figure and it is important that other genotoxic agents are investigated.

Multiple staining techniques are recommended for this type of investigation as they offer definite advantages for the identification of unusual chromosome configurations, autosomal univalents and meiotic aneuploidy. However, it should be pointed out that although ambiguities in the interpretation of meiotic figures are minimized by triple or quadruple staining they cannot be eliminated completely and air-dried preparations will still have inherent drawbacks. The main problem is that the chiasma frequency of the human male is high (approximately 50 per cell) and quadrivalents will not usually be in easily scored configurations such as large open rings or chains.

Sperm Count 10^6 per ml		Somatic Chromosome Analyses			Spermatogenesis Severely Impaired	Meiotic Analysis
Before Treatment	After Treatment	Karyotype	No. of cells Analysed	Abnormal cells		
0.02-0.8	0.08	46,XY	15	0	+	-
-	-	-	-		-	-******
9.7 - 13	0.0	46,XY	15	0	+	-
0.0***	0.0***	46,XY	50	2****	+	-
-	-	46,XY	10	0	+	-
0.7 - 14	38-64	46,XY	50	3*****	-	+*******

***** 1 with dicentrics and fragments, 2 with a chromatid break.

****** Poor technical quality of Diak/MI

******* 1 quadrivalent in 43 Diak/MI

Quadrivalents frequently resemble closely apposed bivalents and can in fact be difficult to evaluate even when banding techniques are employed. Extreme caution must therefore be taken with any analysis and as interpretations tend to be conservative the frequency of translocations will be a minimum estimate. Furthermore small inversions and deletions would be likely to escape detection. Nevertheless where meiotic studies are to be performed the multiple staining techniques should be used in preference to conventional staining methods because of their greater powers of resolution.

3. Cytogenetic Analysis of Human Sperm Chromosomes

Rudak et al. [23] reported the first direct analysis of the chromosome constitution of human sperm in 1978. They fertilized zona-free hamster eggs with hüman sperm and analyzed 60 sperm chromosome complements from one male. This technique is a fundamental breakthrough in meiotic research since it allows precise determination of the frequency of chromosomally abnormal male gametes. It also provides a potential method for assessing the effects of various agents on the chromosome constitution of human sperm.

In order to apply this technique to study the cytogenetic effects of various agents on human sperm, it is first necessary to determine the normal "background" frequency of chromosomal abnormalities in sperm from normal men. Rudak et al., reported a 5% (3/60) frequency of aneuploidy in the sperm chromosome complements of one normal male.

TABLE 4

ABNORMAL SPERM CHROMOSOME COMPLEMENTS (7%)

ANEUPLOID (28 5.3%)		CHROMOSOME BREAKS AND FRAGMENTS (9 1.7%)
HYPERHAPLOID (18 3.4%)	HYPOHAPLOID (10 1.9%)	
24,Y,+1	22,X,-13	*23,X,csg(3)(p21), inv 3(p11q11)
24,X,+B	22,X,-E	23,X,csbCq (2 cells)
24,X,+C (2 cells)	22,Y,-19	23,T,csb7q
24,Y,+C (4 cells)	22,Y,-G	*23,Y,inv(3)(p11q11), csb(7)(q11)
24,Y,+16	22,X,-21	23,Y,csb(14)(q2)
24,X,+g	22,-C or -Y	23,X,csbXq
24,X,+21	21,X,-1,-13	23,X,+ace
24,Y,+21	21,-E,-C or -Y	23,Y,+ace
24,XX	21,X,-18,-22	
24,Y,+2B,-D	21,X,-3,-7,-13 +dic(3;7),+2ace	
24,X,+mar,+ace		
*24,X,+E,inv del 3, inv(P11q11),del(p2)		
24,Y,+16,quadriradial (B;F), ctg2q		
26,X,+9,+10,+13		

*This male was found to carry a pericentric inversion of chromosome 3 in lymphocytes (ascertained through sperm chromosomes).

cause of differential viability. 48.6% (18/37) of the abnormal sperm complements were hyperhaploid and 27.0% (10/37) were hypohaploid. These numbers are not significantly different from the one to one ratio theoretically expected (chi square test). It is interesting that we have not observed any 24,YY complements although various studies [25, 26] have shown that 1-5% of sperm have two fluorescent Y bodies. This may indicate that the fluorescent spots seen in sperm may not always reflect the presence of a Y chromosome.

The frequency of X-bearing sperm to Y-bearing sperm in this sample was 55.4% to 44.6% (290X:233Y). Two abnormal sperm complements (22,-C or -Y and 21,-E,-C or -Y) were not included in this calculation. This is significantly different from the expected one to one ratio ($p < 0.5$, chi square test). It is interesting that Pawlowitzki and Pearson [27] found a very similar ratio of 54.6% X to 43.9% Y based on the presence of a Y body in 10,692 sperm. This may indicate that meiosis does not produce an equal number of X and Y-bearing sperm or alternatively, in our *in vitro* system X-bearing

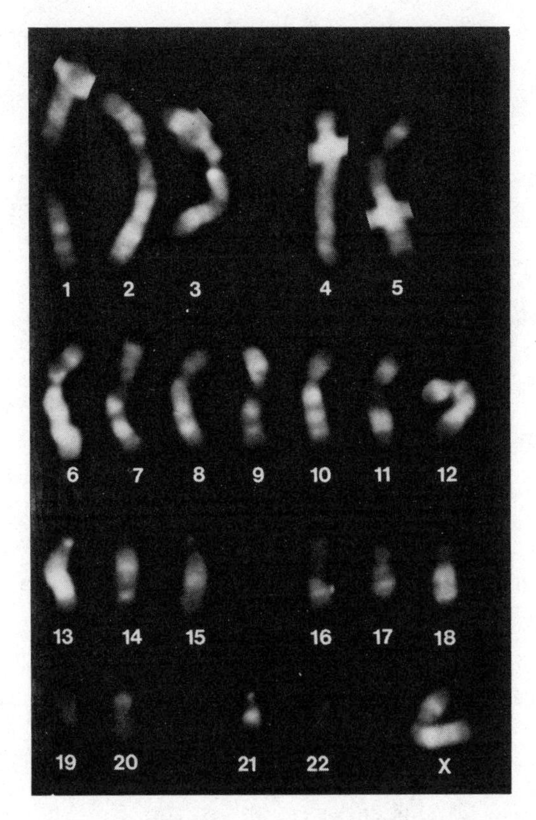

Fig. 5. Q-banded karyotype of a normal 23,X sperm complement.

In a preliminary report [24] we found 9.2% chromosome abnormalities in 240 spermatozoa from 18 normal men. We have subsequently increased this series to 525 spermatozoa from 23 normal men. We will discuss the results of this larger series in some detail.

Q-banding of chromosomes was successful in 56% of spreads to allow identificiation of individual chromosomes. In other complements the chromosomes could only be characterized as to group by routine Giemsa staining. A karyotype of a normal 23,X sperm complement is shown in Fig. 5.

Thirty-seven (7.0%) of the sperm chromosome complements were abnormal: 18 (3.4%) were hyperhaploid, 10 (1.9%) were hypohaploid and 9 (1.7%) had a chromosome break or fragment. Details of the abnormal sperm complements are outlined in Table 1. The aneuploidy in sperm chromosome complements seems to be relatively evenly distributed in all chromosome groups. This would argue that all chromosomes are susceptible to non-disjunction and that certain chromosomes are seen more frequently in spontaneous abortions and newborns be-

sperm may have an advantage in fertilizing zona-free hamster eggs. However selection of sperm seems unlikely since it has been demonstrated that the chromosome constitution of mouse sperm does not affect their ability to fertilize eggs [28, 29].

It is important to extend this research to other populations to determine the variation in the frequency and type of chromosome abnormalities in human sperm. It will then be possible to assess the effects of genotoxic agents.

One serious limitation of this technique is that it requires sperm capable of fertilizing eggs. Many agents suspected of causing sperm chromosome damage may also affect sperm production, motility or capacitation so that the sperm is not capable of fertilization. For example, we have found that only 2 of 14 radiotherapy patients have had sperm with an adequate fertilization of hamster eggs (>10%) to attempt sperm chromosome analysis. This limitation may exclude from study precisely those agents most likely to cause sperm chromosome damage. On the other hand, some agents may cause sperm chromosome damage and yet preserve fertilizing and these would have the greatest potential danger.

In summary the technique of human sperm chromosome analysis by fertilization of zona-free hamster eggs allows precise determination of the chromosome constitution of human sperm. This technique is still in its infancy and research on normal males is required to obtain more base-line information on the frequency and type of chromosome abnormalities in human sperm. In the future this technique may be valuable in assessing the cytogenetic effects of various agents on the male gamete.

CONCLUSIONS

We hope that this paper illustrates the potential of the currently available techniques for analyzing chromosome damage in germ line cells. The pachytene spreading and sperm chromosome techniques, which are relatively recent innovations, enable us to look at aspects of meiosis which have hitherto been impossible to investigate and the sequential staining methods are of great help in interpreting air-dried metaphase I cells. The latter method could also be used for metaphase II cells which are often difficult to analyze in conventionally stained preparations and this would enable us to estimate the rate of metasphase I non-disjunction directly.

Considerable technical progress has undoubtedly been made since the effect of genotoxic agents on human meiosis was last reviewed (Hultén and Luciani [30]) and the main problem now facing us is not so much that the techniques are not good enough but that suitable material is difficult to obtain. This is likely to remain a serious problem in the evaluation of environmental effects in the human but where investigators do have access to material they are urged to use as many of the available techniques as they can in order to make their study as informative as possible.

ACKNOWLEDGMENTS

Support from the M.R.C. (Grant Numbers G.80/0640/4CA) for research on human meiosis is gratefully acknowledged. We should also like to thank Mr. Paul Leedham for his help in obtaining the ovarian material and Dr. J. Berthelsen of the Rigshospitalet, Copenhagen for providing the material from the gonadally irradiated males.

REFERENCES

1. M. J. Moses, S. J. Counce, and D. F. Paulson, Synaptonemal complex complement of man in spreads of spermatocytes with details of the sex chromosome pair, Science, 187:363-365 (1975).
2. J. M. Fletcher, Light microscope analysis of meiotic prophase chromosomes by silver staining, Chromosoma, 72:241-248 (1979).
3. J. Navarro, F. Vidal, M. Guitart, and J. Egozcue, A method for the sequential study of synaptonemal complexes by light and electron microscopy, Hum. Genet., 59:419-421 (1981).
4. A. J. Solari, Synaptonemal complexes and associated structures in microspread human spermatocytes, Chromosoma, 81:315-337 (1980).
5. B. M. N. Wallace and M. A. Hultén, Triple chromosome synapsis in oocytes from a human foetus with trisomy 21, Ann. Hum. Genet., 47:271-276 (1983).
6. M. A. Hultén and J. Lindsten, The behavior of structural aberrations at male meiosis, in: "Human population cytogenetics" (P. A. Jacobs, W. H. Price, and P. Law, eds.), pp. 24-61, Pfizer Medical Monographs 5, Edinburgh University Press, Edinburgh (1970).
7. M. A. Hulten and J. Lindsten, Cytogenetic aspects of human male meiosis, Adv. Hum. Genet., 4:327-387 (1973).
8. M. J. Moses, P. A. Poorman, L. B. Russell, N. L. A. Cacheiro, and A. J. Solari, The synaptonemal complex and a tandem duplication in the mouse, J. Cell Biol., 75:135a (1977).
9. P. A. Poorman, M. J. Moses, L. B. Russell, and N. L. A. Cacheiro, Synaptonemal complex analysis of mouse chromosome rearrangements, I. Cytogenetic observations on a tandem duplication, Chromosoma, 81:507-518 (1981).
10. M. E. Dresser and M. J. Moses, Silver staining of synaptonemal complexes in surface spreads for light and electron microscopy, Exp. Cell Res., 121:416-419 (1979).
11. E. P. Evans, G. Breckon, and C. E. Ford, An air-drying method for meiotic preparations from mammalian testes, Cytogenetics, 3:289-294 (1964).

12. M. Hultén, J. Lindsten, L. Ming Pen-Ming, and M. Fraccaro, The XY bivalent in human male meiosis, Ann. Hum. Genet., 30:119-123 (1966).
13. D. A. Laurie, M. A. Hultén, and G. H. Jones, Chiasma frequency and distribution in a sample of human males: chromosomes 1, 2, and 9, Cytogenet. Cell Genet., 31:153-166 (1981).
14. N. Saadallah and M. A. Hultén, Chiasma distribution, genetic lengths, and recombination fractions: A comparison between chromosomes 15 and 16, J. Med. Genet., 20:290-299 (1983).
15. M. A. Hultén, Chiasma distribution at diakinesis in the normal human male, Hereditas, 76:55-78 (1974).
16. J. G. Brewen, R. J. Preston, and N. Gengozian, Analysis of x-ray induced chromosomal translocations in human and marmoset spermatogonial stem cells, Nature, 253:468-470 (1975).
17. Paris Conference 1971, Standardization in human cytogenetics, Birth Defects, Original Article Series, VIII. The National Foundation, New York (1972).
18. European Collaborative Study Report (August 1981), Preliminary results presented at the 6th International Congress of Human Genetics, September 13-18, 1981, Jerusalem, Israel.
19. P. A. Jacobs, Mutation rates of structural chromosomal rearrangements in Man, Am. J. Hum. Genet., 33:44-54 (1981).
20. A. A. Awa, T. Honda, S. Neriishi, H. Shimba, T. Amano, and H. B. Hamilton, An interim report of the cytogenetic study of the offspring of atomic bomb survivors in Hiroshima and Nagasaki, Abstract of the paper presented at the 6th International Congress of Human Genetics, September 13-18, 1981, Jerusalem, Israel.
21. A. A. Awa, Review of Thirty years study of Hiroshima and Nagasaki Atomic bomb survivors, II. Biological effects, B. Genetic effects. 2. Cytogenetic study, J. Radiat. Res. Supp., 75-81 (1975).
22. Hiroshima and Nagasaki, The physical, medical, and social effects of the atomic bombings, Edited by the Committee for the compilation of materials on damage caused by the atomic bombs in Hiroshima and Nagasaki, Hutchinson (1981).
23. E. Rudak, P. A. Jacobs, and R. Yanagimachi, Direct analysis of the chromosome constitution of human spermatozoa, Nature, 274:911-913 (1978).
24. R. H. Martin, C. C. Lin, W. Balkan, and K. Burns, Direct chromosomal analysis of human spermatozoa, Preliminary results from 18 normal men, Am. J. Hum. Genet., 34:459-468 (1982).
25. A. T. Sumner, J. A. Robinson, and H. J. Evans, Distinguishing between X, Y, and YY bearing human spermatozoa by fluorescence and DNA content, Nature New Biol., 229:231-233 (1971).
26. M. Klasen and M. Schmid, An improved method for Y-body identification and confirmation of a high incidence of YY sperm nuclei, Hum. Genet., 58:156-161 (1981).
27. I. H. Pawlowitzki and P. L. Pearson, Chromosome aneuploidy in human spermatozoa, Humangenetik., 16:119-122 (1972).

28. A. Gropp, in: "Proc. Symp. Institut. National de la Santé et de la Recherche Medicale." (A. Boué and C. Thibault, eds.), pp. 255-268, Paris (1973).
29. C. J. Epstein and B. Travis, Preimplantation lethality of monosomy for mouse chromosome 19, Nature, 280:144-145 (1979).
30. M. A. Hultén and J. M. Luciani, On the possibilities of detecting chromosome aberrations induced by environmental agents in human germ line cells, in: "Genetic damage in Man caused by environmental agents" (K. Berg, ed.), pp. 143-186, Academic Press (1979).

EPIDEMIOLOGIC CONSIDERATIONS IN ASSESSING ADVERSE REPRODUCTIVE OUTCOMES FOLLOWING GENOTOXIC EXPOSURES*

Zena Stein

New York State Psychiatric Institute and Sergievsky Center
Columbia University
630 West 168 Street
New York, New York 10032

Observational studies of human populations, which are the essence of epidemiology, have inherent limitations when compared to experimental studies of laboratory animals. Our methods often appear laborious, our investigations expensive, our inferences weak. These shortcomings are even more apparent in studying toxic exposures than in most other areas of epidemiologic research, for the real-life situations are in this case particularly complex. The fullest exploitation of human observational studies is often greatly restricted by concern for confidentiality on the part of exposed and affected persons, for parsimony by health authorities, for safeguards on the part of industry, and for political considerations on the part of government agencies.

In order to make optimal use of epidemiology we must therefore be aware from the beginning of its limitations. Accepting both the

---

*This paper draws freely on the volume: Assessment of Health Effects at Chemical Disposal Sites, proceedings of a symposium held on June 1-2, 1981 at the Rockefller University, New York City. Edited by Dr. William W. Lowrance. Copyrighted and published by The Rockefeller University, 1981. In particular, the chapter entitled Epidemiologic Considerations in Assessing Health Effects at Toxic Waste Sites pages 125-145 is reproduced in its entirety at the beginning and end of this paper. I thank Drs. Jennie Kline, Patrick Shrout, Dorothy Warburton, and Maureen Hatch for permission to use this material, and Dr. William W. Lowrance and the Rockefeller University to reprint it in this form.

inherent and imposed restrictions which attend observational studies is part of the burden of epidemiology; it calls for much ingenuity and experience fully to explore the causal connections between an exposure and a health effect. Ultimately, however, it is observational studies which provide the most direct tests of whether an exposure is associated with a health effect in humans.

In this paper we consider some of the issues bearing on the design of epidemiologic studies. We limit ourselves mainly to studies which evaluate effects of environmental chemical exposures on human reproductive function, although many of these issues also arise in considering the effects of non-chemical exposures and a range of other health outcomes. We open with some definitions, dealing first with the independent (hypothetical causal) variable and then with the dependent variable (health outcome). We select four endpoints for fuller discussion. Next, we discuss how real-life situations will influence both the choice of study design and the interpretation of findings. Then we distinguish between three types of studies: those where information on the independent and dependent variables is readily to hand (Level I); those where additional information on either or both of these variables can be generated fairly quickly from special studies (Level II); and those where longterm follow-up of both the exposed populations will be needed (Level III). We consider the types of inferences that are likely to be drawn from these studies and ways in which strategies might be strengthened.

THE INDEPENDENT VARIABLE: THE EXPOSURE

For the purposes of this paper we assume a starting point of some unintended chemical contamination of the environment, suspected of causing adverse effects on reproduction. This is not always the starting point for public health action. In practice, the first alert often is a report of untoward events clustered in time and place (such as three births with neural tube defects in quick succession when the expectation in that population over that time period is less than one). With such clusters the initial task is to confirm that the events have indeed occurred, to define the population at risk, and to assess whether the observed number of events is in fact in excess of that expected. Confirming the suspected excess, investigators still may have no clue as to the likely causal factor. Environmental exposure may be suspected. Often the frequency of such suspected exposure among those affected will be compared with that of a sample of comparable unaffected individuals (a case-control study). Suggestive findings may serve as the beginning point for a further effort to explore whether a particular exposure is associated with other health effects as well.

From the perspective of this meeting, however, let us ignore the cluster situation and assume that the starting point is a group of residents who are aware that they may be exposed to some potentially harmful chemical in the environment and they are wondering what, if any, the health effects are. In its way this situation is as difficult as the cluster situation. With clusters of adverse outcome, as we have noted above, we need to search for the cause; with exposures, we need to search for the health effects, often with very little to guide us either from laboratory or clinical science as to likely effects.

It is seldom easy to characterize the exposure. In studies of waste sites, precise description of chemical constituents and identification of relevant exposure is often difficult: not only may there be multiple contaminants, several equally suspect, but products formed through the interaction of these individual contaminants may be the responsible agents. The time of onset of relevant exposure to a chemical, a combination of chemicals, or a by-product, may not be exactly known, or not known at all. For current exposure, it may be necessary to measure levels of the chemical in the environment and also to document the likely routes of exposure (through inhalation, contamination of drinking water or vegetation, or direct contact in homes or schoolyards). It is unlikely that routine monitoring data from previous years will be available to permit estimates of past exposure and comparison with current exposure. Often we have only an imprecise description of the exposure, based on sampling of chemicals currently in the environment, and have little information on the stability of these concentrations over time or on the by-products that may have been formed.

Even if the nature, concentration, timing, and route of exposure are known, we still may be uncertain as to the potential hazard to reproduction if we lack data on the probable dose, time, and duration of exposure of each individual. Identifying groups of residents who have "probably" been exposed is much less satisfactory and seriously weakens the epidemiologic inquiry.

With a few exceptions (such as a study of men exposed 20 years ago to plutonium, for whom precise dose estimates may be calculated, using both past and present biologic measurements), some misclassification of the exposure status of the individual is virtually always likely. As we discuss in more detail later, such misclassification inevitably reduces the power of the study to detect associations between specific exposures and given outcomes. Epidemiologists must resort to constructing less-than-perfect indices of exposures, which they then attempt to relate to health phenomena.

THE INDEPENDENT VARIABLES: REPRODUCTIVE DYSFUNCTION

Any or several of the endpoints listed in Table 1 can be examined. These endpoints were chosen for reasons that are partly

TABLE 1. Analytic Indices of Reproductive Dysfunction

Sexual dysfunction
decreased libido
impotence
Sperm abnormalities
decreased number
decreased mobility
decreased morphology
Sub-fecundity
abnormal gonads, ducts, or external genitalia
abnormal pubertal development
infertility (of male or female origin)
amenorrhea
anovulatory cycles
delay in conception
Illness during pregnancy and parturition
toxemia
hemorrhage
Early fetal loss (to 28 weeks)
Perinatal death
late fetal loss (after 28 weeks) and stillbirth
intrapartum death
death in first week
Decreased birthweight
Gestational age at delivery
prematurity
postmaturity
Altered sex ratio
Multiple births
Birth defects
major
minor
Chromosome abnormalities (detected in early fetuses, through amniocentesis, in perinatal deaths, in livebirths)
Infant mortality
Childhood morbidity
Childhood malignancies
Age at menopause

pragmatic (they are probably capable of being measured by an epidemiologic study) and partly theoretical (they have been shown to vary with environmental pollutants). In the section that follows, we have chosen to discuss four endpoints: fertility, spontaneous abortion, congenital malformations and birthweight. It will become clear that each of these must be approached with due respect for epidemiological pitfalls and the same is true for most of the others listed.

No one has used all the outcomes listed in Table 1 to test for reproductive dysfunction. On the other hand, the choice as to which to study often seems to have been quite arbitrary although the selection of the appropriate endpoint(s) for study may be the most crucial scientific decision made. In a later section we discuss some of the issues to be weighed in choosing the dependent variable: the frequency of the outcome in the unexposed; biological coherence; accessibility of data; and the use of single or multiple endpoints. Certainly we will not be able to search for effects on all possible outcomes without considering the statistical data-gathering and financial costs.

FOUR SELECTED ENDPOINTS

Infecundity and Infertility. Infecundity is the inability to conceive, and can result from physical or psychosocial factors which prevent intercourse, as well as from failure of conception itself. A common operational definition of an infecund couple is a failure to conceive after one year of unprotected intercourse. About 10% of U.S. couples fall within this definition. Infecundity can be male-related or female-related, or it can result from incompatibility in a couple.

Infecundity is a difficult endpoint to investigate and infertility (the failure to produce a live offspring) is the more usual endpoint studied. Fertility is probably influenced by many factors such as social class, age, contraceptive practices, religion, and previous reproductive history; local, regional and temporal variations are considerable. Therefore in investigating a suspected "epidemic" of infertility, selection of an appropriate control group in an unexposed population is almost always necessary and requires great care. Although birth certificates provide a starting point, they will usually need to be supplemented by interviews.

Despite the difficulties of studying infecundity and infertility, it has been reported as an outcome of some chemical exposures. A self-report of reduction in fertility by workers in a pesticide plant in California making dibromochloropropane (DBCP) led to discovery of a high frequency of azospermia among exposed men (Whorton et al., 1977). A study done using questionnaires mailed to women physicians in England reported an increase in involuntary sterility of more than two years duration in women working as anesthesiologists (Pharoah et al., 1977).

Spontaneous Abortion. Spontaneous abortion (defined here as termination of pregnancy before 28 weeks gestation) is the commonest unsuccessful outcome of prenancy, occurring in 15-20% of all recognized pregnancies. There is evidence that very early abortions, before pregnancy is recognized, may occur with a frequency equal to recognized abortions, but these losses are difficult to study without a prospective design and use of very early pregnancy tests.

Because spontaneous abortion is so frequent, it is the outcome having the greatest power to detect a change, given a limited population, as we discuss later. Other considerations mentioned below also make spontaneous abortion a useful indicator for studying reproductive failure.

Under normal circumstances spontaneous abortion acts to screen out abnormal pregnancies: about 90% of all chromosomally abnormal conceptions and about 50% of fetuses with malformations are lost as spontaneous abortions. About one-quarter of spontaneous abortions have abnormalities of the conceptus which are so gross as not even to be compatible with continued development. Since most severe abnormalities of the fetus are lost as spontaneous abortions, one can expect that any agent that increases the rate of abnormalities in term pregnancies will have an even greater effect on the rate of loss of pregnancies as spontaneous abortions. This is especially true for increases in chromosome abnormalities. For example, a doubling of the rate of trisomy at conception would increase the rate of Down syndrome from 0.14% to 0.29% at birth but would increase the rate of spontaneous abortion from 15% to 17.5% and the rate of trisomy among early abortions from 25% to 40%.

Spontaneous abortion has the additional advantage of providing a specimen that can be examined in various ways to provide information about the mechanism of the loss. At present this examination can consist of a chromosomal analysis through culture of fetal tissue, and of a detailed morphologic classification. In the future biochemical or immunological analysis may also be informative. Abortions also can be classified usefully by gestational age. About half of first trimester spontaneous abortions are chromosomally abnormal (Warburton et al., 1980), and these may be further divided into types of abnormalities (trisomies, triploidy, tetraploidy) occurring through different mechanisms. Trisomy is usually the result of meiotic non-disjunction, triploidy often results from dispermy, and tetrapoloidy results from failure of the first mitotic division in the zygote (Hassold et al., 1980). The ability to distinguish among these different etiological classes, particularly between chromosomally normal and chromosomally abnormal, is an aid to causal inference. Since it is unlikely that one agent will produce all types of abortions (although this is of course not impossible), exposures which are equally associated with all classes of abortions must be suspected of being the result of recall bias or confounding.

Against these advantages of studying spontaneous abortion must be set the disadvantages. First, the only sources of data available on this outcome are medical records and personal interviews. The first source will seriously underestimate the frequency of spontaneous abortion, because many early abortions do not receive medical attention or are not recorded. In a New York population we estimated that 40% of early spontaneous abortions were not brought to medical attention. Interview data are subject to recall bias and to inaccuracies of self diagnosis. However, several studies have indicated that women's self reported histories are generally reliable, at least to the extent that they can be checked in medical records (Kline et al., 1978). On the other hand, reports by husbands of their wives' spontaneous abortions are much less reliable. One approach to studying spontaneous abortion frequency is to use both medically verified and all reported abortions in separate analyses.

The increase in precision gained by direct study of the products of conception must be weighed against the considerable expense. Thus special arrangements will be needed to ensure speedy retrieval and storage of the specimen, and skilled laboratory workers are needed to examine them. In the end, successful karyotyping may be achieved in only one-third to one-half of cases.

Because of these difficulties in studying spontaneous abortion, epidemiological studies have proved difficult to interpret and results often inconsistent, especially in relation to environmental exposures. For example, increases in spontaneous abortion rates have been reported to occur among female anesthesiologists in England and the United States (Knill-Jones et al., 1972, 1975; American Society of Anesthesiologists, 1974), in the wives of vinyl chloride workers (Infante et al., 1976), in 2,4,5-T exposed regions in Alsea, Oregon (U.S. EPA, 1979), and at Love Canal (Vianna, 1980). Each of these studies can be faulted on one or more grounds: the anesthetist studies for the use of questionnaires with poor response rates, the vinyl chloride studies for using husbands' recall of wives' miscarriages (and possible recall bias), the Alsea studies for using only hospitalized spontaneous abortions and not testing for similar hospitalization rates in cases and controls, and the Love Canal study for the use of inappropriate control data collected ten years previously in a different population.

Congenital Malformations. Birth defects are much less common than the other reproductive outcomes we are considering; the usual estimate of the frequency of major malformations in term births is

2% to 3%. On the other hand, it is the reproductive outcome which is the most distressing to families, it is the most costly to health resources, and it is also the most visible outcome to the community as a whole.

The rarity of individual malformations argues for the usefulness of grouping together various entities for analysis. Classifying by the organ system involved is convenient but since malformations of the same system may have many different causes, it is not necessarily a useful strategam. By contrast, classification of malformations into etiological groups - those due to chromosomal anomalies, those due to single mutant genes, those due to known environmental factors, and those of multiple or unknown origin - would have advantages. Some malformations, such as those due to known inherited mutant genes, can be exluded from analysis, since they cannot be due to the exposure. The timing of the exposure may also exclude certain kinds of abnormalities: exposures after conception cannot be the cause of most chromosomal abnormalities for instance. Classifying congenital malformations by the time in embryonic development when the defects originate may be useful, when exposure is restricted to a defined period of pregnancy. However, understanding of the pathogenesis of human malformations is seldom sufficient to permit reliable classification along such lines.

Information on congenital malformations may be available from birth certificates, but this is usually very incomplete. Hospital and physician records are a better source, but still are likely to be incomplete. Classification of malformations into etiological groups usually will require extensive history-taking and examination of the patients by physicians having experience in dysmorphology.

The frequency and kinds of malformations which are detected will vary not only with expertise of observers but also with parental race, age, and social class, and with the age and sex of the child. The frequency of detection of some malformations, such as congenital heart disease, will increase with age of the subjects, but the frequency of others which are lethal early in life will decrease with age. All of these factors must be taken into account in comparing groups.

Maternal drug ingestion (thalidomide, anticonvulsants) during pregnancy has been demonstrated to lead to increased rates of specific malformations, and some studies have suggested increases due to environmental exposures, such as to anesthetic gases, lead, or 2,4,5-T. In some of these cases the definition of congenital malformation or "birth defects" was very broad, and appeared to include abnormal palmar creases, asthma, behavioral problems, and skin rashes. While these outcomes may well be sequelae of exposure to toxic chemicals, they should perhaps be analyzed separately and certainly not included uncritically under the category of "birth defects."

Birthweight. Low birthweight is strongly associated with neonatal death, congenital malformations, developmental delay, and chromosomal abnormalities. About 7% of American infants weigh less than 2500 grams at birth, and low birth weight is the most important contributor to infant mortality in our society. Although the probability of survival for infants of given birthweights varies considerably with social class and location, this variability is probably largely a function of the level of medical care.

Birthweight is routinely recorded on birth certificates and hospital records in most parts of the world, and the data are not subject to the same kinds of biases of reporting as are other reproductive outcomes. For this reason the reported increase in low birthweight infants is the most convincing reproductive effect described from Love Canal (Vianna, 1980) and from the areas surrounding lead smelters in Sweden (Nordstrom, 1979).

Maternal age, parity, race, and smoking habits are known risk factors for low birthweight and must be taken into account in analyzing comparison groups. If gestational age is known, one can distinguish between babies born before term but with weight appropriate to gestational age, from those with evidence of poor fetal growth.

The advantages of using low birthweight as an adverse reproductive endpoint are the ready availability of the data and the lack of bias in its collection. Because of the relatively high frequency of 'low birthweight' and because birthweight itself can usefully be analyzed as a continuous variable, this outcome has considerable advantages in statistical analysis.

ISSUES RELATING TO THE DESIGN OF THE STUDY

An investigation of whether living near a dump site is associated with adverse health effects will encounter both technical and human problems. Populations living around a dump site usually are small, thus limiting both the range of outcomes and the size of effects that can be studied. Individuals living in an area are not usually homogeneous either with respect to characteristics that can influence reproductive dysfunction independently of exposure (age, smoking, alcohol drinking) nor with respect to the type, level, duration, or timing of exposure. The exposure itself often is poorly defined. Many of the effects of interest are either rare (such as neural tube defects) or unlikely to have been routinely recorded prior to the investigation (such as spontaneous abortions). To these technical difficulties are added the challenges of carrying out a study in a highly charged atmosphere of anger and fear accompanying

TABLE 2. Frequency of Selected Reproductive Endpoints (Adapted from Conference on the Evaluation of Human Populations Exposed to Potential Mutagenic and Reproductive Hazards; see Acknowledgement)

Event	Frequency per 100	Unit
Azoospermia (absence of sperm)	1	Men
Birthweight* less than 2500 grams	5-15	Livebirths
Failure to conceive after one year of unprotected intercourse	10-15	Couples
Spontaneous abortion within 8-28 weeks of gestation	10-20	Pregnancies
Chromosomal anomaly among spontaneously aborted conceptions (8-28 weeks)	30-40	Spontaneous abortions
Chromosomal anomalies among amniocentesis specimens (unselected women over 35 years old)	2	Amniocentesis specimens
Stillbirth	2-4	Stillbirths + livebirths
Birth defects	2-3	Livebirths
Chromosomal anomalies in livebirths	0.2	Livebirths
Neural tube defects	0.01-1.0	Livebirths + stillbirths
Severe mental retardation	0.4	Children to age 15

*More usefully analyzed as a continuous variable, as discussed.

suspicion of adverse health effects. This atmosphere may make it difficult or impossible to exploit the "natural experiment" fully. Data obtained through personal interview may be doubted unless these can be validated by other evidence.

In this section we discuss how these real-life problems influence both the design of the study and the inferences that can be drawn after the study is completed. We have organized this discussion around the concept of statistical power. Statistical power, defined broadly, is the ability of a study to detect and estimate accurately a change in the frequency (or mean value) of an event. More precisely, statistical power is the probability that an effect of a specific size will be detected when it is present in the universe from which the sample was drawn.

In experimental work the primary question is whether the independent and dependent variables are associated to an extent unlikely to have arisen by chance, assuming the null hypothesis. And this is probably also the first question raised by the community concerned with the possible effects of living near a dump site: Is exposure associated with an increased frequency of ill health?

Viewing a negative result, however, the community has a second question (one raised with less urgency in experimental work): How likely is it that exposure is associated with ill health even though this investigation failed to detect an effect? In the case of exposure to toxic wastes, the answer to this question may have far-reaching implications, both for public health action and eventually for regulatory policy.

For this reason, we have focused our discussion on those aspects of study design which influence statistical power - the ability to detect an effect which is in fact present. We have supposed that a particular type of exposure leads to an increase in the frequency of one or several adverse reproductive outcomes, for example, those given in Table 2.

Two epidemiological strategies might be taken to search for association: one may compare the frequency of adverse reproductive outcomes among people who are exposed with that of people who are unexposed (a cohort or prospective study), or one may compare the frequency of exposure among people affected by the outcome with that of people unaffected by the outcome (case-control or case-referent study). Given identical sample sizes, in general the cohort design is more powerful when exposure is less common than the outcome, and the case-control design is more powerful when the outcome is less common than the exposure.

Four Parameters Influencing Statistical Power

1. The number of persons studied. In general, power increases with increase in sample size. The exception to this generalization is important from the view of efficiency in data collection: in a cohort study there is little increase in statistical power achieved by adding numbers of the unexposed, when the unexposed population is more than three times that of the exposed population, and similarly, in a case-control study, little is gained by selecting more than three controls for every case (Fleiss, 1981).

2. The frequency of response. The response variable is the variable which is not determined at the outset by the study design. In a cohort study, the adverse reproductive outcome is the response variable; in a case-control study, exposure is the response variable. The closer the frequency of the response variable is to 50% in the unexposed or unaffected study group, the greater will be the statistical power of the study.

3. The size of association between exposure and outcome. Obviously, the greater the effect of exposure (say a ten-fold increase as compared to a doubling in the frequency of outcome), the greater will be the likelihood of detecting the increase.

4. Statistical significance level. The formal statistical test of an association is designed to control two kinds of errors: the error of rejecting the null hypothesis of "no association" when the null hypothesis is true (Type 1 error), and the error of not rejecting the null hypothesis when it is in fact false (Type II error). The test statistic allows the probability of a Type I error to be estimated; this probability is often called the alpha level. Before an association is said to be significant, the probability of its occurring as a result of chance sampling fluctuations (that is, the probability of a Type I error) must be less than some preordained value, called the statistical significance level. All things held constant, the lower the significance level, the less statistical power there is for the test. In setting the significance level, however, other technical aspects of the statistical test need to be considered. For example, if the direction of an association can be predicted, it may be appropriate to use a one-tailed test. Significance levels for one-tailed tests result in more statistical power than significance levels for two-tailed tests.

We should not assume uncritically that the same criteria of significance apply in investigations of the type we are discussing as in laboratory studies. For instance the asymmetry between errors of false acceptance and false rejection of the null hypothesis may be inappropriate in some situations. One can imagine situations where an increase in adverse outcome, although not statistically significant at the .05 level, is sufficient to warrant further investigation. Likewise, if the chances of detecting a fairly large effect (such as three-fold increase in a common event) would be mised one time out of five, one might feel very uneasy in proclaiming that no adverse effects are present.

When the setting is a toxic waste site, statistical power to detect modest increases (say two-fold) in even the most common adverse reproductive outcomes, such as spontaneous abortion, is likely to be weak because the communities are usually small and adequate numbers of exposed individuals can rarely be found.

The ability of an investigation to detect the true effect will depend not only on the four factors listed above but also on appropriate and precise definition of exposure, appropriate choice and careful measurement of outcome, and on comparability of the study groups with respect both to potentially confounding variables and to methods of data collection. We now must discuss each of these issues.

Definition of Exposure. We have assumed that in the population, exposure of a particular type is associated with an increased risk of one or several adverse reproductive outcomes. It might be that any exposure to either parent preceding conception is associated with adverse effects regardless of the dose, duration, or timing (whether long before conception or around the time of conception, for example). In contrast, the relevant exposure may be limited to a subgroup of the "ever"-exposed population. For instance, exposures to the woman at least ten years prior to conception have been implicated in studies of the relation of maternal irradiation and triploidy among aborted conceptions (Alberman et al., 1972). Since power increases as the frequency of exposure increases, investigators may choose to define all individuals as exposed who were ever exposed to the suspect agents. If in fact only a subgroup of the "ever" exposed group is at risk of adverse effect, then this definition of exposure will dilute the exposed group with individuals who are uninformative about the effects of exposure, thus reducing the chances of detecting a true association.

Often there is little evidence to go on in defining the relevant exposure. Insofar as the etiology of the suspected outcome is known, some specificity may be given at least to defining the timing and possibly the relevant parent exposed. Thus, for example, most chromosomal anomalies occur either prior to conception, in the germ cells of either parent, or at the time of fertilization. If this is the outcome to be studied, then the exposed group will be limited to individuals exposed prior to or at the time of conception. This added precision in defining exposure will add to the true power of the study, although at first glance it may appear to detract since in most circumstances the number of individuals defined as "exposed" will be decreased when the definition is refined.

In the absence of information permitting some definition of exposure based on prior experience or theory, several types of exposures may be explored in the same study. The effects of the dimensions of exposure, including parent exposed, dose, timing, and duration, can be examined both separately and in combination. This approach has the advantage that the relevant definition of exposure may be uncovered. It has the disadvantage that many comparisons may be made on the same data and will lead to an increased probability of making at least one Type I error. A common way to avoid these errors is to use a significance level for each of the multiple tests formed by dividing the desired significance level for each of the multiple tests (say, .05) by the number of statistical tests to be performed. As discussed above, however, the use of this smaller significance level will result in a loss of power.

Selection of Outcome(s). It is possible, though unlikely, that investigators will be convinced from the start that only one particular reproductive outcome is likely to result from the suspect toxic exposure. In this situation there is obviously no point in examining other outcomes. Usually, however, it will be unclear which

TABLE 3. Estimates of Sizes of Samples of Exposed and Unexposed Needed to Detect, with 80% Statistical Power, a Doubling in the Frequency of Selected Reproductive Events (Computations Assume a Two-Tailed Test at Alpha = .05). (Power Calculations Based on Jacob Cohen, Statistical Power for the Behavioral Sciences, Academic Press, New York, 1969.)

Event	Rates in the unexposed sample	Necessary sample sizes of exposed and unexposed
Chromosome anomalies in livebirths	0.002	11,468 livebirths
Congenital malformations	0.02	1,109 livebirths
Spontaneous abortion	0.15	118 pregnancies
Chromosome anomaly among spontaneous abortions	0.35	31 spontaneous abortions

of the many endpoints that could be studied are likely to follow exposure. Choice will have to be made to examine one or several outcomes depending on availability of data, validity of outcome measurement, frequency of the event, and biological plausibility. If the first approach is to draw on already available data (Level I), investigators may be limited to outcomes which are routinely collected in record systems such as vital certificate data or special ongoing surveillance programs. The first of these offer data on perinatal mortality, birthweight, and sex of liveborn infants. Special systems such as the Center for Disease Control Congenital Malformation Registry offer data on malformations among livebirths for certain U.S. locations. For many reproductive outcomes, data are not available in routinely collected records; thus if spontaneous abortions, or changes in semen, are to be examined, special studies need to be mounted.

It is intuitively obvious that given a fixed sample size, it will be easier to detect changes in the frequency of more common outcomes than in the frequency of rarer outcomes. A doubling of a rare event, such as a neural tube defect, is more likely to arise by chance than a doubling in the frequency of a more common event, such as spontaneous abortion. Table 3 sets out, for cohort studies, the sample sizes of exposed and unexposed individuals which would need to be studied in order to detect a doubling in the frequency of several different outcomes. (In this connection, it may also be useful to search for sub-populations in which the outcomes are concentrated. An obvious sub-population, if the outcome is chromosomal anomaly, is to use spontaneous abortion rather than birth, because chromosomal anomalies are far more concentrated in these conceptions (Table 3).)

When investigators cast a wide net and examine several different endpoints, as often happens, statistical problems arise like those discussed in connection with multiple exposure variables. Investigators may undertake to search for effects on birthweight, neural tube defect, congenital heart defects, and spontaneous abortions, setting the probability of a falsely positive finding (Type I Error) to be .05 for each outcome. In this case, with each test of an association there will be a one-in-twenty chance that an association will be detected, despite the fact that there is no association in the population. Clearly, the more endpoints included in the search, the better the chance of finding at least one that (simply by chance) appears to be associated with the exposure. As before, there is no entirely satisfactory way of avoiding this problem, because if the investigator anticipates the difficulty by setting a more stringent test for statistical significance (say .01 instead of .05), he then raises the chance of missing an effect that is really there - that is, he loses statistical power.

Some endpoints (such as all chromosomal trisomies) seem biologically similar and may usefully be grouped together in analysis. Such a grouping would then be regarded as a single outcome and the number of endpoints studied would be decreased. On the other hand, grouping outcomes which are likely to be distinct with respect to underlying etiologic mechanisms is likely to decrease the chances of detecting an effect on a specific endpoint. Again, in statistical considerations, the more closely we are able to define the relevant endpoints, the greater the power we have to discover an association of given size between the exposure and outcome.

These are some of the reasons we call for caution in choosing outcome events. The decision over whether to examine one isolated outcome, a category of outcomes, or a number of discrete outcomes, preferably is made at the outset of the investigation, using the best biological and clinical evidence available to formulate hypotheses. This will place researchers in the best position to make optimal use of the study population without biasing the results. It is seldom that investigators of chemical exposures enjoy the luxuries of preliminary data exploration or of split samples, strategies available in the laboratory or with large populations.

POTENTIALLY CONFOUNDING VARIABLES

The aim, at the conclusion of a study, is to describe a relationship between exposure and outcome which is, as far as possible, unlikely to be explained by extraneous differences between the two study groups. In order to accomplish this aim, two sources of variation need to be controlled: variation in the characteristics of the study groups which relate to the *a priori* chance of exposure or to outcome, and variation in the quality of data collected for the two study groups.

Comparability of Study Groups. In a case-control study, case and control populations should be comparable in their a priori chances of exposure and in characteristics other than exposure which relate both to the outcome under study and to the exposure variable. Similarly, in a cohort study, exposed and unexposed groups should be comparable in all known risk factors for the outcome. For some outcomes such as birthweight, several characteristics of the woman - pre-pregnancy weight, race, and cigarette smoking - are known to have an important effect. These characteristics may or may not be independent of exposure status. If they are not independent, in order to avoid confounding they will need to be controlled either in design or in analysis. Even if these characteristics are independent of exposure status, statistical control may prove useful by removing unwanted sources of variation and increasing the precision of the estimate of association.

Therefore it is prudent in designing a study to select a comparison group, either unexposed people or the controls in a case-control study, thought to be comparable to the study group with respect to known risk factors for the disease under investigation. Data on each of the potentially confounding variables should be gathered during the investigation so that comparability of the two study groups can be demonstrated and the consistency of association of the potentially confounding variables across various strata can be assessed.

Recall Bias in Assessment of Exposure. This is the most serious disadvantage of a case-control study. Recall bias - the implication that exposure data may be influenced by experience of adverse outcomes - must be suspected whenever interview data from affected and unaffected individuals are depended on as a primary source of information on exposure. Recall bias may produce either spuriously positive findings (cases over-report exposure, controls under-report exposure, or both) or spuriously negative findings (cases under-report exposure). Recall bias may be assessed in several ways: (1) by comparing self-reports of exposure with records of exposure collected prior to or independently of the outcome; (2) by measuring exposure biochemically; or (3) by comparing the frequency of self-reported exposure among several "worried" groups thought to have etiologically distinct conditions (such as chromosomally normal and chromosomally abnormal spontaneous abortions, neural tube defects, and Down's syndrome).

Bias in the Measurement of Outcome. The ascertainment of outcome may be biased with respect to exposure either in case-control studies or in cohort studies. Quite obviously this can occur if data on outcome are obtained solely by interview. For most reproductive outcomes, validation of reports of outcome will be possible through vital certificate data and hospital records. For some outcomes, such as spontaneous abortion, women may not have sought medical attention and care should be taken to evaluate the quality of interview data obtained. Even in studies where ascertainment occurs through routine record systems, bias may inadvertently occur if medical service utilization differs for exposed and unexposed populations.

Each of these issues - comparability of study groups, appropriate and precise definition of exposure status and appropriate selection and careful measurement of outcome, will usually be grappled with to a different extent depending on the time, effort, and funding devoted to the study.

THREE LEVELS OF EFFORT

Studies which draw on already accessible data (Level I studies) will usually provide the least precision in definition of exposure, the smallest range of outcomes to be examined, and the fewest data by which the comparability of study groups can be tested. Estimation of exposure status may be based on the proximity of homes to the toxic waste sites or to contaminated water sources. In such cases, exposure is defined at the ecologic or group level. Rarely will it be possible to distinguish the parent exposed, the duration, or likely dose. With respect to outcome, such studies may draw on vital certificate data or special registries of malformations; thus in most circumstances the range of outcomes will be limited to birthweight, perinatal mortality, sex of offspring, and possibly malformations among births. In most states, birth certificates provide data on a few potentially confounding variables, such as maternal age, but data on personal habits such as cigarette smoking or alcohol drinking are likely to be unavailable. Causal inference is likely to be weak. Both positive and negative findings may need a second level of study. Despite limitations, such studies often provide the most efficient first search for a few select gross outcomes.

A second level of effort (Level II studies) involved the collection of more precise data on exposure to individuals, more careful examination of outcomes (including some not available in routine record systems), and collection of information on potentially confounding variables. As this level of effort, the range of outcomes to be examined can be expanded to include outcomes identified in medical records (spontaneous abortions, malformations), through interview with the study subjects (spontaneous abortion, sexual dysfunction), or through biological studies of subjects (semen characteristics, hormone levels). Precision in the definition of exposure status with respect to duration and timing of likely exposure can be gained through personal interviews. Biological, soil, home, and water measurements of current exposure levels can improve precision in the estimates of dose and in the understanding of major routes of exposure. Such studies will almost invariably require personal contact with the people under study, necessitating care and often lengthy arrangements with the community and political organizations. However, when properly carried out, a Level II study will offer data at the individual level. Inferences from both positive and negative findings are likely to be more secure in this type of study than in ecologic studies carried out at Level I, although it may not always be possible to rule out all selection factors which could produce spuriously positive findings.

Sometimes an exposure will be judged sufficiently suspicious as to justify a third level of effort (Level III studies), prospective follow-up of a cohort of exposed and unexposed individuals. In some countries (such as in the Canadian system for linking occupational and other environmental exposures with later health outcomes in selected individuals) such a follow-up can be achieved without contact with study subjects. In the United States a similar set of records on exposure and outcomes does not exist. At present, Level III study requires contact with the exposed and unexposed subjects at the start of the study, and follow-up contacts to ascertain outcome. Provided that the entire cohort is followed successfully, Level III studies may provide the strongest evidence for causal inference, since exposure is defined prior to the occurrence of outcome, and selection factors are exluded by essentially complete follow-up of the cohort. These studies are, of course, difficult to carry out. More importantly, however, many years are likely to pass before the study is completed and decisions consequent on the findings of the study made.

In considering these three levels of effort, we have identified several components which add to interpretability of studies: (1) assignment of exposure status to individuals, rather than to groups; (2) precision in the description of the duration, timing, dose and route of exposure; (3) selection of the most likely adverse outcomes, rather than only those available in routine records; (4) data allowing examination of the comparability of study groups; and (5) ascertainment of exposure status independently of outcome, and vice versa.

In Level II and Level III studies, nearly all of these aims can be achieved, although in Level II studies it may be difficult to ascertain exposure and outcome status independently. Level I studies tend to fall short in each of these areas, although they offer the most efficient, least intrusive examination of the research question.

We may ask, then, what kinds of information might be integrated into routine record systems, particularly birth certificates, which would improve the power of Level I studies. Assignment of exposure status at the individual rather than group level might be improved if the birth certificate cited, for each parent separately, the number of years he/she resided in the current home. Precision in the description of the duration and timing of exposure would also be improved with this information. Data on the likely route and dose of exposure will be available only if routine environmental monitoring has been carried out in the area and these data are accessible for merging with information on place of residence. A system for routinely updating birth certificates with data on malformations diagnosed within a month of birth would, in theory, improve the utility of birth certificates for information on malformations. Expansion of reporting to include all fetal deaths would extend the range of outcomes which could be examined from routine records, although many fetal deaths will remain unknown to medical facilities. Data on maternal smoking and race, which are not available on all birth certificates, relate to several reproductive outcomes; comparability of study groups could more easily be demonstrated if these data were available from birth certificates. Since birth certificates usually are filled out prior to suspicion that exposure may be causing adverse outcomes, independence in ascertaining exposure and outcome status is easily achieved.

RECOMMENDATIONS RELATING TO LEVELS OF ANALYSIS

Level I data (in particular, birth and death certificates) are the main source of information available for ongoing surveillance. Every effort should be made to improve the quality of these records, extend their range, and improve the flexibility for analysis.

Level II data which rely on individual exposure measures, clinical records, and interviews, in practice are extremely hampered by restrictions on investigators. The multiplicity of health authorities, each with its own institutional review boards, makes these studies extremely difficult to pursue. It might be advantageous to set up special procedures to enable record studies and personal interviews to be carried out speedily in response to environmental emergencies.

Level III studies should be interdisciplinary, linking information regarding exposures (for instance, to substances suspected of being mutagenic or carcinogenic) to current symptoms or signs in individuals (such as chromosome anomalies) and to later health outcomes (birth defects or cancer in offspring).

INTERCONNECTIONS

We are aware of potentially toxic substances in our environment. We have _in vitro_ tests of mutagenicity for substances that enter the body. We have evidence of mechanisms of biological response to exposures. And we have measurable adverse health outcomes. The problem is to thread these elements together so as to enable us to use their connections, each with the other, to develop warning systems, to provide individual risk estimates, to interpret the process. In large part this is a problem of organization: we need cross-disciplinary working groups that together can produce the linkages that are lacking.

Both this symposium and one referred to in our acknowledgement have convinced us that neither the laboratory nor the epidemiological approaches alone will enable us to overcome the problems each faces. For the next step, we require a synthesis.

As we and others have suggested, it is now possible to conduct prospective studies in which exposure is measured, *in vitro* mutagenicity assessed, circulating lymphocytes evaluated, and populations followed for a long time. However, this is laborious and slow, though it is ultimately a most valid approach. We can do better if the problem of linkage is viewed as a formidable, but not insuperable, challenge.

ACKNOWLEDGMENT

Many issues argued here derive from a document prepared for a Conference on the Evaluation of Human Populations Exposed to Potential Mutagenic and Reproductive Hazards, Washington, D.C., January 25-27, 1981, sponsored by the March of Dimes Birth Defects Foundation, the Centers for Disease Control, the National Institute for Occupational Safety and Health, and the Environmental Protection Agency. That document was drawn up by a Committee, chaired by Dr. Warburton and myself, that included Drs. L. Edmonds, M. Hatch, L. Holmes, J. Kline, R. Miller, P. Shrout, D. Whorton, and A. Wyrobek. Others present at the Workshop made significant contributions and suggested modifications.

The authors of the present paper takes personal responsibility for its contents while acknowledging her debt to the earlier paper.

The research has been supported by grants from the National Institutes of Health (1 R01 HD 12207) and the Environmental Protection Agency (R870355 01).

REFERENCES

E. Alberman, P. E. Polani, J. A. Fraser Roberts, C. C. Spicer, M. Elliott, and E. Armstrong, Parental exposure to x-irradiation and Down's syndrome, Annals of Human Genetics, 36:195-208 (1972).

Anesthesiology, Report of the ad hoc committee on the effect of trace anesthetics on the health of operating room personnel, American Society of the Anesthesiologists, Anesthes., 41:321-340 (1974).

A. Bloom (ed.), Guidelines for Studies of Human Populations Exposed to Mutagenic and Reproductive Hazards, March of Dimes Birth Defects Foundation, New York (1981).

J. Fleiss, Statistical Methods for Rates and Proportions, 2nd edition, John Wiley and Sons, New YOrk (1981).

P. F. Infante, J. K. Wagoner, and R. J. Waxweiler, Carcinogenic, mutagenic, and teratogenic risks associated with vinyl chloride, Mut. Res., 41:131-42 (1976).

T. Hassold, P. Jacobs, J. Kline, Z. Stein, and D. Warburton, Effect of maternal age on autosomal trisomies, Ann. Hum. Genet., London, 44:29-36 (1980).

J. Kline, P. Shrout, Z. Stein, M. Susser, and M. Weiss, An epidemiological study of the role of gravity in spontaneous abortion, II, Early Hum. Dev., 1(4):345-356 (1978).

R. R. Knill-Jones, L. V. Rodriguez, D. D. Moir, and A. A. Spence, Anaesthetic practice and pregnancy: Controlled survey of women anesthetists in the United Kingdom, Lancet, I:1326-1328 (1972).

R. R. Knill-Jones, B. J. Newman, and A. A. Spence, Anesthetic practice and pregnancy, Lancet 2:807-809 (1975).

I. Nordstrom, L. Beckman, and I. Nordenson, Occupational and environmental risks in and around a smelter in northern Sweden, I. Variations in birth weight, Hereditas, 88:43-46 (1978).

P. D. D. Pharoah, E. Alberman, and P. Doyle, Outcome of pregnancy among women in anesthetic practice, Lancet, I:34-36 (1977).

U.S. Environmental Protection Agency, Epidemiology Studies Division, Six years spontaneous abortion rates in three Oregon areas in relation to forest, 2,4,5-T spray practices (1979).

N. J. Vianna, A. K. Polani, R. Regal, S. Kim, G. E. Haughie, and D. Mitchell, Love Canal studies of human reproductive loss, In supra, No. 182 (1980).

D. Warburton, Z. Stein, J. Kline, and M. Susser, Chromosome abnormalities in spontaneous abortions; Data from the New York City study, in: Human Embryonic and Fetal Death (I. H. Porter and E. B. Hook, eds.), Academic Press, New York, 261-287 (1980).

D. Whorton, R. M. Krauss, S. Marshall, and T. Milby, Infertility in male pesticide workers, Lancet, 2:1259-61 (1977).

A REFERENCE POPULATION:
FOR THE EVALUATION OF NATURAL OCCURRING
INTERINDIVIDUAL VARIATION IN THE RESPONSE
TO GENOTOXIC AGENTS

Åke Nordén and Ronald W. Pero

Department of Medicine and
Department of Biochemical and Genetic Extoxicology
University of Lund
Lund, Sweden

and

Unit of Community Health Care Sciences
Dalby, Sweden

1. INTRODUCTION

A reference population is theoretically a "normal population" with which any other population suspected to be exposed to genotoxic or other unwarranted environmental factors can be compared for the evaluation of health hazards.

By using an entire population of adequate size rather than a sample it is hoped that interindividual variation of the response to environmental factors should be represented as far as this is possible.

The health status of the reference population must be known through continuous data recording of health events over extended periods of time. The influx into and outflow from the population should be known so that individuals moving away can be reached and their health assessed when needed.

The reference population thus defined also offers clinical material for comparative investigations of diagnostic procedures.

2. THE DALBY COMMUNITY CARE RESEARCH CENTER

The Dalby Community Care Research Center opened in 1968 in the South of Sweden with the intentions to formulate the role of primary care. Data recording of medical events was organized originally for a population of 8000, later for about 20,000 at present providing information over a period of 14 years. Already in 1947 part of the population - Lundby - was subjected to studies covering more than 99% of a population of 2500 [1]. The intensions were to explore "in an unselected population the frequency of mental variants." Follow-up studies in 1957 and 1972 have been almost as complete as the original.

The Dalby Community Care Research Center started as a joint project between the Swedish Government and the local county council. The center therefore was in charge both of routine primary care and research. To-day primary care organization in Sweden is to a great extent based on experience gathered at Dalby. Since 1982 the Dalby Unit is part of the University of Lund.

3. SUMMARY OF RESEARCH PROGRAMMES 1982

3.1. Prospective study of an aging propulation including all of those who in 1969 and 1970 reached retiring age (67 years). They have been examined at two-year intervals by physical and laboratory investigations. Their social and economical situation is repeatedly being assessed. A nutritional study by the use of the "double portion technique" has been undertaken in a sample of the population [2]. The food has been analyzed chemically and provides material for further studies of essential and toxic components. The death rate between 67 and about 80 years of age corresponds to that of Sweden. Close to 90% have cared for themselves at home so far.

3.2. Upper respiratory infections represent the most common problem in primary care. It has been found that bacterial infections rather than viral are frequent causes of sickness. Studies of the use of antibiotics earlier suspected not to be indicated could be shown to be used adequately by primary care physicians.

3.3. Urinary tract infections have lately focused on infections caused by *Staphylococcus saprophyticus* - a pathogen in the urinary tract predominantly in women. The unique pathogenicity for the urinary tract may be due to the presence of receptors not found in other tissues.

3.4. Care of the chronically ill child is a new program for the assessment of the home situations of children with severe diseases - diabetes, leukemia, malformations, renal insufficiency, etc. This group of children is increasing as medical techniques develop.

3.5. Child health program includes health control at 4-years of age and the health control of school children. The problem for small children is predominantly otitis media - a clinical trial of pneumococcus vaccins is in progress. In schoolchildren serious diseases are being observed more often than expected. Four boys with Crohn's disease were diagnosed when 0.4 cases had been expected.

The abuse of alcohol and cannabis are subject to special prevention programs.

3.6. The prevalence of handicap in the population is the subject of an inventory by the district nurses - a research program for the nurse within her own district. It should make the nurse an investigator of problems which occur in the population for which she is responsible.

3.7. Diabetes. In 1968 an organization was introduced for the care of the diabetic through a specially trained nurse. This is now a universally accepted model in Sweden. It means that the majority of diabetic patients (= type II) can be adequately cared for within the primary care organization. The effect of antidiabetic drugs is part of the program and also studies of the diabetic diet. By the use of a continuous blood glucose analyzer the effect of dietary components on blood-glucose, insulin, and other hormones is under study.

3.8. Rheumatoid arthritis. The physical therapists at the center are organized to take care of the majority of patients with rheumatoid arthritis in cooperation with the social service. The program also includes clinical pharmacological studies of drugs.

3.9. Hypertension. A hypertension clinic with a nurse has been organized but is at present in a special research program being reinvestigated. Reports suggest that in perhaps 30% of all hypertensives alcoholism is a factor of pathogenetic importance.

3.10. Care of the alcoholic. The routine use of laboratory techniques for the diagnosis of unhealthy alcohol consumption has been assessed. An organization for the care of the chronic alcoholic as part of the primary care resources has been developed. The use of a former patient as a social worker has made it possible to keep efficient and continuous contact with even severe abusers.

3.11. Opthalmological program originally to demonstrate the prevalence of visual defects in the population. By the introduction of automatic visual field analysis the natural history of glaucoma has been studied. Previous assumption that the increased intraocular pressure is the primary cause has been reevaluated. Circulatory failure of the blood supply to the optic nerve may be the first phase of the disease. Low pressure glaucoma is being observed with increasing frequency. Current therapy in order to lower intraocular pressure has to be reevaluated.

3.12. The clinical pharmacology of primary care program was initially focusing on the effect of food on absorption of drugs. It was found that there was both a great interindividual variability and a variability between different drugs. Present studies based on absorption tests aim at an "optimalization" of the prescription of drugs in diseases like diabetes, hypertension, rheumatoid arthritis, and arthrosis.

3.13. Data recording of disease events illustrates the disease pattern in different age groups and give information on leading disease conditions. Rapid changes have occurred in the population from a rather old population with a dominance of cardiovascular diseases and diabetes to a young population with a dominance of infectious diseases. Trends in the requirement for medical care since 1968 have recently been analyzed. It has been found that demands are less than previously assumed. It may mean that current plans for the training of doctors, and the need of hospital beds will have to be reevaluated.

3.14. Studies of DNA-repair were originally taken up in order to better understand the aging process. One hypothesis for the genetic control of aging suggested that progressively deficient DNA-repair systems had a decisive effect on the outcome of aging [3].

Insight into factors controlling aging might also throw light on the mechanisms behind the long incubation times for cardiovascular diseases, type II diabetes and malignant diseases.

A method designed to estimate an individual's sensitivity to mutagenic exposures was established based on *in vitro* N-acetoxy-2-acetylaminofluorene (NA-AAF) stimulation of unscheduled DNA synthesis (UDS) in resting lymphocytes from peripheral blood [4]. It was found that mutagen sensitivity increased with age, was higher in males than in females, and was also higher in subjects with higher blood pressure.

In order to further relate mutagen sensitivity thus determined to the general health status, 266 individuals were selected to represent the normal distribution of age, sex, and blood pressure in the general population by Dalby, Sweden [5]. The general health status was described by collecting, from the national health insurance records, information regarding all disease episodes during a five-year period.

It was found that in subjects with high mutagen sensitivity the prevalence of cardiovascular diseases was higher than in subjects with low mutagen sensitivity ($p < .025$), infectious diseases were also more prevalent in the high mutagen sensitivity group ($p < 0.005$). There were more cases of malignant diseases in the high mutagen sensitivity group than in the group with low sensitivity but the numbers were too few to reach statistical significance. These data were taken as evidence in support of the hypothesis that lymphocyte sensitivity to the induction of DNA damage by NA-AAF indicates the relative risk of individuals' developing the major diseases important in general health care [5].

REFERENCES

1. E. Essen-Möller, Individual traits and morbidity in a Swedish rural population, Acta Psychiatrica et Neurologica Scandinavica, Suppl. 100:1-160 (1956).
2. B. Borgström, Å. Norden, B. Åkesson, M. Abdulla, and M. Jägerstad, Nutrition and Old Age, Scand. J. Gastroenterology 14, Suppl. 52:1-299 (1979).
3. F. M. Burnet, A genetic interpretation of ageing, Lancet 2: 480-483 (1973).
4. R. W. Pero, C. Bryngelsson, F. Mitelman, T. Thulin, and Å. Nordén, High blood pressure related to carcinogen-induced unscheduled DNA synthesis, DNA carcinogen binding and chromosome aberrations in human lymphocytes, Proc. Nat. Acad. Sci. USA, 73:2496-2500 (1976).
5. R. W. Pero and Å. Nordén, Mutagen sensitivity in peripheral lymphocytes as a risk indicator, Environmental Research, 24: 409-424 (1981).

INVESTIGATIONS OF SMOKING AND RELATED HEALTH COMPLICATIONS AND GENOTOXIC HAZARDS IN A PREVENTIVE MEDICAL POPULATION PROGRAM IN MALMÖ, SWEDEN

Erik Trell, Lars Janzon, and Rolf Korsgaard

Departments of Preventive Medicine
and Community Medicine
University of Lund
Malmö General Hospital
Malmö, Sweden

Department of Tumour Cytogenetics
The Wallenberg Laboratory
University of Lund
Lund, Sweden

Malmö is a Swedish city of 230,000 inhabitants, served by one general hospital. Since 1975, there has been an institute of Preventive Medicine integrated within the regular hospital servies at the Medical Department of Malmö General Hospital. The design of the department, as outlined in Fig. 1, seeks to avoid a conveyor-belt type of isolated "health check-ups," and instead create an ambulatory ward for individual risk factor assessment and intervention. The investigative units can be used in the afternoon as outpatient clinics for further investigation and treatment of the risk factors identified in the screening, e.g., hyperlipidemia and hypertension as outlined in Fig. 2.

Whole birth year cohorts of Malmö residents are invited to the screening. The capacity is 10,000 mammographic and 5000 internal medical screenings each year with subsequent follow-up intervention with regard to the risk factors identified. The program is directed towards early detection and prevention of the new major internal medical non-infectious diseases and health hazards of Western Societies, e.g., arterial hypertension and other cardiovascular diseases, diabetes mellitus, overweight, cancer, alcoholism, and complications of tobacco smoking. Furthermore, continuous testing and incorporation of new methods for identification and intervention of current risk factors to the human organism in our modern environment are important aims of the program.

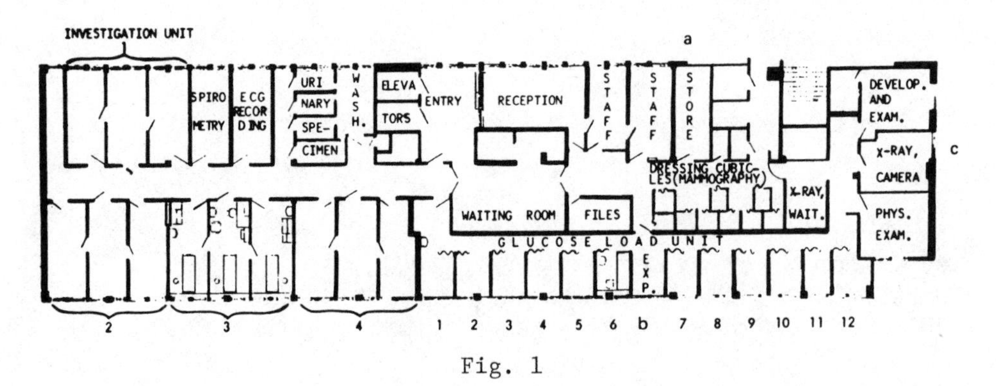

Fig. 1

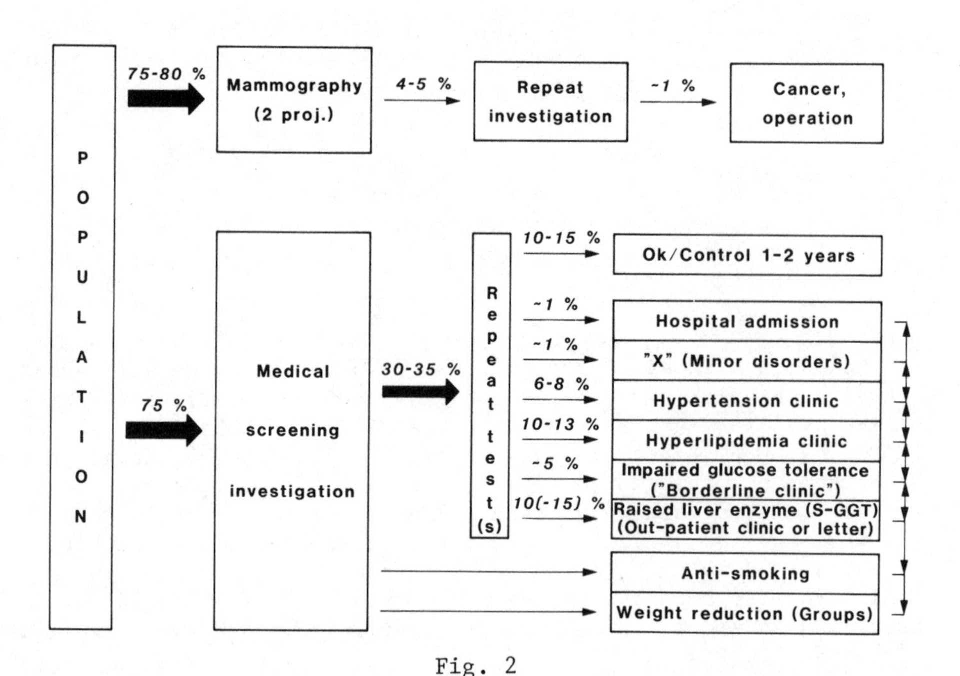

Fig. 2

The screening is monitored and recorded by an interactive computer system (Fig. 3), which allows on-line registration and representation of demographic information, test results and questionnaires, generation of test protocols and other listings, as well as storing, retrieval and statistical processing of all the accumulated information in the data bank [1].

The screening questionnaire is answered, checked, and recorded by the screening subjects by a simple "yes"-, "no"-, "don't know"-,

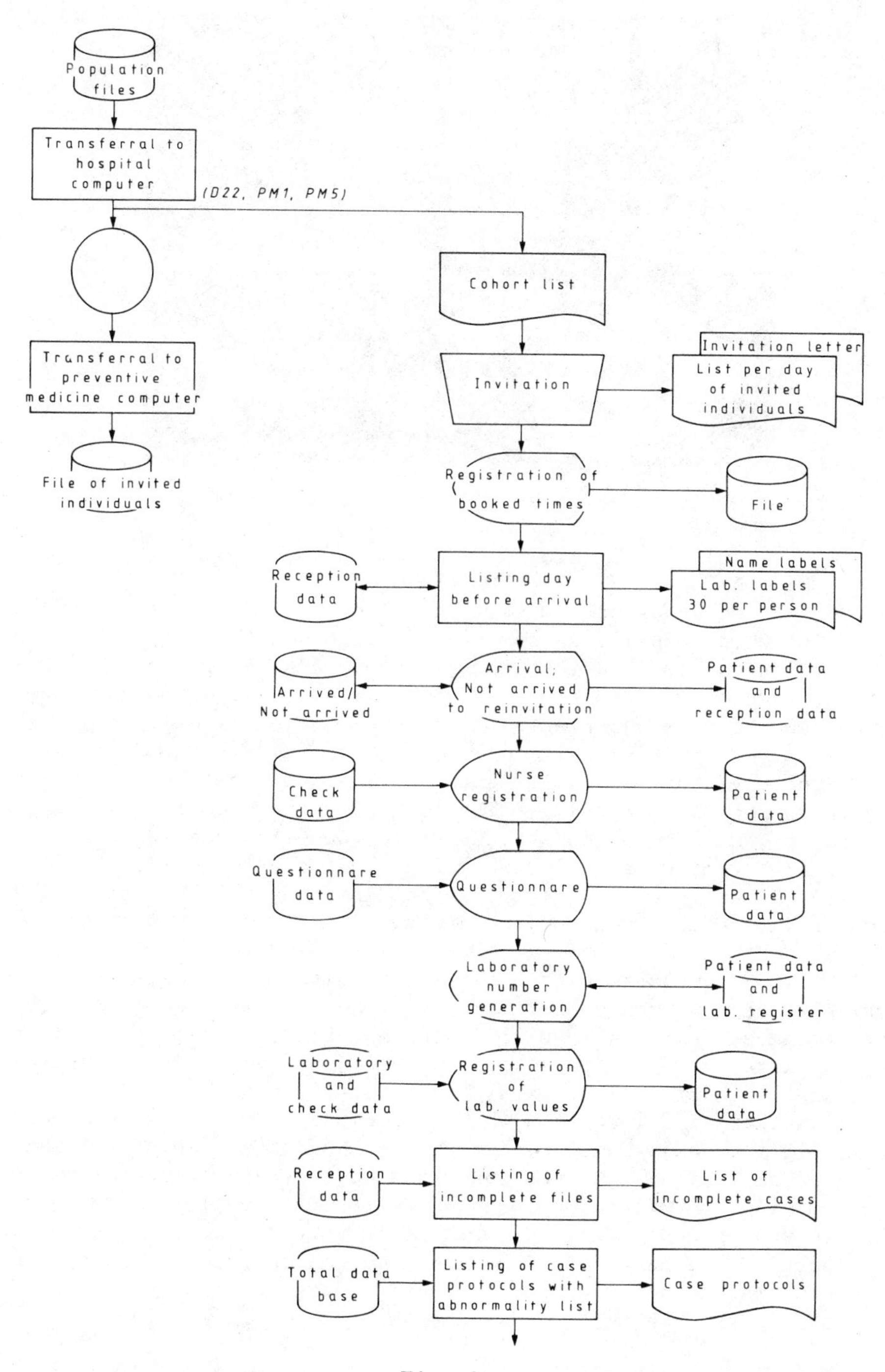
Population files
Transferral to hospital computer
(D22, PM1, PM5)
Transferral to preventive medicine computer
File of invited individuals
Cohort list
Invitation
Invitation letter
List per day of invited individuals
Registration of booked times
File
Reception data
Listing day before arrival
Name labels
Lab. labels 30 per person
Arrived/Not arrived
Arrival; Not arrived to reinvitation
Patient data and reception data
Check data
Nurse registration
Patient data
Questionnare data
Questionnare
Patient data
Laboratory number generation
Patient data and lab. register
Laboratory and check data
Registration of lab. values
Patient data
Reception data
Listing of incomplete files
List of incomplete cases
Total data base
Listing of case protocols with abnormality list
Case protocols

Fig. 3

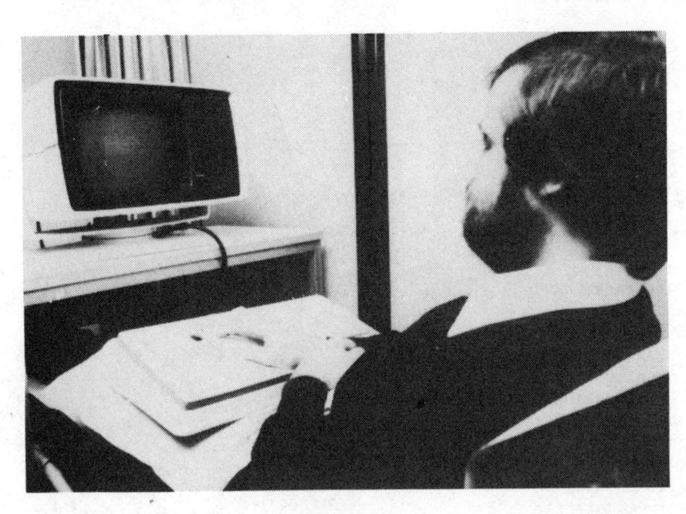

Fig. 4

and "back" -hole keyboard mask (Fig. 4) over the standard keyboard [2]. The Questionnaire is comprised of 260 questions and provides information on hereditary factors, alcohol use, diet, physical exercise, symptoms, medication, and other items related to environmental and genotoxical agents.

Probably, the most important and widespread of these factors in modern societies is smoking. Table 1 shows the questions on smoking habits, knowledge and attitudes used on our screening questionnaire. The following tables summarize some of the results of the statistical analysis of interrelationships between these answers and the motivation and success in reducing or quitting consumption [3].

Table 2 demonstrates the overall smoking frequencies in a strictly age- and sex-uniform sample of the middle-aged male screening attendees (born in 1930-1931 and investigated in 1978). Fifty percent were smokers. More interestingly, however, Table 3 indicates that knowledge of smoking hazards and complications were highest in the ex-smokers and lowest in the light tobacco consumers, but paradoxically increasing with higher smoking quantity. This may mean that formal insight into smoking risks may have influenced some to stop smoking, but is of no consequence for more dedicated smokers. This is supported by Table 4, which illustrates that an asserted knowledge of the association between smoking and, for instance, certain cancers did not increase the desire or success in reducing or eliminating the use in these male health screening attendees.

Table 5 shows that symptoms of complications such as dyspnea and chronic bronchitis increased with increasing smoking category,

TABLE 1. Smoking Questions in Screening Questionnaire

QUESTIONNAIRE

Did you ever smoke daily for at least 6 months?
Do you smoke at present?
Do you inhale?
Have you been smoking more than 10 years?
Have you been smoking more than 5 years?
Have you been smoking more than 1 year?
Have you considerably reduced your tobacco consumption during the last 6 months?
Did you quit smoking more than 5 years ago?
Did you quit smoking between 1 and 5 years ago?
Did you quit smoking last year?
Do you smoke cigarettes every day?
Do you smoke more than 40 cigarettes daily?
Do you smoke more than 30 cigarettes daily?
Do you smoke more than 20 cigarettes daily?
Do you smoke more than 10 cigarettes daily?
Do you smoke more than 3 cigars daily?
Do you smoke 2-3 cigars daily?
Do you smoke 1 cigar daily?
Do you smoke a pipe?
Do you smoke more than 1 packet of pipe-tobacco/week?
Would you like to quit smoking?
Do you believe in an association between smoking and cancer of the throat, lungs and/or mouth?
Do you believe in an association between smoking and cancer of the urinary bladder?
Do you believe in an association between smoking, pneumonia and/or chronic bronchitis?
Do you believe in an association between smoking and myocardial infarction?
Do you believe in an association between smoking and generally increased disability?
Do you consider air pollution a greater threat to the individual health than smoking?
Do you believe hereditary factors are more important than smoking for diseases that generally are considered caused by smoking?
Do you believe that disorders associated with smoking won't improve by quitting?
Do you believe that disability associated with smoking will improve by quitting?
Do you believe that the health benefits from quitting are as yet unproven?
Do you use sleeping pills more than 3 times a week?
Have you during the last year been on sick leave at least 3 times?
Have you been undergoing treatment for psychiatric disorders?
Do you get short of breath when you walk together with somebody in your own age with a normal pace?*
Have you during at least the last 2 years during at least 3 months each year had a cough every morning and evening?**
Have you during the last 2 years during at least 3 months every year been coughing phlegm, both mornings and evenings?**
Do you have episodes with shortness of breath and wheezing?***
Do you consider yourself completely healthy?

*Question on dyspnoea.
**Questions on chronic bronchitis.
**"YES" to both was required as a criterion for chronic bronchitis.
***Question on asthma.

TABLE 2. Smoking Categories in 47-Year-Old Men Attending Health Screening (n = 1037)

	n	% of all subjects
Non-smokers	230	22
Ex-smokers	290	28
Smokers	517	50
All	1 037	100
Pure cigarette smokers		% of all smokers
<10	38	7
10–20	162	31
>20	121	24
All	321	62
Pure pipe smokers		% of all smokers
≤1 packet/week	18	3
>1 packet/week	24	5
All	42	8
Pure cigar smokers		% of all smokers
	22	4
		% of all smokers
Mixed smokers	132	26

Non-smoker: never smoked.
Ex-smoker: quit smoking and not smoked for at least one month.
Smoker: has smoked regularly at least one gram of tobacco/day for at least one year.
Pure cigarette smoker: smokes at present 10; 10–20; 20 cigarettes/day.
Pure cigar smokers: (too few to be grouped).
Pure pipe-smokers: smokes at present <1 pack/week; >1 pack/week.
Mixed smokers: (combination of varying amounts of pipe tobacco, cigarettes and/or cigars).

and also that the number considering themselves as healthy decreased while overt diseases showed somewhat inconsistent patterns. Furthermore, as illustrated in Table 6, the presence of a smoking-related symptom like dyspnea influenced both the desire and the ability to quit or reduce smoking. This supports the concept that in dedicated smokers indication and counselling of actual individual complications related to smoking is a more worthwhile anti-smoking strategy than general information on smoking hazards.

Previous studies indicate that carboxyhaemoglobin in blood may be a better prospective risk indicator than sole quantity of smoking. It may better reflect the amount of inhaled tobacco compounds

TABLE 3. The Percentage of Smokers, Ex-Smokers and Non-Smokers Believing in an Association between Smoking and Lung-Cancer, Urothelical Cancer, Pneumonia-Bronchitis, Myocardial Infarction, Arteriosclerotic Leg Disease, General Disability Together with the Percentage Who Believe Air Pollution and Hereditary Factors are More Important than Smoking for these Diseases

Percentage yes answers	Never-smokers $n=230$	Ex-smokers $n=290$	Cigarette smokers 10 $n=38$	10–20 $n=162$	20 $n=121$	Pipe smokers $n=42$	Cigar smokers $n=22$	Mixed smokers $n=132$	All smokers $n=517$	All individuals $n=1\ 037$
Do you believe in an association between smoking and cancer of lung, mouth, and/or throat?	44	64	34	44	50	43	36	45	44	50
Cancer of the bladder	9	10	11	7	12	2	0	7	8	9
Pneumonia–Bronchitis	38	49	29	31	36	29	27	36	33	39
Myocardial infarction	34	51	29	35	44	17	36	40	37	40
Arteriosclerotic leg disease	14	18	11	18	14	12	27	17	16	16
Increased disability in general	38	50	24	32	29	24	28	36	31	38
Do you consider air pollution more important than smoking for the development of these diseases?	18	29	18	30	36	36	18	30	31	27
Do you believe hereditary factors are more important than smoking for these diseases?	11	15	11	12	21	14	9	15	15	14

TABLE 4. The Influence of Asserted Knowledge as to the Health Hazards of Smoking on the Wish to Quite and Success to Reduce the Daily Tobacco Consumption

Do you believe in an association between smoking and	Do you want to quit		Have you reduced your tobacco consumption?	
	Yes %	No %	Yes %	No %
Lung cancer				
Yes	55	45	10	90
No	54	46	6	94
Cancer of the bladder				
Yes	56	44	15	85
No	54	46	9	91
Chronic bronchitis				
Yes	53	47	8	92
No	55	45	11	89
Myocardial infarction				
Yes	53	47	11	89
No	55	45	9	91
Intermittent claudication				
Yes	65	35	10	90
No	52	48	10	90
Increased disability in general				
Yes	61	39	13	87
No	51	49	8	92

than the number of cigarettes smoked due to variations in inhalation habits, tobacco brand, host factors etc. [4]. Figure 5 shows the results of determinations of carboxyhaemoglobin in blood (COHb%) after overnight smoking abstinence in the sample of middle-aged male screening attendees statistically analyzed with regard to smoking habits, attitudes and knowledge. As seen, few never- or ex-smoking individuals had values in excess of 1% (and these all had some haematological abnormality), while in the smokers the values showed great interindividual variability. Most values were above 1.0%, however, and these can all be utilized as a biochemical information, feedback and control instrument in individual anti-smoking motivation and argumentation.

TABLE 5. The Percentage of Smokers, Ex-Smokers and Non-Smokers with Dyspnoea, Chronic Bronchitis, Asthma, Intermittent Claudication, Previous Myocardial Infarction Together with the Percentage Using Sleeping Pills at Least 3 Times/Week, the Percentage with Earlier Psychiatric Treatment, the Percentage with at Least 3 Episodes of Sick Leave Last Year, and the Percentage Who Consider Themselves Healthy

% yes answers	Never-smokers $n=230$	Ex-smokers $n=290$	Cigarette smokers 10 $n=38$	Cigarette smokers 10–20 $n=162$	Cigarette smokers 20 $n=121$	Pipe smokers $n=42$	Cigar smokers $n=22$	Mixed smokers $n=132$	All smokers $n=517$	All individuals $n=1\ 037$
Dyspnoea	1	3	0	6	6	5	0	3	4	3
Chronic bronchitis	2	5	8	10	12	7	5	10	10	7
Asthma	0	3	3	2	7	0	5	5	4	3
Intermittent claudication	0	5	5	1	5	2	1	2	3	3
Previous myocardial infarction	0	1	3	1	1	0	0	1	1	1
% using sleeping pills at least 3 times/week	1	2	3	1	7	2	0	2	3	2
% earlier treated by psychiatrist	6	13	8	9	14	14	9	17	13	11
% on sickleave at least 3 times last year	4	7	18	14	12	5	5	14	13	9
% who consider themselves as healthy	76	72	87	71	64	76	82	67	72	72

TABLE 6. The Influence of Dyspnoea on the Wish to Quit, Success in Reducing the Tobacco Consumption and Self-Concept of Health

	Do you want to quit?		Have you reduced your tobacco consumption?		Do you consider yourself as healthy?	
History of	Yes %	No %	Yes %	No %	Yes %	No %
Dyspnoea						
Yes	64	36	27	73	32	68
No	54	46	9	91	72	28

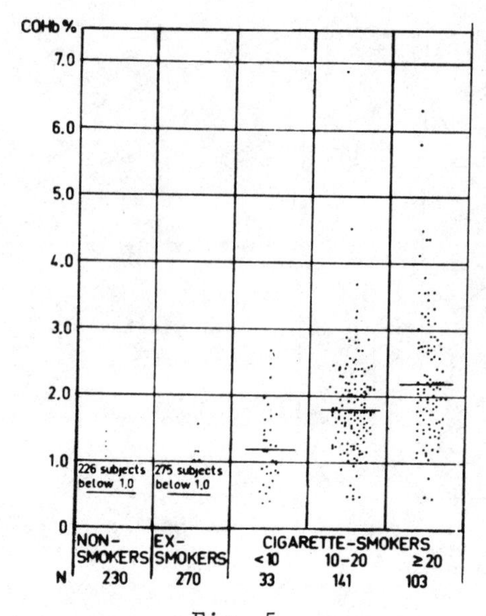

Fig. 5

Measurement of carbon monoxide in expired air (CO) is an even better, instant method for the same purpose, and is now performed directly by the nurses in all smokers as part of the screening investigation. The results are regularly discussed in relation to smoking habits and hazards, with the aim of getting a decision from the screening participant with regard to quitting or reducing consumption. Figure 6 shows the form (in Swedish) used by the nurses to note down the results of the CO measurement and of their discussion with the screening subject (antirökundersökning = anti-smoking investigation;... första kontakt = first visit;... uppmanad sluta roka = asked to quite smoking;... ja = yes;... nej = no;... respons och positivit beslut = response and positive decision;... tveksamt =

Antirökundersökning

Personnummer

Malmö allmänna sjukhus
Förebyggande medicin
Claesgatan 7
214 01 MALMÖ

Första kontakt

Rökkonsumtion

.... cigarretter/dag ☐ rullar cigarretter själv .... tobakspkt (50 g)/vecka

☐ röker pipa .... tobakspkt/burk (50 g)/vecka

.... cigarrer/dag

CO-värde:

Uppmanad att sluta röka	☐ Ja		☐ Nej
Respons och positivt beslut	☐ Ja	☐ Tveksamt	☐ Nej
Egen bedömning betr genomförande	☐ Ja	☐ Tveksamt	☐ Nej

Anm

Uppföljning betr rökvanor göres efter 1 år | Datum | Sign

Uppföljningsresultat efter 1 år

☐ Rökfri sedan ....

☐ Rökfri period under året fr o m .... t o m .... (.....v/mån)

☐ Har ej försökt ☐ uppger sig ej kunna ☐ vill ej

☐ Neddragen rökkonsumtion jämfört med den för 1 år sedan

Rökkonsumtion

.... cigarretter/dag ☐ rullar cigarretter själv .... tobakspkt (50 g)/vecka

☐ röker pipa .... tobakspkt/burk (50 g)/vecka

.... cigarrer/dag

Noterade förbättringar (medicinska värden, prestationer, aktivitet m m)

Beslut inför nästa 1 års-period

Datum | Sign

82.04 Blankett nr 957

Fig. 6

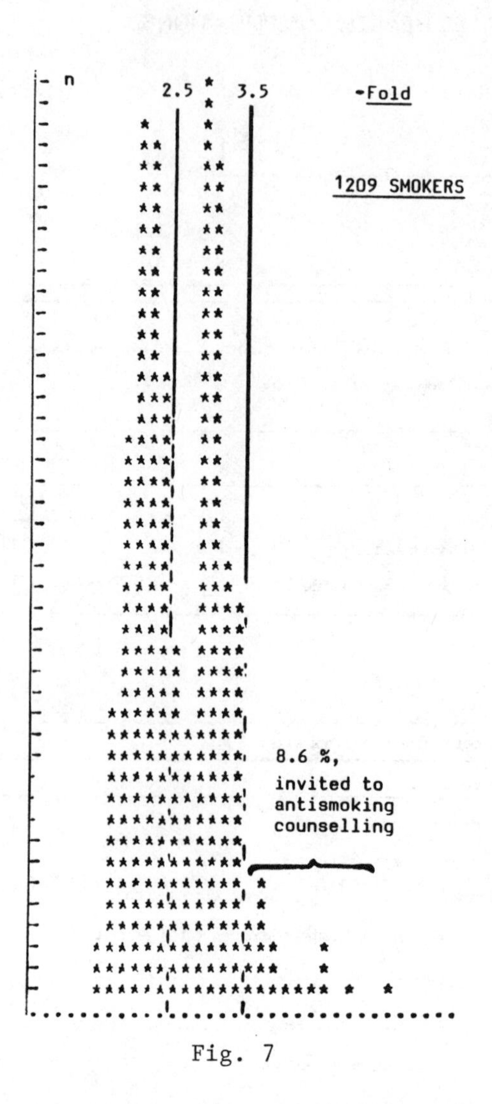

Fig. 7

doubtful;... egen bedömning betr. genomförande = own estimation of success;... uppföljning betr. rökvanor göres efter 1 år = follow-up of smoking habits after 1 year;... datum = date.)

However, there may also be an individual susceptibility to different smoking-related complications. Respiratory tract cancers associated with smoking show strong hereditary influences and there is evidence that the intracellular enzyme system, aryl-hydrocarbon hydroxylase plays an important role in this. The mixed-function oxygenase aryl-hydrocarbon hydroxylase (AHH) is believed to activate polycyclic aromatic hydrocarbons, regarded as the most important proximate carcinogens of tobacco smoke, to reactive intermediates-ultimate carcinogens- in the cells into which the tobacco compounds are absorbed, e.g., the bronchial epithelium. This metabolism initially involves the microsomal oxidation of the double bonds and the formation of an epoxide. Epoxides are chemically labile, highly energetic and form covalent bonds with certain proteins, RNA and, most importantly, DNA. The diol-epoxides and the dihydro-diol-epoxides of unsubstituted polycyclic aromatic hydrocarbons are also of major importance and could be ultimately reactive forms of these carcinogenic compounds.

Previous studies have indicated a genetic heterogeneity of AHH inducibility and an overrepresentation of cases with high AHH inducibility in retrospective studies of smoking-related lung [5], larynx [6], and oral cancers [7] as well as leukoplakias [8]. This prompted us to perform AHH inducibility assays, using techniques described in References 5-8, in consecutive population subsamples of middle-aged male smokers, ex-smokers, and non-smokers. Figure 7 shows the results in 1209 smokers, and Fig. 8a-b the corresponding findings in 304 ex-smokers and 218 non-smokers. In all samples there was an apparently tri-modal distribution with low (1-2.5), intermediate (2.6-3.5) and high (over 3.5 fold) induction levels on the order of 46-54, 39-45 and 7.6-10.5%, respectively. This is about the same frequency as in previously described normal reference materials [5-8].

In separate studies on the same subjects we have demonstrated a very good reproducibility, on the order of $r = 0.9$, both for inter-observer and intraindividual, as well as temporally of our assay method. Therefore, we have invited all the smoking individuals with high induction levels to a group-mediated anti-smoking program, where the high induction levels form a biological imperative rather than feed-back in the stop smoking counselling. So far, the results have been favorable, with the majority of the invited individuals responding to the invitation and trying to stop or reduce their consumption. Long-term results are still lacking, however. The smoking habits thus do not seem to influence the AHH-induction, but the AHH induction levels may be utilized to influence the smoking habits. In collaboration with a Swedish program group for genetic toxicology, genotoxic effects of smoking are now evaluated in small, but well characterized and matched normal population subsamples from the preventive medical screening investigation material. Some of the effects studied and methods applied in the pilot samples are: Plasma tests such as xenobiotics analyses, analyses of plasma proteins and other plasma components; erythrocyte tests including haemoglobin alkylation and glutathionperoxidase activity determinations; lymphocyte studies such as DNA repair synthesis, carcinogen binding, DNA and chromosome aberrations, Sister chromatid exchange, point mutations, aneuploidy, etc.; granulocyte studies, and urine tests. The department of preventive medicine here serves as a reference base from which methods can be compared

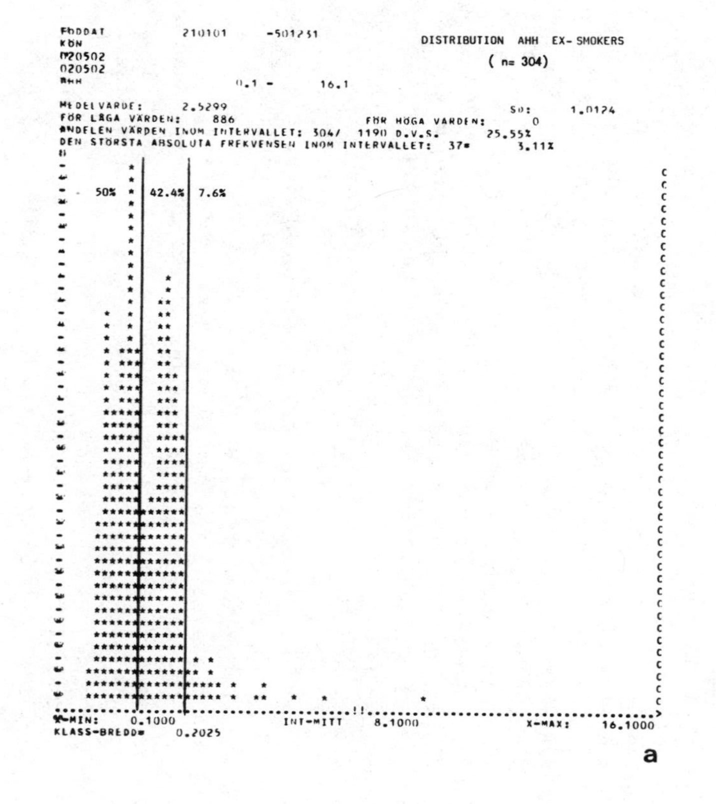
DISTRIBUTION AHH EX- SMOKERS
(n= 304)
50%
42.4%
7.6%
a

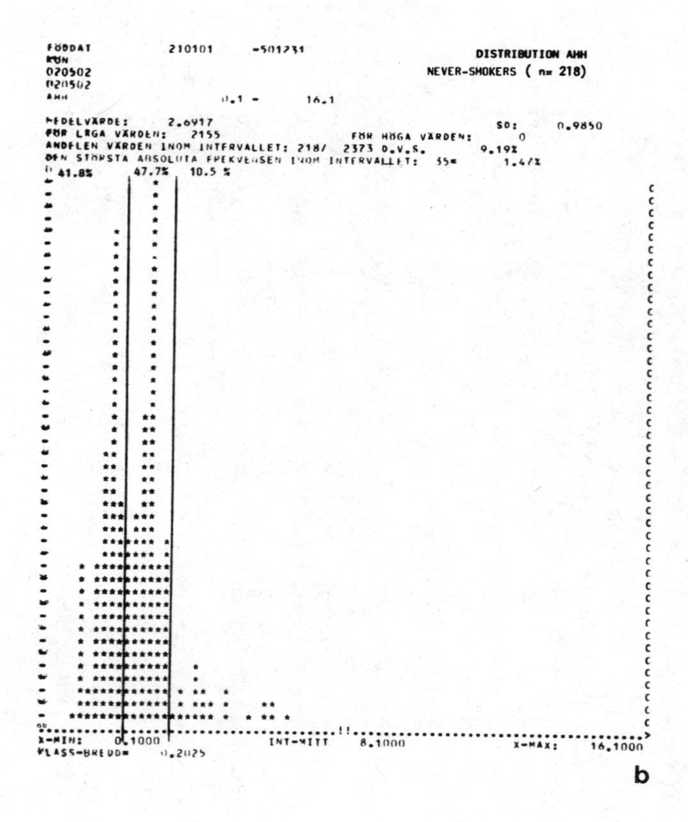
DISTRIBUTION AHH
NEVER-SMOKERS (n= 218)
41.8%
47.7%
10.5 %
b

Fig. 8

and results can be related to background data, in particular smoking habits. The blood samples are obtained at the department, prepared in its laboratory and the samples are distributed by ordinary air mail to special research laboratories all over Sweden. So far, no results are available, but the methods are of great potential interest to future preventive medicine and environmental health activities.

REFERENCES

1. E. Trell, N. Dahlberg, C. Larsson, J.-O. Krantz, A. Laser, and B.-G. Peterson, Interactive Computer Program for Monitoring Multiphasic Health Screening, Computer Progr. Biomed., 12:262 (1980).
2. Idem, Interactive Computer Program for Self-Distributed Medical Questionnaire in a Population Health Screening, Computer Progr. Biomed, 14:257 (1982).
3. L. Janzon, S.-E. Lindell, and E. Trell, Smoking and Disease, Scand. J. Soc. Med., 9:127 (1981).
4. L. Janzon, S.-E. Lindell, E. Trell, and P. Larme, Smoking habits and carboxyhaemoglobin, A cross-sectional study of an urban population of middle-aged men, J. Epidemiol. Commun. Health, 25:271 (1981).
5. G. Kellerman, C. R. Shaw, M. Luyten-Kellerman, Aryl hydrocarbon hydroxylase inducibility and bronchogenic carcinoma, New Engl. J. Med., 289:934 (1973).
6. E. Trell, R. Korsgaard, I. Mattiasson, B. Hood, P. Kitzing, and B.-G. Simonsson, Arylhydrocarbonhydroxylase inducibility and laryngeal carcinomas, Lancet, ii:140 (1976).
7. E. Trell, G. Björlin, L. Andreasson, R. Korsgaard, and I. Mattiasson, Carcinoma of the oral cavity in relation to arylhydrocarbonhydroxylase inducibility, smoking, and dental status, Int. J. Oral Surg., 10:93 (1981).
8. L. Andreasson, G. Björlin, R. Korsgaard, I. Mattiasson, E. Trell, and L. Trell, Leukoplakia of oral cavity, smoking, and arylhydrocarbon hydroxylase inducibility, Postgrad Med., 58:138 (1982).

CONTRIBUTORS

James W. Allen, Health Effects Research Laboratory, U.S. Environmental Protection Agency, Research Triangle Park, North Carolina 27711.

Bo Andersson, Institute of Forensic Medicine, Karolinska Institute, Stockholm, Sweden.

J. George Bekesi, The Mt. Sinai Medical Center, One Gustave L. Levy Place, New York, New York 10029.

John R. Bend*, National Institute of Environmental Health Sciences, P. O. Box 12233, Research Triangle Park, North Carolina 27709.

William L. Bigbee, Biomedical Sciences, Division, Lawrence Livermore Laboratory, P. O. Box 5507, Livermore, California 94550.

Arthur D. Bloom*, Department of Pediatrics, College of Physicians and Surgeons of Columbia University, 630 West 168 Street, New York, New York 10032.

Peter J. Bokos, Interventions, 1313 S. Michigan Avenue, Chicago, Illinois 60605.

Alan R. Boobis, Department of Clinical Pharmacology, Royal Postgraduate Medical School, London W12 OHS, United Kigdom.

Ernest Borek, AMC Cancer Research Center and Hospital, Lakewood, Colorado 80214.

Elbert W. Branscomb, Biomedical Sciences Division, Lawrence Livermore Laboratory, P. O. Box 5507, Livermore, California 94550.

Jan Carlstedt-Duke, Department of Medical Nutrition, Karolinska Institute, Huddinge University Hospital F69, S0141 86 Huddinge, Sweden.

Larry D. Claxton, Health Effects Research Laboratory, U.S. Environmental Protection Agency, Research Triangle Park, North Carolina 27711.

James E. Cleaver, Laboratory of Radiobiology, School of Medicine, University of California at San Francisco, San Francisco, California 94143.

F. Bernard Daniel, Health Effects Laboratory, U.S. Environmental Protection Agency, 26 West St. Clair, Cincinnati, Ohio 45268.

Donald S. Davies, Department of Clinical Pharmacology, Royal Postgraduate Medical School, London W12 OHS, United Kingdom.

Frederick J. de Serres*, Office of the Director, National Institute of Environmental Health Sciences, P. O. Box 12233, Research Triangle Park, North Carolina 27709.

Robert M. Donahoe, Department of Psychiatry, Emory University, Atlanta, Georgia 30322/Georgia Mental Health Institute, Atlanta, Georgia 30306.

Lars Ehrenberg, Institute of Radiobiology, University of Stockholm, Stockholm, Sweden.

Deborah C. Eltzroth, Department of Psychiatry, Emory University, Atlanta, Georgia 30322/Georgia Mental Health Institute, Atlanta, Georgia 30306.

Arthur Falek, Department of Psychiatry, Emory University, Atlanta, Georgia 30322/Georgia Mental Health Institute, Atlanta, Georgia 30306.

Alf Fischbein, The Mt. Sinai Medical Center, One Gustave L. Levy Place, New York, New York 10029.

N. E. Garrett, Northrop Services, Inc., Environmental Sciences, Research Triangle Park, North Carolina 27709.

Mikael Gillner, Department of Medical Nutrition, Karolinska Institute, Huddinge University Hospital F69, S0141 86 Huddinge, Sweden.

Hira L. Gurtoo, Department of Health, State of New York, Roswell Park Memorial Institute, 666 Elm Street, Buffalo, New York 14263.

Jan-Åke Gustafsson, Department of Medical Nutrition, Karolinska Institute, Huddinge University Hospital F69, S0141 86 Huddinge, Sweden.

Lars-Arne Hansson, Department of Medical Nutrition, Karolinska Institute, Huddinge University Hospital F69, S0141 86 Huddinge, Sweden.

Beryl Hartley-Asp, Pharmacology Department, AB Leo, Box 941, Helsingborg, Sweden.

David G. Hoel*, National Institute of Environmental Health Sciences, P. O. Box 12233, Research Triangle Park, North Carolina 27709.

Bertil Högberg, Department of Medical Nutrition, Karolinska Institute, Huddinge University Hospital F69, S0141 86 Huddinge, Sweden.

Benkt Högstedt, Department of Clinical Genetics, University Hospital, Lund, Sweden.

Felicia Hollingsworth, Department of Psychiatry, Emory University, Atlanta, Georgia 30322/Georgia Mental Health Institute, Atlanta, Georgia 30306.

Ernest B. Hook, Birth Defects Institute, New York State Department of Health, Albany, New York 12237.

Shiu L. Huang, Northrop Services, Inc., Environmental Sciences, Research Triangle Park, North Carolina 27709.

Maj. A. Hultén, Regional Cytogenetics Laboratory, East Birmingham Hospital, Bordesley Green East, Birmingham B9 5ST, United Kingdom.

Lars Janzon, Departments of Preventive and Community Medicine, University of Lund, Lund, Sweden.

Ronald H. Jensen, Biomedical Sciences Division, Lawrence Livermore Laboratory, P. O. Box 5507, Livermore, California 94550.

Sören Jensen, Institute of Radiobiology, University of Stockholm, Stockholm, Sweden.

G. Clare Kahn, Department of Pathology and Laboratory Medicine, Hospital of the University of Pennsylvania, 3400 Spruce Street, Philadelphia, Pennsylvania 19104.

Rolf Korsgaard, Departments of Preventive and Community Medicine, University of Lund, Lund, Sweden.

Robert S. Lake, Schering-Plough, Lafayette, Pharmaceutical Research Division, Box 32, Lafayette, New Jersey 07848.

Bo Lambert, Department of Clinical Genetics, Karolinska Hospital, Stockholm, Sweden.

Roger Larsson, Institute of Forensic Medicine, Karolinska Institute, Stockholm, Sweden.

D. A. Laurie, Regional Cytogenetics Laboratory, East Birmingham Hospital, Bordesley Green East, Birmingham B9 5ST, United Kingdom.

Phaik-Mooi Leong, Department of Nutrition and Food Science, Massachusetts Institute of Technology, Cambridge, Massachusetts 02139.

Sven Lindstedt*, Institute of Clinical Chemistry, Sahlgrenska Hospital, 413 45 Göteborg, Sweden.

Göran Löfroth, Department of Radiobiology, Wallenberg Laboratory, University of Stockholm, Stockholm, Sweden.

Johan Lund, Department of Medical Nutrition, Karolinska Institute, Huddinge University Hospital F69, S0141 86 Huddinge, Sweden.

John J. Madden, Department of Psychiatry and Department of Biochemistry, Emory University, Atlanta, Georgia 30322/Georgia Mental Health Institute, Atlanta, Georgia 30306.

Nils Mandahl, Department of Tumor Cytogenetics, University of Lund, Lund, Sweden.

R. H. Martin, Alberta Childrens Hospital, Medical Genetics Clinic, S. W. Calgary, Alberta, Canada.

Jun Minowada, Department of Health, State of New York, Roswell Park Memorial Institute, 666 Elm Street, Buffalo, New York 14263.

Felix Mitelman, Department of Clinical Genetics, University Hospital, Lund, Sweden.

Peter Moldeus*, Institute of Forensic Medicine, Karolinska Institute, Stockholm, Sweden.

Martha M. Moore, Health Effects Research Laboratory, U.S. Environmental Protection Agency, Research Triangle Park, North Carolina 27711.

Åke Nordén*, Research Division 2, EB-Block, Lasarettet, 221 85 Lund, Sweden.

Magnus Nordenskjöld, Department of Clinical Genetics, Karolinska Institute, Stockholm, Sweden.

Martin Nygren, Department of Organic Chemistry, University of Umeå, Umeå, Sweden.

Robert Olin, Royal Institute of Technology, Stockholm, Sweden and ASTRA Pharmaceuticals, Södertälje, Sweden.

Beverly Paigen, Department of Health, State of New York, Roswell Park Memorial Institute, 666 Elm Street, Buffalo, New York 14263.

Ronald W. Pero, University of Lund, Wallenberg Laboratory, Fack 7031, S 220 07 Lund, Sweden.

Lorenz Poellinger, Department of Medical Nutrition, Karolinska Institute, Huddinge University Hospital F69, S0141 86 Huddinge, Sweden.

R. Julian Preston, Oak Ridge National Laboratory, P. O. Box Y, Oak Ridge, Tennessee 37830.

Claes Ramel, Department of Genetics, Wallenberg Laboratory, University of Stockholm, 106 91 Stockholm, Sweden.

Christoffer Rappe, Department of Organic Chemistry, University of Umeå, Umeå, Sweden.

N. Saadallah, Regional Cytogenetics Laboratory, East Birmingham Hospital, Bordesley Green East, Birmingham B9 5ST, United Kingdom.

Irving J. Selikoff, Department of Community Medicine, The Mt. Sinai School of Medicine, 5th Avenue and 100th Street, New York, New York 10029

H. A. J. Schut, Department of Pathology, Medical College of Ohio, 3000 Arlington Avenue, Toledo, Ohio 43614.

David A. Shafer, Department of Psychiatry, Emory University, Atlanta, Georgia, 30322/Georgia Mental Health Institute, Atlanta, Georgia 30306.

Y. Sharief, Northrop Services, Inc., Environmental Sciences, Research Triangle Park, North Carolina 27709.

William Sheridan, National Institute of Environmental Health Sciences, P. O. Box 12233, Research Triangle Park, North Carolina 27709.

Thomas R. Skopek, Department of Molecular Biophysics and Biochemistry, Yale University, New Haven, Connecticut 06520.

Zena Stein, New York Psychiatric Institute, Columbia University College of Physicians and Surgeons, 722 West 168 Street, New York, New York 10032.

G. D. Stoner, Department of Pathology, Medical College of Ohio, 3000 Arlington Avenue, Toledo, Ohio 43614.

Gary H. Strauss, Health Effects Research Laboratory, U.S. Environmental Protection Agency, Research Triangle Park, North Carolina 27711.

H. Eldon Sutton, The University of Texas at Austin, Department of Zoology, Austin, Texas 78712.

William G. Thilly, Department of Nutrition and Food Science, Massachusetts Institute of Technology, Cambridge, Massachusetts 02139.

Rune Toftgård, Department of Medical Nutrition, Karolinska Institute, Huddinge University Hospital F69, S0141 86 Huddinge, Sweden.

Margareta Törnqvist, Institute of Radiobiology, University of Stockholm, Stockholm, Sweden.

Erik Trell, Departments of Preventive and Community Medicine, University of Lund, Lund, Sweden.

B. M. N. Wallace, Department of Genetics, University of Birmingham, Birmingham B15 2TT, United Kingdom.

Michael D. Waters, Health Effects Research Laboratory, U.S. Environmental Protection Agency, Research Triangle Park, North Carolina 27711.

Andrew Wyrobek, Biomedical Sciences Division, Lawrence Livermore Laboratory, P. O. Box 5507, Livermore, California 94550.

Gösta Zetterberg, Institute of General Genetics, University of Uppsala, Box 7003, 750 07 Uppsala, Sweden.

---

*Session Chairmen.

INDEX